NEUROBIOLOGY OF INVERTEBRATES

NEUROBIOLOGY OF INVERTEBRATES

Edited by

J. SALÁNKI

Director of the Biological Research Institute of the
Hungarian Academy of Sciences (Tihany)

1828—1968

AKADÉMIAI KIADÓ · BUDAPEST · 1968

PROCEEDINGS OF THE SYMPOSIUM HELD AT THE BIOLOGICAL
RESEARCH INSTITUTE OF THE HUNGARIAN
ACADEMY OF SCIENCES (TIHANY) SEPTEMBER 4—7, 1967

ISBN-13: 978-1-4615-8620-3   e-ISBN-13: 978-1-4615-8618-0
DOI: 10.1007/978-1-4615-8618-0

JOINT EDITION PUBLISHED BY AKADÉMIAI KIADÓ, BUDAPEST
AND PLENUM PRESS, NEW YORK

SOFTCOVER REPRINT OF THE HARDCOVER 1ST EDITION   1968

PREFACE

In September 1967 a Symposium on Neurobiology of Invertebrates was held at Tihany, in the Biological Research Institute of the Hungarian Academy of Sciences, coinciding with the 40 years anniversary of this Institute. Its Department of Experimental Zoology, representing the most important basis for researches in Hungary on the nervous system of invertebrates, organized the meeting.

The Symposium covered both morphological and functional aspects of invertebrate neurobiology from the viewpoints of elementary processes as well of regulatory mechanisms. The complex approach of identic or similar problems is a generally accepted trend in biological research — this tendency is well reflected in the 34 papers presented by participants of the Symposium coming from different countries of Europe and the United States. The volume contains all but one of the lectures held at the meeting; the paper of A. K. Voskresenskaya included in the Proceedings could not be read because of the tragic death of the author, some weeks before the Symposium. The volume is separated into 4 parts according to the 4 days program, however this division is rather tentative because of overlappings in the different fields. Discussion following the lectures are also published in short, however it was impossible to give a full picture in this respect.

The Biological Department of the Hungarian Academy of Sciences deserves special gratitude for the financial support of the Symposium. I wish to express also my thanks to all contributors and to my colleagues for their help in organizing the Symposium and for their technical assistance in preparing this volume.

János Salánki

CONTENTS

CONTENTS

PROCESSES ON CELLULAR LEVEL

NEUROHUMORS AND ENZYMES

INTEGRATION

LIST OF PARTICIPANTS

Ádám, Gy.
Hungary
  Department of Animal Physiology,
  Eötvös Loránd University
  Budapest
  VIII., Puskin u. 3.

Arvanitaki-Chalazonitis, A.
France
  Institut de Neurophysiologie et
  de Psychophysiologie
  Marseille
  31, chemin J. Aiguier

Bierbauer, J.
Hungary
  Department of Histology and Embriology,
  University Medical School
  Budapest
  IX., Tűzoltó u. 58.

Boettiger, E. G.
USA
  University of Connecticut
  Storrs, Conn.

Chalazonitis, N.
France
  Institut de Neurophysiologie et
  de Psychophysiologie
  Marseille
  31, chemin J. Aiguier

Corning, W. C.
USA
  Department of Psychology
  Fordham University
  Bronx, N. Y.

Cottrell, G. A.
Great Britain
  Wellcome Laboratories of
  Pharmacology Laboratory
  University of St. Andrews
  St. Andrews

Csillik, B.
Hungary
  Department of Anatomy,
  University Medical School
  Szeged

David, O. F.
USSR
  Sechenov Institute of Evolutionary
  Physiology and Biochemistry,
  Academy of Sciences of the USSR
  Leningrad, K-223
  Thorez pr. 52.

Fehér, O.
Hungary
  Department of Animal Physiology,
  József Attila University
  Szeged

Gerasimov, V. D.
USSR
  A. A. Bogomolets Institute
  of Physiology,
  Academy of Sciences of the USSR
  Kiev-24
  ul. Bogomoltsa 4.

GLAIZNER, B.
Great Britain
  Department of Zoology,
  University of Southampton
  Southampton

GUBICZA, A.
Hungary
  Biological Research Institute,
  Hungarian Academy of Sciences
  Tihany

HÁMORI, J.
Hungary
  Department of Anatomy,
  University Medical School
  Budapest
  IX., Tűzoltó u. 58.

HIRIPI, L.
Hungary
  Biological Research Institute,
  Hungarian Academy of Sciences
  Tihany

HUGHES, G. M.
Great Britain
  Department of Zoology,
  University of Bristol
  Bristol

KERKUT, G.
Great Britain
  Department of Physiology and
  Biochemistry,
  University of Southampton
  Southampton

KISS, I.
Hungary
  Biological Research Institute,
  Hungarian Academy of Sciences
  Tihany

KOMISSARCHIK, Ya. Yu.
USSR
  Institute of Cytology,
  Academy of Sciences of the USSR
  Leningrad, F-121
  ul. Maklina 32.

KOSTYUK, P. G.
USSR
  A. A. Bogomolets Institute
  of Physiology,
  Academy of Sciences of the USSR

  Kiev-24
  ul. Bogomoltsa 4.

KUZIEMSKI, H.
Poland
  Department of Physiology,
  Medical Academy
  Gdansk

LÁBOS, E.
Hungary
  Biological Research Institute,
  Hungarian Academy of Sciences
  Tihany

LUKACSOVICS, F.
Hungary
  Biological Research Institute,
  Hungarian Academy of Sciences
  Tihany

MADARÁSZ, I.
Hungary
  Department of Physiology,
  University Medical School
  Szeged

MANDELSTAM, YU. E.
USSR
  Sechenov Institute of Evolutionary
  Physiology and Biochemistry,
  Academy of Sciences of the USSR
  Leningrad, K-223
  Thorez pr. 52.

MIROLLI, M.
USA
  NIMH
  Division of Special Mental
  Health Research
  Washington, D. C.

ZS.-NAGY, I.
Hungary
  Biological Research Institute,
  Hungarian Academy of Sciences
  Tihany

NISTRATOVA, S.
USSR
  Institute of Developmental Biology,
  Academy of Sciences of the USSR
  Moscow
  ul. Vavilova 26.

NOLTE, A.
GFR
    Zoologisches Institut
    44 Münster (Westf.)
    Badestrasse 9.

PLOTNIKOVA, S. I.
USSR
    Sechenov Institute of Evolutionary
    Physiology and Biochemistry,
    Academy of Sciences of the USSR
    Leningrad, K-223
    Thorez pr. 52.

S.-RÓZSA, K.
Hungary
    Biological Research Institute,
    Hungarian Academy of Sciences
    Tihany

RÖHLICH, P.
Hungary
    Department of Histology and
    Embryology,
    University Medical School
    Budapest
    IX., Tűzoltó u. 59.

SAKHAROV, D. A.
USSR
    Institute of Developmental Biology,
    Academy of Sciences of the USSR,
    Moscow
    ul. Vavilova 26.

SALÁNKI, J.
Hungary
    Biological Research Institute,
    Hungarian Academy of Sciences
    Tihany

SHISHOV, B. A.
USSR
    Helminthological Laboratory,
    Academy of Sciences of the USSR
    Moscow, B-71
    Leninski pr. 33.

SVIDERSKY, V. L.
USSR
    Sechenov Institute of Evolutionary
    Physiology and Biochemistry,
    Academy of Sciences of the USSR
    Leningrad, K-223
    Thorez pr. 52.

SZENTÁGOTHAI, J.
Hungary

Department of Anatomy,
University Medical School
Budapest
IX., Tűzoltó u. 58.

VARANKA, I.
Hungary
    Biological Research Institute,
    Hungarian Academy of Sciences
    Tihany

VEPRINTSEV, B. N.
USSR
    Institute of Biological Physics,
    Academy of Sciences of the USSR
    Moscow region
    Putchino on OKA

VERZÁR, F.
Switzerland
    Institut für Gerontologie
    Basel
    Nonnenweg 7.

VINNIKOV, Ya. A.
USSR
    Sechenov Institute of Evolutionary
    Physiology and Biochemistry,
    Academy of Sciences of the USSR
    Leningrad, K-223
    Thorez pr. 52.

WALKER, R. J.
Great Britain
    Department of Physiology and
    Biochemistry,
    University of Southampton
    Southampton

WELLS, M. J.
Great Britain
    Department of Zoology,
    University of Cambridge
    Cambrigde

WILLOWS, A. O. D. *USA*
    Friday Harbor Laboratory
    Friday Harbor
    Washington 98250

ZEIMAL, E. V.
USSR
    Sechenov Institute of Evolutionary
    Physiology and Biochemistry,
    Academy of Sciences of the USSR
    Leningrad, K-223
    Thorez pr. 52.

OPENING ADDRESS

J. Salánki

Director of the Biological Research Institute
of the Hungarian Academy of Sciences
Tihany, Hungary

It is a great pleasure for me to great the participants in this Symposium here in the Biological Research Institute of the Hungarian Academy of Sciences. As the director of this Institute and the head of the Organizing Committee I particularly want to welcome those colleagues who have come from abroad, especially since, for most of them, this is their first visit to Hungary. I extend our best wishes to them and hope that it will be a pleasant sojourn and one worth remembering.

As you know, the Symposium has been organized in connection with the anniversary of this Institute. It was just 40 years ago, that the Biological Research Institute in Tihany was built and opened. During its 40 years of existence, the Institute has played a very important role in promoting and encouraging biological research in this country. For varying lengths of time, most of the prominent Hungarian biologists of today, even those who now live and work in other countries, have been members of the staff of the Institute or have worked here several times as visiting researchers. In the earlier years, the work of the Institute was rather diversified, including physiology, bacteriology, genetics, botany, hydrobiology, biochemistry, and other subjects. However, during the last five years, after some reorganization, two departments came into existence: the Department of Experimental Zoology and the Department of Hydrobiology. The Department of Experimental Zoology works on problems of invertebrate neurobiology. This was one of the reasons it was decided to organize this kind of Symposium in connection with the anniversary of the Institute.

The other and more important reason for organizing this Symposium was the great development and increasing number of new discoveries in the field of the morphology, physiology, and neurochemistry of the nervous system of invertebrate animals.

Although neural regulation and its fundamental elements are based on similar structures and principles in the whole of the animal kingdom, some special problems do exist which necessitate the organization of meetings of scientists working on invertebrate brains and their structural and functional properties.

It is quite obvious that the neurobiology of invertebrates is a very wide field of research. Scientists from many different disciplines work in this field, but their specialization of interests may mean that they work apart from one another. At the same time, a great effort is being made to remove the barriers between these specializations and to use a combination of morphological, physiological, and chemical methods in approaching and solving problems. This was the reason for organizing this Symposium which is centered on the neurobiology instead of the neuromorphology, neurophysiology, or neurochemistry of the invertebrates. We hope that the wide-ranging interests of the participants will contribute to the success of our discussions.

At the same time, we hope that this Symposium will render a service not only through formal discussions of mutual problems, but also by enabling informal discussions to take place outside the conference room. We also hope that this Symposium will create an opportunity for personal contact between scientists of different countries and will serve to increase understanding not only with regard to special biological problems, but also in many other respects.

Symposium on Neurobiology of Invertebrates 1967 (17—25)

INTRODUCTORY REMARKS

TECHNICAL PROBLEMS IN THE STUDY
OF NEURON NETWORKS

J. SZENTÁGOTHAI

Department of Anatomy, University Medical School
Budapest, Hungary

Most neurons of highly differentiated character are relatively easily accessible with specific stains such as the Golgi procedures and the methylene blue stains. Many types of such neurons were studied in invertebrates and vertebrates in the classical period of neurohistology and have also been studied more recently with renewed interest, due to the great advance in our understanding of the functions of individual neurons. In addition, the interconnections between the more differentiated neurons can be investigated by the use of degeneration studies, provided they are sufficiently long. This line of research was less successful up until now in invertebrates because of the lack of reliable specific staining techniques for degenerated terminals. This difficulty has now been surmounted by the possibility of combining degeneration experiments with observations under the electron microscope (EM). The characteristic osmiophilic stage of degenerating axon terminals, well established by now in the vertebrate nervous system, has been shown by Gray (1964) and Hámori and Horridge (1966) to also be characteristic for the degeneration of invertebrate axons and synaptic terminals. There is thus no obstacle, in principle, to unraveling the intricate connections between different neurons, both in the vertebrate and in the invertebrate nervous system, provided, of course, that (1) the neurons are separated from one another by sufficiently large distances, to render possible the use of degeneration methods, or (2) the shape and arborization pattern of the neurons is sufficiently differentiated — as, for example, in the cerebellar cortex — so that various types of local connections can be recognized with the aid of Golgi techniques. The situation immediately becomes very difficult if these two fundamental conditions are not satisfied. Although, for example, in the spinal cord, most of the neurons are fairly highly differentiated and it is very easy to distinguish dendrites, axons, initial axon collaterals, and terminal axon ramifications in the Golgi picture, the meshwork established between these various elements is of rather uniform structure. This is one of the main reasons that, in spite of a really spectacular development in the unit level physiology of the spinal cord and considerable advances made with the aid of modern degeneration techniques, we do not know much more about the synaptology properties of the spinal neuronal network than was known by Cajal (1909).

The situation becomes even more difficult if the neurons of any nervous network are lowly differentiated, i.e., axons and dendrites are not readily distinguished. Such networks occur in abundance both in invertebrates and in vertebrates. In the higher vertebrates, they are largely confined to the peripheric autonomic plexus, as those of the intestine, where only one type of neuron, the Dogiel type 1, has clearly separable dendrites and one axon. In the more numerous Dogiel type 2, no such differentiation of cell processes is generally possible (Fig. 1a). The true nerve networks, or, better, feltworks, of lowly differentiated nerve cells are of course well known in many phyla of invertebrates. Figures 1b and c show a characteristic example of such a feltwork in echinoids *(Strongylocentrotus purpuratus)*, where the various kinds of spinous appendages are surrounded at their base by a muscle cone. Each of these cones is surrounded by a nervous ring — belonging to the basiepithelial nerve plexus — built up of very lowly differentiated nerve cells, unipolar, bipolar, or sometimes stellate, but without the slightest possibility for distinguishing the various kinds of processes, such as dendrites and axons. It is only with considerable difficulty that these nerve elements can be specifically stained with the usual methods. The Bielschowsky type stains, particularly the Gross–Schultze modification, and perhaps gold-chloride methods, are so far the only methods that can give satisfactory results.

It would be a great mistake, though, to simply dismiss these nerve feltworks as nets of diffuse conductivity and of low integrative capacities. Such nerve nets show quite baffling integrative properties, and, although quite understandably, observations are highly controversial, there can be little doubt that their study will be most rewarding in the future. I would even venture to predict that it is these seemingly diffuse neuron networks that will offer us some basic clues for the understanding of neural organization as a whole. First, although primitive, or, more correctly, lowly differentiated in structure, the gross arrangement of these feltworks are a highly organized structure above the unit level, as in echinoids, where they appear in well-defined rings with denser accumulations of nerve cells along both borders, with systematic oblique interconnections between the opposite marginal cell accumulations. (We could also speak of the rather intricate and, unfortunately, ill-understood mode of connections with the true epithelial nerve network of the basiephithelial plexus.) Secondly, the higher integrative centers of both invertebrates and vertebrates, although built up in most cases of highly differentiated neurons, have certain important properties in common with the so-called 'primitive nerve nets', such as (1) large numbers of neurons with (2) relatively short axons having (3) an extreme abundance of reciprocal interconnections. The gross arrangement of the interneuron feltwork may be rigidly organized in geometric fashion as we know it in the cerebellar cortex, or it may show a certain trend toward becoming diffuse, as is the case in the cerebral cortex due to the

---

Fig. 1. a — Myenteric plexus of the small intestine. Feltwork of Dogiel type 2 cells; cat. b — Strongylocentrotus purpuratus, nerve ring surrounding muscle cone of spinous appendage. c — Same as b with higher power. All figures from Gross–Schultze stained material

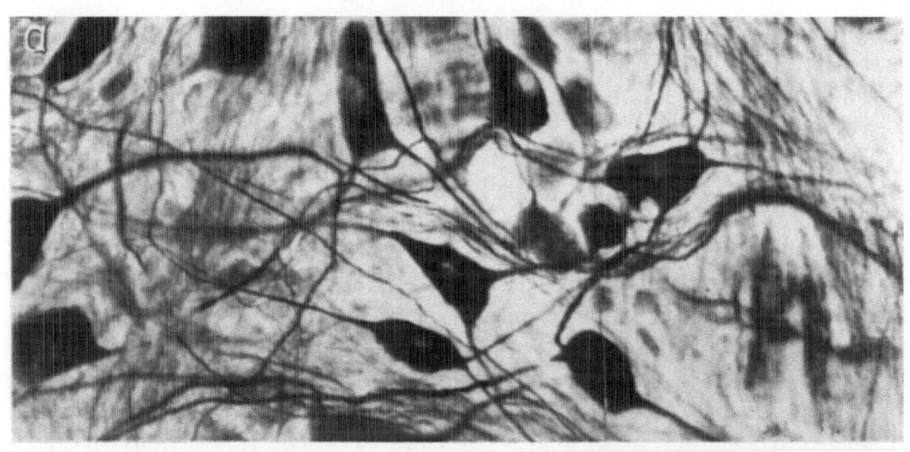

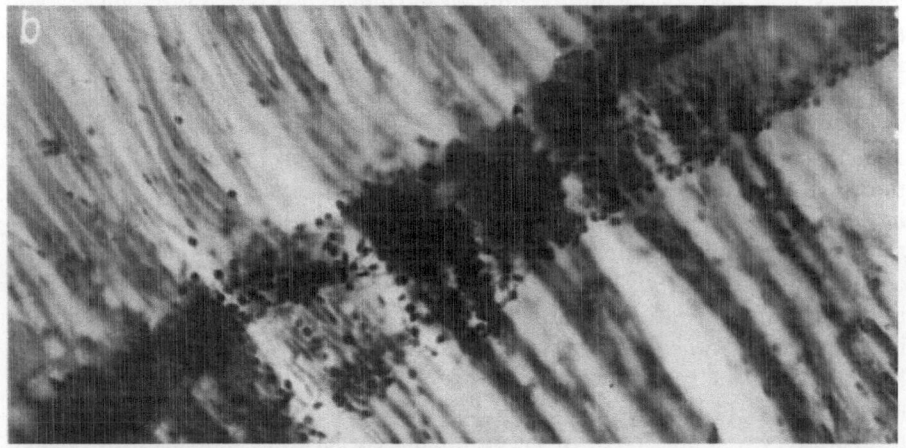

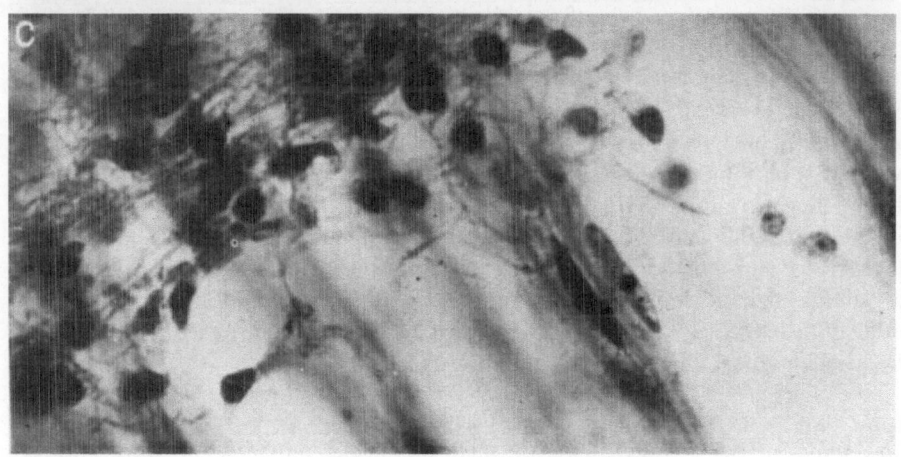

large number of Golgi 2nd type neurons with a bushy arborization of their axons, or it may become an almost random network, as in some parts of the hypothalamus (Szentágothai *et al.* 1962). But in all cases, the above three principles remain essentially unchanged.

These brief considerations may suffice to show how important the study of these nerve feltworks actually is, and how important it is not to be satisfied with the formal statement that nerve elements with ill-defined processes are interconnected in some way or other. Indeed, suitable combinations of conventional light microscopy methods including specific nerve stains and degeneration methods with EM studies both normal and using degeneration are extremely powerful tools the possibilities of which have so far been exploited on only the most superficial level. It is my object to call the attention of neurobiologists, particularly those working in the field of functional neuroanatomy, to certain technical possibilities that can be used with advantage in the analysis of neural networks. It is obvious that there are no simple rules that could be applied in stereotyped fashion, the diversity of such networks requiring careful adaptation of certain basic principles to the varying circumstances.

The mode of approach that I am advocating is the method of 'persisting elements' in completely or partially isolated tissue regions containing fragments of the nervous network under study. The most radical procedure along this line involves the tissue culturing of pieces of nervous tissue. As is known, this kind of work has, in recent years, yielded most important fundamental results concerning the relation between impulse transmission and the structural development of neuron connections (Crain *et al.* 1964). However, with tissue culturing, much of the original topographic relations of the neurons is lost. If isolation or partial isolation is carried out *in loco* with preservation otherwise of the usual environment and/or blood supply, the topographical relations of the several nerve elements having their perikarya inside the isolated fragment, and hence not undergoing degeneration, are very well preserved. It was customary earlier to let animals with isolated neuronal tissue slabs survive for at least two months (Szentágothai 1958, 1962, 1964, 1965) in order to make absolutely sure that all nerve elements of extrinsic source were degenerated. Later, however, when EM studies had made it clear that degeneration of the true synaptic terminals is much quicker than expected on the basis of light-microscope studies, it turned out that there was no need to wait for such long periods. In some regions — for example, in the spinal dorsal horn — degeneration of synapses is so rapid that two to three days of postoperative duration are sufficient to have all synaptic terminals of extrinsic origin not only degenerated but mostly removed or taken up into the glia. This is of considerable advantage, since in such a short time, the original topography of the elements is unlikely to be greatly changed because of regeneration, transneuronal atrophy of the persisting nerve elements, or glial hypertrophy.

As examples for this mode of approach, two figures are shown from a recent study aiming at an identification of Renshaw cell synapses on motoneurons and an analysis of the intrinsic synaptic relations of the dorsal horn. Both studies,

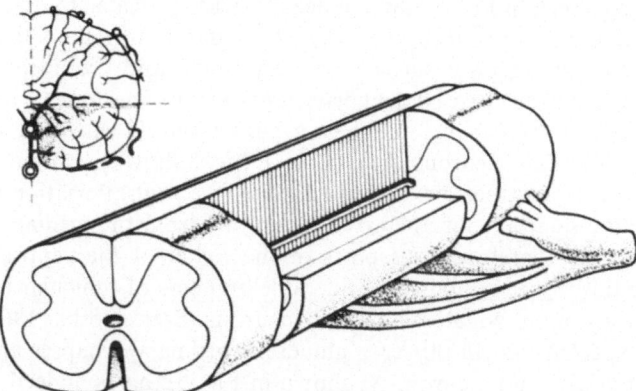

FIG. 2. Diagram illustrating the operation procedure to stain an isolated ventral horn preparation. Small diagram above shows the two planes of cut (dashed line) and that in consequence of injuring the arteries of the anterior fissure most part of the anterior horn usually undergoes necrosis. Further explanation in the text.
(From Szentágothai 1958)

made by Réthelyi and Szentágothai, will be reported in more detail elsewhere, and only some preliminary information will be given here to illustrate the possibilities offered by this experimental approach.

Figures 2 and 3 show the result of the repetition of an older experiment by Szentágothai (1958), now supplemented with EM analysis of motoneurons. In order to study the mutual interconnections between spinal motoneurons and Renshaw neurons, as postulated by Eccles *et al.* (1954), a chronically isolated ventral horn preparation has been developed. This is accomplished by splitting the enlarged part of the lumbar cord in the length of about 1.5–2 segments. One half is left to ensure approximately satisfactory movement for one hindleg, and the other half is split horizontally into a dorsal and a ventral quadrant; the dorsal with the entering dorsal roots is removed, and the remaining ventral quadrant is crushed by a watch-makers forceps at the upper and lower ends of the split. Thus (Fig. 2) an isolated ventral quadrant is gained, which is supplied by vessels entering through the remaining ventral rootlets. The fact that most of the grey matter is supplied by the deep fissure arteries (Fig. 2, upper diagram) is of advantage, since only a small halfmoon-shaped strip of grey matter remains, in most cases intact of the ventro-lateral border of the ventral horn including the region at the site of the emergence of the ventral rootlets where the Renshaw cells are now known to be localized (Thomas and Wilson 1965). In the original study of Szentágothai (1958), in such chronic preparations no boutons of ordinary size were found on the surface of motoneurons, only an extremely delicate network of fibres that was barely visible with the highest light-microscopy power. It was concluded, therefore, that the terminals of Renshaw axons on motoneurons ought to be of unusually small size. As seen in Fig. 3, this is indeed the case provided that nothing of the central part of the ventral horn survives. The synaptic terminals that persist in contact with the motoneuron surface are very small, and are deeply embedded into the motoneuron surface, and the postsynaptic membrane appears to be thickened below the entire contact region. These synapses should thus be the inhibitory synapses of the Renshaw neurons. As a control, Fig. 3 shows part of a motoneuron surface from a similar experiment in which, due to a slight deviation of the knife when making the sagittal split, the arteries entering from the median fissure were not destroyed, and therefore large parts of the center of the ventral horn were preserved. As this region contains considerable numbers of funicular interneurons having numerous initial collaterals to motoneurons (Szentágothai 1967, Fig. 5) it ought to be expected that in this case numerous ordinary synapses are preserved on the surface of the motoneuron. As shown in Fig. 3, this is, indeed, the case.

Figure 4 shows the repetition for EM investigation of another chronic isolation experiment, designed for the study of intrinsic interneuronal connections of the substantia gelatinosa (Szentágothai 1964). For the details of the experiment, the reader is referred to the original article. In this case — taking advantage of the very rapid degeneration and removal of axon fragments in this region — the animal was left to survive for only two days after the experiment instead of two months, as originally used for light-microscopy purposes. As seen from this figure, so soon after the isolation a few degenerated axon fragments can already be detected incorporated into glial processes. The rather small unmyelinated axons, also having synapses with dendrites or dendritic spines, cut lengthwise in this longitudinally oriented section of the neuropile correspond to the intrinsic longitudinal axon system of the substantia gelatinosa, as described earlier (Szentágothai 1964). This picture does not change significantly if the animals are allowed to survive for one to two months, apart from a considerable hypertrophy of the glia that obscures the original topographic relationships of the nerve elements. This example is presented to show that for EM purposes the total or partial isolation of neural tissue slabs need not be maintained for longer periods, which is a great advantage avoiding disturbing influences from regeneration, glial hypertrophy and transneuronal atrophy. The synapses persisting in this isolated dorsal horn preparation can thus be separated from the many and various kinds of synapses established in the substantia gelatinosa by extraneous elements.

FIG. 3. a, b and c — Small synaptic terminals of Renshaw cell axons (Rt) persisting on motoneuron cell body surface (Mcb) of chronically (2 months) isolated ventral horn preparation. d — Same kind of preparation with preservation of part of the ventral horn center. Persisting large axon terminals (Lat) in contact with motoneuron cell body (Mcb) are endings of initial collaterals of funicular neurons that have survived in the center of the ventral horn. — Dog, scale 1 μ

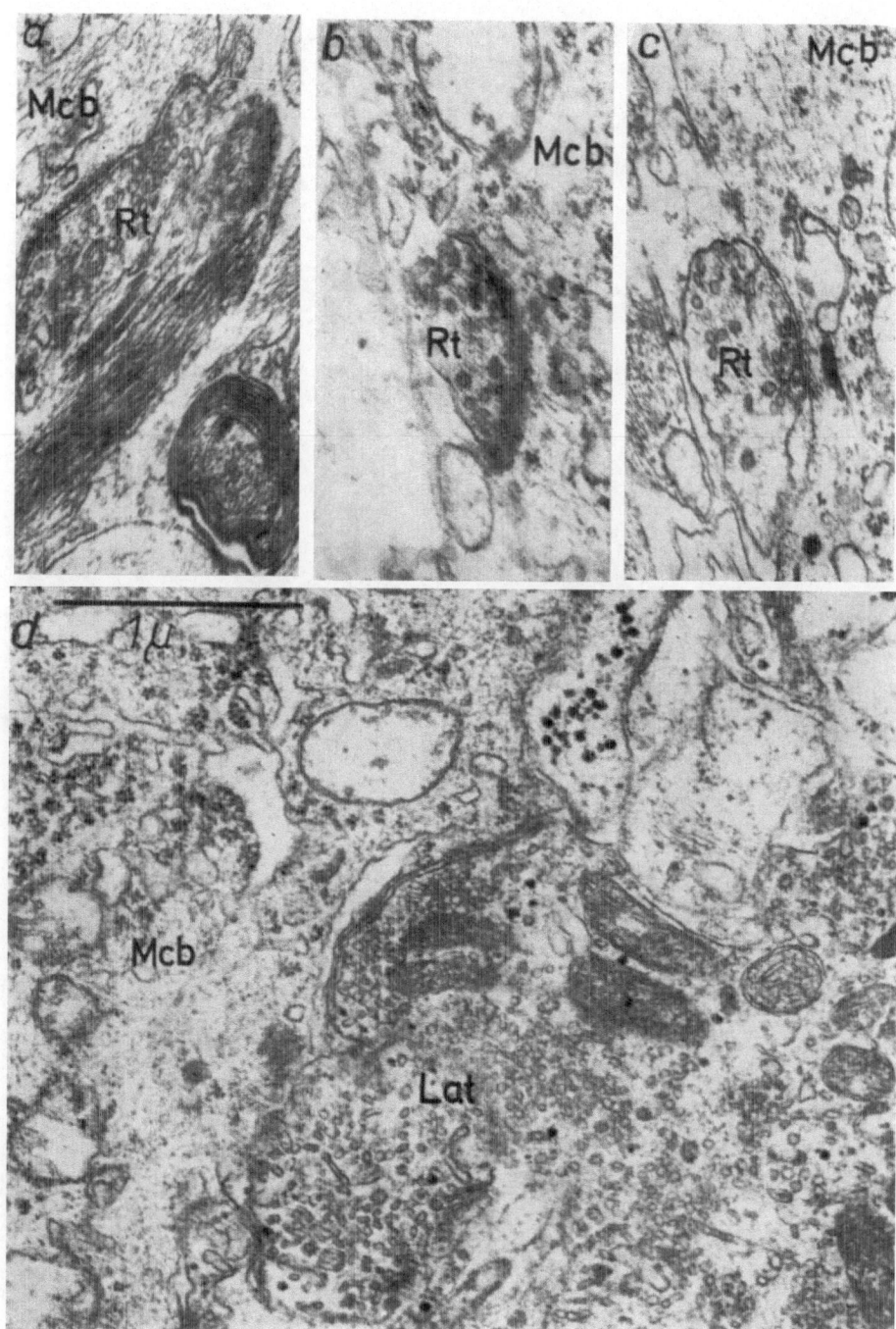

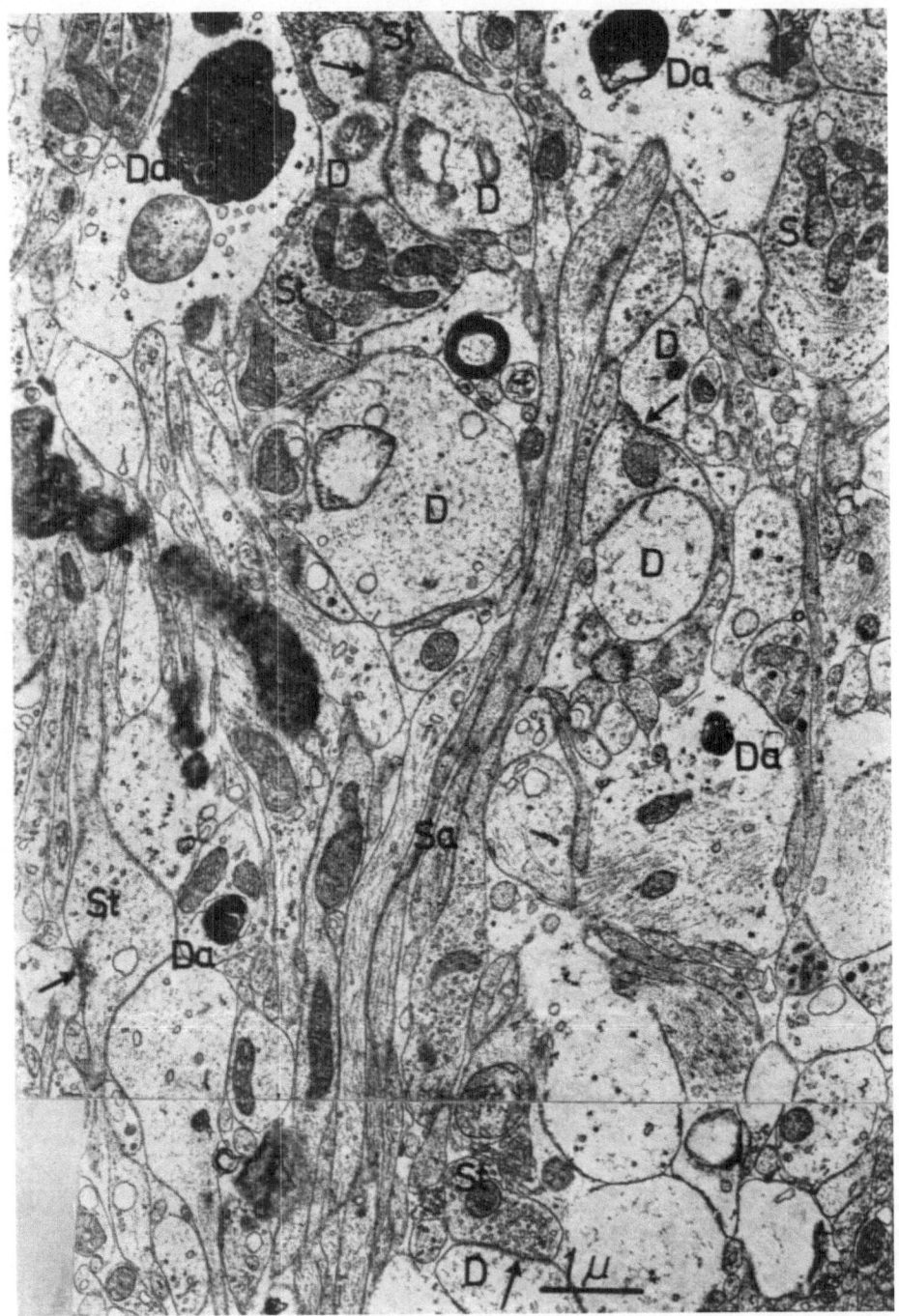

SUMMARY

Attention is drawn to the difficulties in dealing with feltworks built up of short neurons or short connections established by initial collaterals. The importance of such neuron networks both in lower and in higher animals justifies the introduction of new experimental-histological techniques for their study. An attempt is made to show by two simple examples dealing with the spinal cord how the technique of neural isolation of tissue fragments with preservation of their blood supply can be exploited with advantage on the level of the electron microscope for identification of the synapses of certain short interneuronal connections.

REFERENCES

CAJAL, S. R. y (1909): *Histologie du Systeme Nerveux de l'Homme et des Vertébres*. Maloine, Paris, Vol. 1.

CRAIN, S. M., BUNGE, R. P., BUNGE, M. B. and PETERSON, E. R. (1964): Bioelectric and electron microscope evidence for development of synapses in spinal cord cultures. *J. Cell Biol.* 23 114—115.

ECCLES, J. C., FATT, P. and KOKETSU, K. (1954): Cholinergic and inhibitory synapses in a pathway from motoraxon collaterals to motoneurons. *J. Physiol. (Lond.)* 126 524—562.

GRAY, E. G. (1964): The fine structure of normal and degenerating synapses of the central nervous system. *Archieves de Biologie (Liége)* 75 285—299.

HÁMORI, J. and HORRIDGE, G. A. (1966): The lobster optic lamina. III. Degeneration of retinula cell endings. *J. Cell. Sci.* 1 271—274.

SZENTÁGOTHAI, J. (1958): The anatomical basis of synaptic transmission of excitation and inhibition in motoneurons. *Acta morph. Acad. Sci. Hung.* 8 287—309.

SZENTÁGOTHAI, J. (1962): On the synaptology of the cerebral cortex. In: *Structure and Function of the Nervous System*. (Ed.: S. A. Sarkissov) Medgiz, Moscow, 6—14.

SZENTÁGOTHAI, J. (1964): Neuronal and synaptic arrangement in the substantia gelatinosa Rolandi. *J. comp. Neurol.* 122 219—240.

SZENTÁGOTHAI, J. (1965): The synapse of short local neurons in the cerebral cortex. In: *Modern Trends in Neuromorphology*. (Ed.: J. Szentágothai) *Symp. Biol. Hung.* 5 Akadémiai Kiadó, Budapest, 251—276.

SZENTÁGOTHAI, J. (1967): Synaptic architecture of the spinal motoneuron pool. In: *Recent Advance in Clinical Neurophysiology. Electroenceph. Clin. Neurophysiology*, (Ed.: L. Widén, Suppl. 25.) Elsevier, Amsterdam, 4—19.

SZENTÁGOTHAI, J., FLERKÓ, B., MESS, B. and HALÁSZ; B:, *The Hypothalamic Control of the Anterior Pituitary*. Akadémiai Kiadó, Budapest, 1962.

THOMAS, R. C. and WILSON, V. J. (1965): Precise localization of Renshaw cells with a new marking technique. *Nature (Lond.)* 206 211—213.

FIG. 4. Neurally isolated substantia gelatinosa after two days survival. Fragments of degenerated axons (Da) can be recognized as dark bodies incorporated into light glial profiles. Persisting small intrinsic axons (Sa) and their synaptic thickening (St) can be seen, some in contact with dendrites (D). Sites of synaptic contacts are indicated by arrows. — Cat, scale 1 μ

NEUROMORPHOLOGY

Symposium on Neurobiology of Invertebrates 1967 (29—48)

STRUCTURAL, CYTOCHEMICAL AND FUNCTIONAL ORGANIZATION OF STATOCYSTS OF CEPHALOPODA

Ya. A. Vinnikov, O. G. Gasenko, A. A. Bronstein, T. P. Tsirulis, V. P. Ivanov and G. A. Pyatkina

Laboratory of Evolutionary Morphology, Sechenov Institute
of Evolutionary Physiology and Biochemistry,
Academy of Sciences of the USSR
Leningrad, USSR

In the process of animal evolution, well-developed organs responding to changes in position with regard to the earth's gravity field have appeared not only in vertebrates which belong to Deuterostomia, but also in some specimens of Protostomia. They are particularly well impressed in Cephalopoda, which possess specialized statocysts that contain maculae with otoliths and cristae with cupules. Physiological investigations (Boycott, Young, etc.) have shown that the macula of a statocyst is a static receptor organ signaling the orientation of the head and eyes of an animal, whereas cristae arranged in three planes respond to angular accelerations. Hence, statocysts of Cephalopoda are in their anatomy and function, analogous to the vestibular apparatus of vertebrates.

A question arises in this connection: what are the likenesses (if any) and what are the differences between the ultrastructural and cytochemical organization of the vestibular organ in vertebrates and the statocysts in Cephalopoda, which have been developed independently in the process of evolution? While a considerable number of electron-microscope and histochemical investigations (Wersall 1956; Vinnikov *et al.* 1965) have been concerned with the vestibular apparatus of vertebrates, the statocysts of cephalopods have received little attention.

Besides the histological study of these organs in *Octopus* by Young (1960), there is only a brief preliminary communication of Barber (1960) on the ultrastructural organization of maculae and cristae of *Octopus vulgaris*.

MATERIALS AND METHODS

In this study, we used 36 octopi *(Octopus sp.)* and 40 squids *(Ommastrephes sloanei pacificus)* inhabiting the sea of Japan not far from Putyatin Island. In decapitated animals, the cartilaginous capsule of the statocyst was revealed under the binocular microscope and its cavity embedded in a fixative fluid. A subsequent additional fixation and histochemical treatment were performed after maculae and cristae had been revealed.

To obtain histological preparations, fluids of Bouin, Carnoy and Zenker were used. For histochemical studies, the isolated statocysts fixed in 1% formol and 5% neutral formol were transferred to standard incubation media to reveal succinic-dehydrogenase activity according to Nachlas *et al.* (1957) and acetylcholinesterase activity according to Koelle (1951). After the completion of the reaction, the material was additionally fixed in formol and investigated on total preparations or sections obtained on the freezing microtome after the receptor organs had been embedded in gelatin.

For electron-microscope studies, maculae and cristae were fixed for 1–2 h in 1% OsO$_4$ in seawater and embedded in butylmethacrylate or araldite. The material was also fixed in 5% glutaraldehyde prepared on seawater with additional fixation in 1% OsO$_4$ and embedding in araldite. It was then sectioned on the LKB–4800 ultratome and, after contrasting with plumbum citrate, observed in the JEM–6c electron microscope.

RESULTS

Paired statocysts of octopi and squids are located in the base of the skull in front of a funnel leading to the mantle cavity. They are represented as a membraneous oval sac filled with endolymph. In octopi this sac is immersed in a perilymphatic cavity limited by the cartilaginous capsule. It is supported by connective tissue linings pierced by vessels and nerves. In squids, there is no perilymphatic cavity and the membraneous sac adjoins cartilaginous walls of the capsule. The statocysts are lined internally (except for receptor regions of maculae and cristae) with one-layer flat epithelium lying on the fibrous connective tissue lining with carti-laginous light stratum. In connective tissue, one can find single muscle fibers, hair receptor cells, and nerve endings.

Both in octopi and squids, the macula is an oval thickening on the anteromedial wall of the statocyst, i.e., its location corresponds to that of the saccular macula in vertebrates. In octopi of 2–4 kg, the macula has the shape of a regular ellipse (1.5–2.0 mm) with a conical otolith above it. The squid macula (1.0–0.8 mm) has a rough surface. The overcovering otolith is shaped like a holed disk. It should be pointed out that in squids, in addition to the main macula, we find two (upper and lower) small additional maculae covered by a layer of small otoconia and known as maculae neglectae. They lie medially at a right angle to the main macula especially the lower one, occupying a spatial position of the utricular macula of vertebrates.

Fig. 1—11. The structural organization of the macula receptor layer in cephalopods. A — axon; ANF — afferent nerve fiber; BB — basal body; Ot — otolith; C — cupule; ENF — efferent nerve fiber; ENE — efferent nerve ending; F — foot of the basal body; K — kin-ocilia; M — mitochondria; Mv — microville; MER — membrane of endoplasmic reticulum; O — otoko- nia; OM — otolith membrane; Ph — phalanx of the supporting cell; RC — receptor cell; R — rootlet; S — stereocilia-like process; SC — supporting cell; UNC — unipolar nerv ecell; N — nucleus

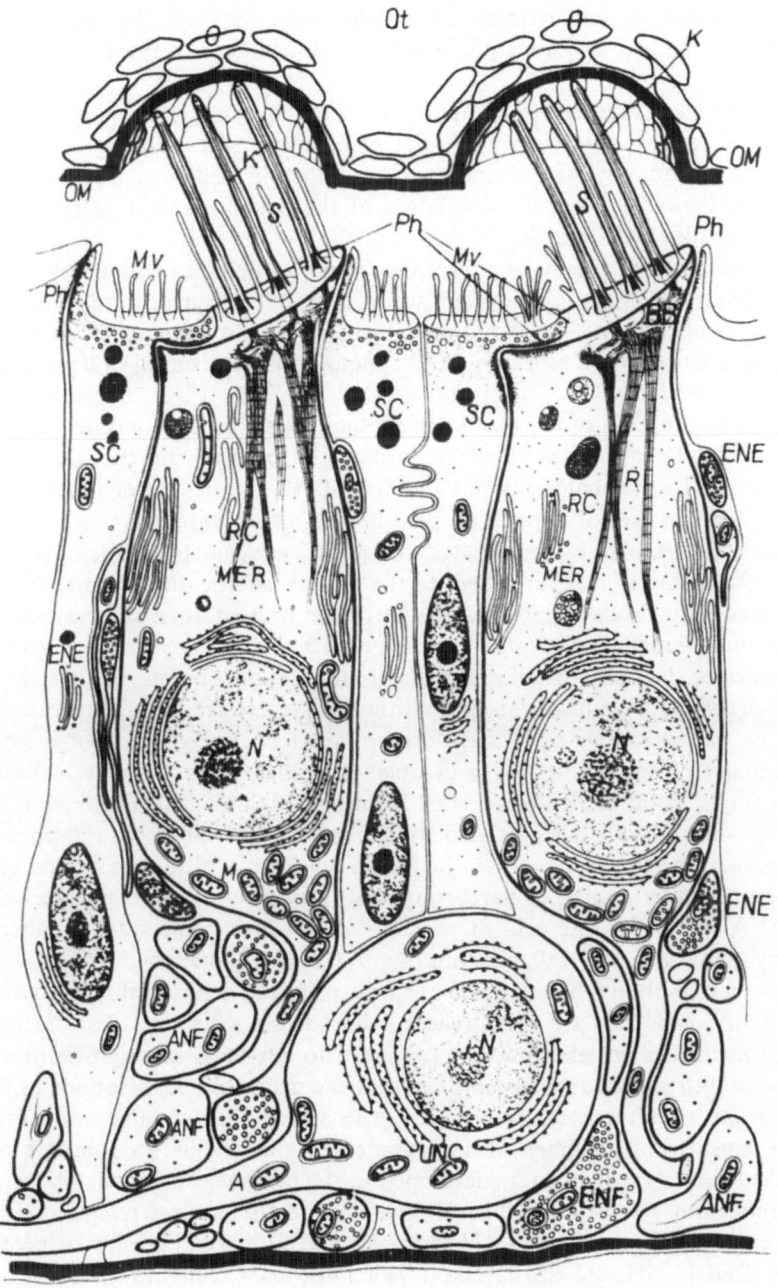

In octopi, there is a nerve trunk along the long axis of the macula that runs into central ganglia. In squids, the nerve trunk is formed outside the macula.

The macula of a statocyst is composed of three layers: (a) primary receptor cells with hairs; (b) underlying layer of large neurons; and (c) a layer of nerve fibers. Scattered among these elements are supporting cells running from the surface to the base of the macula. The edges of the macula consist of supporting cells which gradually diminish.

Sensory hair cells of the macula have the form of a cylinder with a conical top (Fig. 1). Because of this, the apical surface of each cell is bent toward the central axis of the macula, along the length of which nerves are running. Such a slope probably may be regarded as a sign of morphological polarization of the receptor cell.

Hairs of receptor cells of the macula in cephalopods consist of motile kinocilia- and stereocilia-like processes (Fig. 2). It must be pointed out that unlike vertebrates, we find in cephalopods not one but 70–120 kinocilia per sensory cell, containing a typical set of 9 pairs of peripherical and 2 unpaired central fibrils. Resembling the maculae of vertebrates, the apexes of kinocilia, of length 5–7 μ and 0.3 μ in diameter, are attached to the beams of the otolith membranes. The latter contain dome-like niches in the upper part of each cell, into which bundles of hairs enter. Such a structure not only holds a large otolith above the macular surface, but also promotes a direct stimulation of sensory cells by the action of shearing forces of the otolith. At the sites where kinocilia come out into the surface of a receptor cell, there are rounded or polygonal cavities about 0.5 m in diameter surrounded by a cytoplasmic roller firing 4–5 stereocilia-like processes (Figs. 1 and 2). Kinocilia start from basal bodies which have similarly oriented feet-like cones. Cross-striated roots come out spirally from the opposite side of the basal bodies just beneath the feet. They form a plexus above the basal bodies from which a number of narrowing roots repeating with a major period of about 500 Å run inside the cell. Stereocilia-like processes, the total number of which may amount to 160–180, are thin (about 0.1 μ in diameter) evaginations of the cytoplasm resembling the microvilles that surround each kinocilium as rosettes. Strereocilia-like processes are about twice as short as kinocilia and, like the stereocilia in the macula of vertebrates, they probably do not show any significant relation to the otolith membrane. Receptor cells of the macula in cephalopods exhibit a great variety of ultrastructures. Besides the above cross-striated roots, they contain a considerable number of membranes of the rough endoplasmic reticulum. The latter is most strongly developed in the perinuclear region. The endoplasmic reticulum forms a peripheral zone around the nuclear region (Fig. 1). In this area, a large amount of RNA, protein, and functional groups of protein molecules (COOH, SH, SH–SS) can be revealed histochemically. In the cytoplasm of the supranuclear region of hair cells there are numerous mitochondria, structures of the Golgi apparatus, vacuoles of various size, multivesicular bodies, and lysosomes. A round nucleus is a bit shifted to the base of the hair cell, which

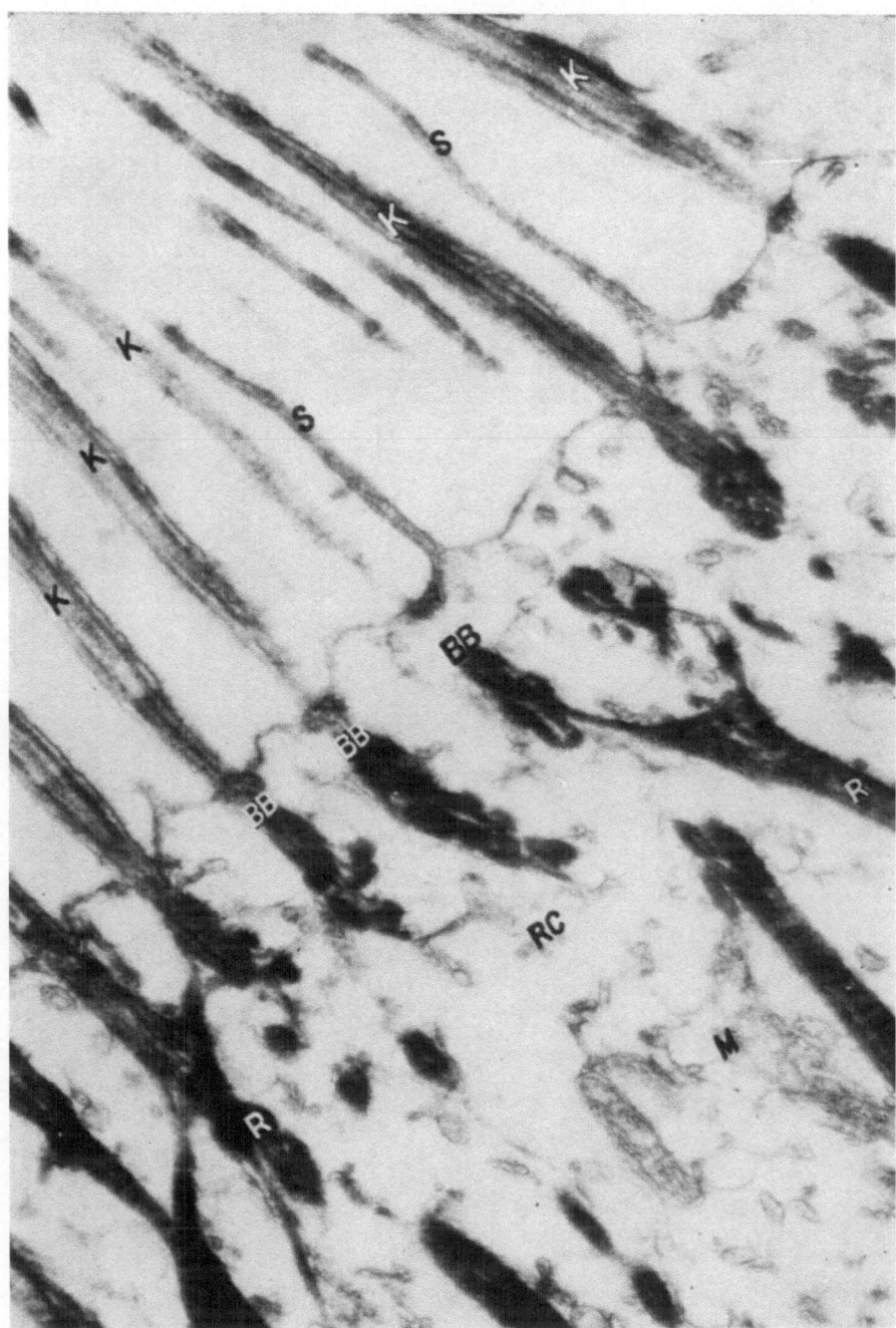

FIG. 2. The top of a hair cell in the macula of the Octopus. × 40,000

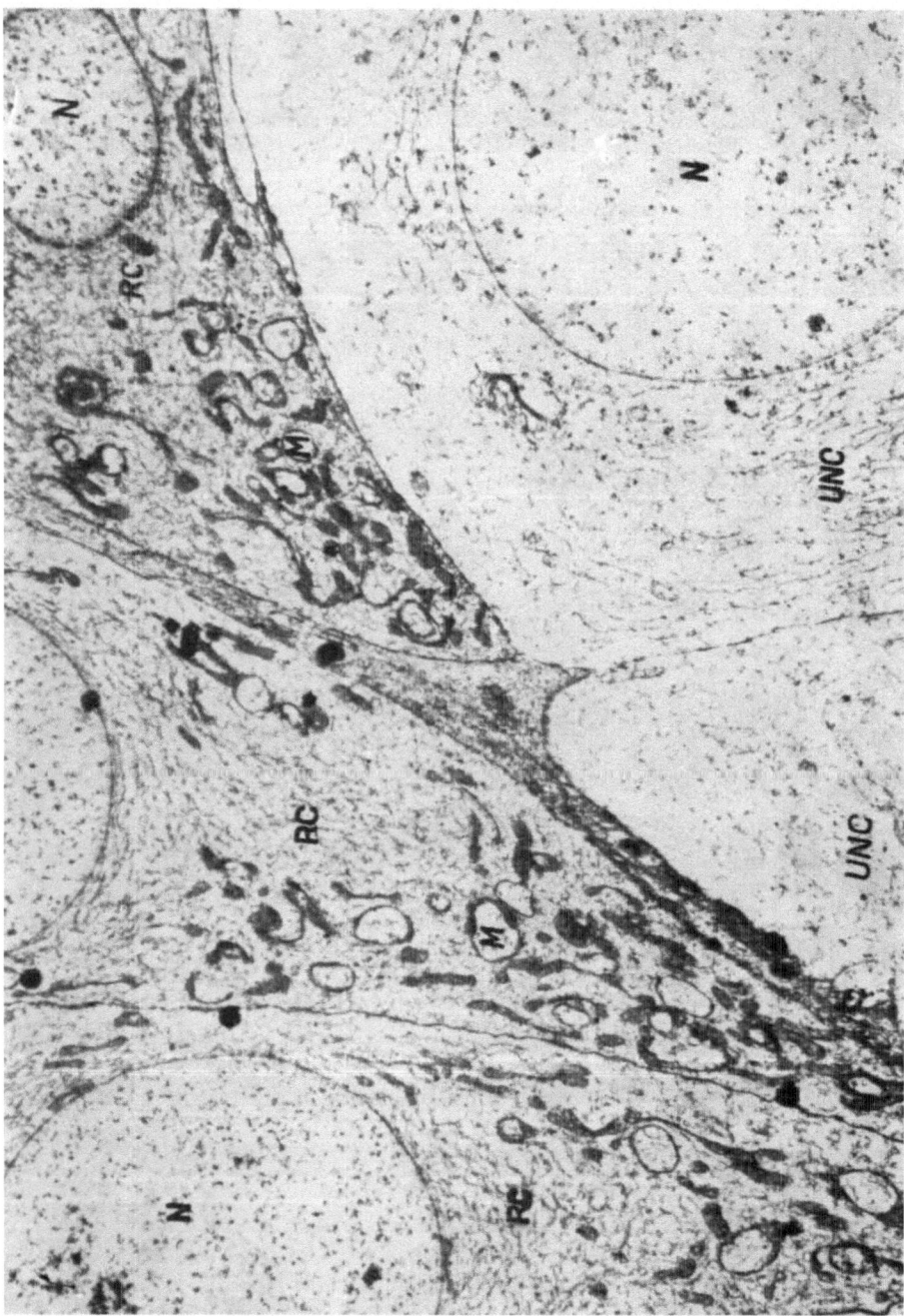

FIG. 3. Basal parts of receptor cells at the site of their transition into axons and bodies of unipolar neurons

gradually narrows making a characteristic cone that continues as an axon penetrating deep between the underlying bodies of nerve cells (Fig. 3). The cone and the axon are characterized by a great number of mitochondria. Due to this, when revealing the activity of oxidative enzymes in the light microscope, one may simultaneously observe an outline of the lower part of the receptor cell and the axon. We never observed dendrite emerging from the body of a receptor cell, as had been suggested by Young (1960). Our data show that large oval cells lying in the second layer of the macula are unipolar neurons. The suggestion that they are multipolar (Young 1960) or glial cells (Barber 1965) seems to be erroneous. These neurons have a characteristic axonal cone giving off a thick unipolar process (Fig. 4). The branching of this process has not yet been observed. The cytoplasm of a nerve cell is abundant in mitochondria, elements of the Golgi apparatus, vacuoles, and, particularly in the membranes of the endoplasmatic reticulum (Figs. 3 and 4), presenting in the light microscope a typical picture of Nissl substance containing RNA and proteins with sulfhydril and carboxyl groups.

Axons of various diameter can be found in the nerve fiber layer. Among these axons, however, are distinctly visible 'light' fiber-containing fibrils, large mitochondria, and a small number of synaptic vesicles, as well as 'dark' fibers exclusively rich in synaptic vesicles (200–300 Å in diameter) (Fig. 4). 'Dark' and 'light' nerve fibers along the whole length of the layer may establish synaptic relationships. Hence, the layer of macular nerve fibers may be regarded as a neuropile. The 'dark' fibers proceed outside the neuropile and, passing around the unipolar neurons, end in synapses on the lateral surface of hair receptor cells. The histochemical treatment of the macula, sensitive to the presence of acetylcholinesterase, has shown the cholinergic nature of these fibers (Fig. 5). Some cholinergic nerve fibers may go beyond the limits of the neuropile, forming nerve endings in a wall of the statocyst membraneous sac. The above results permit a suggestion that the 'dark' fibers which are abundant in synaptic vesicles and show a high acetylcholinesterase activity are efferent, as is the bundle of Rasmussen of the vertebrate labyrinth.

Supporting cells of the macula form linings between receptor and nerve cells on one side and nerve fibers on the other (Fig. 1). However, all the above elements, including hair receptor cells, can make contacts in certain regions (Fig. 3). At the top of the receptor layer supporting cells make up phalanges dividing apexes of receptor cells, the phalanx portion of the supporting cells form a mushroom-like widening which covers the lower part of the sloping surface of the receptor cell to which it is connected by desmosomes. On the opposite side, the phalanx makes a protrusion which laterally covers the bulging portion of the top of another receptor cell (Fig. 1). Due to this fact, only that part of the surface of the receptor cell which is uncovered by the phalanx and bearing hairs is facing the otolith membrane and immersed in endolymph. The surface of the phalanx is covered with branching microvilli which may even penetrate between hairs of the sensory cells. In the middle of a supporting cell there is an elongated nucleus of high electron density. The cytoplasm of the supporting cells is filled up with numerous granules including

YA. A VINNIKOV et al.

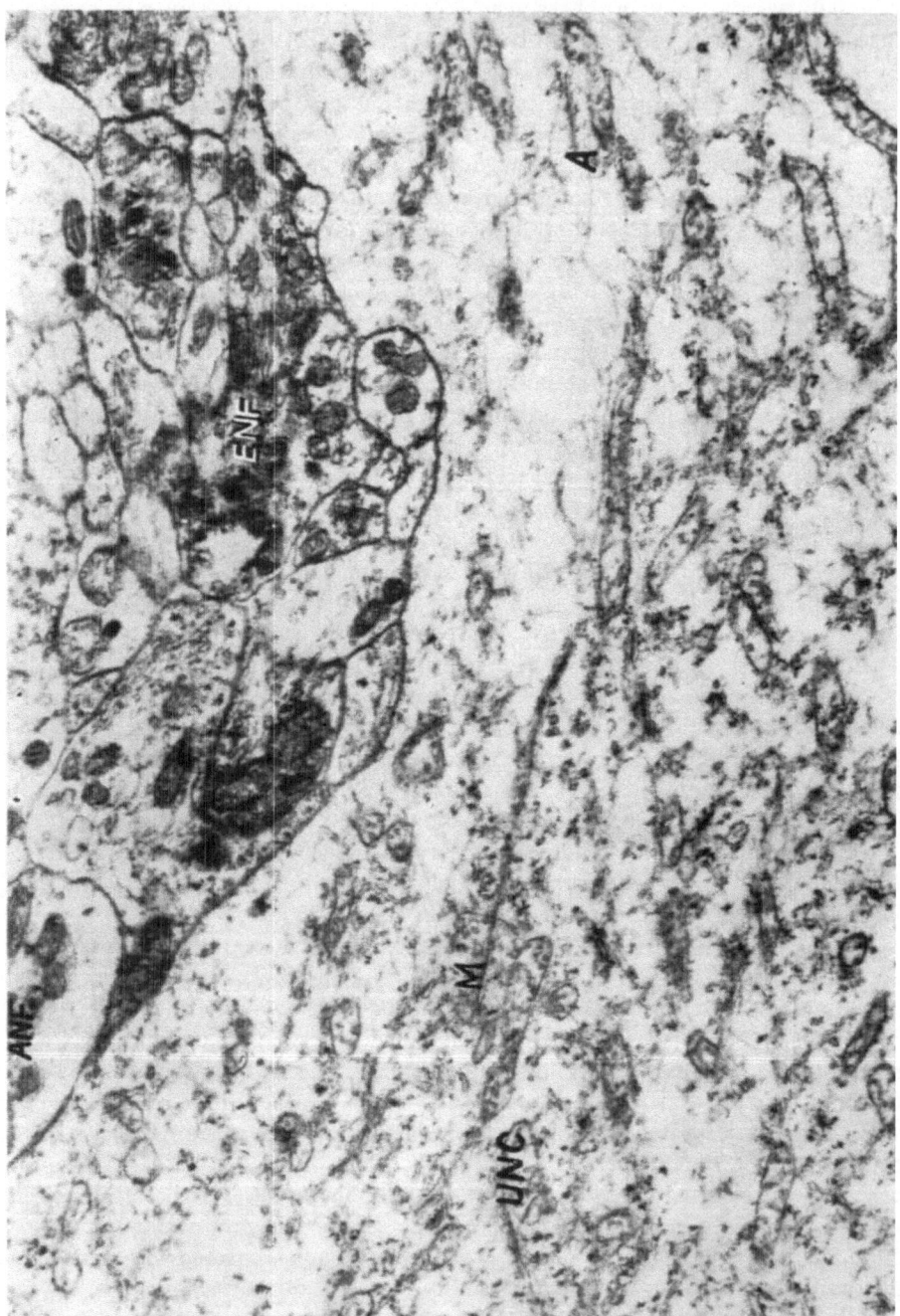

FIG. 4. The process emerging from the unipolar neuron in the Octopus macula. × 20,000

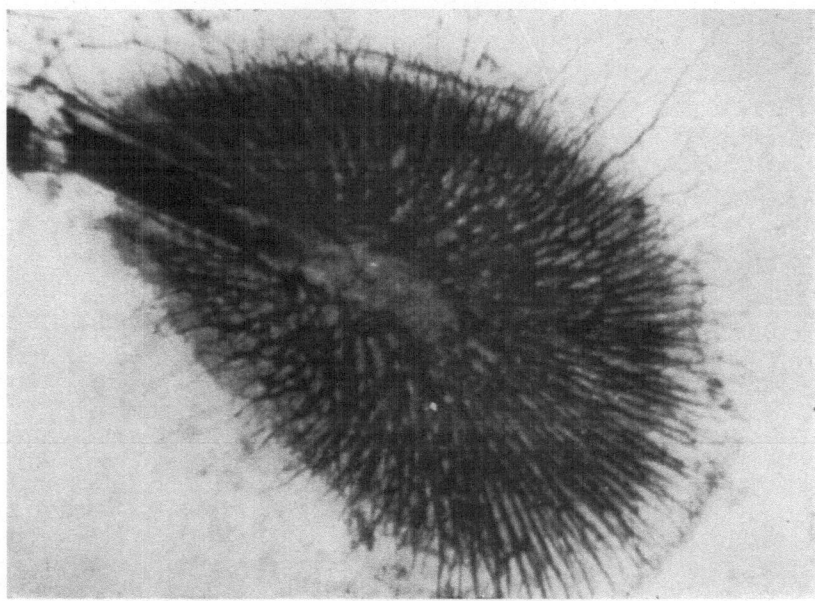

FIG. 5. Acetylcholinesterase activity in the nerve trunk and emerging nerve fibers in the Octopus macula. Plane preparation, by Koelle's method

secretory granules, mitochondria, elements of the Golgi apparatus, and nonfibrillar bundles.

Cristae of the statocysts in Cephalopoda have the appearance of a long comb 1–0.9-mm wide covered by a transparent cupule. They lie on a pad of cartilage and, as in vertebrates, are located in three perpendicular planes. At the anterior surface of the statocyst, a crista makes a semicircle in the horizontal plane. This transverse horizontal crista has a separate nerve trunk. One crista bends backward along the lateral surface; this is a longitudinal horizontal crista, to which a separate nerve is attached. Toward the caudal end, a crista rises sharply upward, forming a vertical portion supplied with a nerve. Thus, in cephalopods, unlike vertebrates, two cristae do not have a vertical but instead a horizontal orientation. In octopi the main parts of the crista are divided into three parts. In squids, such division does not exist. Adjacent to the cristae lie finger-like cartilagenous protrusions — anticristae (Owsjannikov and Kovalevsky 1867, Hamlyn-Harris 1903). In the octopus, there is only one anticrista, while in the squid their number are equal to 11. Variation in the distribution of anticristae in cephalopods support the statement of Owsjannikov and Kovalevsky (1867) and also a suggestion of Young (1960) that they change the flow of endolymph to provide a differentiated stimulation of various parts of the crista when the animal is moving in different directions. Cristae in the statocysts of cephalopods have the same structure irrespective of their position in the organ. Each of its sections in the octopus and squid is a sep-

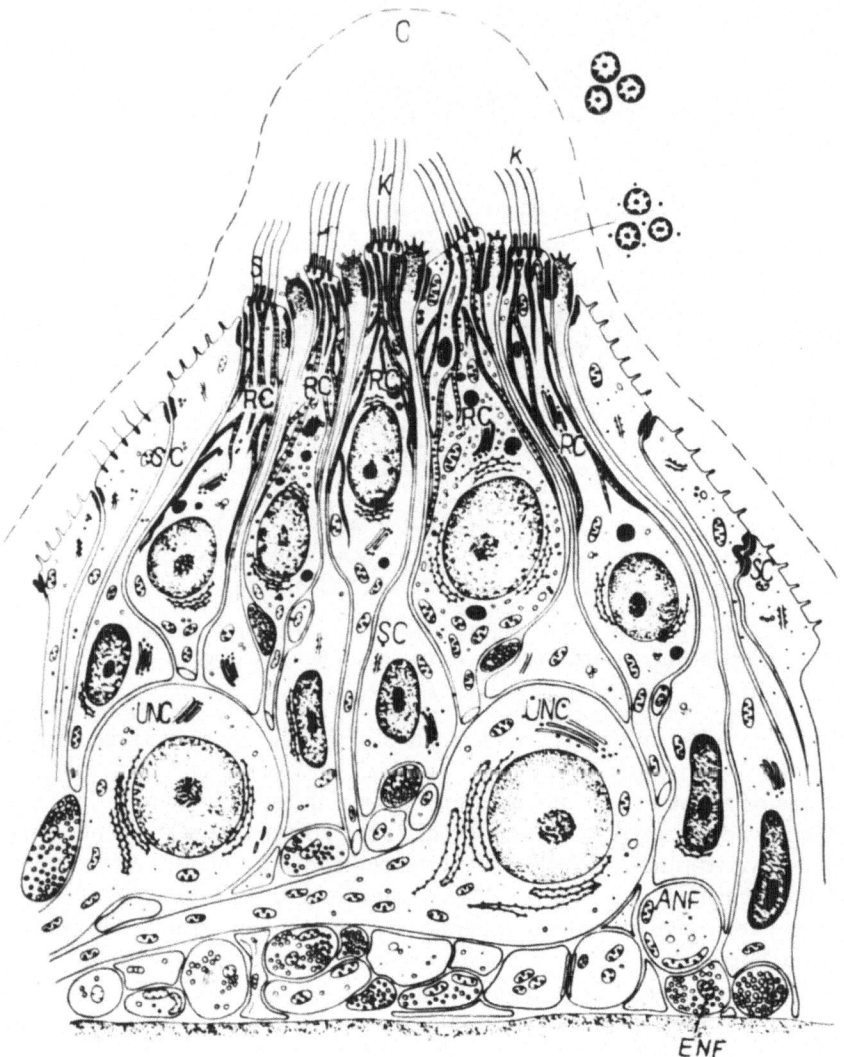

FIG. 6. The structural organization of the receptor layer in the cristae of cephalopods

arate functional unit enclosed in a transparent cupule. The main part of a crista consists of a series of large hair receptor cells occupying the apex. Beneath this there are few large unipolar neurons. The base of the crista is bedded with a layer of nerve fibers. All these elements are held up by supporting cells proceeding as a shaft lining from the cartilaginous bad to the apex of the crista between receptor cells. Other supporting cells have microvilles arranged symmetrically on each other forming a tilted lining along the lateral surface of the crista.

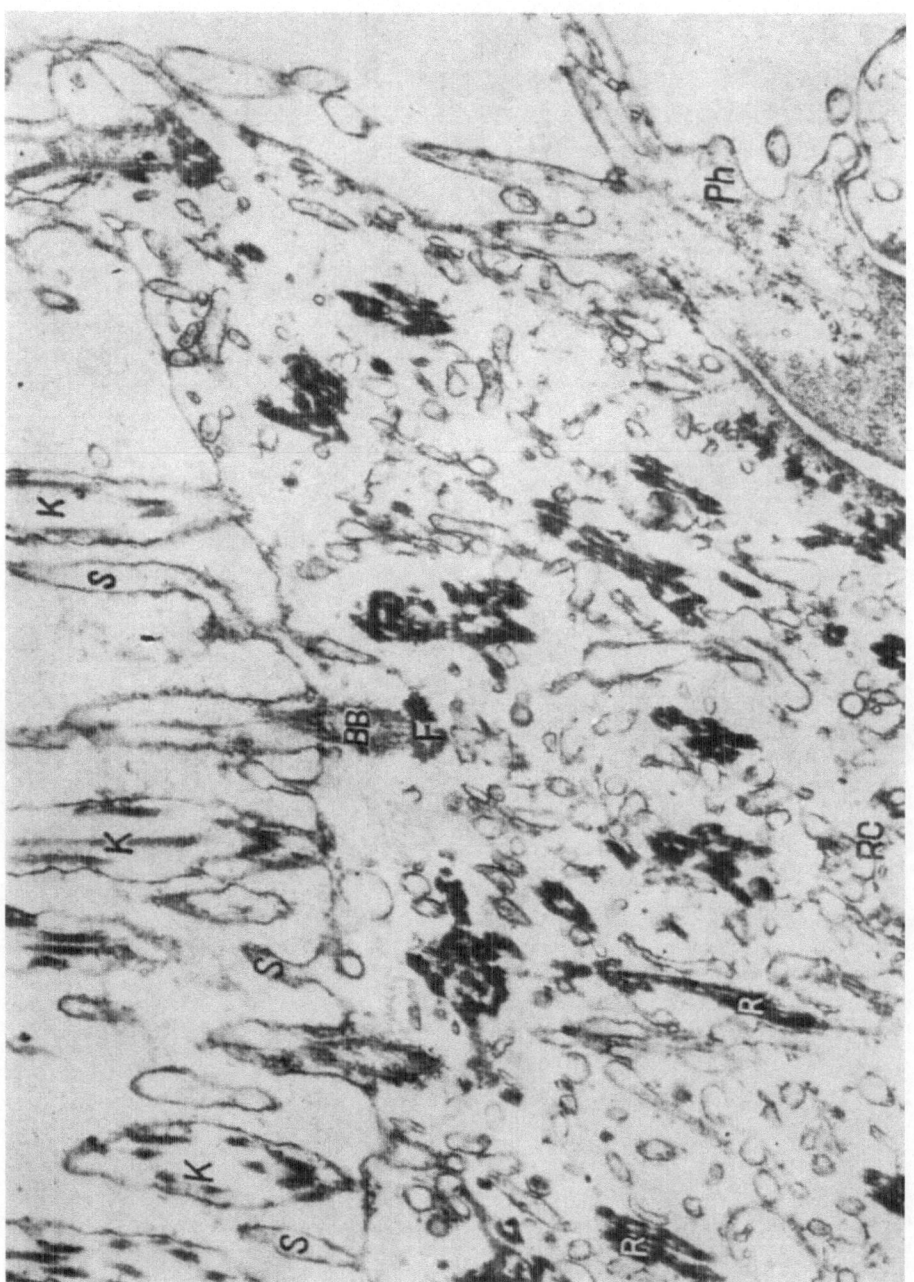

FIG. 7. The top of the lateral hair cell of the squid crista. × 37,000

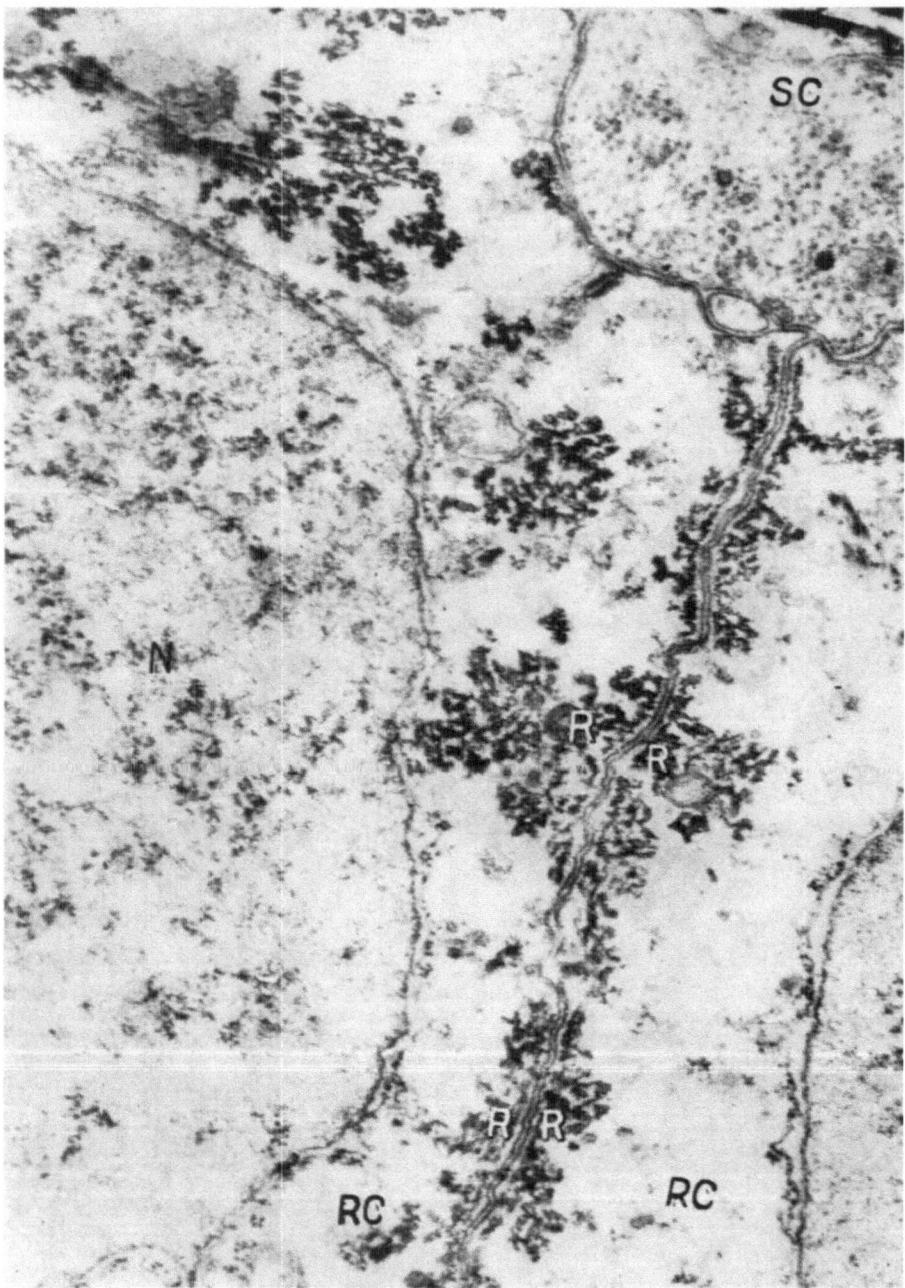

FIG. 8. Transverse section across the body of a receptor cell in the squid crista. The contact of rootlets with the plasmic membrane. × 48,000

Observation of the transverse and longitudinal sections across the squid crista shows that it consists of five rows of hair receptor cells (Fig. 6). A sharp asymmetry in the dimensions of these cells is very distinct. From one of the lateral surfaces of the crista there arise two rows of lateral hair cells in the shape of narrowing cylinders from the base of which an axon is formed. The third central row in the crista consists of flattened, rather narrow hair cells, the nuclei of which lie higher than the nuclei of other hair cells. The fourth row is composed of the largest cylindrical hair cells, $1/2$–2 times the diameter of all the adjoining receptor cells of the crista and having a dense osmiophilic cytoplasmic matrix. The fifth row consists of lateral hair cells similar in their shape and dimensions to the receptor cells of the first and second rows. The phenomenon described is particularly distinct in the squid crista, but can also be found in octopi. The asymmetry of the dimensions of hair cells has been noted by Young (1960), who did not, however, describe, a number of flattened hair cells at the apex of the crista. This is probably due to the fact that these cells are not clearly visible in the light microscope. Young suggested that in the center of the octopus crista, there lie large hair cells forming one or two rows in the neighboring sections. Unfortunately, such has not yet been observed.

Each of the receptor cells of a crista carries 100–120 kinocilia (Fig. 7). The base of each kinocilium is surrounded by a cytoplasmatic circumferential fall that gives rise to several stereocilia-like processes. The total number of these processes in a cell may be as large as 200. In their structure and dimensions they do not differ from the previously described similar formations in the macula. As in the macula, the apical surface of the lateral receptor cells of the crista are cut out at a sharp angle (Fig. 7). Their slanted surface is turned toward a row of central hair cells with club-shaped tops protruding above other receptor cells. Hairs occupy a strictly vertical position in relation to the cupule. Bundles of lateral hair cells are turned toward them. Slopes of the apex and of the hairs are particularly pronounced in the large lateral hair cells that adjoin only on one of the lateral sides of the crista some central hair cells. It may be concluded that in the crista of cephalopods as well as in vertebrates, we find morphological polarization of receptor cells which in its character rather resembles the polarization of receptor cells in the lateral line of vertebrates (Flock and Wersall 1962). It may be suggested that the sloping of the cupule in either direction only induced excitation in two rows of lateral hair cells: in the first and second ones or in the fourth and fifth ones. Hair cells lying in two opposite rows must have been inhibited. However, taking into account an asymmetrical arrangement of some large hair cells in the crista, it can be suggested that a slope of the cupule in both possible directions in statocysts is not unambigous from the functional point of view. As for the central row of hair cells, their elements interact synergically with any pair of the lateral hair cell rows. As in the macula kinocilia, basal bodies of all hair cells in the crista have roots (Fig. 7) with a regularly repeating period of 500 Å. They form a plexus in the region of the basal bodies, and then as separate processes penetrate deep into the cytoplasm toward the cytoplasmic membrane. In a cross section of hair cells of the

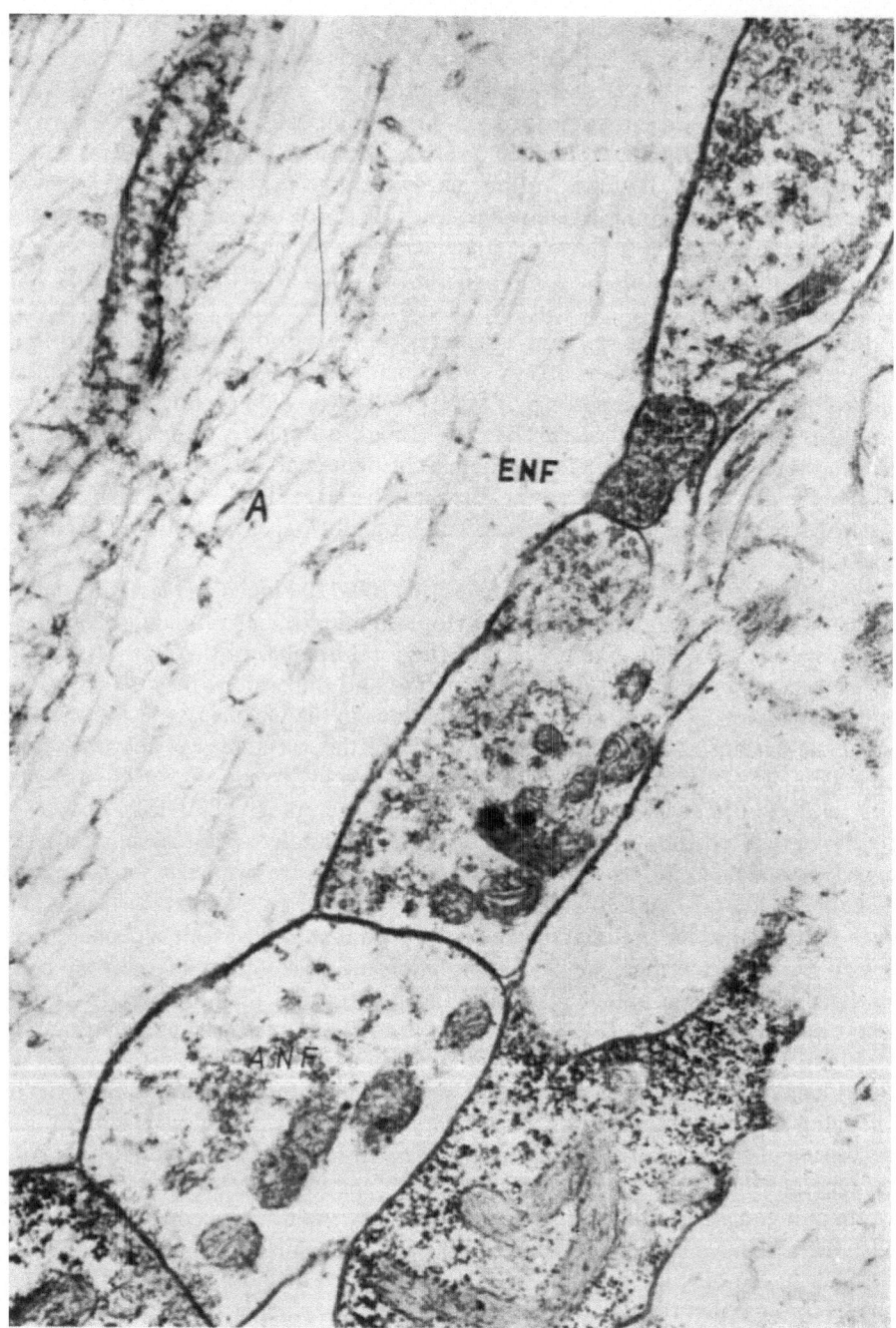

FIG. 9. Nerve fiber layer of the Octopus crista. × 31,000

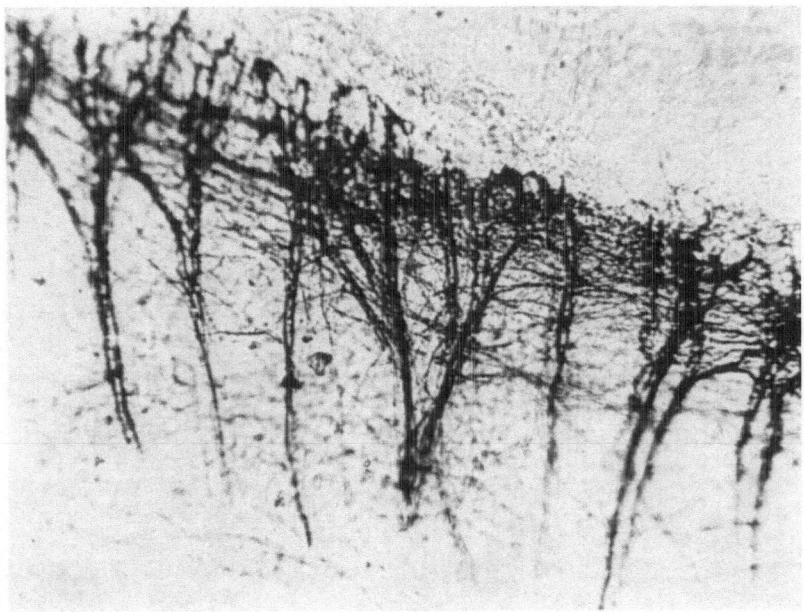

FIG. 10. Acetylcholinesterase activity in efferent nerve fibers of the Octopus crista
by Koelle's method. Micrograph

crista it can be seen that bundles of roots are distributed in a bead-like way along
the whole perimeter of the cell, making an immediate contact with the inner surface
of its plasmic membrane (Fig. 8). The intimate connection of roots with basal
bodies of kinocilia, the formation of a well-pronounced plexus joining all basal
bodies, and, finally, a close contact with a plasmic membrane of the receptor cell
suggest that all these structures participate in the transmission of excitation in
hair cells.

The ultrastructural organization of the cytoplasm in the hair cell of the crista
on the whole resembles the above-described structure of the macula hair cell.
They too are rather rich in membranes of the endoplasmatic reticulum, mitochondria, osmiophilic granules, and multivesicular bodies. A considerable amount
of RNA, as well as a number of protein, thiol, and carboxyl groups, were revealed
histochemically. All these facts may be regarded as evidence showing a considerable intensity of energy and functional processes in hair cells.

As in the macula, large oval cells underlying hair cells turned out to be unipolar
neurons by their ultrastructural and cytochemical organization. Neither in the
light nor in the electron microscope have we observed any signs of branching of
dendrites or processes approaching the surface of the crista and carrying hair
bundles, as had been suggested earlier (Young 1960).

Nerve fiber layers of the crista are less developed than similar layers of the macula. However, they too have the character of a neuropile and contain interacting

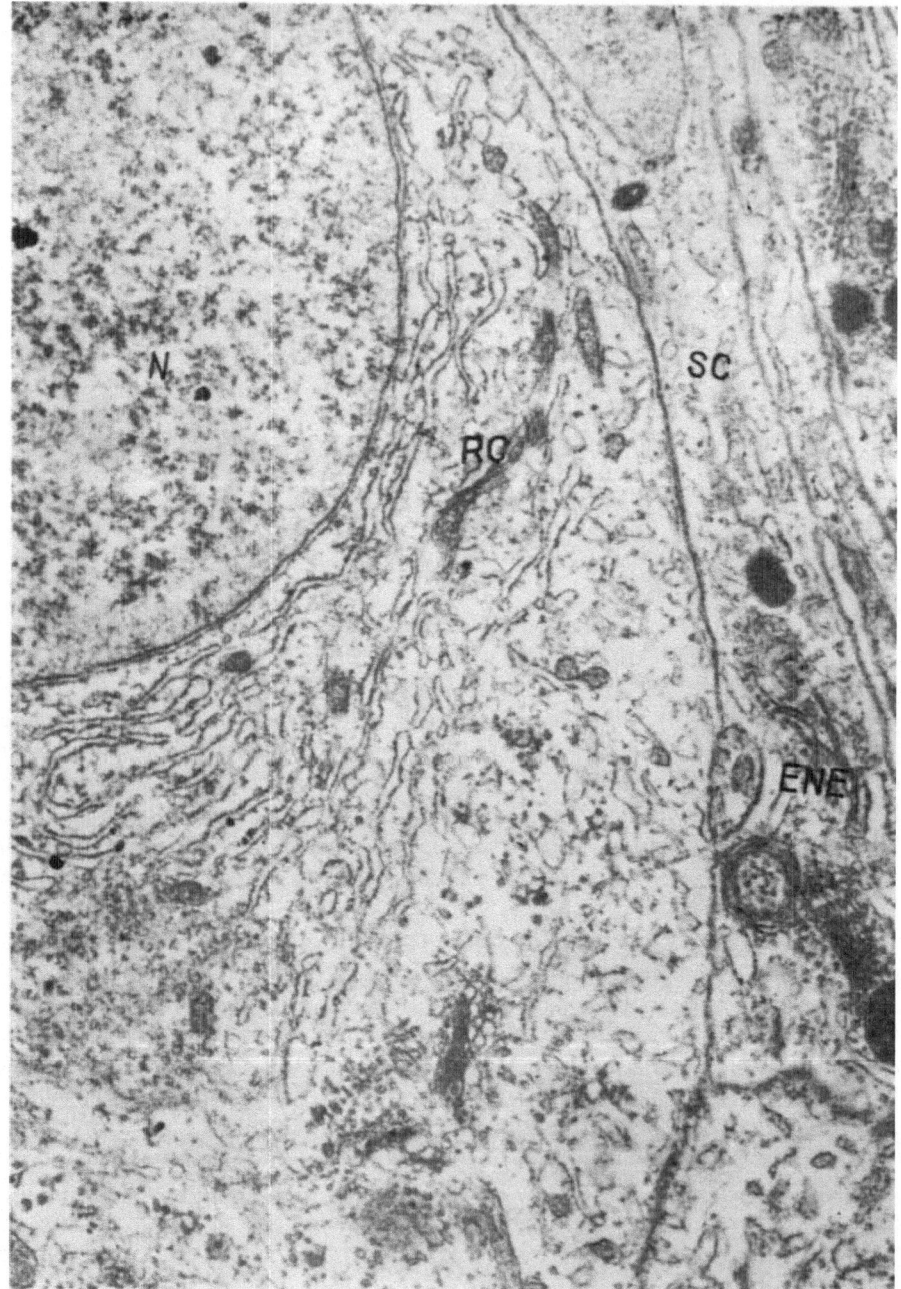

FIG. 11. Hair cells of the Octopus crista with adjoining nerve ending. × 20,000

'dark' and 'light' fibers of various diameters (Fig. 9). In a layer of nerve fibers, histochemical reactions reveal a complex system of cholinergic fibers (Fig. 10) which, as in the macula, penetrate the layer of receptor cells coming into contact with their bases and lateral surface (Fig. 11). The electron-microscope observations show that around receptor cells there are 'dark' nerve endings filled with synaptic vesicles which seem to be providing efferent innervation of the crista.

Around the apex of the crista receptor cells supporting cells form locked phalanges pierced by microvilles (Fig. 6). However, in deeper parts of the crista, supporting cells acquire a rod-like shape. They are filled with tonofibrils and can be found only at the joint of several receptor cells (Fig. 8). Because of this, plasmatic membranes of neighboring receptor cells can maintain close contact.

CONCLUSION

Similar biophysical models of animal organs perceiving the orientation of a body in the field of gravity were formed by nature in the process of evolution. Despite differences in the origin and mode of evolution of cephalopods and vertebrates, their statocysts and vestibular apparatus show a striking likeness at the organ level of organization as well as at cellular and subcellular ones. Indeed, both have organs which comprise maculae with otoliths and cristae with cupules. Receptor cells of these organs, both in cephalopods and vertebrates possess two kinds of hairs, some of which belong to kinocilia linked directly with the otolith membrane or cupule. The likeness between the ultrastructural organization of equilibrium organs of cephalopods and vertebrates is also confirmed by the fact that their receptor cells exhibit a morphological polarization, reflecting the capability of these mechanoreceptors to perceive not only the intensity but also the direction of a stimulus. In both organs there is an efferent cholinergic innervation. The differences detected in the ultrastructure and cytochemistry of statocysts and organs of the vestibular apparatus are only variants of mechanisms perceptive of gravitational forces or energy. Finally, it should be pointed out that there is an obvious resemblance of the maculae and cristae of statocysts in both squids and octopi. The differences discussed above are characteristic for the organ level. They seem to reflect differences in the character and rapidity of animal movement, with an obvious prevalence of the reactive mode of movement in squids (Akimuschkin 1963). This fact seems to be in accord the observation of Kreps et al. (1967) concerning a larger concentration of some phospholipids in the squid brain than in the brain of an octopus.

SUMMARY

The ultrastructural and cytochemical organization of statocysts in the octopus, *Octopus sp.*, and the squid, *Ommastrephes sloanei pacificus*, in many respects resemble the vestibular apparatus of vertebrates. Receptor cells in the maculae and cristae of cephalopods are crowned with bundles of hairs consisting of 160–180 short stereocilia-like microvilles and 70–120 kinocilia. As in the case of vertebrates, kinocilia of these cells are connected with the otolith membrane or the cupule. Since the tops of receptor cells and hairs are bent in a strictly definite direction, the cells are morphologically polarized. Basal bodies of kinocilia in statocysts form cross-striated rootlets. In hair cells of the squid crista, these rootlets are attached to the plasmatic membrane of the cell. The cytoplasm of hair cells contain a well-developed endoplasmic reticulum and a great number of mitochondria and osmiophilic granules. Axons emerge from the basal part of receptor cells. Large cells underlying hair cells in the maculae and cristae are unipolar neurons. Hair cells are innervated by fibers containing synaptic vesicles and acetylcholinesterase. Those are efferent fibers. Supporting cells are distributed among receptor cells, neurons, and nerve fibers. Their tops are typical phalanges.

REFERENCES

AKIMUSCHKIN, I. I. (1963): *Cephalopods of the USSR Seas.* Izd. AN SSSR, Moscow (in Russian).

BARBER, V. (1965): Preliminary observations on the fine structure of the *Octopus* statocyst. *J. Microscopic.* **4** 547–550.

BOYCOTT, B. (1961): The functioning of the statocysts of *Octopus vulgaris. Proc. roy. Soc. B.* **152** 78–87.

DIJKGRAAF, S. (1961): The statocyst of *Octopus vulgaris* as a rotation receptor. *Publ. Staz. zool. Napoli* **32** 64–87.

FLOCK, A. (1964): Structure of the macule utriculi with special reference to directional interplay of sensory responses as revealed by morphological polarization. *J. Cell. Biol.* **17** 413–431.

FLOCK, A. and WERSALL, J. (1962): A study of the orientation of the sensory hairs of the receptor cells in the lateral line organ of fish with the special reference to the function of the receptors. *J. Cell. Biol.* **15** 19–28.

HAMLYN-HARRIS, R. (1903): Die Statocysten der Cephalopoden. *Zool. Jb.* **18** 327–358.

KLEIN, K. (1931): Die Nervendigungen in der Statocysts von Sepia. *Z. Zellforsch.* **14** 481–516.

KOELLE, G. (1951): Elimination of enzymatic diffusion artifacts in histochemical localization of cholinesterases and survey of their cellular distributions. *J. Pharmacol. exp. Ther.* **103** 153–171.

KREPS, E. M., KRASILNIKOVA, V. I., PATRIKEEVA, M. V., SMIRNOV A. A. and CHIRKOVSKAYA, E. V. (1967): The comparative investigation of the phospholipids in the nervous system of Cephalopoda. *J. evol. biochim. physiol.* **3** 101–109 (in Russian).

MATURANA, H. and SPERLING, S. (1963): Undirectional responses to angular acceleration recordered from the middle cristal nerve in the statocyst of *Octopus vulgaris. Nature* **197** 815–816.

NACHLAS, M., TSOY, K., DE COUZA, E. CHENG, C. and SELIGMAN, A. (1957): Cytochemical demonstration of succinic dehydrogenase by the use of a new-p-nitrophenil-substituted ditetrazole. *J. Histochem. Cytochem.* **5** 420.

OWSJANNIKOW, P. A. and KOVALEVSKY, A. (1867): Über das Centralnervensystem und das Gehororgan der Cephalopoden. *Med. Sci. St. Petersbourg.* **2.**

SPOENDLIN, H. (1964): Organization of the sensory hairs in the gravity receptors in utricule and saccule of the squirrel monkey. *Z. Zellforsch.* **62** 701—716.

VINNIKOV, Ya. A., GASENKO, O. G., TITOVA, L. K. and BRONSTEIN, A. A. (1963): Morphological and histochemical studies on the animal labyrinth in the changed gravity field. *Izv. Akad. Nauk SSSR, biol.* **2** 222 (in Russian).

VINNIKOV, Ya. A., GASENKO, O. G., TITOVA, L. K., BRONSTEIN, A. A., OSIPOVA, I. V., GOVAR-DOVSKY, V. I., ARONOVA, M. Z. and ZHINKIN, I. L. (1965): Electron-microscopic studies on the utricule of some vertebrates in a state of the relative quiet and after acceleration. In: *Funkcionalnaja evolutia nervnoi sistemi. Nauka* 170—182 (in Russian)

WERSALL, J. (1956): Studies on the structure and innervation of epithelium on the cristae ampullares in the guinea pig. *Acta otolaring.* Suppl. 126.

YOUNG, J. (1960): The statocysts of *Octopus vulgaris. Proc. roy. Soc. B.* **152** 3—29.

YOUNG, J. (1964): *A Model of Brain.* Oxford, Clarendon Press.

DISCUSSION

J. Szentágothai: Are efferent nerve endings situated only on the receptor cell body or can they form synapses along axons of these cells?

Ya. A. Vinnikov: Efferent nerve endings can form synapses along axons as well as on the cell body. This fact is apparently very important, because efferent inner-vation seems to regulate not only the function of the body of the cell but the func-tion of the axon too.

B. Csillik: Which technique did you use for the histological demonstration of cholinesterase? How long was the incubation period? It is a well-known fact that too long an incubation often results in an artificial staining (impregnation) of nervous elements by a simple attachment of copper thiocholin sulphide to nerve fibers. Do you have any proof to exclude such a possibility?

Ya. A. Vinnikov: We used the histochemical method of Koelle and Friedenwald in the modification of Gerebtzoff. We determined the localization of nonspecific cholinesterase by using the butyrylcholinesterase technique. This is the reason we think that in our experiments the localization of specific acetylcholinesterase was determined.

J. Salánki: Did you find or do you suggest that there are any other synapses present besides the cholinergic ones?

Ya. A. Vinnikov: I have no reason to deny the existence of the noncholinergic synapses, but we have not investigated them as yet.

P. Röhlich: You mentioned that the ciliary rootlets of the hair cells contact the inner side of the cell membrane. Do you attribute a significant role to the rootlet in the transmission of signals from the cilia to the cell membrane, or can the rootlet be regarded merely as a supporting structure?

Ya. A. Vinnikov: There is no doubt that some information is transmitted along the root, but it is not clear what kind of information is transmitted. Neither do I know the direction of information transmission: from basal bodies to the membrane, or in the opposite direction. This question needs electrophysiological investigation.

F. G. Boettiger: Would you discuss more fully how directional sensitivity is determined in these statocysts on the cellular level?

Ya. A. Vinnikov: We think that in cephalopods polarization is due to the sloping arrangement of the cells, or, more precisely, of their apexes, in cristae and maculae. During a movement at the cupule or otolith, only some of the lateral hair cells are excited, and cells which are situated on the opposite side are inhibited. This is, however, only an assumption. Electrophysiological investigations are required to support this assumption.

Symposium on Neurobiology of Invertebrates 1967 (49—57)

STRUCTURAL AND FUNCTIONAL PECULIARITIES OF THE SYNAPTIC TRANSMISSION IN INSECTS

Yu. E. MANDELSTAM

Sechenov Institute of Evolutionary Physiology and Biochemistry,
Academy of Sciences of the USSR
Leningrad, USSR

Questions concerning the role of the cholinergic mechanism in the synaptic transmission in insects have been discussed for a number of years now (Roeder, Kennedy and Sampson 1947; Harlow 1958; Voskresenskaya 1959; Colhoun 1963).

Biochemical investigations of the insect nervous system have revealed some components of the cholinergic system (acetylcholine, cholinesterase, cholin-acetylase).

Physiological tests, however, have yielded unconvincing evidence concerning the mediatory role of acetylcholine which had been shown for vertebrates. This can be explained by the fact that structural and morphological peculiarities of the nervous organization in insects make it hard to use pharmacological methods for studying the nature of synaptic transmission.

By Gerebtzoff's modification of the Koelle–Friedenwald acetylthiocholine method (1959), we showed that the cholinesterase activity of the Asiatic locust *(Locusta migratoria)* nervous system is localized in the neuropile. The enzymic activity was not detected in the nerve cell bodies (Fig. 1a). Later on, the histochemical method of determining cholinesterase localization was used to estimate the penetration of anticholinesterase compounds into the nervous system of the locust.

Anticholinesterase compounds of similar chemical structure with or without a charge were injected into the cavity of the locust abdomen by means of a microsyringe. Lethal doses were much higher for ionized compounds than for the unionized ones.

In the *in vitro* experiments, reverse correlations were observed. A compound with a charged sulphur atom showed a 3000-fold increase of affinity to cholinesterase as compared to a compound without a charge (Kabachnik, Bretskin and Michelson 1964).

In another variant of these tests, equal weight quantities of these compounds were introduced into locusts of two different groups in doses resulting in a 100% lethality. Ten minutes after the injection of the unionized compound, the activity of the enzyme was practically unnoticeable (Fig. 1b). In the case of the ionized compound, after the same period of time, the cholinesterase activity did not yet

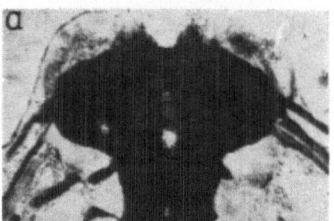

normal distribution

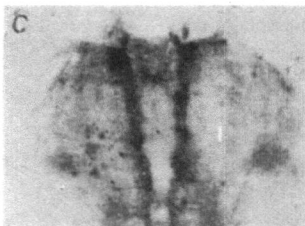

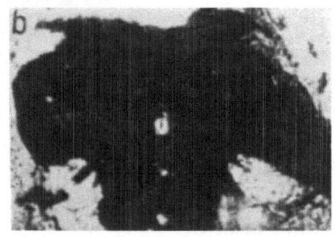

after injection of ГД-7

$$C_2H_5O\diagdown_{p}\diagup^0$$
$$CH_3\diagup \diagdown S\text{-}(CH_2)_2\text{-}SC_2H_5$$

ГД-7

after injection of ГД-42

$$C_2H_5O\diagdown_{p}\diagup^0 \qquad \overset{\oplus}{S}\diagup^{CH_3}\cdot[CH_3SO_4]^-$$
$$C_3\diagup \diagdown S\text{-}(CH_2)_2\diagup \diagdown C_2H_5$$

ГД-42

FIG. 1. Distribution of cholinesterase activity in metathoracic ganglia of the *Locusta migratoria* L. *a)* normal distribution; *b), c)* ten minutes after injection of lethal doses of anticholinesterase compounds

caterpillar pronymphal stage

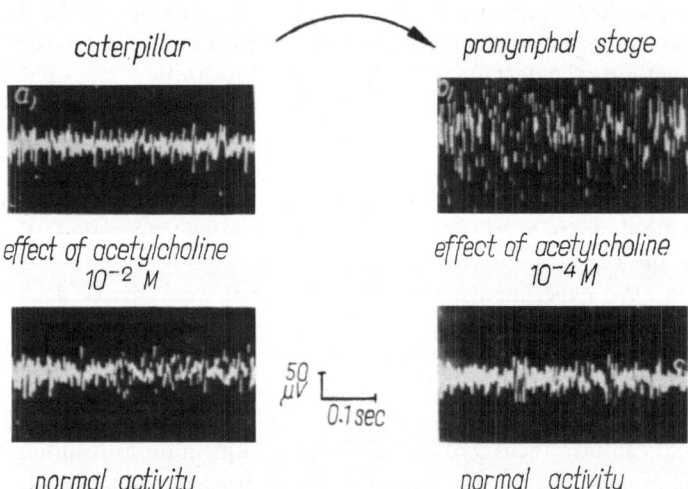

effect of acetylcholine
10^{-2} M

effect of acetylcholine
10^{-4} M

$50\,\mu V$ $\llcorner$
$0.1\,sec$

normal activity

normal activity

FIG. 2. Bioelectrical activity of the nerve cord of *Pieris brassicae* L.
a) caterpillar, *b)* pronymphal stage

differ from the normal activity (Fig. 1c). The first signs of inhibition were noted only a few hours later.

There most likely exists in the nervous system of insects some selective barriers that impede the penetration of ionized anticholinesterase compounds as well as acetylcholine at the sites of synaptic contacts.

In our further studies we attempted to investigate the sensitivity of the insect nervous system to acetylcholine during ontogenesis. Lepidopterae were a suitable subject for such investigations.

After the fourth molt and during the pupal stage the sensitivity of the *Pieris brassicae* nerve cord to acetylcholine was controled daily. Insects were grown in chambers with an artificial climate under the same thermal and light regimens. Only at the pronymphal stage was acetylcholine at a concentration of 10^{-4}–10^{-5} M able to increase the spontaneous electric activity of the nerve cord (Fig. 2). The pronymphal stage lasted for 1–2 days up to the onset of the pupal stage.

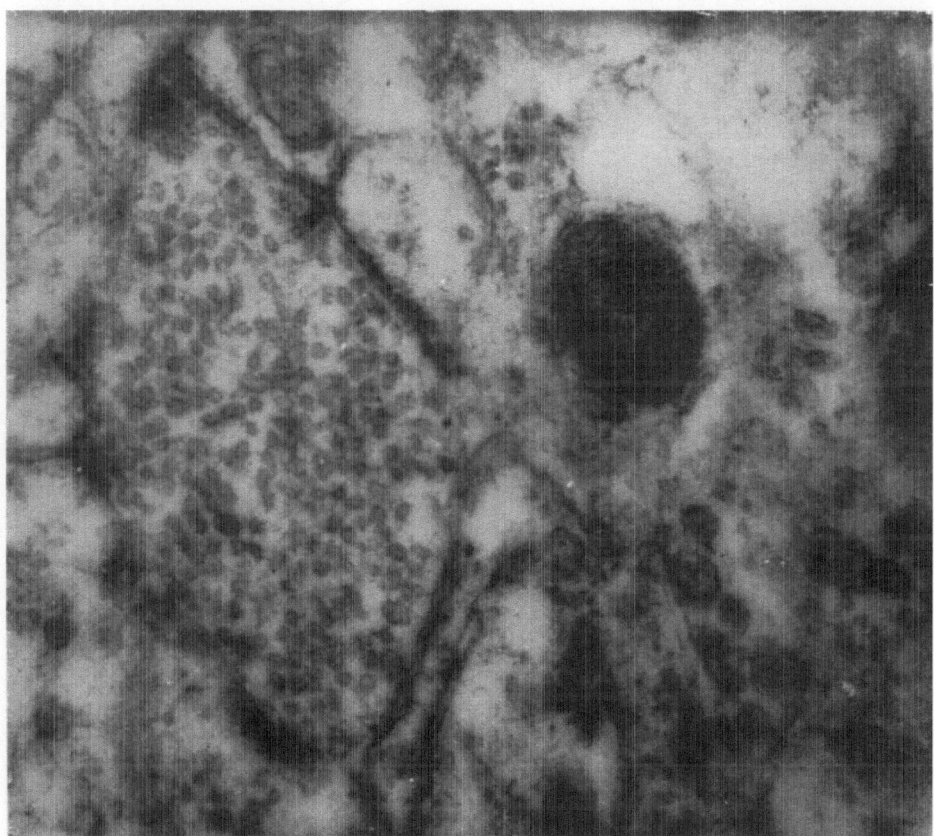

FIG. 3. Localization of cholinesterase activity in neuropile of the locust methathoracic ganglia. × 70,000 Karnowsky's method (1964)

In caterpillars and pupas, acetylcholine even at a concentration of 10^{-2} M did not alter the level of the spontaneous bioelectrical activity of the nerve cord (Mandelstam and Tyschenko 1964).

Thus, at certain stages of insect development acetylcholine penetrates into the nerve cord despite the presence of a quaternary nitrogen atom. This fact is in good agreement with changes occurring in the gangliar sheaths of the cabbage butterfly during its metamorphosis (Heywood 1965).

The experimental data obtained on the nervous system of locust imago, pine silk worm, and cabbage butterfly caterpillars suggest the excitatory effect of anticholinesterase and cholinomimetic compounds (nicotine, arecoline, pilocarpine, etc.) and demonstrate as well the part played by cholinergic mechanisms in the central nervous system of insects.

This evidence is confirmed by electron-microscope data on the cholinesterase localization in the nervous system of the locust. The enzymic activity was determined directly in the sites of synaptic contacts (Fig. 3).

Much less is known about the nature of neuromuscular transmission in insects. Recently, Kerkut and his collaborators, applying pharmacological techniques, obtained many new and interesting facts on this subject (Kerkut, Shapiro and Walker 1965).

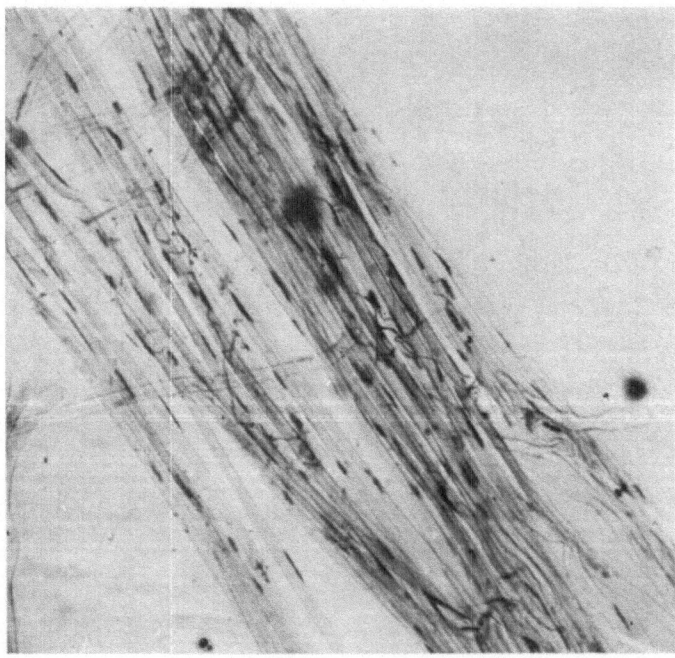

FIG. 4. Motor endings in the bumblebee intersegmentary muscles.
Methylene blue. × 100

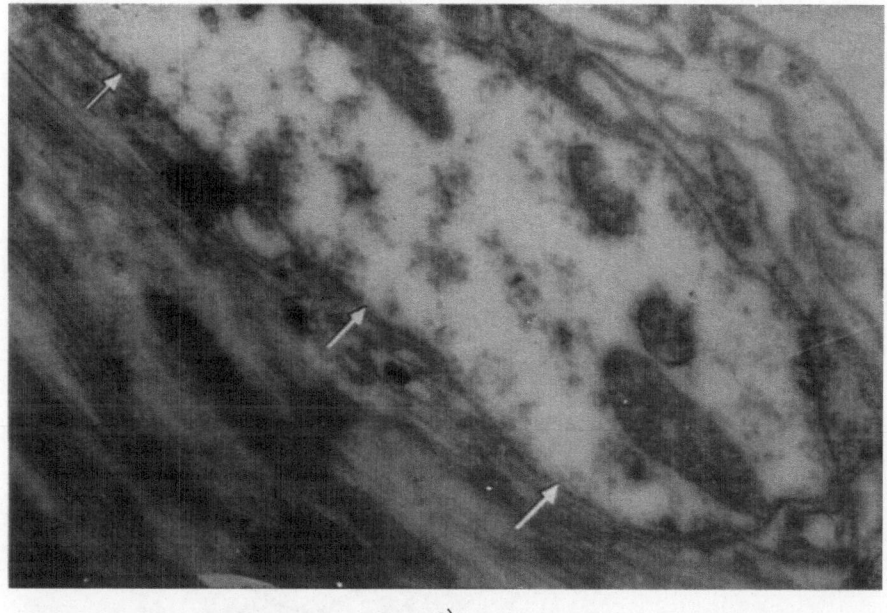

a)

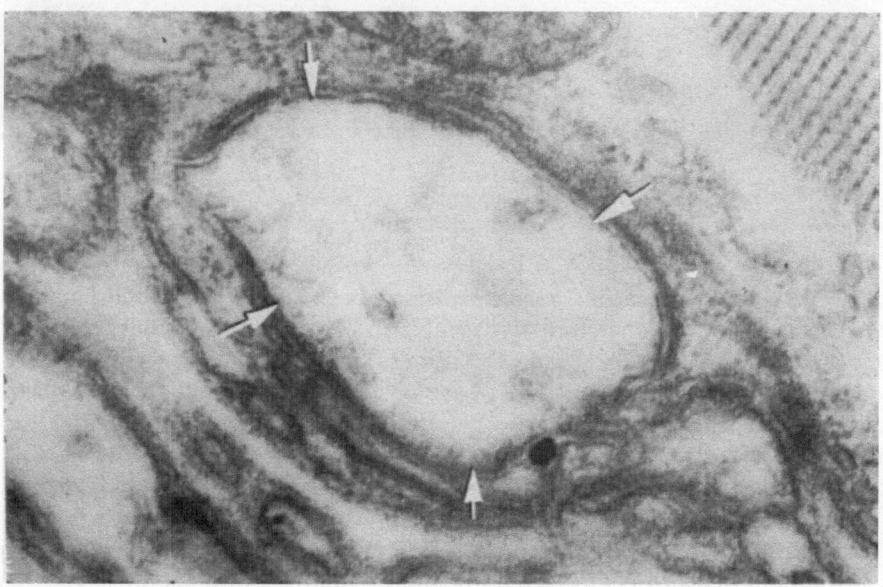

b)

FIG. 5. Structure of neuromuscular synapsis of the bumblebee. a) Intersegmentary muscle, × 20,000; b) longitudinal flight muscle. × 70,000. The arrows show the synaptic cleft

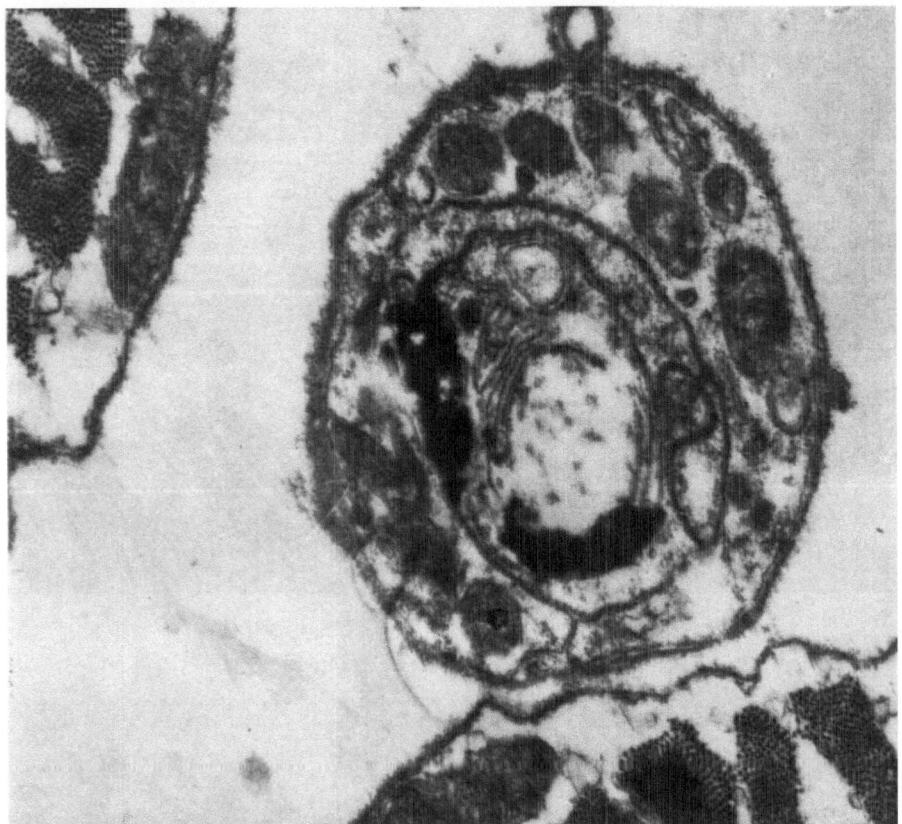

FIG. 6. Localization of cholinesterase in the intersegmentary muscle axon
of the bumblebee. × 26,000

In the study of this problem, a great deal of methodological difficulty is caused
by the presence in insects of a polyneuronal innervation of heterogenous functional
significance. It would therefore be desirable to compare muscles possessing only
one type of innervation. According to Tiegs (1955) and Ikeda and Boettiger (1965),
longitudinal wing muscles in insects have only fast innervation. On the preparations stained with methylene blue, one can see an immensely dense and well-
developed net of motor endings. In the slow intersegmentary muscles of the
bumblebee we have revealed endings that resemble the endplate of vertebrates
(Fig. 4).

As was shown by Auber (1960), each ending is formed by a single axon only.
Thus, a homogeneous character of the motor endings allows us to suggest that
they are similar from the functional point of view. The electron-microscope study
has shown that this type of ending is formed mainly by lemnoblasts.

Despite functional differences, the ultrastructures of neuromuscular synapses in the longitudinal wing and on the intersegmentary muscles of the bumblebee are, on the whole, similar.

The width of the synaptic cleft is 160–200 Å. The synaptic vesicles are 400–500 Å. Folds in the postsynaptic region are nearly absent (Fig. 5a). It must be pointed out, however, that in wing muscles, synapses occur which are of the type of circumference (Fig. 5b) described by Smith (1960) for the beetle wing muscles.

The peculiarity of the motor endings in the intersegmentary muscle of the bumblebee and the small diameter of its fibers enabled us to use it as a suitable subject for histochemical determinations of cholinesterase localization. In light and electron microscopy, the cholinesterase activity was revealed along the terminal nerve endings, and as far as we can judge, the activity of the enzyme is connected with mesaxon membranes (Fig. 6). No enzymic activity was revealed in the membrane of the neuromuscular synapses.

SUMMARY

It was shown by a histochemical method of cholinesterase localization that ionized anticholinesterase compounds penetrate into the nervous system of locust imago much slower than the unionized ones.

During lepidopterae metamorphosis, some changes occur in the barriers of their nerve system: the ionized compounds penetrate to the sites of the synaptic contacts. The data on the ability of the anticholinesterase and cholinomimetic compounds to excite the nerve cord and those on the cholinesterase localization in synapses allow us to conclude that the cholinergic mechanisms play an important part in the nervous system of insects.

In the intersegmentary muscle of the bumblebee, the cholinesterase activity is found along nerve courses, while it never occurs in the neuromuscular synapses.

REFERENCES

Auber, J. (1960): Observations sur la innervation motorice des muscles des insectes. *Z. Zellforsch.* **51** 705–724.

Colhoun, E. H. (1963): The physiological significance of acetylcholine in insects and observation upon other pharmacologically active substances. In: *Advances in Insect Physiology* **1** 1–45.

Gerebtzoff, M. A. (1960): *Cholinesterases*. Pergamon Press, London.

Harlow, A. (1958): The action of drugs on the nervous system of the locust *Locusta migratoria*. *Ann. appl. Biol.* **46** 55–73.

Heywood, R. B. (1965): Changes occuring in the central nervous system of *Pieris brassicae* L. (Lepidoptera) during metamorphosis. *J. insect. physiol.* **2** 413–430.

Ikeda, K. and Boettiger, E. G. (1965): Studies on the flight mechanism of insects. II. The innervation and electrical activity of the fibrillar muscles of bumblebee, Bombus. *J. insect. physiol.* **2** 779–798.

Kabachnik, M. I., Brestkin, A. P. and Michelson, M. J. (1964): *Mechanism of the Physiological Action of Phosphoorganic Compounds.* "Nauka", Moscow (in Russian).

KARNOVSKY, M. S. (1964): The localization of cholinesterase activity in rat cardiac muscle by electron microscopy. *J. Cell. Biol.* **23** 217—232.

KERKUT, G. A., SHAPIRO, A. and WALKER, R. J. (1965): The effect of acetylcholine, glutamic acid and GABA on the contractions of perfused cockroach leg. *Comp. Biochem. Physiol.* **16** 37—48.

MANDELSTAM, YU. E. and TYCHENKO, V. P. (1968): Cholinergic mechanism in Lepidopteran central nervous system. *J. evol. physiol. biochem.* (in press) (in Russian).

ROEDER, K. D., KENNEDY, N. K. and SAMPSON, E. A. (1947): Synaptic conduction to giant fibers of the cockroach and the action of anticholinesterase. *J. Neurophysiol.* **10** 1—10.

SMITH, D. S. (1960): Innervation of the fibrillar flight muscle of an insect: Tenebrio molitor (Coleoptera). *J. biophys. biochem. Cytol.* **8** 447—466.

TIEGS, O. W. (1955): The flight muscles of insects; their anatomy and histology, with some observation on the structure of striated muscle in general. *Phil. Trans.* **238** 221—359.

VOSKRESENSKAYA, A. K. (1959): *Functional Properties of the Neuromuscular Apparatus in Insects.* Izd. Akad. Nauk USSR, Moscow–Leningrad (in Russian).

DISCUSSION

G. A. Kerkut: Did you manage to localize cholinesterase histochemically at the insect nerve-muscle junction?

Yu. E. Mandelstam: For the determination of cholinesterase localization we applied Koelle's method with Gerebtzoff's modification, as well as the method of Karnovsky and Barnett. However, with neither of these methods could we demonstrate cholinesterase activity in the investigated insect endplate.

B. Csillik: The structure of the insect endplates you have shown is similar to that of gamma-innervated endings ('small nerve system endings') of vertebrates, which lack junctional folds, or, if any, contain only a very few of them. However, in such gamma endings, acetylcholinesterase is located in pre- and postsynaptic membranes, as shown last year by Dr. Joó and myself. Acetylcholinesterase is absent from insect endplates, but a nonspecific esterase (arylesterase?) has been demonstrated there by Dr. Hámori. It would be of utmost importance to show the fine structural localization of these nonspecific esterases in insect endplates; our recent studies with Miss Elizabeth Knyihár have proved that in mammalian endplates, nonspecific esterases occupy the middle dense layer of the synaptolemma, within the synaptic cleft. In view of the possible role of amino acids in neuromuscular impulse transmission, the fine structural localization of nonspecific esterases in the synaptic cleft might throw new light upon the molecular events that take place during endplate activity.

Yu. E. Mandelstam: Of course the demonstration of the nonspecific cholinesterase activity in the insect endplate is intriguing. But, as shown in our experiments, the histochemical reaction proved to be negative, and since we were primarily interested in the localization of the cholinesterase we cannot discuss this problem at the present.

G. A. Cottrell: I wish to ask how you fixed your tissue for electron miscroscopy, because I believe that the appearance of small vesicles and granules varies with the method of fixation?

Yu. E. Mandelstam: By measuring the size of synaptic vesicles, the tissue was fixed in 2.5% glutaraldehyde in phosphate buffer, followed by a fixation with 1% OsO_4.

J. Hámori: The similarity between the vertebrate endplates (V. E.) and those of the insect intersegmentary muscle (I. E.) seems to be only superficial. Though they may look similar in 'macroscopic' or in low-power light-microscopic pictures, they differ considerably under the electron microscope. (a) As the author has mentioned, the width of the synaptic cleft varies in I. E. from 100 to 200 Å (in our insect material, between 80 and 150 Å), whereas in the characteristic V. E., the cleft is between 300 and 600 Å. (b) Subsynaptic apparatus characteristic for many differentiated V. E. has been never observed in I. E. (c) In I. E., there is usually a rich, well-developed system of sarcoplasmic reticulum, or 'rete synapticum' (after Edwards) connecting the postsynaptic membrane with the neighboring mitochondria and myofibrils. A corresponding system in V. E. has never been found. (d) Insect muscle fibers always have a multiple (and, in many cases, polyneural) innervation, whereas in the majority of vertebrate skeletal muscles, the proportion of endplate to muscle fibers is 1 : 1. All these facts may add to the basic difference in their physiological properties, i.e., that for the muscles of insects, the initiation of a local depolarization area is characteristic at each myoneural junction, whereas in vertebrates − with the only exception being the slow system of the frog − a propagated change of membrane potential occurs.

Yu. E. Mandelstam: The morphological properties of insect muscle innervation as well their synaptic electrogenesis are well known. When I referred to the nerve endings of the insect intersegmentary muscle as being of similar type as the vertebrate endplates, I noted only the similarity of these nerve endings as seen in preparations stained with methylene blue.

In the slide the multiterminal character of this nerve innervation is demonstrated − in a nerve fiber a couple of motoric nerve endings can be seen.

Furthermore the definition 'nerve ending of endplate type' has already been accepted in the morphological literature for certain types of endings which had been observed in insect muscle several years ago.

Symposium on Neurobiology of Invertebrates 1967 (59—68)

THE STRUCTURE OF THE SYMPATHETIC NERVOUS SYSTEM OF INSECTS

S. I. PLOTNIKOVA

Sechenov Institute of Evolutionary Physiology and Biochemistry,
Academy of Sciences of the USSR
Leningrad, USSR

The development of the vegetative nervous system in the phylogenetic line of Protostomia has been studied by Zawarzin and his histological school (Zawarzin 1924, 1941; Tsvileneva 1950, 1951, 1966; Kolmogorova 1959). Zawarzin (1924, 1941) proved that in this line insects have the most differentiated part of the vegetative nervous system, i.e., the system of the ventral unpaired nerve (or median nerve) which may be regarded as analogous to the sympathetic nervous system of vertebrates.

Numerous investigations of Voskresenskaya and her co-workers (1950, 1959, 1960, 1964) have shown that the unpaired nerve exerts a considerable influence on the functional properties of somatic motor apparatus and that this influence can be imitated by adrenomimetic substances and eliminated by adrenolytics.

In light of these studies, it would be interesting to investigate the structure, conduction pathways, and the nature of nervous elements of the vegetative centers in insects with an established function of the unpaired nerve.

Conduction paths were studied on total preparations of the *Locusta migratoria migratoria* L. nerve cord that were stained with methylene blue. Following a technique of Zawarzin, detected neurons were drawn with the help of a drawing apparatus. The monoaminergic nature of nervous elements in some representatives of Protostomia *(Locusta migratoria migratoria* L., *Allolobophora caliginosa* F., *Lumbricus terrestris* L., *Dendrocoeleum lacteum* M., *Actinia equina* L., *Bunodactis stella* V.) was revealed by a specific sensitive fluorescent method of Falck–Hillarp (Falck 1964) in the modification of Govyrin (1965). Sections obtained on the freezing microtome were lyophilized for 8 h according to the method of Eränkö (1954). The histochemical reaction with formaldehyde was carried out at 80 °C. The sections were embedded in polysterol through xylol. The results of the reaction were observed on an ML–2 microscope.

OBSERVATIONS

A more detailed study of the structure of the unpaired nerve centers has been made on metathoracic ganglia of the locust nerve cord. As can be seen on the preparations stained with methylene blue, this is a composite ganglion, consisting of four neuromeres: one thoracic (Fig. 1; I) and three abdominal (Fig. 1; II, III, IV). Each neuromere gives off one unpaired nerve [Fig. 1; (1, 2, 3, 4)]. Effector neuron dendrites of the unpaired nerve system are located in the motor neuropile of each neuromere. Sensory fibers end in the ventral region. The bodies of effector neurons in the unpaired nerve are situated on the ventral as well as on the dorsal side of the ganglion. On the dorsal side of each neuromere there is a group of small neurons [Fig. 1 (5)], sending axons into the unpaired nerve. The bodies of two larger symmetric effector neurons [Fig. 1 (6, 6^1)] lie in the ventral region of each neuromere, whereas their dendrites are branching in the region of the motor neuropile. These dendrites form a medial nucleus in the medial zone of the neuropile [Fig. 1 (7)] and two lateral ones, in the lateral region [Fig. 1 (8)]. The axons of all these neurons enter corresponding unpaired nerves.

Through the distribution of dendrites in the unpaired nerve nuclei the system is assumed to include some neurons the axons of which go to the periphery as components of somatic nerves. For each abdominal neuromere of the metathoracic ganglion, such neurons are assumed to comprise (a) two large symmetric neurons [Fig. 1 (9, 9^1)] the dendrites of which end in the medial nucleus while axons go toward the periphery as components of somatic nerves, and (b) two pairs of symmetric neurons [Fig. 1 (10, 10^1)] with dendrites arranged in the region of lateral nuclei and axons that come out as components of somatic nerves innervating sheaths of nerves and trachea. Judging by the distribution of synapses in medial and lateral nuclei, these elements make up a unified system with the neurons of the unpaired nerve.

To study the nature of nervous elements of the unpaired nerve, a reaction for catecholamines was carried out on longitudinal sections across the dorsal region of the metathoracic ganglion. A specific intense green fluorescence was observed in medial [Fig. 2 (1)] and lateral nuclei of the unpaired nerves. Catecholamines were found to be localized in dendrites.

This reaction thus establishes the adrenergic nature of nervous elements making nuclei of unpaired nerves.

Unpaired nerve centers of the nerve cord and the suboesophageal ganglion are linked to the tritocerebral part of the insect brain — the supraoesophageal gan-

---

Fig. 1. Distribution of effector elements of unpaired nerves in the motor neuropile of metathoracic ganglion. I, II, III, IV — neuromeres; 1, 2, 3, 4 — unpaired nerves, 5 — bodies of small neurons (represented only in the second neuromere), 6, 6^1 — effector neurons of the unpaired nerve, 7 — medial and 8 — lateral nuclei of unpaired nerves; 9, 9^1 — effector neurons with dendrites ending in unpaired nerves and axons entering the periphery as components of the segmented nerves; 10, 10^1 — neurons with axons coming into somatic nerves; methylene blue

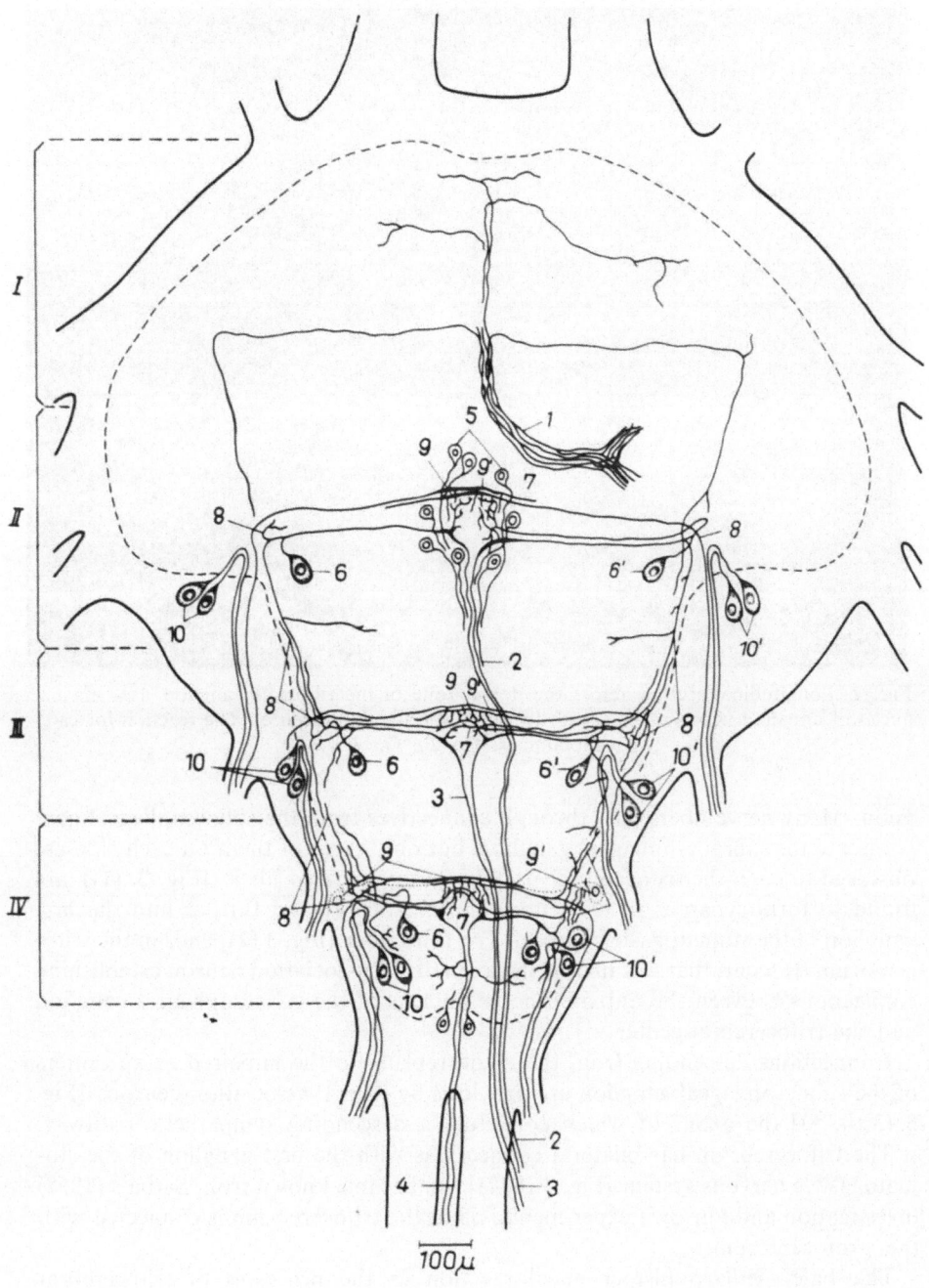

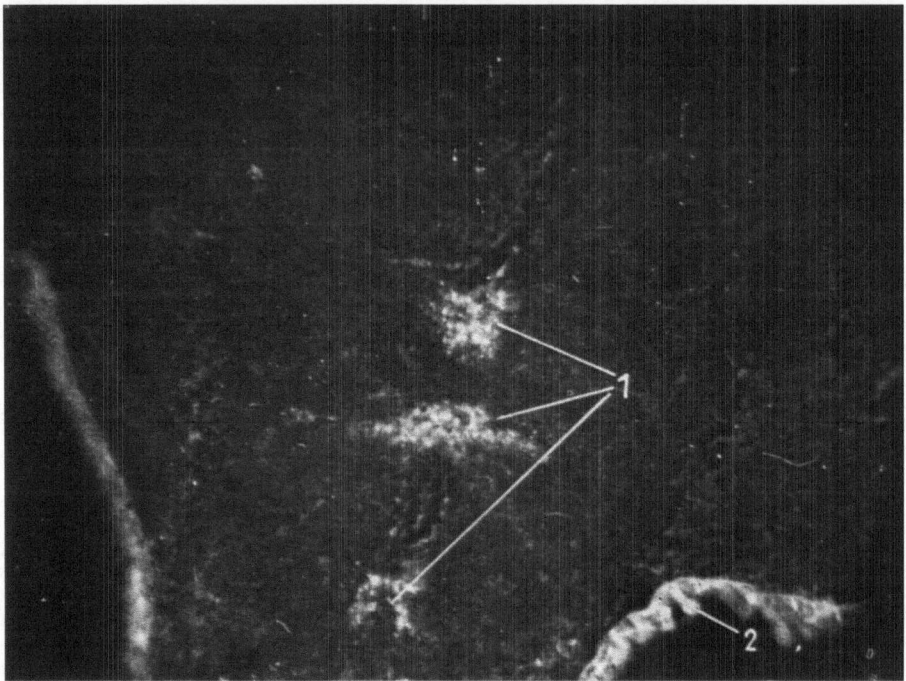

Fig. 2. Longitudinal section across the dorsal zone of metathoracic ganglion. 1 — medial nuclei of unpaired nerves; 2 — sheath (with nonspecific fluorescence). The reaction for catecholamines. Ob 20, Oc 3

glion. Many nerve fibers pass through connectives from the suboesophageal ganglion into the supraoesophageal ganglion, but only three of them on each side are observed to enter the tritocerebral part of the brain. Two fibers [Fig. 3 (1)] are found to form synapses in the tritocerebrum and proceed further into the first ganglion of the stomatogastric system. The third fiber [Fig. 3 (2)] ends in the tritocerebrum. It seems that this fiber is the axon of the association neuron establishing connections between the unpaired nerve nucleus of the suboesophageal ganglion and the tritocerebral center.

Connections descending from the tritocerebrum to the unpaired nerve centers of the suboesophageal ganglion are provided by several association neurons [Fig. 3 (3, 4, 5)] the axons of which constitute a descending sympathetic pathway.

The tritocerebrum has bilateral connections with the first ganglion of the stomatogastric nervous system [Fig. 3 (6, 7)]. As became known from Satija's (1958) investigation and our own experimental data, the tritocerebrum is connected with the protocerebrum.

The Falck–Hillarp histochemical reaction in the processes of tritocerebral vegetative neurons indicates the presence of catecholamines [Fig. 4 (1)].

The presence of catecholamines in the unpaired nerve nuclei and tritocerebrum

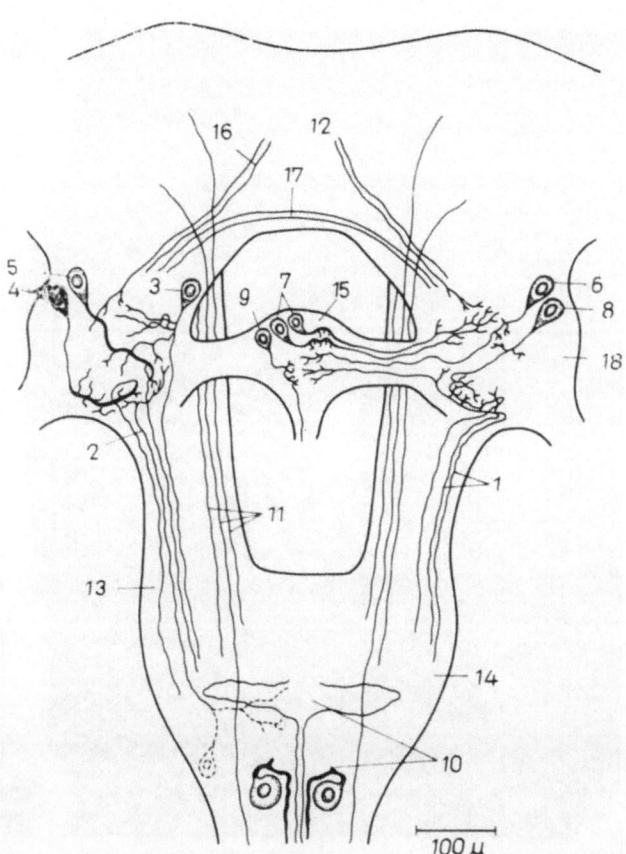

FIG. 3. Diagram of conduction pathways of Aeschna nymph tritocerebrum. (So as not to overcrowd the diagram, nervous elements are shown only on one side of the center.) 1 — Vegetative nerve fiber going thorugh the tritocerebrum into the first ganglion of the stomatogastric system; 2 — nerve fiber ending in the tritocerebrum; 3, 4, 5, 6, 7, 8 — associaton cells; 9 — effector neuron of the frontal ganglion; 10 — unpaired nerve nucleus in the suboesophageal ganglion; 11 — somatic fibers of the connective; 12 — supraoesophageal ganglion; 13 — connective; 14 — suboesophageal ganglion; 15 — frontal ganglion; 16 — fibers linking the tritocerebrum and protocerebrum; 17 — fibers linking the right and left parts of the tritocerebrum; 18 — tritocerebrum

of insects is another piece of evidence revealing the sympathetic nature of these centers.

Hence, the tritocerebrum cephalizes the unpaired nerve nuclei, the stomatogastric nervous system, and establishes connections between these systems and the protocerebral parts of the supraoesophageal ganglion, thus representing the higher vegetative center of the insect.

Other less higher Protostomia also have higher vegetative centers. This becomes obvious with a comparison of our own evidence of the vegetative conduction

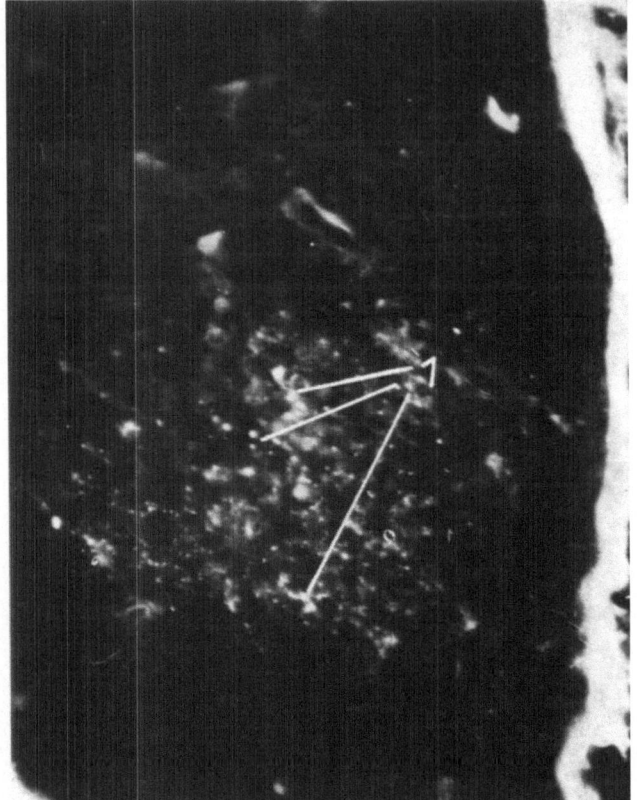

FIG. 4. Longitudinal section across the locust tritocerebrum.
1 — Cut off process of sympathetic neurons. The reaction
for catecholamines. Ob 20, Oc 3

FIG. 5. Diagram of catecholamine distribution in the nervous system of Protostomia. *a* —
Distribution of nervous elements, containing monoamines in *Actinia*. *B* — Nervous conduction
paths in *Dendrocoelum lacteum* M. (according to Hanström). *b* — Distribution of nervous
elements containing monoamines in *Dendrocoelum lacteum* M. *C* — Diagram of vegetative
conduction pathways in the earthworm (according to Newyvaka). *c* — Distribution of nervous
elements containing monoamines in earthworm nervous system. *D* — Nervous conduction
pathways of the crayfish (according to Orlov). *E* — Vegetative conduction pathways of Aeschna
nymph. *F* — Vegetative conduction pathways of Asiatic locust. *f* — Distribution of monoamines in vegetative nervous elements of the Asiatic locust tritocerebrum and metathoracic
ganglion. 1 — Higher vegetative centers. 2 — Vegetative nerve cord centers. 3 — Supraoesophageal ganglion. 4 — Suboesophageal ganglion. 5 — Sensitive nerve cells of earthworm
pharynx. 6 — Sensitive nerve cells of earthworm dermomuscular tube

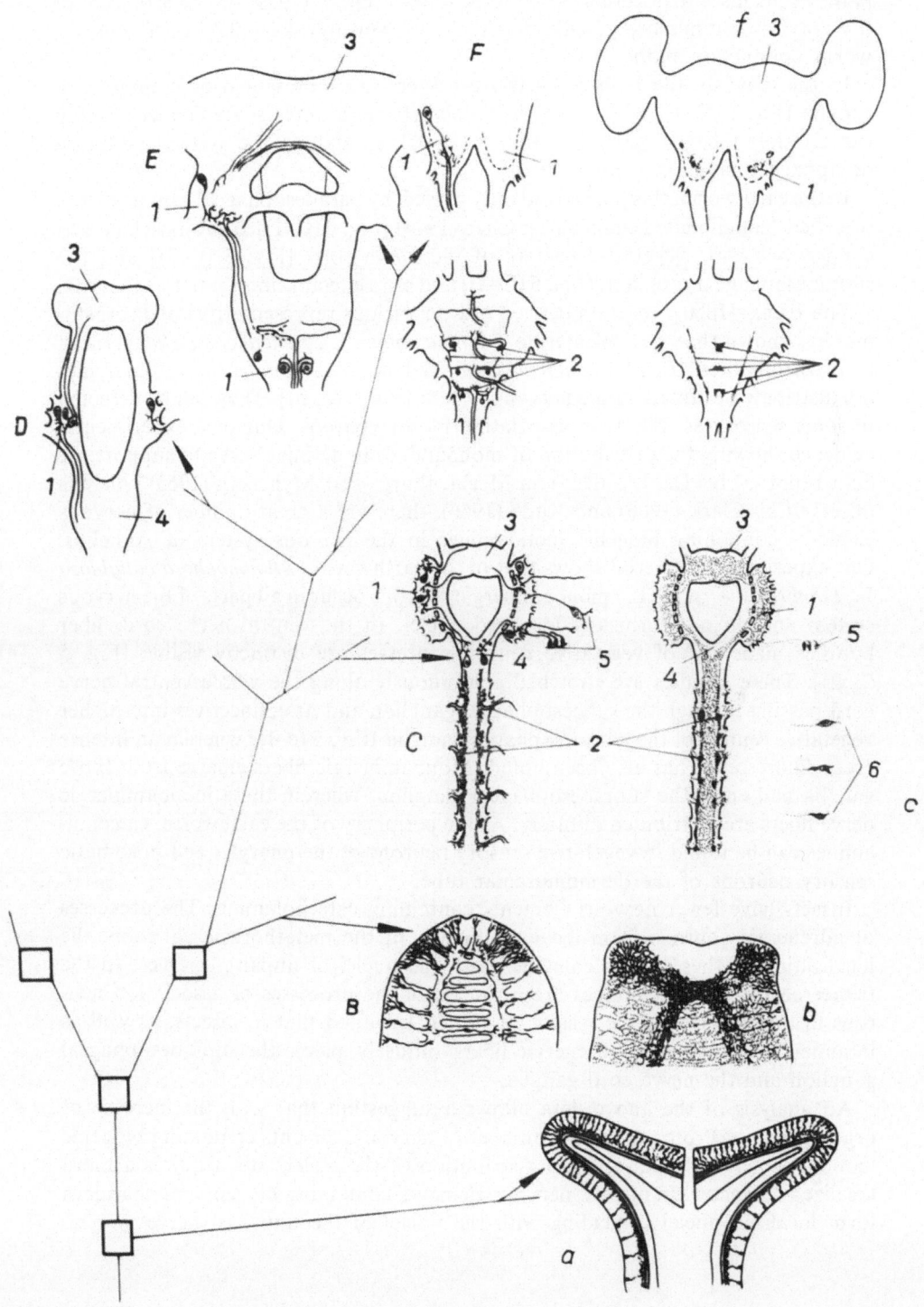

S. I. PLOTNIKOVA

paths in the insect tritocerebrum with the data of Orlov (1929) on the structure of the crayfish commissure ganglia and those of Newmyvaka (1960) on the earthworm conduction paths.

In the crayfish, the higher vegetative centers are represented as commissural ganglia [Fig. 5 (D–1)] which, like the tritocerebrum in insects, are connected with the supraoesophageal ganglion [Fig. 5 (D–3)], stomatogastric system, and suboesophageal ganglion [Fig. 5 (D–4)].

In the earthworm, this part is evidently played by paraoesophageal ganglia or the so-called 'ganglionated chain', after Chen (1944) [Fig. 5 (C–1)]. They have bilateral connections with vegetative centers of the nerve cord [Fig. 5 (C–2)] and the supraoesophageal ganglion, [Fig. 5 (C–3)] and send effector fibers into the pharynx.

The Falck–Hillarp reaction carried out on various representatives of Protostomia has shown that in Coelenterata *(Actinia equina* L., *Bunodactis stella* V.*)* and in plathelminthes *(Dendrocoelum lacteum* M.*)* neurons comprising monoamines are distributed diffusely in the nervous system [Fig. 5 (a, b)]. They can be detected in some sensory as well as in association motor neurons. Our own experimental evidence showing the distribution of monoamines in actiniae serve to support the data reported by Dahl, Falck, von Mecklenburg, and Myhrberg (1963). As was observed by Clark (1966) and Rude (1966), there are a great number of nervous elements containing biogenic monoamines in the nervous system of Annelids. Our experimental material shows that in the earthworm *(Allolobophora caliginosa* F., *Lumbricus terrestris* L.*)* monoaminergic neurons occur in all parts of the nervous system and do not form any localized nuclei. In the ventral nerve cord, fiber bundles made out of vegetative neuron processes are distinctly visible [Fig. 5 (c, 2)]. These bundles are stretched continuously along the whole ventral nerve cord passing through the suboesophageal ganglion and its connectives into higher vegetative centers of the paraoesophageal ganglia [Fig. 5 (c–1)] wherein an intense green fluorescence has also been noted. Monoaminergic fibers emerge from those ganglia and enter the supraoesophageal ganglion, wherein these monoaminergic nerve fibers are distributed diffusely. At the periphery of the earthworm, catecholamines can be found in vegetative sensory neurons of the pharynx and in somatic sensory neurons of the dermomuscular tube.

Insects have fewer nervous elements containing catecholamine. The processes of adrenergic neurons form localized nuclei. In the metathoracic ganglion, the localization of these nuclei coincides with the nuclei of unpaired nerves. In the tritocerebrum, catecholamines are arranged in the processes of association neurons in the unpaired nerve system. It should be noted that in insects, as well as in annelids, a number of adrenergic fibers diffusely pierce the supraoesophageal ganglion and the nerve cord ganglia.

An analysis of the above data allows a suggestion that with the increase of organization of Protostomia, the number of nervous elements containing biogenic monoamines is reduced, and the distribution of these elements acquires a more localized character. In insects, nervous elements containing biogenic monoamines form localized nuclei coinciding with the nuclei of the unpaired nerve system.

SUMMARY

Neuronal relationships have been studied in unpaired nerve centers of the nerve cord and in the cephalizing higher vegetative center of the tritocerebrum. The evidence of neurohistological analysis has been compared with the histochemical reaction for monoamines. In insects, catecholamine-containing nervous elements make localized nuclei coinciding with the nuclei of the unpaired nerve system, which is additional evidence in favor of its sympathetic nature. As has been demonstrated on representatives of the phylogenetic line of Protostomia, the lowering of organization leads to an increase in the number of nervous elements containing monoamines and their distribution acquires a diffuse character.

REFERENCES

CHEN, T. T. (1944): The morphology of the anterior autonomic nervous system of the earthworm, *Lumbricus terrestris. J. comp. Neurol.* **80** 191—210.

CLARK, M. E. (1966): Histochemical localization of monoamines in the nervous system of the polychaet *Nephtys. Proc. roy. Soc. B.* **165** 308—325.

DAHL, E., FALCK, B., VON MECKLENBURG, C. and MYHBERG, H. (1963): An adrenergic nervous system in sea anemones. *Quart. J. micr. Sci.* **104** 531—534.

ERÄNKÖ, O. (1954): Simple apparatus for freeze-drying of animal tissue. *Acta path. microbiol. scand.* **25** 426—432.

FALCK, B. (1962): Observations on the possibilities of the cellular localization of monoamines by a fluorescence method. *Acta physiol. scand.* **56** suppl. 197 1—25.

GOVYRIN, V. A. (1965): On the absence of direct sympathetic innervation of skeletal muscles. *Dokl. Akad. Nauk SSSR.* **160** 1179—1183 (in Russian).

KOLMOGOROVA, E. J. (1959): Structure of the central parts of the nervous system in Opitorchis felineus. *Zool. J.* **38** 1627—1633 (in Russian).

NEWMYVAKA, G. A. (1948): Materials on comparative histology of the nervous system. Ventral cord of the earthworm. *Sb. pamyati akad. A. A. Zawarzin.* Akad. Nauk SSSR 27—53 (in Russian).

NEWMYVAKA, G. A. (1966): *Nervous System of the Earthworm.* Moscow–Leningrad (in Russian).

ORLOV, J. A. (1924): Die Innervation des Darmes der Insecten (Larven von Lamellicorniern). *Z. wiss. Zool.* **122** 425—502.

ORLOV, J. A. (1925): Über den histologischen Bau der Ganglien des Mundmagennervensystems der Insekten. *Z. mikr.-anat. Forsch.* **2** 39—110.

ORLOV, J. A. (1929): Über den histologischen Bau der Ganglien des Mundmagennervensystems der Crustaceen. *Z. Zellforsch.* **8** 493—541.

RUDE, S. (1966): Monoamine-containing neurons in the nerve cord and body wall of *Lumbricus terrestris. J. comp. Neurol.* **128** 397—405.

SATIJA, R. C. (1958): A histological study of the brain and thoracic nerve cord of Aeschna nymph with special reference to the descending nervous pathways. *Res. Bull. Punjab Univ.* **138** 33—47.

TSVILENEVA, V. A. (1950): Nerve cord of Aeschna (after the method of A. A. Zawarzin). I. The structure of the thoracic ganglia. *Proc. Acad. Sci. of SSSR,* B. **2** 92—128.

TSVILENEVA, V. A. (1951): Nerve cord of Aeschna (after the method of A. A. Zawarzin). II. The structure of the abdominal ganglia. *Proc. Acad. Sci. of SSSR,* B. **2** 66—116 (in Russian).

TSVILENEVA, V. A. (1953): Concerning the nervous structure of the locust nerve cord. *Izv. otd. Est. Nauk Ak. Nauk Tadzhiskoi SSSR* **2** 49—55 (in Russian).

TSVILENEVA, V. A. (1964): The nervous structure of the ixodid ganglion. *Zool. Jb. Anat.* **81** 579—602.

VOSKRESENSKAYA, A. K. (1950): On the "sympathetic" innervation of skeletal muscles in insects. *Physiol. J.* SSSR 36, 2. 176—183 (in Russian).

VOSKRESENSKAYA, A. K. (1959): *Functional Properties of the Neuromuscular System in Insects.* Moscow–Leningrad 1, 188 (in Russian).

VOSKRESENSKAYA, A. K. and SVIDERSKY, V. L. (1960): The role of central and sympathetic nervous system in the function of the tymbal muscles of cicadas. *J. Ins. Physiol.* **6** 26—35.

VOSKRESENSKAYA, A. K. and SVIDERSKY, V. L. (1960): Analysis of the nature of rhythmical trace reactions in the neuromuscular apparatus of the insect wing. *Physiol. J. Akad. Nauk SSSR* **46** 9 1050—1055 (in Russian).

VOSKRESENSKAYA, A. K. and SVIDERSKY, V. L. (1964): Influence of the supraoesophageal ganglia on segmental locomotor systems in insects. *Physiol. J. Akad. Nauk SSSR* **50** 7 835—840 (in Russian).

ZAWARZIN, A. A. (1924): Zur Morphologie der Nervenzentren. Das Bauchmark der Insekten. Ein Beitrag zur vergleichenden Histologie (Histologische Studien über Insekten VI). *Z. wiss. Zool.* **122** 323—324.

ZAWARZIN, A. A. (1941): Essays on evolutionary histology of the nervous system. In: *Izbr. Tr. A. A. Zawarzin Akad. Nauk SSSR* Moscow–Leningrad 1950, 3, 1—419 (in Russian).

DISCUSSION

R. J. Walker: Is there any evidence, using fluorescence microscopy, of yellow cells, indicating the presence of serotonin in the insect nervous system?

S. I. Plotnikova: The use of fluorescence microscopy did not reveal yellow cells at all.

J. Salánki: It is very interesting that there is a concentration of the material with green fluorescence during the phylogeny beginning from coelenterates through worms to the insects. Theoretically it may be either dopamine or adrenaline or noradrenaline. My question is whether these are the same everywhere, or if the substance itself is different according to different phyla? Furthermore, I would also like to ask whether or not there are good histochemical procedures for making a clear distinction between these catecholamines?

S.I.Plotnikova: We have not studied the definition of different catecholamines.

J.Szentágothai: Did you stain the preparations only with methylene blue or also impregnate with silver?

S. I. Plotnikova: They were stained only with methylene blue.

D. A. Sakharov: How did you separate the dendrites of the efferent neurons from the axons?

S. I. Plotnikova: This was not a problem, since in the preparation stained with methylene blue it could clearly be seen that the dendrites of the efferent neurons end in the motor neuropile, while the axons lead through the nerves going from the ganglion to the periphery.

Symposium on Neurobiology of Invertebrates 1967 (69—84)

HISTOCHEMICAL AND ELECTRON-MICROSCOPIC STUDIES ON THE RELATION BETWEEN DOPAMINE AND DENSE-CORE VESICLES IN THE NEURONS OF *ANODONTA CYGNEA* L.

I. Zs.-NAGY

Biological Research Institute of the Hungarian
Academy of Sciences
Tihany, Hungary

On the basis of data from the literature and our own investigations I should like to sum up some properties of the neurons of *Anodonta cygnea:*

1. Falck's method (Falck 1962) gives green fluorescence in the majority of axons and in several nerve cells, and this, as made evident by analyses, originates predominantly from dopamine (Dahl *et al.* 1962, 1966; Zs.- Nagy 1967a). The dopamine can be depleted from the axons with reserpine. Three to four hours after reserpine treatment, practically no green fluorescence remains in the neuropile of the cerebral ganglion. Within 12 h, all cells produce green fluorescence in the cerebral ganglia, which intensifies for 2–3 days; from the second day, more and more green fluorescence is found in the axons as well (Zs.-Nagy 1967a).

2. The second property of the neurons is that the majority of axons and several cells give a specific positivity with Sterba's pseudoisocyanine fluorescence method (Sterba 1961) (Zs.-Nagy 1965). This method, according to Sterba (1964), indicates the carrier protein of the elementary neurosecretory granules.

As determined with the electron microscope the majority of axons contain vesicles which are very similar in their morphology to the elementary neurosecretory granules (Fig. 1) (Zs.-Nagy 1964). The size of these vesicles is, on the average 1000 Å; their distribution is shown in Fig. 2. Fourty-six % are full dense-core vesicles, 28 % are in the process of evacuation, and the rest are empty. A very small percentage of the cells also contain dense-core vesicles ,but these are always full, and somewhat greater in size than in the axons (Fig. 3). Most nerve cells contain a large number of cytosomes (Fig. 4). Some years ago, they had been interpreted as a neurosecretory material on the basis of their paraldehyde fuchsine positivity (Fährmann 1961; Baranyi and Salánki 1963; Antheunisse 1963), but now it is clear that they have nothing to do with the neurosecretion (Nolte *et al.* 1965; Sakharov *et al.* 1965; Gabe 1966). The cytosomes also have an intensive yellow autofluorescence (Zs.-Nagy 1967b).

A relation between dense-core vesicles and catecholamines was presumed in the Molluscs by Gerschenfeld in 1963, but has not yet been proved. However, evidence of such a relation can be of great importance in clearing up the function of dense-core vesicles as well as that of dopamine. For this reason, we set as a goal the clarification of the following questions:

I. ZS.-NAGY

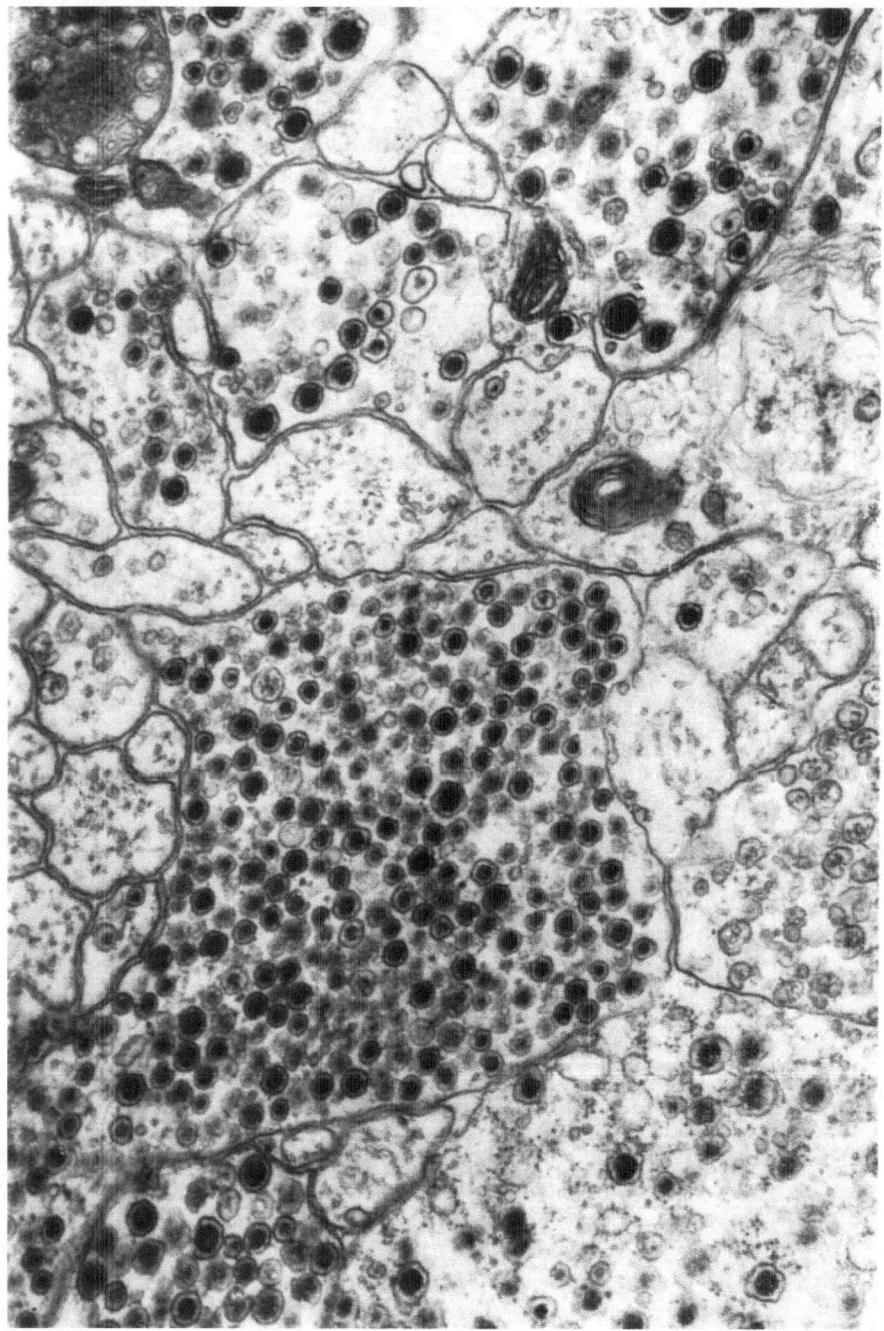

FɪG. 1. Different types of the dense-core vesicles in the axons. Cerebral ganglion. × 30,000

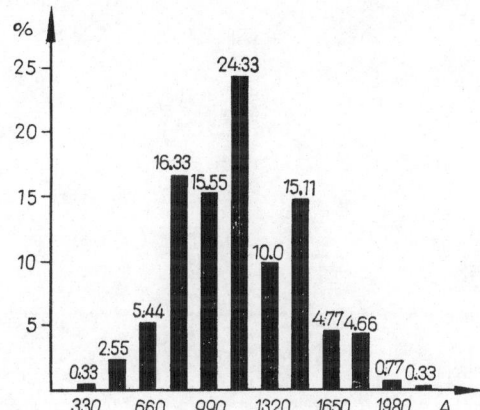

FIG. 2. Size distribution of the vesicles in the axons
of a cerebral ganglion

1. Are the axons that contain dopamine or pseudoisocyanine-positive material identical?

2. How are the dense-core vesicles transformed after the reserpine treatment, which brings about a significant change in the dopamine content of the neurons?

To answer these questions, we photographed the localization of dopamine as detected by the Falck's method (Fig. 5a), then made the pseudoisocyanine-reaction after deparaffinization in the same section and rephotographed the same spots of the section (Fig. 5b); we then compared the two localizations. This demonstrated that the green axons are identical with those of pseudoisocyanine-positivity, although there are some pseudoisocyanine-positive axons in which no green fluorescence was found. The green cells, in which no autofluorescence granules, i.e., no cytosomes, were found also showed pseudoisocyanine-positivity. The pseudoisocyanine-positivity was also examined after reserpine treatment, and we found that after the depletion of dopamine the pseudoisocyanine-positivity of the axons remained, but diminished to a certain extent.

The second question was answered by investigating one of the cerebral ganglia at different points of time after the injection of 10 mg/kg reserpine so as to determine the dopamine content by Falck's method, while the cerebral ganglion of the other side was prepared electron-microscopically.

Three hours after reserpine treatment, when the dopamine is practically indetectable by Falck's method, the dense-core vesicles of the axons (Fig. 6) show very considerable changes: the average diameter (Fig. 7) decreased to 825 Å, the internal part of great electron density remained intact only in about 4% of the vesicles, and 96% of the vesicles were either in the process of evacuation or were empty.

Twelve hours later we tried again, but no green fluorescence had yet appeared in the axons of the neuropile, and the diameter of the vesicles continued decreasing

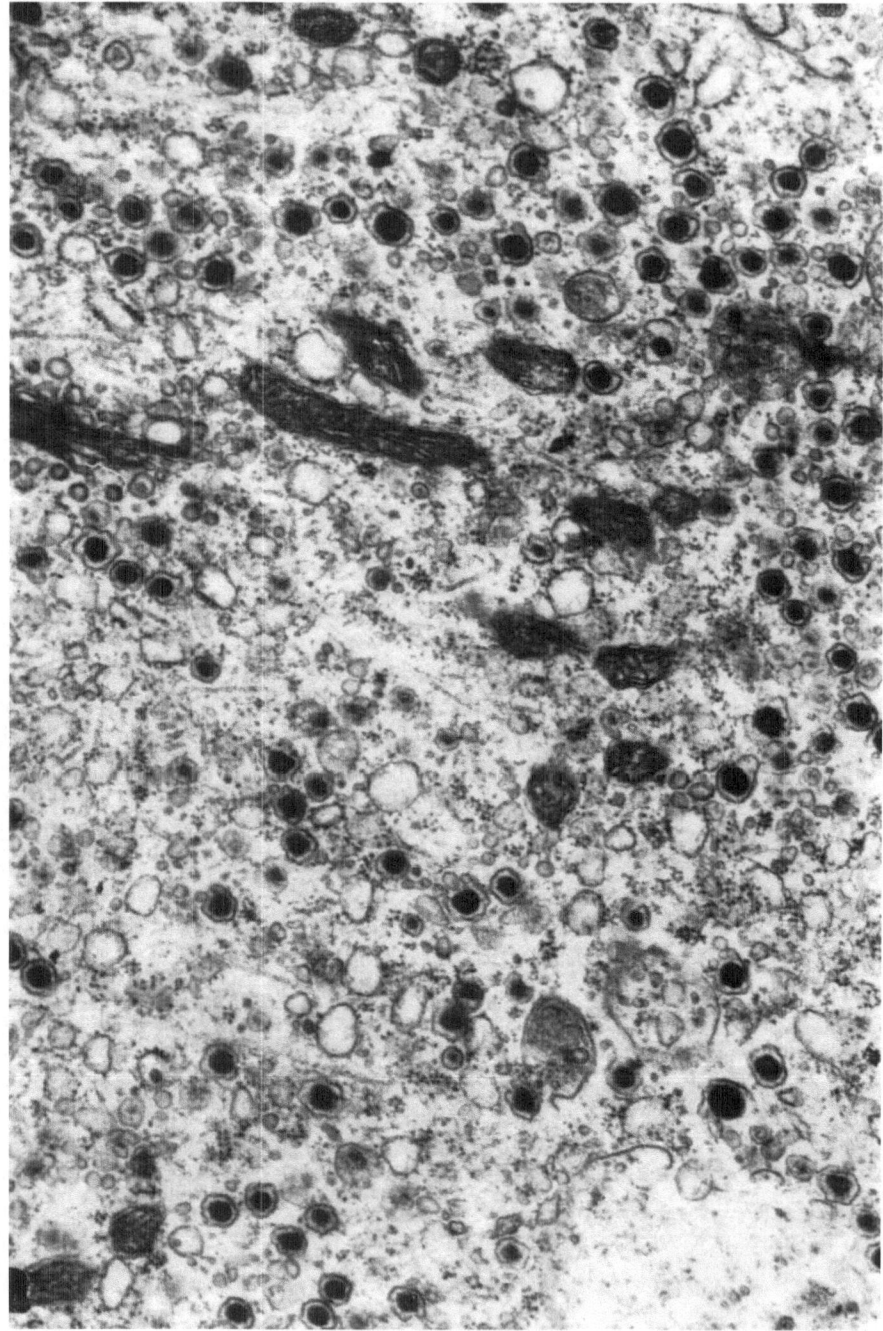

FIG. 3. Part of a perikaryon containing dense-core vesicles. Cerebral ganglion. × 30,000

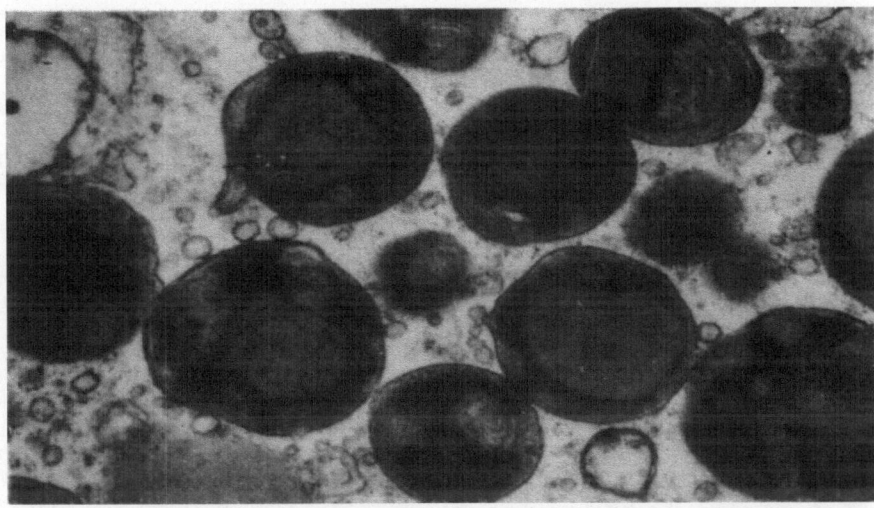

FIG. 4. Cytosomes of smaller diameter in the visceral ganglion. × 30,000

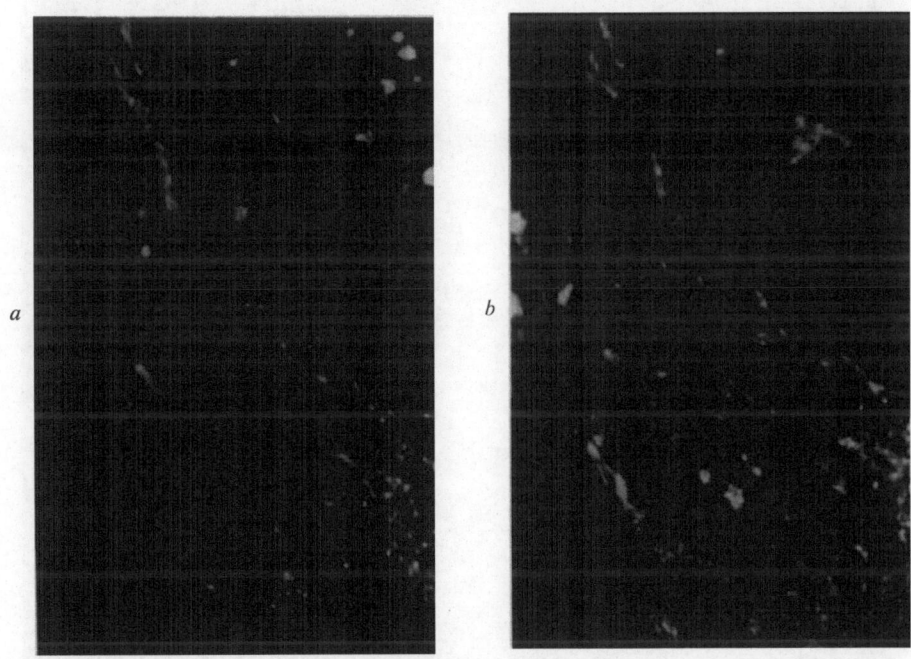

FIG. 5. *a* — Falck-reaction in the visceral ganglion. The green fluorescence indicates the localization of catecholamines in the axons of the neuropile. Uncovered section. × 315. *b* — The field shown in Fig. 5a after Sterba's pseudoisocyanine reaction. The specific yellow fluorescence is localized almost identically with the green fluorescence of Fig. 5a. × 315

I. ZS.-NAGY

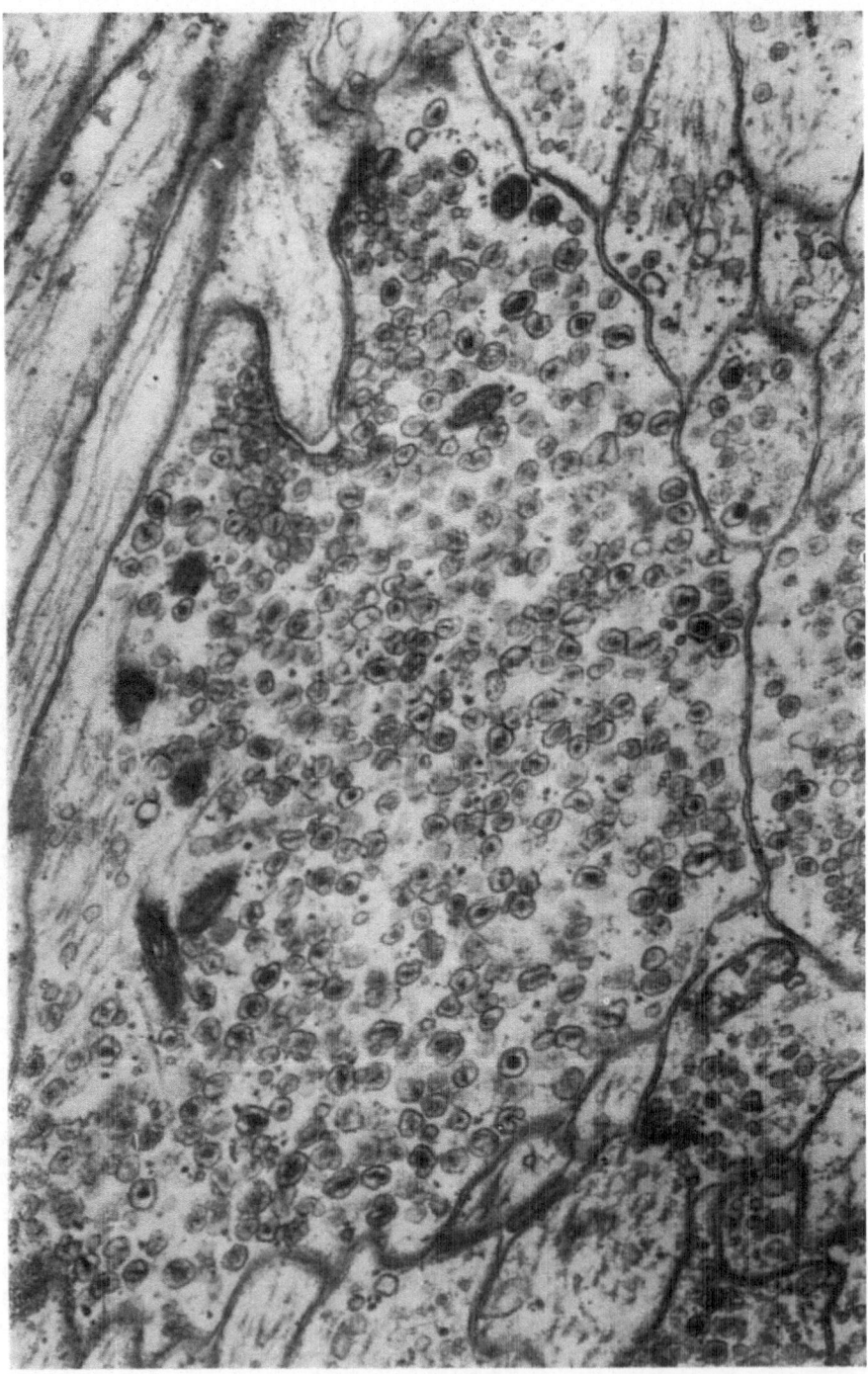

FIG. 6. Effect of reserpine treatment after 3 h on the structure of dense-core vesicles in the axons. Cerebral ganglion. × 30,000

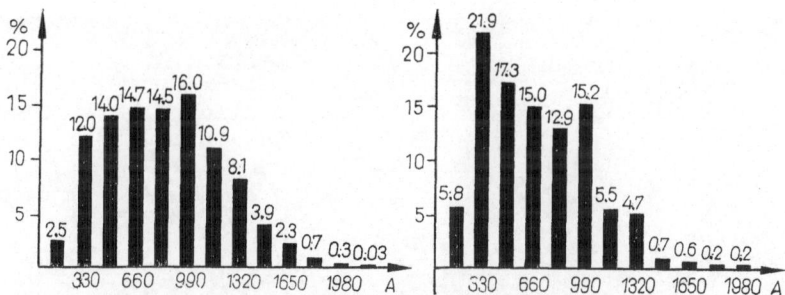

FIG. 7. Size distribution of vesicles in the axons after reserpine treatment. The first diagram shows the distribution after 3 h and the second one after 12 h consequent to reserpine injection

to 670 Å (Fig. 7). At the same time, all cells began fluorescing green, and dense-core vesicles started forming in them, something which is normally characteristic only for a small percentage of nerve cells. In the formation of dense-core vesicles, the rich Golgi-complex plays an important part (Fig. 8), but both the nucleus and the endoplasmic reticulum are strongly activated. Figure 9 shows a very active endoplasmic reticulum, similar to ergastoplasm, which normally occurs very rarely in the cells. It is very conspicuous that the great part of the cytosomal material is used up in this time. In the beginning, this manifests itself in the fact that the cytosomes become smaller, their electron density diminishes, and their membranes break out and the contents seem to stream out into the cytoplasm (Fig. 10). When only very few or no cytosomes are to be found in the cells, the number of dense-core vesicles increases greatly in the perikarya (Fig. 11).

Three days subsequent to the injection of reserpine, the nerve cells contain a great deal of dopamine and there are many dense-core vesicles in their cytoplasm. At the same time, however, the axons also abound both in dopamine and intact dense-core vesicles.

After 6–8 days, the structure of the cytoplasm has not yet been restored with respect to the cytosomes, but the number of dense-core vesicles had decreased in the perikarya by this time, and we can see many full dense-core vesicles in the axons of the neuropile (Fig. 12), the size of which is identical with that of the cytoplasmic dense-core vesicles.

Our results can be interpreted as follows:

It is not yet a settled question as to what extent the pseudoisocyanine-positive axons can be regarded as neurosecretory in the invertebrates, since many authors hold the dense-core vesicles, even on the basis of their ubiquitous presence, as noncholinergic synaptic vesicles (Hagadorn *et al.* 1963; Gerschenfeld 1963; Sakharov *et al.* 1965; Zs.-Nagy 1965). However, the fact that the pseudoisocyanine method is suitable for detecting nerve elements containing elementary neurosecretory granules or dense-core vesicles is very advantageous for us, as no

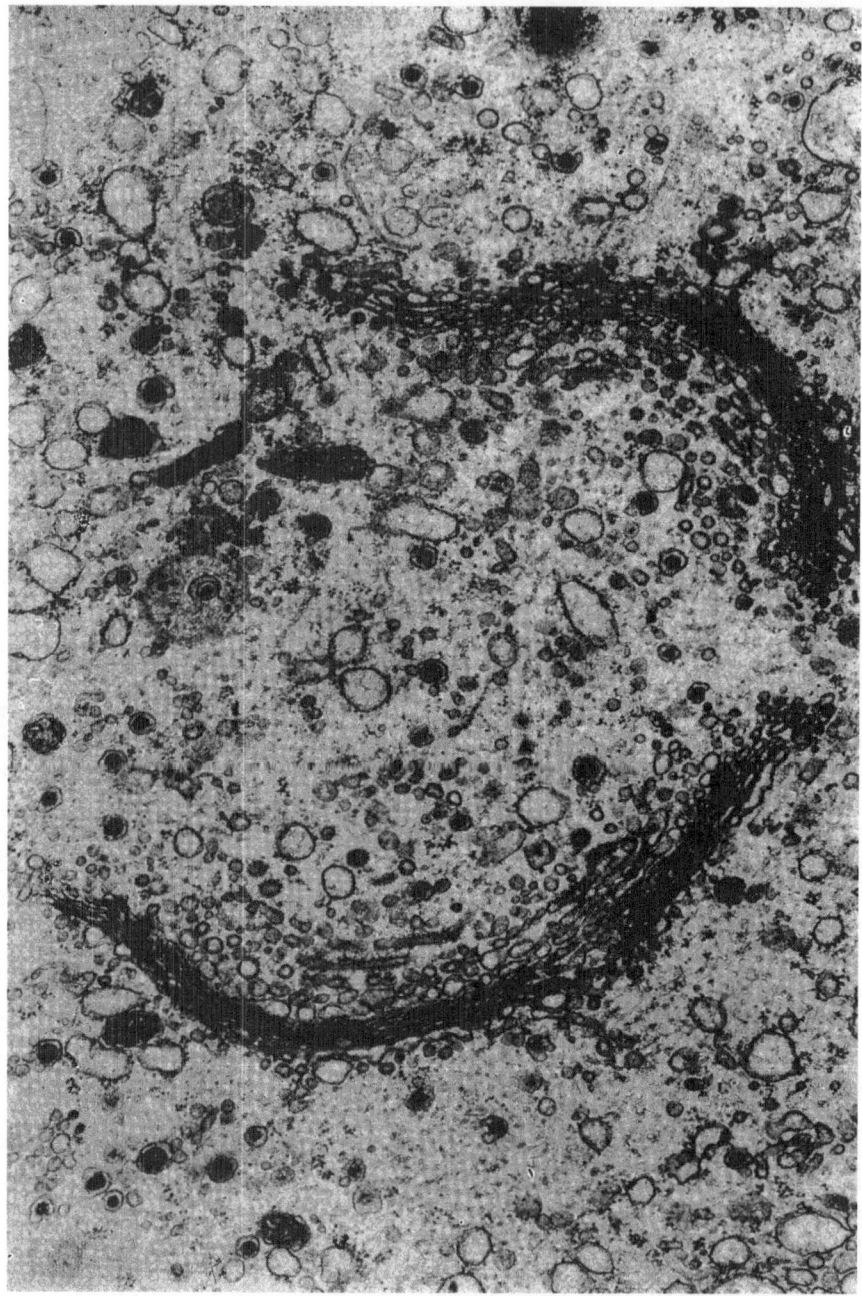

FIG. 8. Golgi-complex in the cerebral ganglion 12 h after reserpine treatment. × 25,000

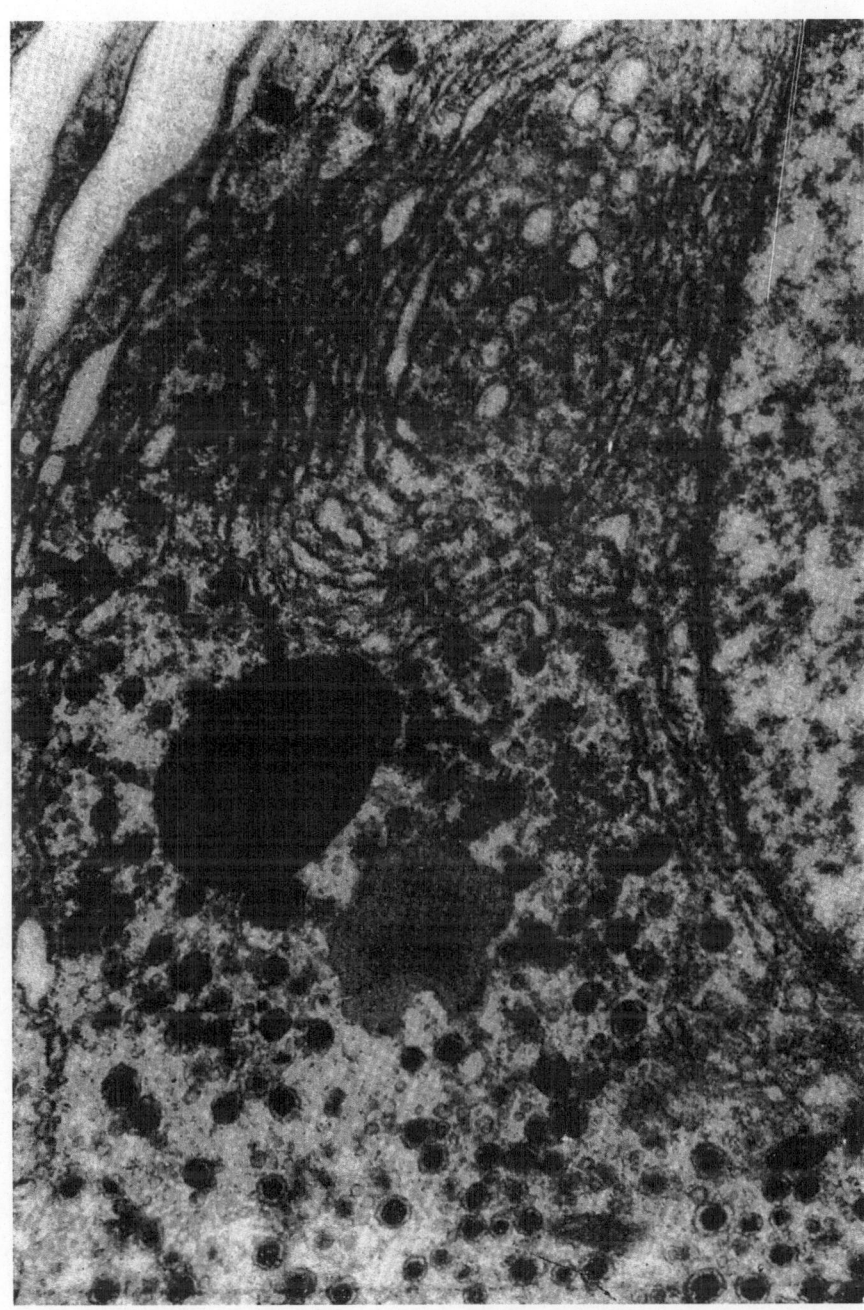

FIG. 9. Ergastoplasm in the cerebral ganglion 12 h after reserpine treatment. × 25,000

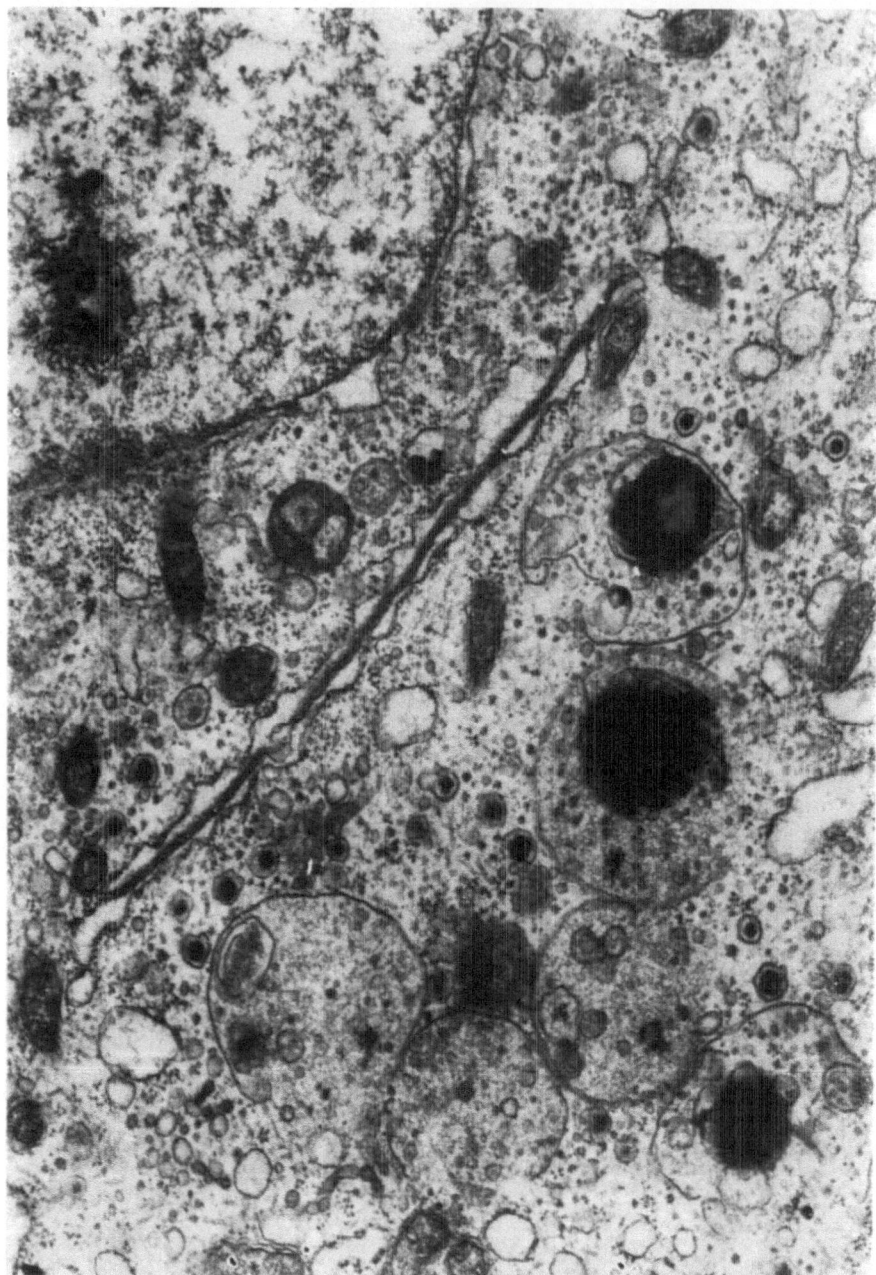

FIG. 10. Cytosomes 12 h after reserpine treatment. Cerebral ganglion. × 30,000

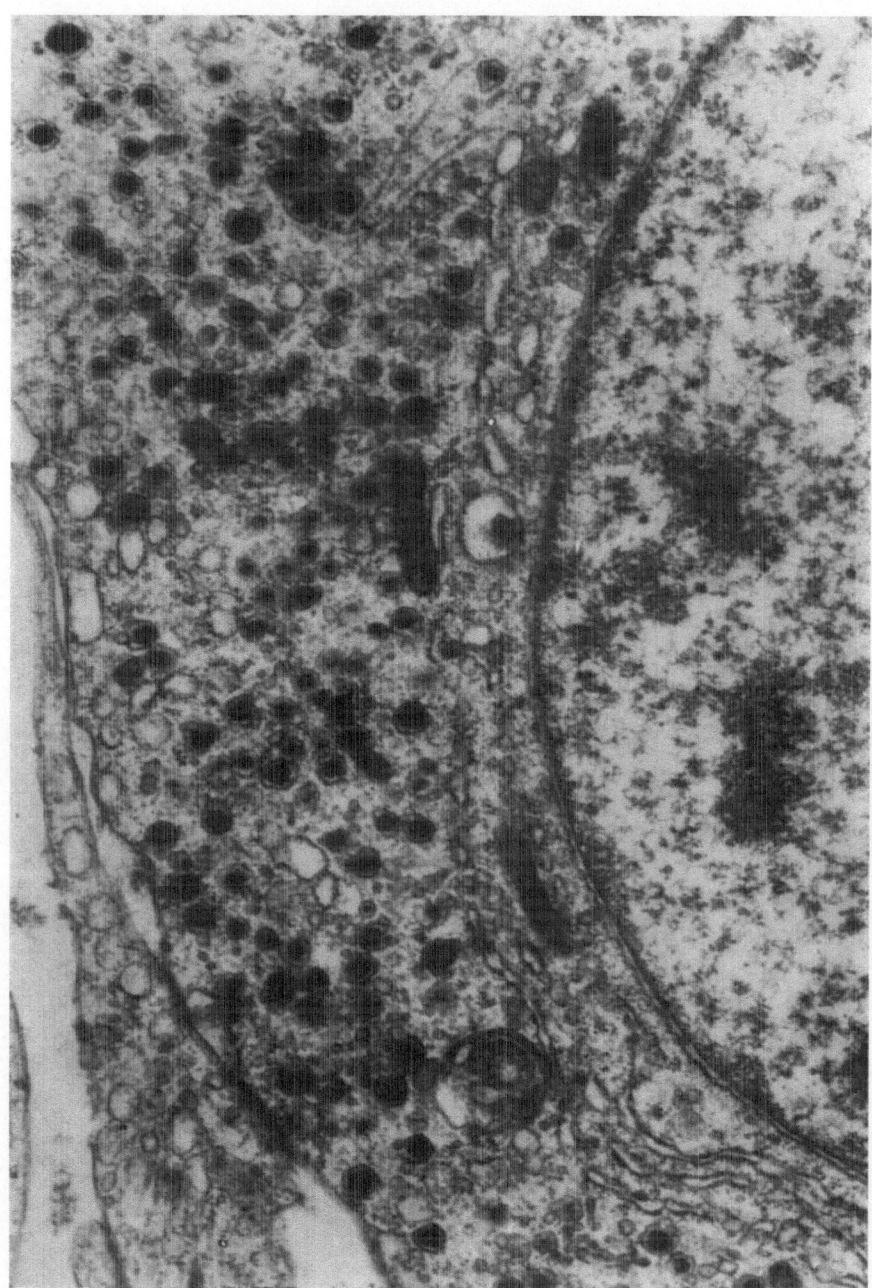

FIG. 11. A number of dense-core vesicles in the perikaryon 12 h after reserpine treatment. Cerebral ganglion. × 25,000

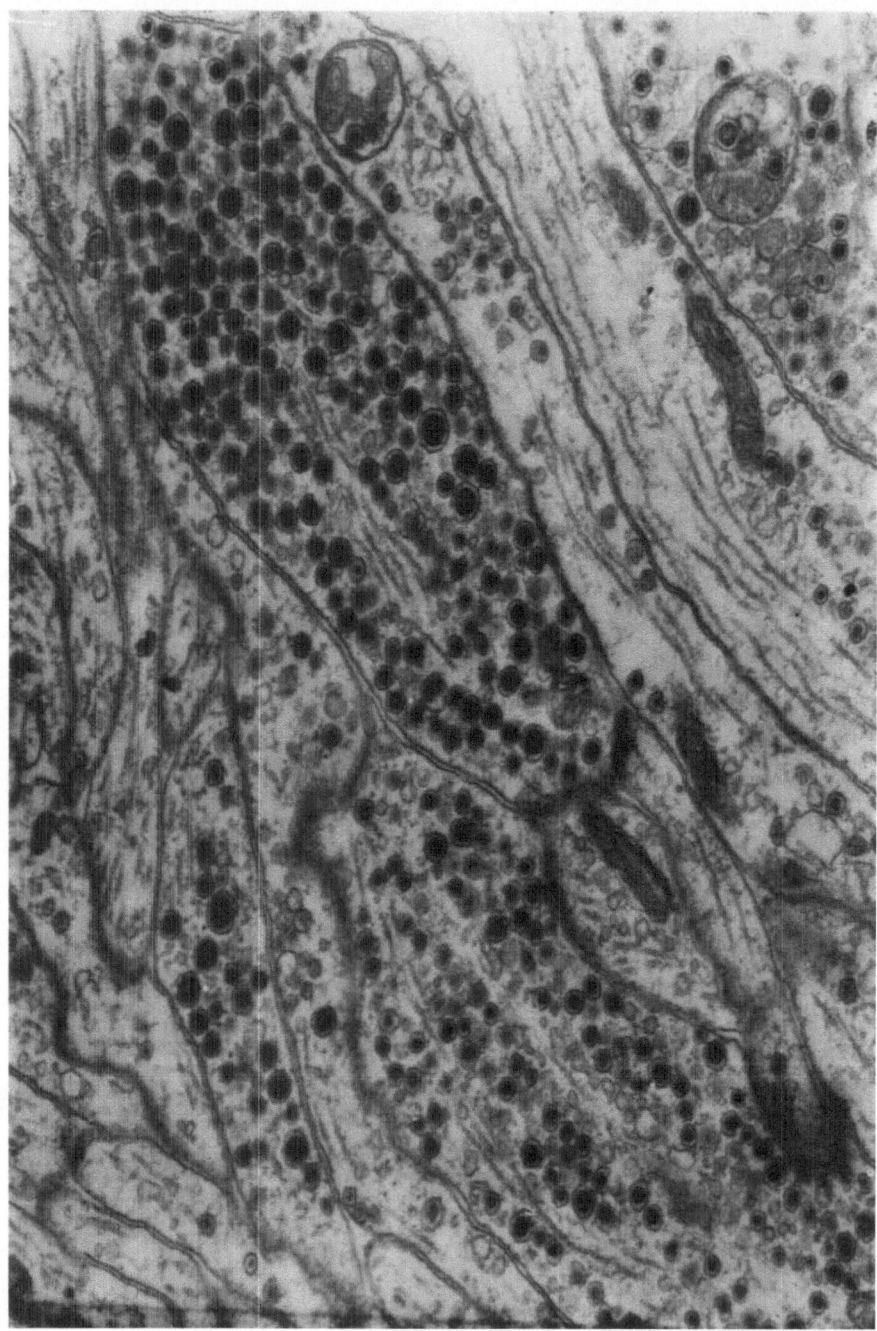

FIG. 12. Picture of the neuropile 8 days after reserpine treatment. Cerebral ganglion. × 25,000

other microscopic method of similar sensitivity is at our disposal. A further advantage of this method is that, following the photographing of the localization of catecholamines, it can be applied to the very same section.

In the axons, in which the localization of dopamine and pseudoisocyanine-positivity is identical, the relation of dopamine and the dense-core vesicles, reacting with the pseudoisocyanine, is obvious. Considering that the evacuation of dopamine after reserpine treatment was followed by a significant evacuation of even the dense-core vesicles on the electron-microscopic level in the axons, the dense-core vesicles seem to be the subcellular storage sites of the dopamine in the neurons of *Anodonta cygnea*.

The question remains of explaining those axons in which catecholamine was not found but pseudoisocyanine-positivity was. The results of reserpine treatment can serve as essential proof, as they showed that after the evacuation of catecholamines the pseudoisocyanine-positivity remained, i.e., the carrier protein did not disappear together with the active substance. Also, it could be seen electron-microscopically that after the depletion of catecholamines the vesicles kept their membrane, but became smaller, with the density of their internal part diminishing. On this basis, it seems that the axons which contained no catecholamine but reacted with pseudoisocyanine can be interpreted as the ones that just depleted their catecholamine-content during their function.

Our experiments show at the same time that the carrier protein reacting with the pseudoisocyanine is, in all probability, localized in the membrane of the dense-core vesicles. Of course, it is not out of question that the dense-core vesicles may not only contain dopamine, but other active substances too, even in the case of mussel. Attention must be called to the fact that in the cerebral ganglia of *Lymnaea stagnalis* (Gastropoda) the great part of the pseudoisocyanine-positive axons never contain catecholamines in histochemically detectable quantities (Sakharov and Zs.-Nagy 1968). It is obvious that the active substance could significantly change in the process of evolution, but the carrier protein, at least with respect to its pseudoisocyanine-positivity, has not changed to such a great extent.

One part of the green cells, which contained autofluorescence granules, i.e. cytosomes, did not react with pseudoisocyanine. This can be interpreted as follows: The formation of catecholamines in the cytoplasm does not start together with that of the carrier protein, or, at an earlier stage, this protein does not react as yet with the pseudoisocyanine. In this stage, the cells only show a week green fluorescence and relatively few dense-core vesicles and many cytosomes are to be found in the cells. Later, more and more dense-core vesicles come into being, while the cytosomes are used up, and it is only at this time that the pseudoisocyanine-positivity appears.

The depletion of the cytosomal material in the production of dense-core vesicles supports the hypothesis of Nolte and her co-workers (1965) according to which the cytosomes represent a 'Stoffwechseldepot'. This hypothesis is also supported by our other investigations (Zs.-Nagy 1967b).

All in all, the investigations presented here prove Gerschenfeld's hypothesis (1963) that the dense-core vesicles may contain catecholamines. In the case of mussels, the dopamine is localized in the dense-core vesicles which originate in the cytoplasm, and from here they go into the axons. These data, together with the fact that the interneuronal mediator function of the dopamine is physiologically demonstrated in certain neurons of Gastropods (Gerschenfeld 1964), show that dense-core vesicles can indeed be interpreted as synaptic vesicles of monoaminergic nature in the Molluscs, and that the morphological signs of their evacuation — the decrease of the outer diameter and the shrinking or disappearance of the internal part to be seen also in normal axons, especially at the synaptic contacts — correspond to the morphological manifestation of the evacuation of catecholamines.

SUMMARY

The localization of axons and nerve cells, the dopamine contents of which are detectable with Falck's fluorescence histochemical method, and the localization of dense-core vesicles (DCV) corresponding to the pseudoisocyanine (PSI)-positive material, have been compared in the same preparations; furthermore, the changes of dopamine content and ultrastructure of the neurons have been examined after treatment with 10 mg/kg reserpine. It has been established that:

1. Axons containing dopamine are identical with those of PSI-positivity. However, there are some PSI-positive axons in which dopamine was not detectable.

2. The complete depletion of dopamine from the axons of cerebral ganglion was obtained 3–6 h after reserpine treatment. After the same length of time, the PSI-positivity of the axons remained, but was to a certain extent diminished.

3. PSI-positivity was found in those dopamine-containing perikarya only, which had no autofluorescence granules.

4. Normally, half of the DCV are in the process of evacuating in the axons and the other half are morphologically intact. At 3 h after reserpine treatment, the morphological signs of the evacuation process (decrease of the outer diameter, shrinking or disappearance of the internal electron-dense part) are to be seen in 96% of the DCV. After 12 h the axons contain no dopamine as yet, and the DCV correspond to the empty form. By this time, however, the production of dopamine has started in the cytoplasm of the nerve cells, and it continues for 2–3 days. At the same time, we can see electron-microscopically that the bulk of the DCV are in the majority of the perikarya, which is normally characteristic only for a small percentage of nerve cells. It is obvious that in the production of the DCV the Golgi complex plays an active part, and the majority of the cytosomal material is used up in this process.

5. Sixty to seventy-five hours subsequent to the injection of reserpine, the nerve cells contain a great deal of dopamine and there are many DCV in their cytoplasm. At the same time, however, the axons are filled with dopamine and intact DCV are numerous.

6. One hundred fifty to two hundred hours after the injection of reserpine, the structure of the cytoplasm is not yet restored, but the neuropile is normal with regard to dopamine-content and DCV.

The localization of dopamine is similar to that of the DCV; during the depletion of dopamine the change of DCV is observed in the axons and together with the appearance of dopamine the DCV in great number are also to be found in all perikarya. This clearly demonstrates that the dopamine is stored in DCV on a subcellular level. Furthermore, it is obvious that the signs of evacuation are the morphological manifestation of the functional release of the dopamine in the DCV.

REFERENCES

ANTHEUNISSE, L. J. (1963): Neurosecretory phenomena in the zebra mussel, *Dreissena polymorpha* Pallas. *Arch. néerl. Zool.* **15** 237—314.

BARANYI, B. I. and SALÁNKI, I. (1963): Studies on neurosecretion in the central nervous system of *Anodonta cygnea. Acta biol. Acad. Sci. hung.* **13** 371—378.

DAHL, E., FALCK, B., LINDQUIST, M. and MEKLENBURG, C. (1962): Monoamines in Mollusc neurons. *Kungl. Fysiografiske sällskapets i Lund Förhand* **32** 89—91.

DAHL, E., FALCK, B., VON MEKLENBURG, C., MYHRBERG, H. and ROSENGREN, E. (1966): Neuronal localization of dopamine and 5-hydroxytryptamine in some Mollusca. *Z. Zellforsch.* **71** 489—498.

FALCK, B. (1962): Observations on the possibilities of the cellular localization of monoamines by a fluorescence method. *Acta physiol. scand.* **56** Suppl. 197.

FÄHRMANN, W. (1961): Licht- und elektronenmikroskopische Untersuchungen des Nervensystems von Unio tumidus (Philipsson) unter besonderer Berücksichtigung der Neurosekretion. *Z. Zellforsch.* **54** 689—716.

GABE, M. (1966): Neurosecretion. *International Series of Monographs in Pure and Applied Biology, Zoology Division*. Pergamon Press Oxford, Vol. **28** 132—136.

GERSCHENFELD, H. M. (1963): Observations on the ultrastructure of synapses in some Pulmonate Mulluscs. *Z. Zellforsch.* **60** 258—275.

GERSCHENFELD, H. M. (1964): A non-cholinergic synaptic inhibition in the central nervous system of a Mollusc. *Nature* **203** 415—416.

HAGADORN, I. R., BERN, H. A. and NISHIOKA, R. S. (1963): The fine structure of the supraoesophageal ganglion of the rhynchobdellid leech, *Theromyzon rude*, with special reference to neurosecretion. *Z. Zellforsch.* **58** 714—758.

NOLTE, A., BREUCKER, H. and KUHLMANN, D. (1965): Cytosomale Einschlüsse und Neurosekret im Nervengewebe von Gastropoden. *Z. Zellforsch.* **68** 1—27.

SAKHAROV, D. A., BOROVYAGIN, V. L. and Zs.-NAGY, I. (1965): Light, fluorescence and electron microscopic studies on "neurosecretion" in *Tritonia diomedia* Bergh (Mollusca, Nudibranchia). *Z. Zellforsch.* **68** 660—673.

SAKHAROV, D. A. and Zs.-NAGY, I. (1968): Localization of biogenic monoamines in cerebral ganglia of *Lymnaea stagnalis. Acta biol. Acad. Sci. hung.* **19** 145—157.

STERBA, G. (1961): Fluorescenzmikroskopische Untersuchungen über die Neurosekretion beim Bachneunauge (Lampetra planeri Bloch). *Z. Zellforsch.* **55** 763—789.

STERBA, G. (1964): Grundlagen des histochemischen und biochemischen Nachweises von Neurosekret (= Trägerprotein der Oxytozine) mit Pseudoisozyaninen. *Acta Histochem.* **17** 268—292.

Zs.-NAGY, I. (1964): Electron-microscopic observations on the cerebral ganglion of the fresh water mussel (*Anodonta cygnea* L.). *Annal. Biol. Tihany* **31** 147—152.

Zs.-Nagy, I. (1965): Fluorescence microscopic examination with pseudoisocyanin on the neurosecretory activities of the fresh water mussel *Anodonta cygnea* L. *Annal. Biol. Tihany* **32** 123—127.

Zs.-Nagy, I. (1967a): Histochemical demonstration of biogenic monoamines in the central nervous system of the lamellibranchs mollusc *Anodonta cygnea* L. *Acta biol. Acad. Sci. hung.* **18** 1—8.

Zs.-Nagy, I. (1967b): Histological, histochemical and electron microscopical studies on the cytosomes of the nerve cells in *Anodonta cygnea* L. (Mollusca, Lamellibranchiata). *Annal Biol. Tihany* **34** 25—39.

DISCUSSION

G. A. Cottrell: (a) Did you measure biochemically the proportion of dopamine depleted by reserpine in your experiments?

(b) Did all the animals survive for more than a few hours after reserpine injection?

I. Zs.-Nagy: (a) No, I did not.

(b) As was shown, I examined the ganglia from 3 h to 8 days after injection of 10 mg/kg reserpine. It is not a lethal dose for mussels. All the animals survived this reserpine treatment. Dahl and his co-workers (1966) gave 30 mg/kg reserpine repeatedly.

B. Csillik: Is the size and shape of dopamine-containing dense-core vesicles dependent of the fixative? In other words, does the possibility exist that the percentage distribution of 'full' and 'empty' vesicles would be different in aldehyde-fixed specimens as contrasted to osmium-fixed samples?

I. Zs.-Nagy: It is possible, but in these experiments I always fixed the ganglia in osmium and so I have no data about the effect of aldehyde fixative.

M. Mirolli: Is the spatial distribution of the broken cytosomes in the neuron's cytoplasm different in normal and reserpinized animals?

I. Zs.-Nagy: We can see some broken cytosomes in normal animals as well, but they have no characteristic distribution in the cytoplasm. In reserpinized animals, however, the majority of cytosomes show the broken and emptying form within 2–3 days.

Symposium on Neurobiology of Invertebrates 1967 (85—94)

SUBMICROSCOPIC ORGANIZATION OF FIBERS OF THE CRAB'S LEG NERVES AND DISTRIBUTION OF A VITAL DYE (HELIOGEN BLUE SBL) ON THEIR ULTRASTRUCTURES

Ya. Yu. Komissarchik and S. B. Levin

Institute of Cytology, Academy of Sciences of the USSR
Leningrad, USSR

Studies have been made of the ultrastructural organization of crab axons and of the distribution of the acid vital dye (Heliogen Blue SBL) on the ultrastructures in the normal state and during physiological excitation. For this purpose, previously described electron-microscopic and cytophotometric methods have been used (Komissarchik and Levin 1964; Levin and Golfand 1964; Komissarchik and Levin 1965; Levin *et al.* 1968).

SUBMICROSCOPIC ORGANIZATION OF MEMBRANOUS STRUCTURES OF LARGE AXON SHEATHS

The axoplasm of all the fibers investigated is surrounded by a 75-Å, three-layered membrane-axolemma. The sheath that envelopes it consists of a periaxonal layer of Schwann cells. The sheath becomes more complicated with an increase in fiber diameter (Figs. 1 and 2). On all the axons investigated, the cytoplasmic membrane of Schwann cells is separated from the axolemma by narrow gaps (the gaps of the axon-Schwann membrane) 120–150 Å thick. Axons of large- and median-sized diameter are, as a rule, surrounded by one layer of Schwann cells. Canals piercing this layer are immensely complicated. (Figs. 2 and 3). Their walls consist of cytoplasmic membranes of neighboring Schwann cells. A great number of electron micrographs show that, on the one hand, these canals fall directly into axon-Schwann gaps, and, on the other hand, into some regions of extracellular space (Figs. 1 and 2). By analogy with myelinated sheaths of nerve fibers in higher animals, these areas are designated as internal and external mesaxons. The width of the gap of these canals is, in the majority of cases, equal to that of an axon-Schwann membrane. The lengths of the canals range widely from 5 to 15 μ. The interrelation of neighboring Schwann cells may be well observed on fibers of medium-sized diameter, cross sections of which reveal no more than two Schwann cells. Thus, Fig. 1 shows two internal and two external mesaxons. The boundaries of the cells are rather complicated forming some additional membranes in certain regions of the sheath. Sheaths of large axons comprise a greater number of cells.

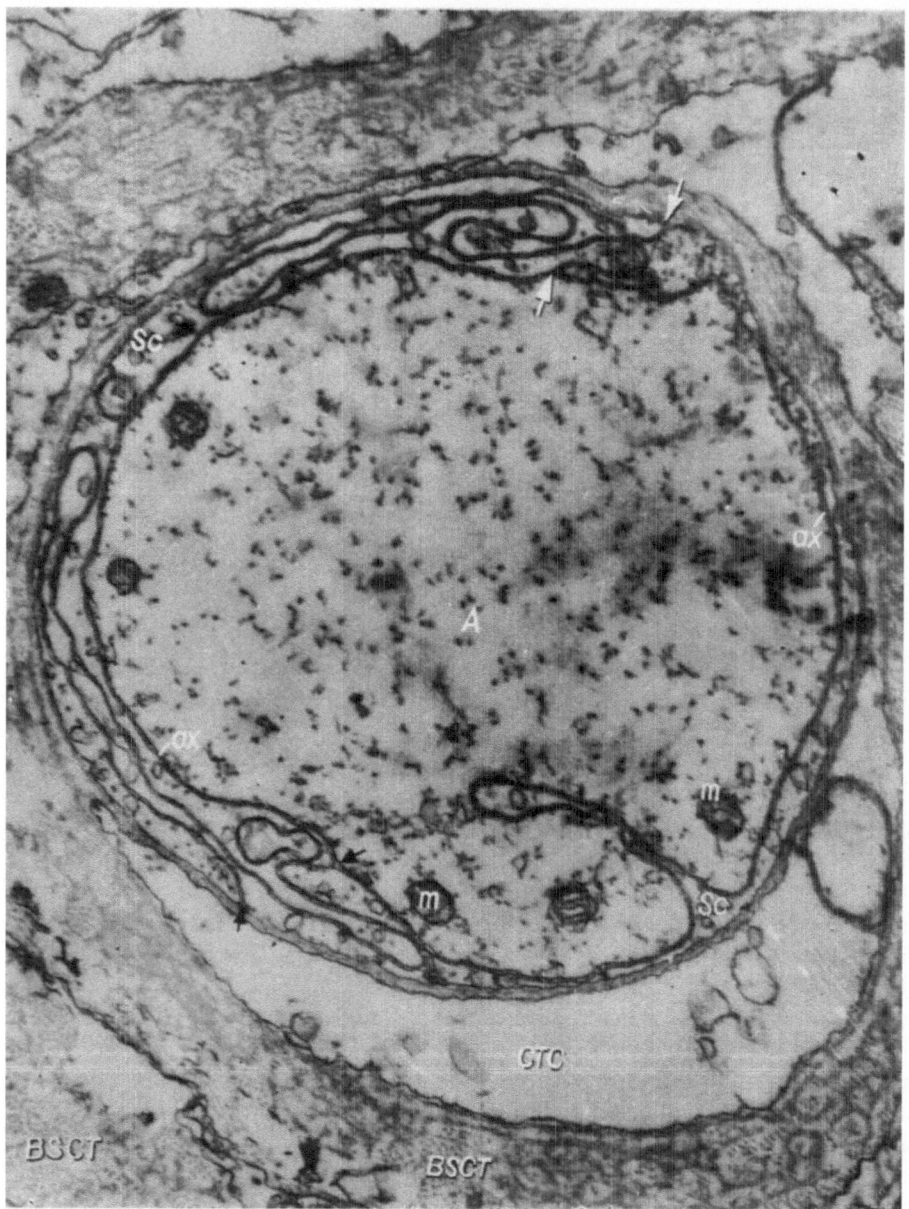

FIG. 1. Cross section of a single axon of medium-sized diameter. The periaxonal layer of the sheath consists of two Schwann cells (Sc). Mitochondria (m) in the axoplasm (A) are arranged near the axolemma (ax). The arrows indicate internal (IMA) and external (EMA) mesaxons. The external sheath of the axon consists of fibrillar bundles of the basic substance of connective tissue (BSCT) and connective tissue cells (CTC). × 28,000

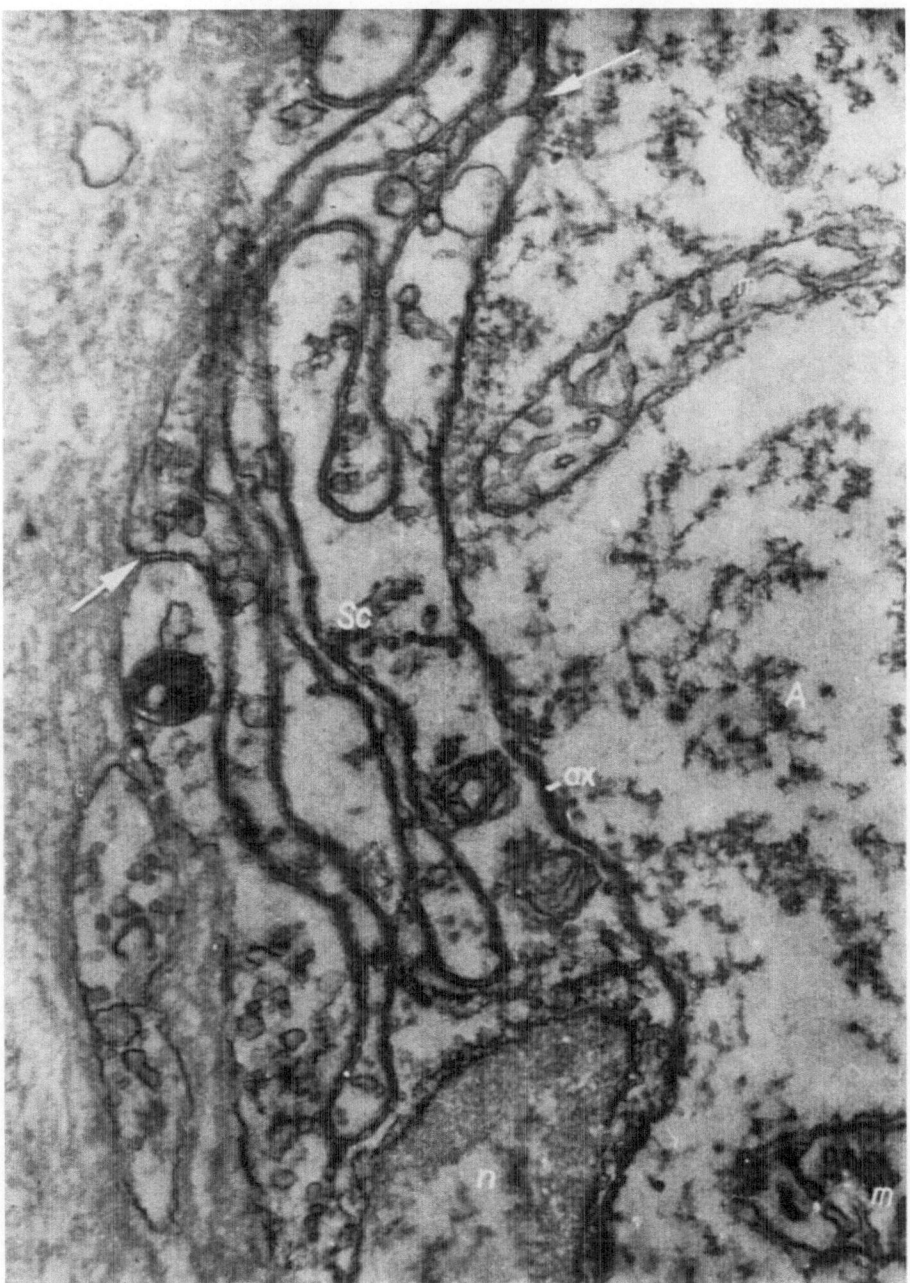

Fig. 2. Cross section of a region of a large axon. A boundary between adjacent Schwann cells is rather complicated. The section is made in the region of the nucleus (n) of a Schwann cell. Designations the same as in Fig. 1. × 45,000

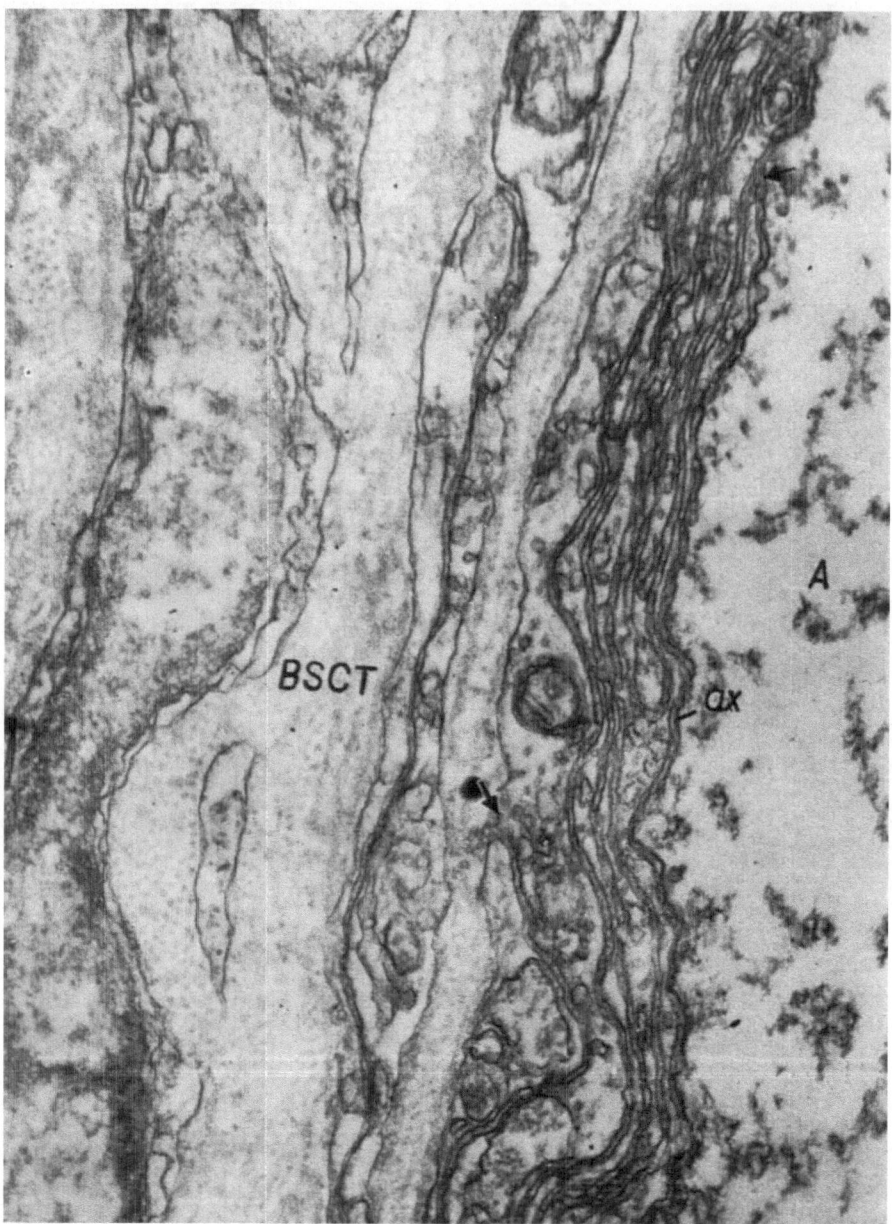

FIG. 3. Cross section of a large axon. Membranous structures of the periaxonal layer of Schwann cells resemble the lipoprotein periodic structures observed in poorly myelinated nerve fibers of higher animals. × 50,000

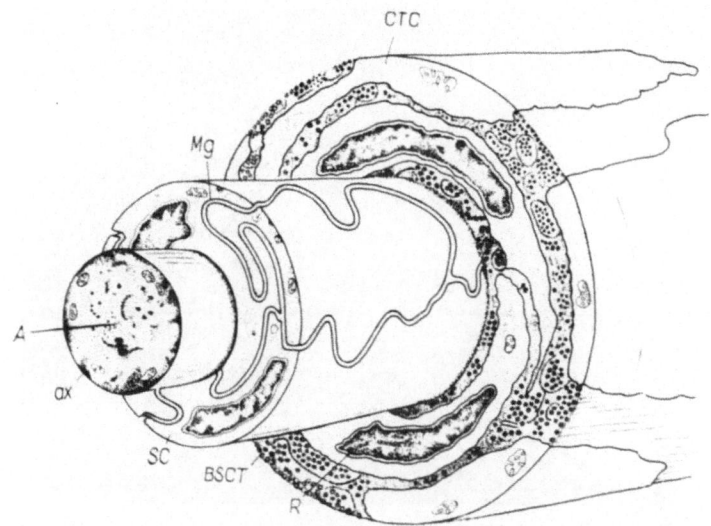

FIG. 4. A three-dimensional representation of the large axon, compiled from electron-microscope micrographs of sections made in different directions. For the sake of clearness, the dimensions of different elements of the fiber are not correlated. Designations the same as in Fig. 1

The distribution of canals between neighboring Schwann cells is somewhat different in such fibers. In some cases, we observed canals in which both ends entered extracellular space. In other cases, one end of a canal goes into an axon-Schwann membrane while the other falls into two canals opening into extracellular space. The relationships in question are demonstrated in Fig. 4. Such pictures may be interpreted as resulting from the overlapping of Schwann cells lying along the fiber.

The described submicroscopic organization of sheaths of large axons in the crab is in principle no different from the structure of the giant axon sheaths of the squid (Villegas and Villegas 1960).

DISTRIBUTION OF PHTHALOCYANINE DYE
ON THE ULTRASTRUCTURES OF VITALLY STAINED AXONS

To reveal organic dyes in the electron microscope, the following criteria need to be met (Isenberg 1957; Bondarev 1957; Finck 1958): The dyes should contain elements of high atomic weight, or their concentration in certain parts of a cell must be sufficiently high. Possessing a relatively high electron optical density,

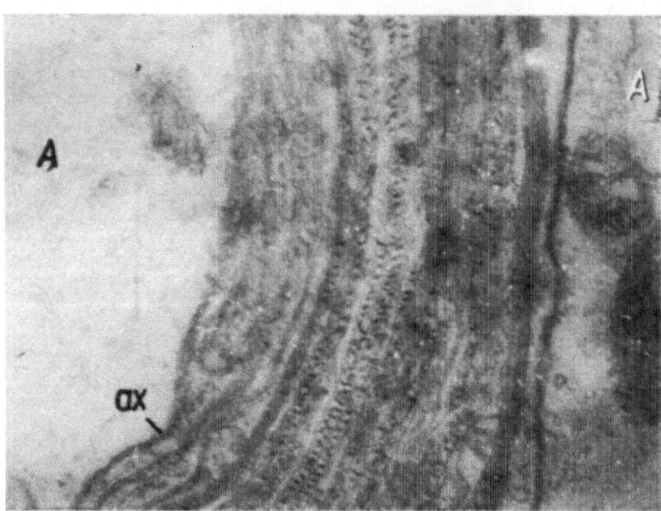

FIG. 5. Cross section of an area of a large unstained axon. Weak
contrast of membranous structures is visible. × 35,000

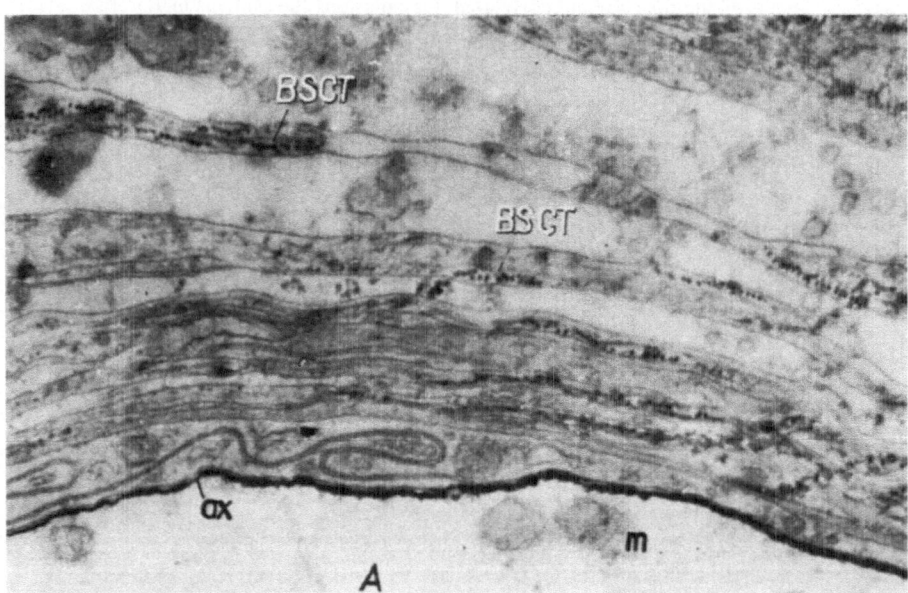

FIG. 6. Cross section of a region of a large axon stained with Heliogen Blue. The general
contrast of membranes is increased. In the same regions of the axolemma there appear accumulations of the dye either as diffusively distributed dense material or as single accumulations.
The dye absorption by fibrils of the basic substance of the connective tissue was found to be
increased. × 30,000

the dye should not be sublimated or destroyed by an electron beam. Moreover, it must be nontoxic. It is important that it should not dissolve in embedding media, which are, for the most part, hydrophobic substances. On the other hand, the dye should not dissolve in water solutions of fixatives or alcohols, nor during ultra-thin sectioning, when sections get into a water surface. It is therefore essential that the dye should be strongly bound with a substrate.

Special experiments have shown that phthalocyanine dye (Heliogen Blue SBL) satisfies the above criteria (Komissarchik and Levin 1964). The molecular weight of this dye is 782, and the maximal size of a molecule is 15 Å. It is a substantive dye and is strongly bound to cellulose fibers and proteins (Vickerstaff 1950).

To study the distribution of this dye on the ultrastructure of crab axons, the latter were kept in a 2% solution of Heliogen Blue in artificial seawater for 1.5–2 h, the time required for diffusion equilibrium.

On the sections of dyed axons, the dye is revealed on membranous structures of sheaths which have increased electron-microscope contrast when compared with pictures presented by such structures of unstained axons (Fig. 5). In both cases, the contrast of the sections was not improved with heavy metal salts. Along with an increase in the contrast of sheath membranes in certain regions of the axolemma, we revealed accumulations of the dye either as diffusely distributed dense material or dense granules passing throughout all the axolemma (Fig.6).

The granules vary in size from 50 to 400 Å. On micrographs made at large magnification, one can see that granules of the dye, distributed for the large part in the axolemma, sometimes exceed its boundaries and enter the axoplasm and gaps of axon-Schwann membranes (Fig. 7). In the majority of cases, dense granules in the axolemma are discretely distributed. It must be pointed out that the above pictures showing the distribution of the dye on the axolemma cannot be revealed along the whole length of the axon in cross and longitudinal sections.

In areas showing a three-layered character of the axolemma it may be seen that its external layer is stained along its whole length, whereas the internal one is stained in certain areas (Fig. 7). The dye is absorbed not only by membranes of the axon sheath. To a considerable extent, it may be bound by the fibrillar substance of connective tissue (Fig. 6). In a number of cases, widened mesaxon gaps were revealed in stained axons. Sometimes they appear as cavities of irregular shape, reaching a size of up to 0.7 μ (Fig. 7).

The increased absorption of the dye by external components of the axolemma and by fibrillar material of connective tissue is in agreement with well-known facts of acid-dye binding by proteins. This is confirmed by experiments with actomyosine. It was found that one actomyosine molecule may bind as many as 720 dye molecules. However, brain phospholipids do not bind the dye.

The uneven distribution of the dye on the axolemma may be indicative of its structural heterogenity and its difference from the cytoplasmic membrane of a Schwann cell.

On the basis of experimental evidence obtained by vital staining of axons with Heliogen Blue SBL, a suggestion can be made that the dye enters the axolemma

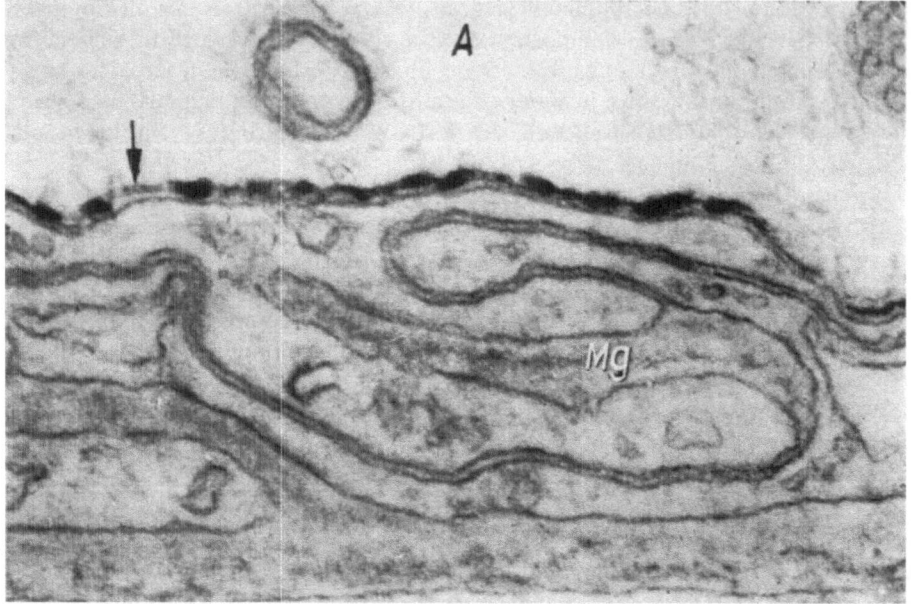

FIG. 7. Cross section of a large crab axon. As in the above micrograph, there is an increase in the contrast of sheath membranes. The dye is localized in axolemma largely as discrete granules. Local widening of mesaxonal gaps (Mg) are visible. In the areas of axolemma which reveal a triple-layered structure (arrow), the dye is localized on internal protein components.
× 90,000

through mesaxonal gaps. Such a supposition is based on the electron-microscope evidence on the distribution of the dye on ultrastructures of the axonal sheath. At the same time, the size and geometry of mesaxonal gaps are such that the total amount of dye bound by fibers may pass through them.

CHANGES IN BINDING HELIOGEN BLUE SBL
DURING THE EXCITATION OF AXONS

The stimulation was carried out by rectangular impulses of threshold strength at 10 Hz during 1–1.5 h in 2% dye. The subsequent treatment of objects for electron-microscope investigations was the same as for nonstimulated axons. The electron-optical density of the axolemma of these fibers was found to be considerably less than in stained nonstimulated axons. In the axolemma, we could not find sites with increased absorption of the dye (Fig. 8). These data show that during prolonged stimulation of the axon there is a decrease in the dye binding by the axolemma.

The decrease in the absorption of Heliogen Blue during the stimulation of axons was also shown by means of the cytophotometry of vital fibers. Short-term stimulation of axons (0.5–8 min) by pulses at 25, 50, and 100 Hz caused a decrease in the amount of the bound dye during the staining of fibers in a 0.01% solution of dye. The stimulation of axons that were dyed and then immersed in seawater, increased the desorption of the dye (Table 1).

This effect is proportional to the frequency of stimulation. The amount of desorbed dye per impulse from 1 cm^2 of the axolemma is about 0.1 pM.

The process undergoes no significant changes when the temperature of the washing solution ranges from 10 to 29 °C. These data also confirm the suggestion

TABLE 1

Increase in desorption of Heliogen Blue in a previously stained axon excited during seawater extraction

Number of measurements	Frequency of stimulation (Hz)	Increase in desorption			
		in $\dfrac{\text{mol} \cdot 10^9}{\text{cm}^2 \cdot \text{min}}$	Significance of differences (P_S)	in $\dfrac{\text{mol} \cdot 10^{13}}{\text{cm}^2 \cdot \text{imp.}}$	Significance of differences (P_S)
11	25	0.19±0.03		1.47±0.23	
10	50	0.31±0.03	0.98	1.13±0.11	0.89
21	100	0.37±0.03	0.97	0.73±0.09	0.99
				mean = 1.11	

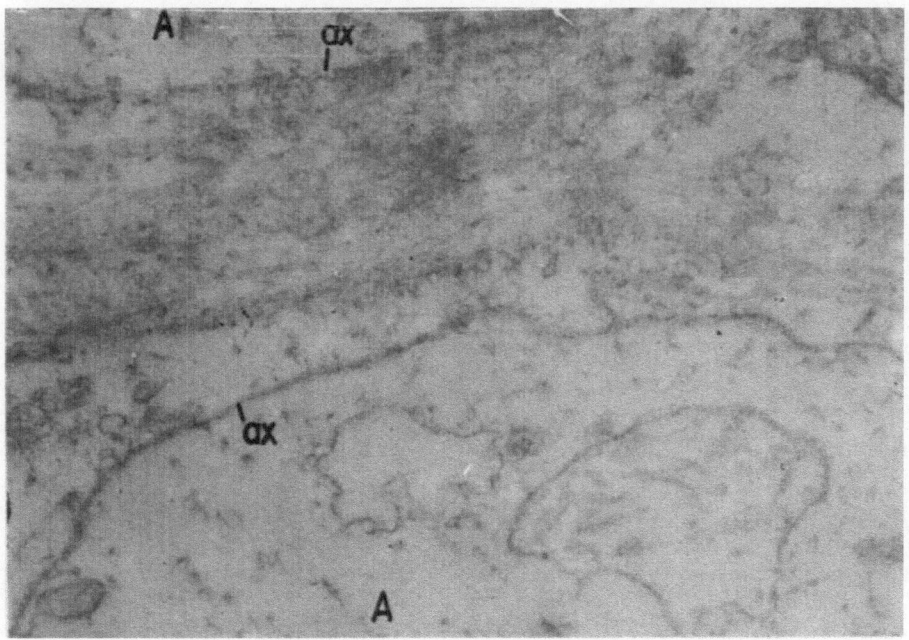

FIG. 8. A part of the cross section of the axon stimulated during staining. No increase in the electro-optical contrast of the axolemma can be observed. × 30,000

that desorption of the dye is connected with the action potential but not with trace biochemical processes.

Hence, electron-microscopic and cytophotometric investigations of the distribution of acid dyes on axon membranes indicate that their structures change during excitation. These changes seem to be connected with nonlipid components of membranes binding the dye. One may not neglect the fact that these changes can play a certain role in the mechanism of the permeability increase of monovalent cations during spikes.

SUMMARY

A submicroscopic organization of membranous structures of the crab large axons and the distribution of the acid vital phthalocyanine dye (Heliogen Blue SBL) on these structures in the normal state and during physiological excitation have been investigated. Evidence is given which shows that the dye enters the axolemma through the water gaps of mesaxons. Selective localization of the dye on membranous structures of the sheath and on fibrils of the basic substance of the connective tissue has been observed. The uneven distribution of the dye on the axolemma indicates its structural heterogeneity as well as its difference from the cytoplasmic membrane of a Schwann cell. The electron-microscopic and cytophotometric methods show that during the stimulation of axons with electrical current of threshold strength the binding of Heliogen Blue by membranes of sheaths decreases. The data show that the dye is bound by protein components of the membrane. A decrease in the affinity of Heliogen Blue to the axolemma may indicate structural changes of its protein components during excitation.

REFERENCES

BONDAREV, V. (1957): Morphology of particulate glycogen in guinea pig liver revealed by electron microscope after freezing and drying and selective staining in block. *Anat. Rec.* **129** 97—108.

FINCK, H. (1958): An electron microscope study of basophilic substances of frozen-dried rat liver. *J. biophys. biochem. Cytol.* **4** 291—300.

ISENBERG, I. (1957): The use of organic dyes in electron microscopy. *Bull. math. biophys.* **19** 279—292.

KOMISSARCHIK, Ya. Yu. and LEVIN, S. V. (1964): Electron microscope studies on the distribution of phthalocyanine dye in vitally stained crab axons. *Cytology (USSR)* **6** 605—609 (in Russian).

KOMISSARCHIK, Ya. Yu. and LEVIN, S. V. (1965): A possibility means for revealing phthalocyanine dye in vitally stained nerve fiber by electron microscopy. *Cytology (USSR)* **7** 104—108 (in Russian).

LEVIN, S. V. and GOLFAND, K. A. (1964): Cytophotometric method of vitally stained cells. *Cytology (USSR)* **6** 525—530 (in Russian).

LEVIN, S. V., ROSENTAL, D. L., GOLFAND, K. A. and KOMISSARCHIK, Ya. Yu. (1968): Changes in sorption of phthalocyanine dye (Heliogen Blue SBL) by axon membranes of the crab during excitation. *Cytology (USSR)* **10** 312–321 (in Russian).

VICKERSTAFF, T. (1950): *The Physical Chemistry of Dyeing*. London.

VILLEGAS, G. and VILLEGAS, R. (1960): The ultrastructure of the giant nerve fibre of the squid axon-Schwann cell relationship. *J. Ultrastruct. Res.* **3** 362—373.

Symposium on Neurobiology of Invertebrates 1967 (95—109)

FINE STRUCTURAL CHANGES INDUCED IN PHOTORECEPTORS BY LIGHT AND PROLONGED DARKNESS

P. RÖHLICH

Department of Histology and Embryology, Medical University School
Budapest, Hungary

Light energy is known to produce some distinct changes in photopigment molecules of invertebrates (Goldsmith 1964); the question may then arise of whether photoreceptor structures containing such pigments (Langer and Thorell 1965, 1966) also show some alterations at the ultrastructural level. Since the normal fine-structure of many invertebrate photoreceptors has been clarified in the last decade (see, for example, Eakin 1965; Röhlich 1967), the basis for experimental investigations has been established. It is surprising, however, that very few results have been reported on the ultrastructural effects of altered light conditions (Röhlich and Török 1962; Röhlich and Törő 1965; Eguchi and Waterman 1967). The present paper will describe some distinct and reproducible changes induced by light and prolonged darkness on the photoreceptor of two invertebrate species.

In the first part of this paper, the effect of intense illumination on the fine structure of the *Daphnia* eye will be reported. The ommatidium of the compound eye consists of a crystalline cone and of eight retinular cells partly surrounding the cone. The retinular cells (Röhlich and Törő 1965) are elongated unipolar cells having their nucleus in their distal part while an axon arises from the opposite end of the cell and runs into the optic ganglion to establish synaptic contacts with the second neuron. The most characteristic structure of the retinular cell is the rhabdomere, a highly regular system of long microvilli, which, together with similar rhabdomeres, form a compact axial structure, the rhabdom (Fig. 1). The microvilli of the rhabdom are arranged in two mutually perpendicular directions (Fig. 2) which are both perpendicular to the incident light. The rhabdom of the arthropod eye was shown by cytophotometric studies (Langer and Thorell 1965, 1966) to contain a rhodopsin-like visual pigment, and hence to represent the specific photoreceptor structure of the arthropod eye.

The animals *(Daphnia magna Str.)* were dark-adapted for 18 h and then exposed to an illumination of 900 lx for 10, 20, 30, and 60 min. The light source was a diffuse daylight in one experiment and a 100-W tungsten bulb in the other. In the latter case, care was taken to filter thermal beams to avoid a temperature increase of the water. Dark-adapted and illuminated animals were fixed in a formaline-glutaraldehyde mixture (Karnovsky 1965) followed by postfixation is osmium.

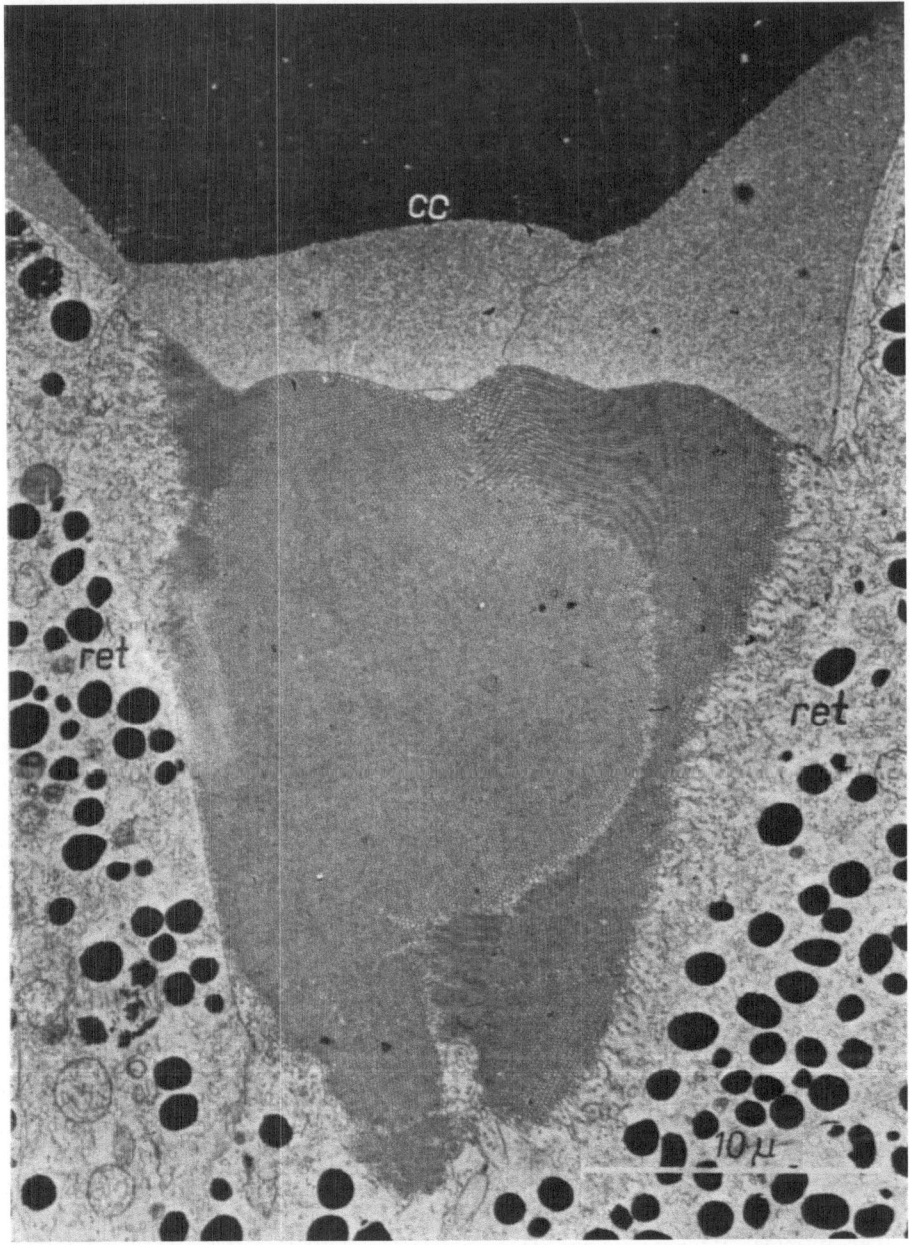

FIG. 1. Detail from an ommatidium of the compound eye of the *Daphnia*. The rhabdom consisting of a highly ordered array of microvilli occupies the axis of the ommatidium. cc — crystalline cone; ret — retinular cell

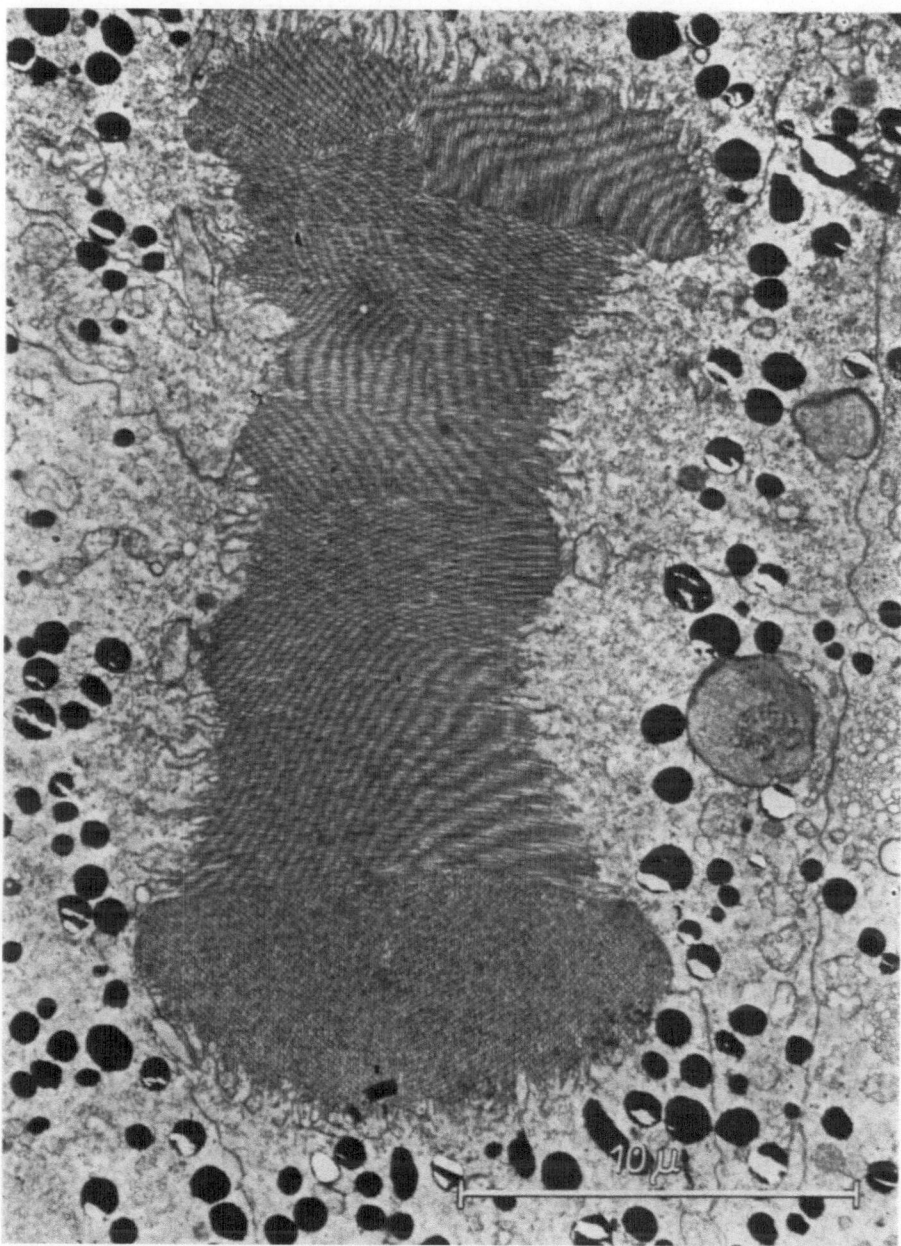

FIG. 2. Cross section of the rhabdom showing the two main mutually perpendicular directions of the microvilli

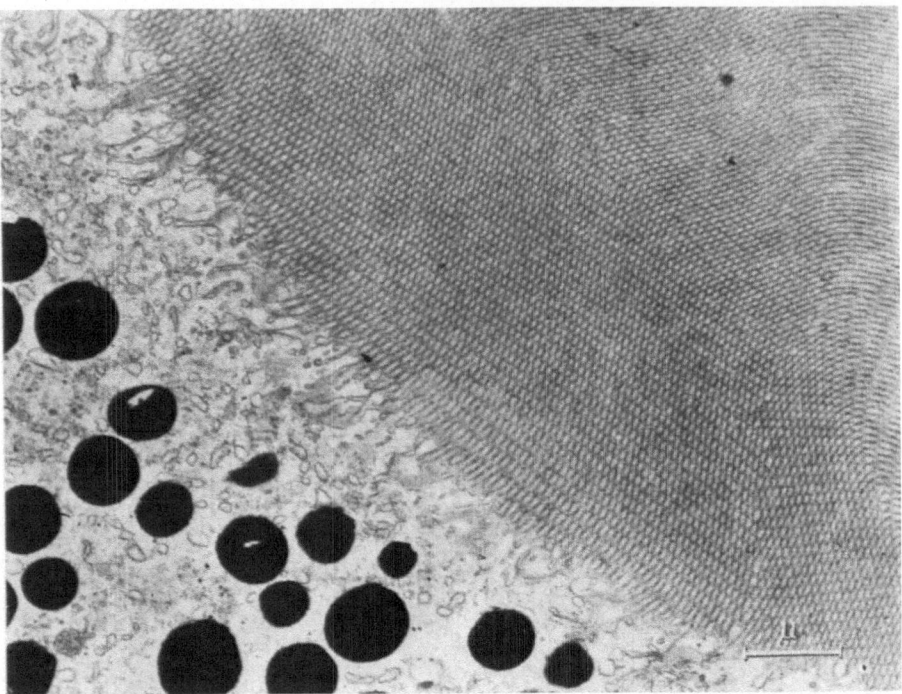

F<small>IG</small>. 3. Detail from the rhabdomere of a dark-adapted *Daphnia*. The number of cell-membrane invaginations and vesicles is fairly low, and the rhabdomere is well delineated toward the cytoplasm

Sections made from araldite-embedded specimens were stained with uranyl acetate and lead citrate and examined in a JEM 6c electron microscope.

Figure 3 shows the rhabdomere of a dark-adapted retinular cell. The regular microvillous system with a sharp delineation toward the cytoplasm is easily seen. A few plasma membrane invaginations are present at the base of the microvilli.

Twenty minutes of illumination induces marked changes in the basal zone of the rhabdomere (Fig. 4). Long membrane infoldings appear and the border between the rhabdomere and the cytoplasm becomes increasingly irregular. Numerous bristle-coated (complex) vesicles which can be derived by pinching off from the membrane of the microvilli become apparent in this region. The other type of vesicle is represented by peculiar flattened vesicles having a diameter of 1000–1500 Å (Fig. 5).

After 60 min of illumination, the highly ordered structure of the rhabdomere becomes increasingly looser and its dimensions are reduced. The reduction in size may be so intense that after 60 min of illumination about half of the rhabdomere may be lost. It seems that the irregular membrane folds and the great number of vesicles develop progressively at the expense of the ordered membranous system of the rhabdomere (Fig. 6).

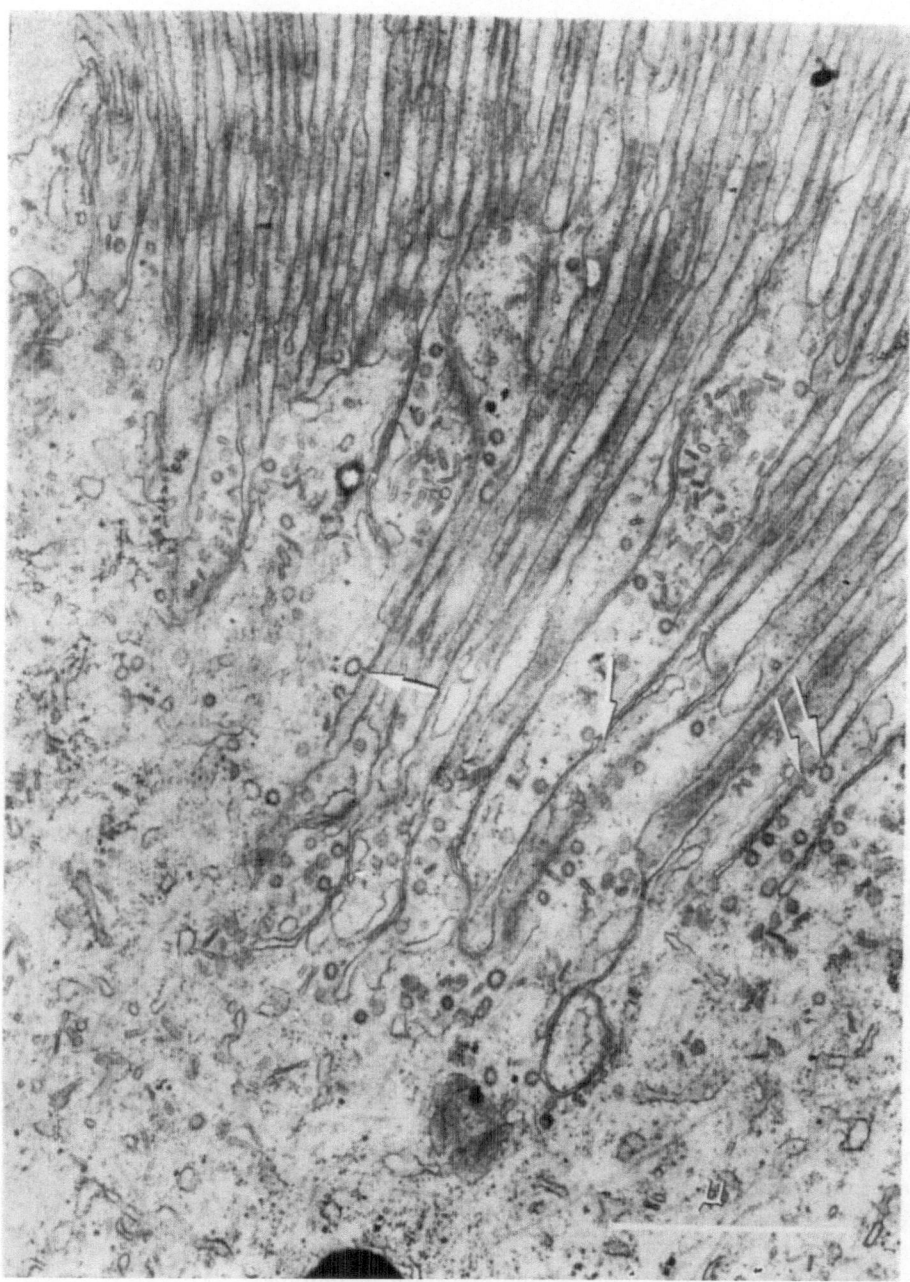

FIG. 4. Detail from a retinular cell of a *Daphnia* illuminated for 20 min. Bristle-coated (complex) vesicles are pinching off (arrows) in great number from long plasma membrane infoldings at the base of the rhabdomere

P. RÖHLICH

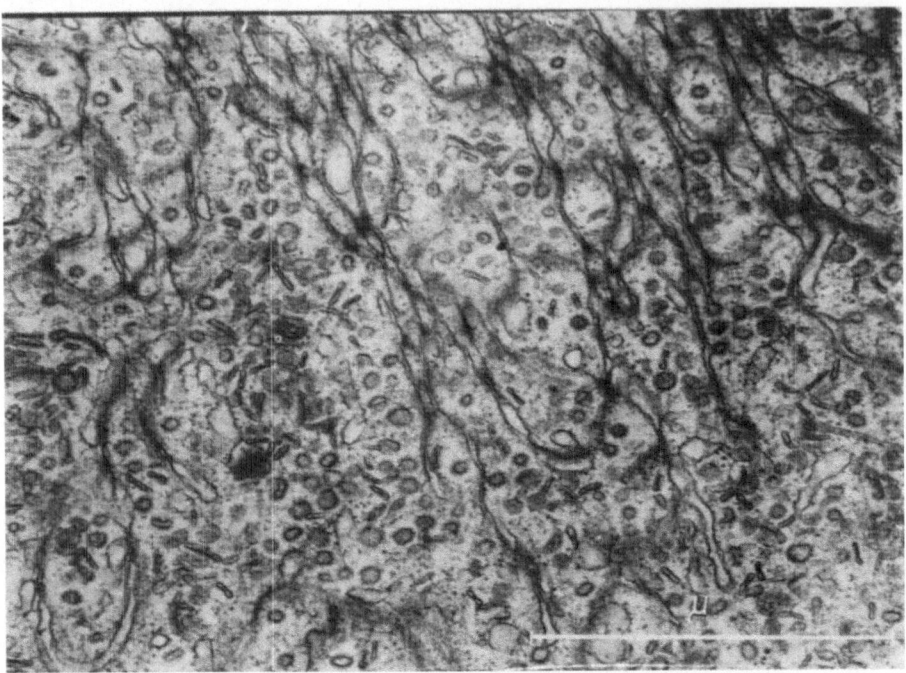

FIG. 5. Bristle-coated and flattened vesicles at the base of the rhabdomere from the same material as Fig. 4

The experiment indicates that light energy induces a gradual instability and disint⸗gration of the photoreceptor of the *Daphnia*. A similar effect produced by the same light intensity was observed by Kuwabara (1966) and by us (Röhlich 1967) on the outer segments of the rods in albino rats (Fig. 7).

The destabilizing effect of light on photoreceptor structures may be explained if the photopigment molecules are assumed to be an integral constituent of the photoreceptor membrane. There are several factors supporting this assumption (see Röhlich 1967). Biochemical data (Hubbard, Bownds and Yoshizawa 1965) indicate that the stability of the photopigment molecule is secured by its chromophore group, the retinal, vitamin-A aldehyde. On the absorption of a photon, the retinal molecule is isomerized into an all-trans form which, because of steric hindrance, hydrolyzes from the protein (opsin) molecule, thus inducing considerable conformational changes and instability of the latter. Therefore — if the photopigment is indeed built into the photoreceptor membrane as is supposed — instability of the photopigment molecule as a consequence of light absorption may lead to the instability of the photoreceptor membrane.

In the second part of this paper, we describe the effect of prolonged darkness on the fine structure of the planarian photoreceptor. The eye of the planarian consists of a pigment cup in which club-like endings of retinular cells are embedded

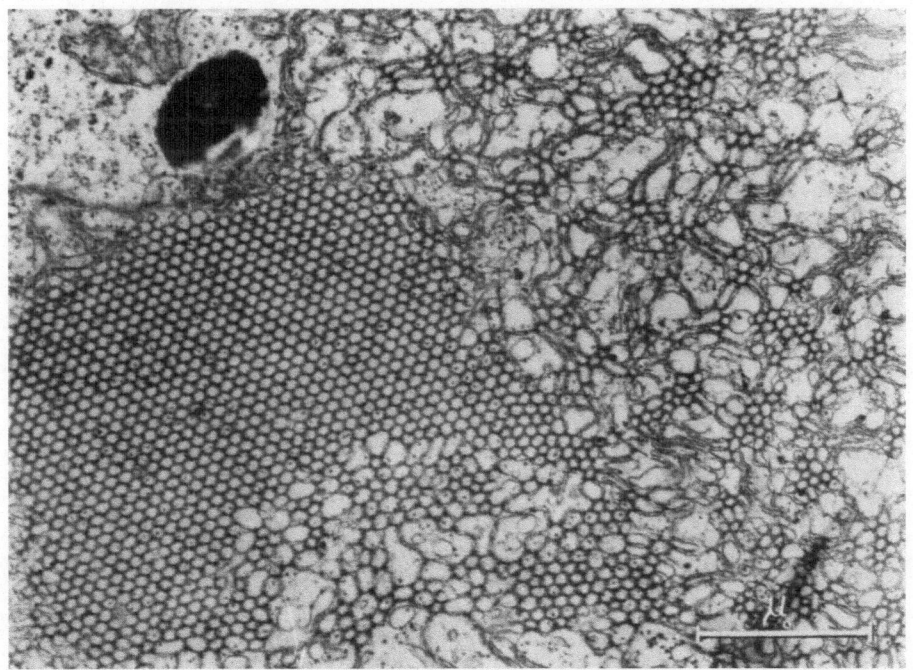

Fig. 6. Disintegrating rhabdomere from a *Daphnia* illuminated for 60 min

(retinal clubs). The retinal clubs are thickened endings of retinular cell processes and are equipped with a highly developed microvillous border (Röhlich and Török 1960, 1961). Since the microvillous border consitutes the major part of the retinal clubs, the area of the eye cup is almost completely occupied by long and slender microvilli (Fig. 8) and may be supposed to represent the specific photoreceptor structure of the planarian eye.

Three groups of planarians *(Dugesia tigrina)* were kept in complete darkness for 7, 14, and 21 days, respectively. One group of experimental animals was, after 28 days in darkness, divided into three groups and exposed to normal light conditions for 6 hours, 2 days, and 7 days, respectively. The controls were animals kept under the same conditions except for a normal laboratory illumination. All groups were fixed in Millonig's osmium tetroxide solution (Millonig 1961) and embedded in araldite. Thin sections were double-stained with uranyl acetate and lead citrate and examined in a JEM 6c electron microscope.

The control animals show the regular and well-developed microvillous system of the retinal club as has been described previously (Röhlich and Török 1961). The earliest changes can be observed in animals kept for 7 days in total darkness: the membranes of the microvilli are broken up at several places − preferentially in the basal half of the microvilli − into anastomosing tubules and vesicles (Fig. 9). After 14 days in darkness, the disintegration of the microvilli extends into large

P. RÖHLICH

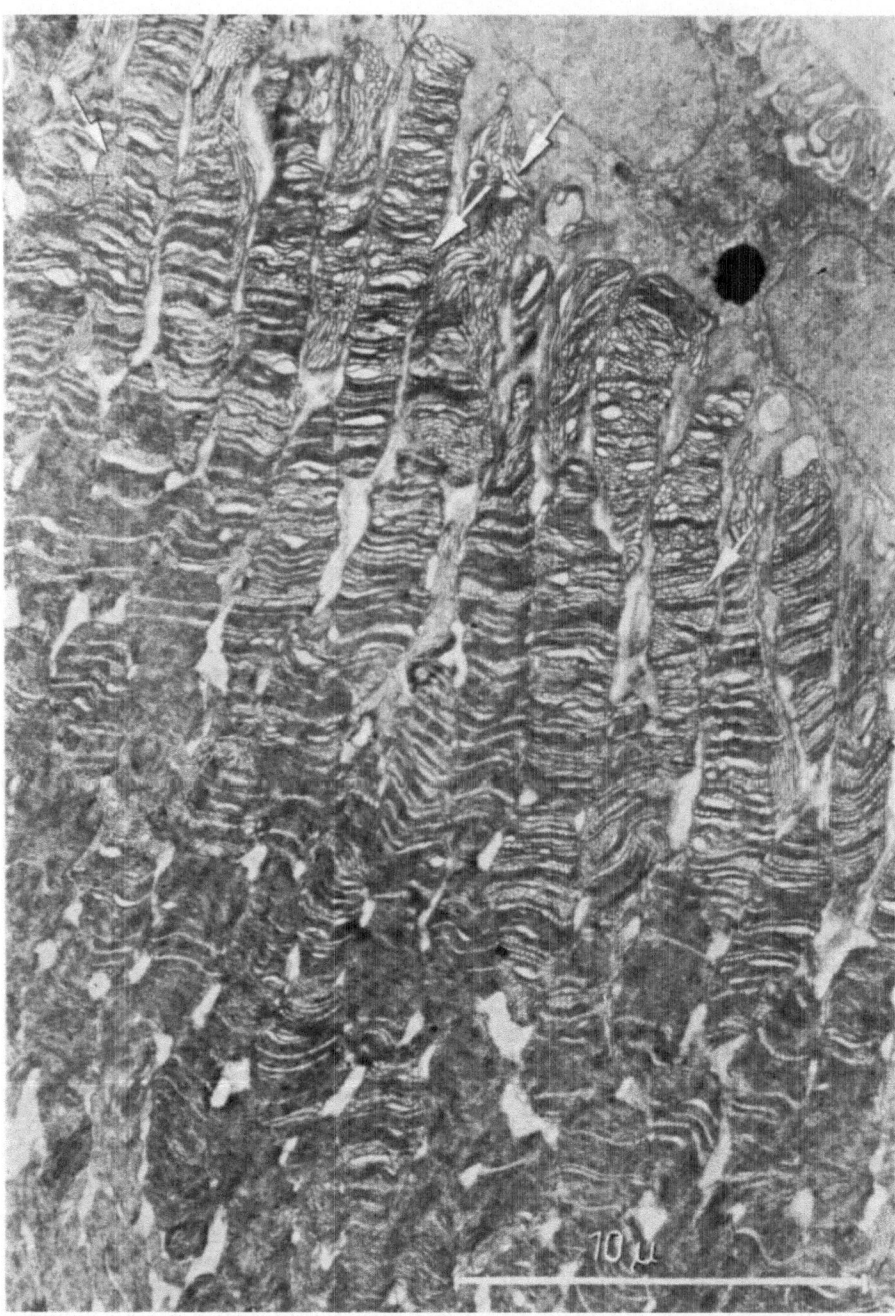

Fig. 7. Outer segments of rods from an albino rat illuminated for 72 h. Many discs of
the outer segments are transformed into tubular structures (arrows)

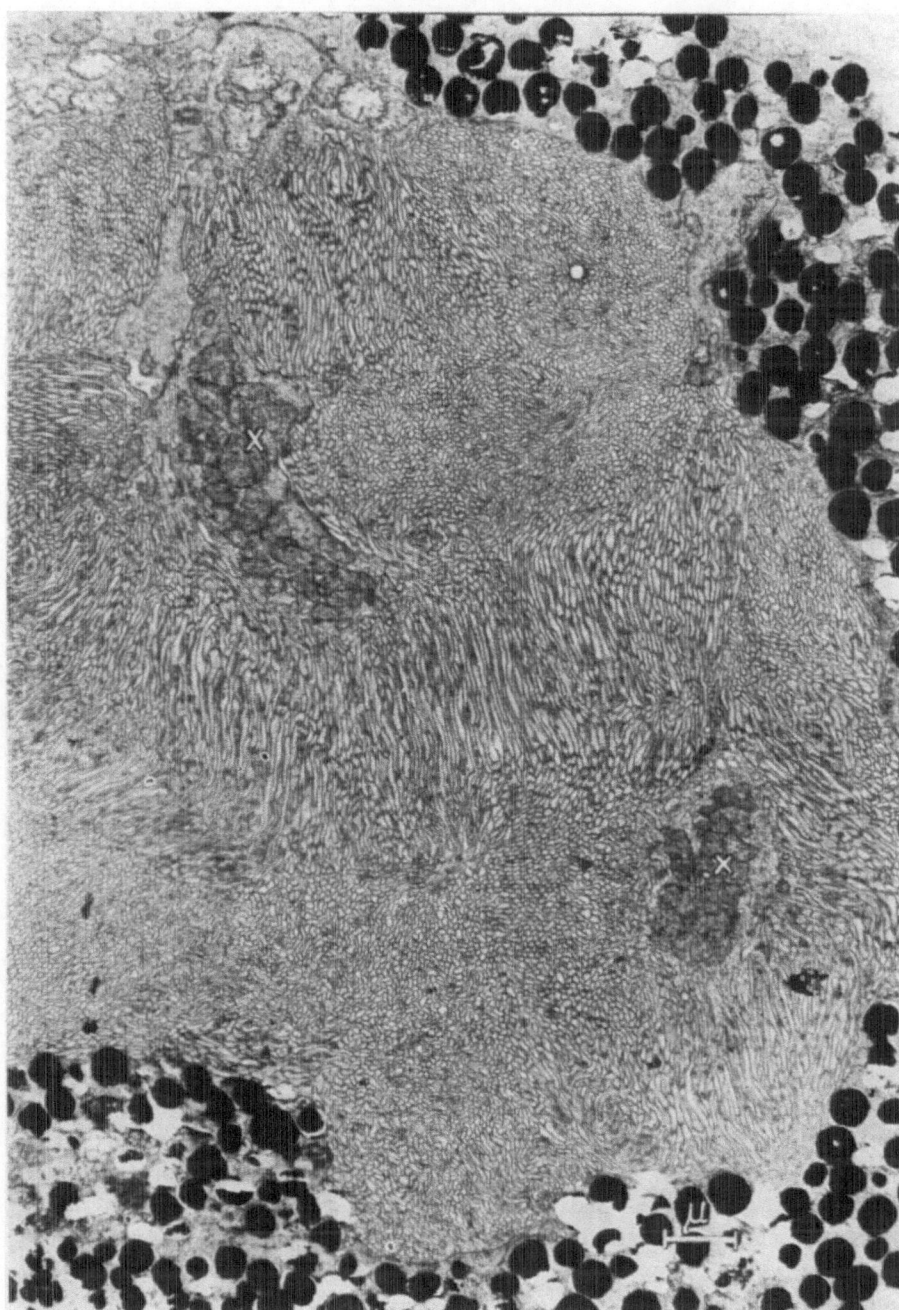

FIG. 8. Detail from the eye of the planarian *Dugesia tigrina*. The pigment cup is completely filled with retinal clubs. x—cytoplasmic core of the retinal club containing aggregated mito-chondria

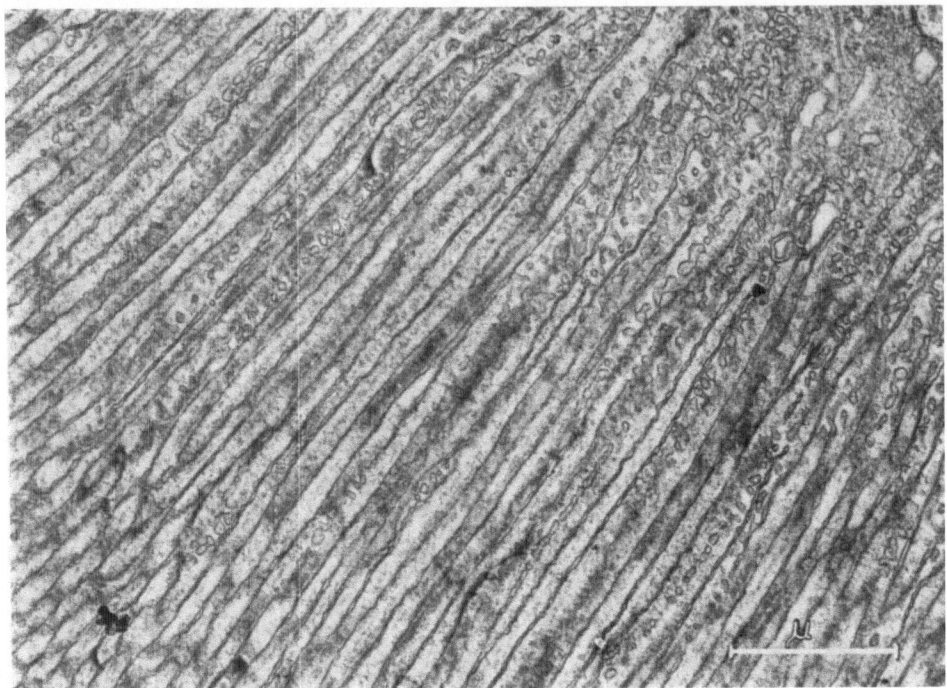

Fig. 9. Detail of a retinal club from a planarian kept in total darkness for 7 days. Adjacent membranes of the microvilli are broken up at several places into anastomosing tubules or rows of vesicles

areas (Fig. 10) and the microvillous border is decreased, resulting in empty spaces (Fig. 12) between the retinal clubs. In the 21-day experimental group (Fig. 11), the retinal clubs show serious alterations: strong decrease in size, disintegration of the microvillous system, dilatation of the endoplasmic reticulum, etc. Intact microvilli are practically absent, and their site is occupied by irregularly folding membranes and vesicles. As was shown earlier (Röhlich 1966), the transformation of photoreceptor membranes into vesicles after osmium fixation can be regarded as a sign indicating the lability of the membrane. When experimental animals were reexposed to normal light conditions, the degenerative process could be stopped and the retinal clubs regenerated rapidly (Fig. 13).

Degenerative changes similar to those described in this paper were also observed in crustaceans (Eguchi and Waterman 1965), where identifiable alterations were found after three months in darkness. It seems, therefore, that the total absence of light induces degeneration of the photoreceptor structure in two large groups of invertebrates. The question of whether this finding could be extended to additional invertebrate phyla awaits further experimental studies.

No explanation can be given at the present time for the mechanism by which the absence of light induces the degeneration of the photoreceptor. It is well known

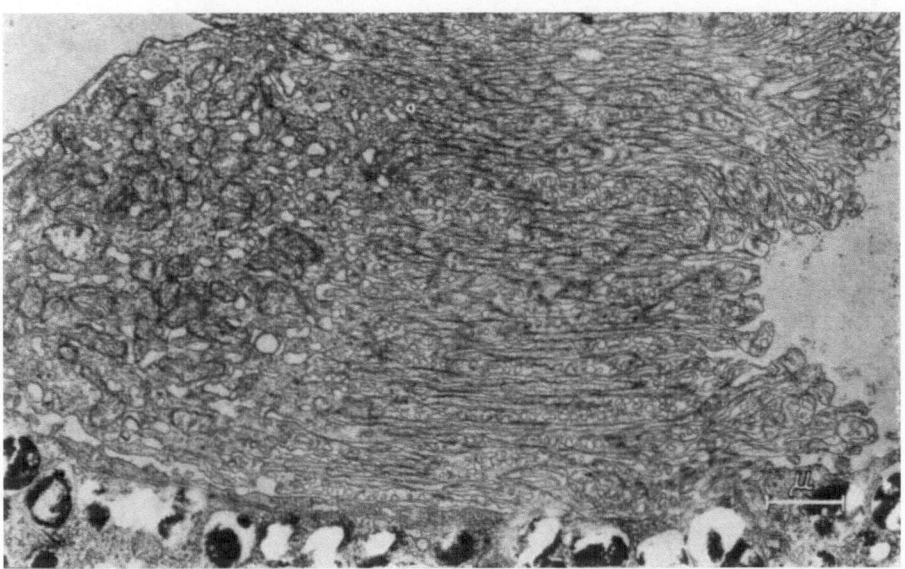

FIG. 10. Retinal club from a planarian kept 14 days in darkness. The disintegration of the membranes of the microvilli shows an advanced stage. Empty spaces appear between the individual retinal clubs

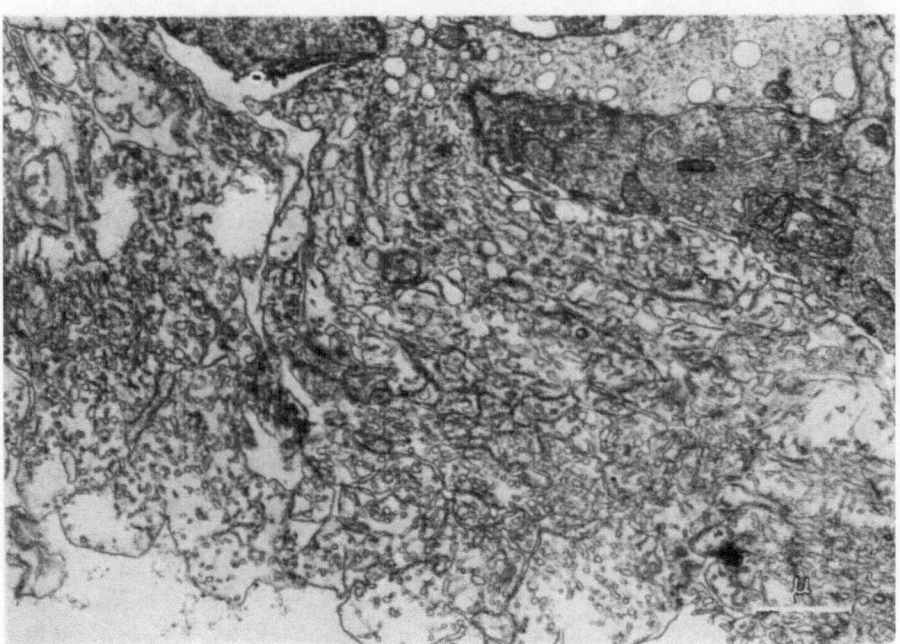

FIG. 11. End stage of degeneration of a retinal club (21 days in darkness). The microvilli are replaced by irregular membrane folds and vesicles

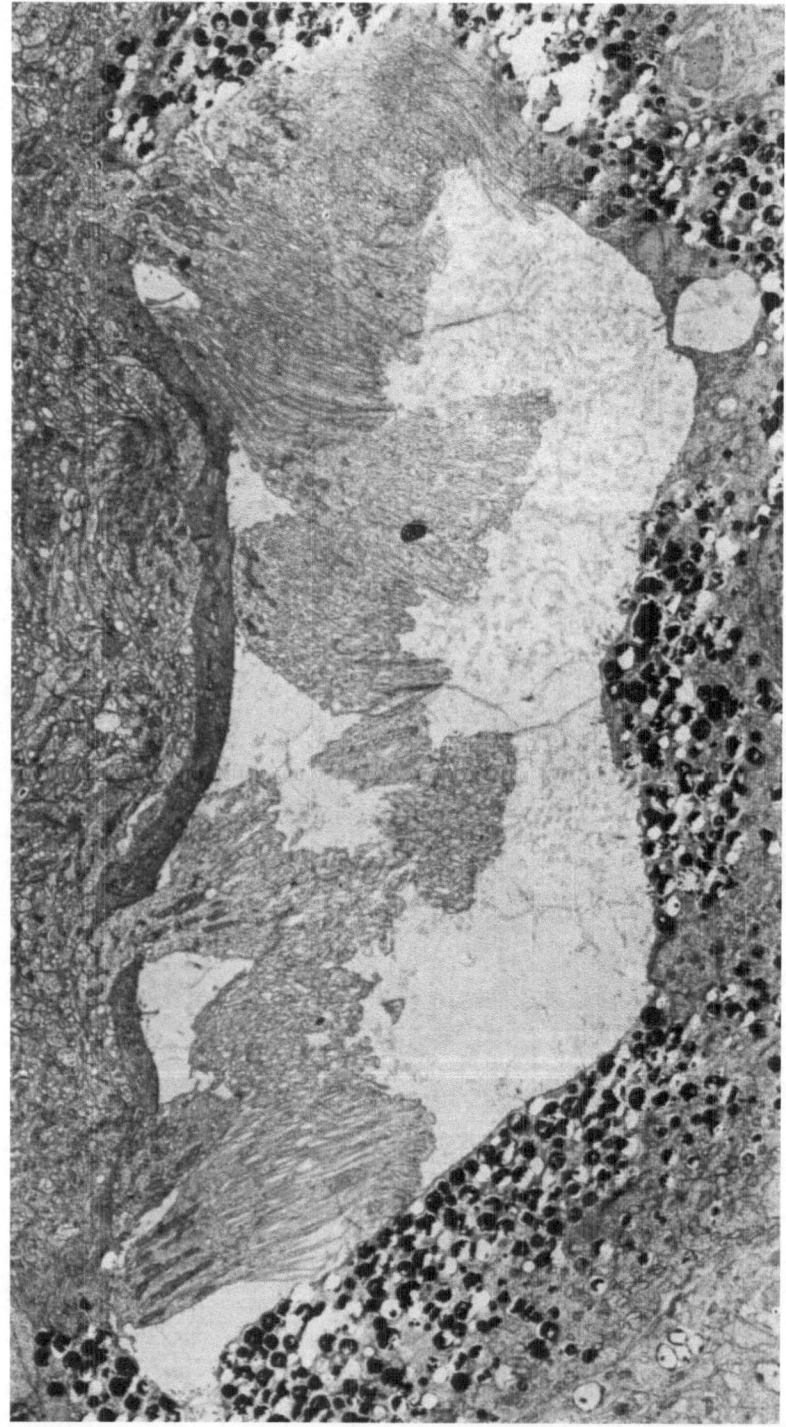

Fig. 12. The eye cup of a planarian after 21 days in darkness. Only a few degenerating retinal clubs are present in the interior of the shrunken eye cup. Large empty spaces indicating the total disappearance of the retinal clubs are found between the individual clubs

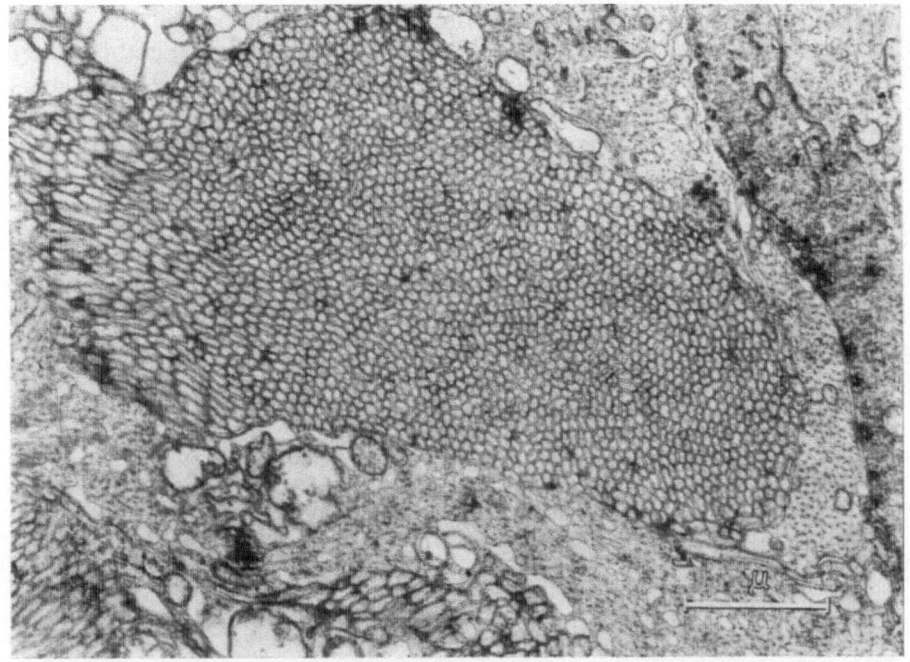

FIG. 13. Regenerating microvilli of a retinal club from a planarian kept in total darkness for 4 weeks and exposed to normal light conditions for 7 days

that total darkness results in degeneration of the plant photoreceptor, the chloroplast. But whereas the chloroplast membrane needs light energy for the synthesis of several of its molecular components, it is not known if light is also required for the building up of the invertebrate photoreceptor membrane.

SUMMARY

Experiments aimed at finding some fine-structural changes of photoreceptors as a consequence of changed light conditions were frequently carried out in the past, but few succeeded in demonstrating significant alterations. The present paper describes some distinct changes in the photoreceptor structures of two invertebrate species.

1. The effect of a 900 lx illumination was studied on the rhabdom of the *Daphnia* eye. The first changes were observed after 10–20 min of illumination: bristle-coated vesicles and peculiar disc-like vesicles appeared in great number at the base of the microvilli. The vesicles were regarded as degradation products of the microvilli membranes. After 60 min, the basal half of the rhabdom was found to have disintegrated. The findings are interpreted as the specific destabilizing effect of light on the photoreceptor membrane.

P. RÖHLICH

2. The effect of prolonged darkness was followed in planarian photoreceptors. The first sign of a degenerative process was observed in the increased osmium-sensitivity of the membranes of the microvilli. Severe degenerative changes developed after two and three weeks in darkness: the microvillous border of the retinal clubs shortened considerably and the microvilli gradually disintegrated. The process ended in complete loss of the photoreceptor ending. By exposing the experimental animals to normal light conditions, the degenerative process could be reversed in several days.

REFERENCES

EAKIN, R. M. (1965): Evolution of photoreceptors. *Cold Spring Harb. Symp* **30** 363—370.

EGUCHI, E. and WATERMAN, T. H. (1965): Fine structure patterns in crustacean rhabdoms. In: The functional organisation of the compound eye. Ed.: C. G. Bernhard, *Wenner Gren Center, Internat. Symp.* **7** 105—124.

EGUCHI, E. and WATERMAN, T. H. (1967): Changes in retinal fine structure induced in the crab *Libinia* by light and dark adaptation. *Z. Zellforsch.* **79** 209—299.

GOLDSMITH, T. H. (1964): The visual system of insects. In: *The Physiology of Insecta* Vol. **I**, Ed.: M. Rockstein, 397—462. Acad. Press, Inc. New York.

HUBBARD, R., BOWNDS, D., and YOSHIZAWA, T. (1965): The chemistry of visual photoreception, *Cold Spring Harb. Symp.* **30** 301—315.

KARNOVSKY, M. J. (1965): A formaldehyde-glutaraldehyde fixative of high osmolality for use in electron microscopy. Abstr. V. Ann. Meeting, Am. Soc. Cell Biol. *J. Cell Biol.* **27** 137A—138A.

KUWABARA, T. (1966): Membranous transformation of photoreceptic organ by light. *Electron Microscopy* Vol. **II** 501—502.

LANGER, H. and THORELL, B. (1966a): Microspectrophotometry of single rhabdomeres in the insect eye. *Exp. Cell Res.* **41** 673—677.

LANGER, H., and THORELL, B. (1966b): Microspectrophotometric assay of visual pigments in single rhabdomeres of the insect eye. In: "The functional organization of the compound eye", Ed.: Bernhard C. G., *Wenner Gren Center, Internat. Symp.* **7** 145—149.

MILLONIG, G. J. (1961): Advantages of a phosphate buffer for OsO_4 solutions in fixation. *J. appl. Physics* **32** 1637.

RÖHLICH, P. (1966): Sensitivity of regenerating and degenerating planarian photoreceptors to osmium fixation. *Z. Zellforsch.* **73** 165—173.

RÖHLICH, P. (1967): Photoreceptorok finom szerkezete normális és kísérleti feltételek között. (Fine structure of photoreceptors in normal and experimental conditions). *Dissertation*, Budapest.

RÖHLICH, P. and TÖRŐ, I. (1965): Fine structure of the compound eye of *Daphnia* in normal, dark- and strongly light-adapted state. *Eye Structure, II. Symp.* Ed.: J. W. Rohen, Schattauer Verlag, Stuttgart, 175—186.

RÖHLICH, P. and TÖRÖK, L. J. (1960): Photoreceptor structures in the planarian eye and their morphogenesis during regeneration. *Proc. Europ. Reg. Conf. Electron Microscopy.* Delft, 822—826.

RÖHLICH, P. and TÖRÖK, L. J. (1961): Elektronenmikroskopische Untersuchungen des Auges von Planarien. *Z. Zellforsch.* **54** 362—381.

RÖHLICH, P. and TÖRÖK, L. J. (1962): The effect of light and darkness on the fine structure of the retinal club in *Dendrocoelum lacteum. Quart. J. micr. Sci.* **104** 543—548.

DISCUSSION

G. A. Kerkut: How quickly do the membranes reform after illumination?

P. Röhlich: Regeneration of the photoreceptor membranes can be observed as early as after 24 h in normal light conditions. It seems that 7 days are enough to produce the total regeneration of the retinal clubs.

J. Salánki: It seems to me that the light intensity you used is high. Could not the reaction be considered a pathological one?

P. Röhlich: To give an idea about the light intensity used in this experiment: it was not more than the illumination produced by diffuse daylight 100 cm inside a window looking to the north. The morphological alterations were surely somewhat exaggerated in our case; similar light conditions may, however, often occur during the life of the animal.

Symposium on Neurobiology of Invertebrates 1967 (111—122)

SYNAPTIC ORGANIZATION OF THE LOBSTER OPTIC LAMINA

J. HÁMORI* and G. A. HORRIDGE†

* Department of Anatomy, University Medical School, Budapest, Hungary
† Gatty Marine Laboratory, University of St. Andrews, Scotland

INTRODUCTION

The most superficial neuropile region of the optic lobe, the optic lamina, or lamina ganglionaris, of arthropods is placed between the retina and the second neuropile layer, the medulla. The retina in the crustacean compound eye is built up of a large number of ommatidia, which are the characteristic light-perceptive cells, each consisting of 6–8 retinula cells in most species. The retinula cell membrane is elaborated into a complex light-sensitive system of tubules called the rhabdomere by which the energy of the photon is transformed to nervous excitation, not necessarily impulses. The optic lamina is the place where the retinula cell axons, after passing through the basement membrane in small bundles, terminate upon the axons of ganglion cells of the lamina. After a decussation in the optic chiasma, the second-order fibers reach the medulla, where they terminate on the third neuron in the chain (Fig. 1). In other words, the optic lamina is the first of a series of relay stations on the optic pathway. Since the functional properties of the retinula cells can be recorded directly with microelectrodes in favorable species, an elucidation of the morphology of the first synapse will assist in the interpretation of the function of the lamina.

STRUCTURE OF THE LAMINA

The lamina of the lobster *Homarus* consists of five distinct layers (Fig. 2). The first, most superficial layer, is of ganglion cell bodies. The second is a glial layer, or sheet, which is pierced at regular intervals to allow retinula fibers and ganglion cell axons to enter the columnar region. Below the glial sheet lie more ganglion cell bodies, forming the third layer. Functionally, the most important layer is the fourth and thickest layer, the columnar region, which is separated from the optic chiasma by a second glial sheet, which is the fifth layer of the lamina. This glial layer has holes at regular intervals through which optic fibers leave the columnar region for the medulla. In the columnar (fourth) region, 100—140 μm thick, all

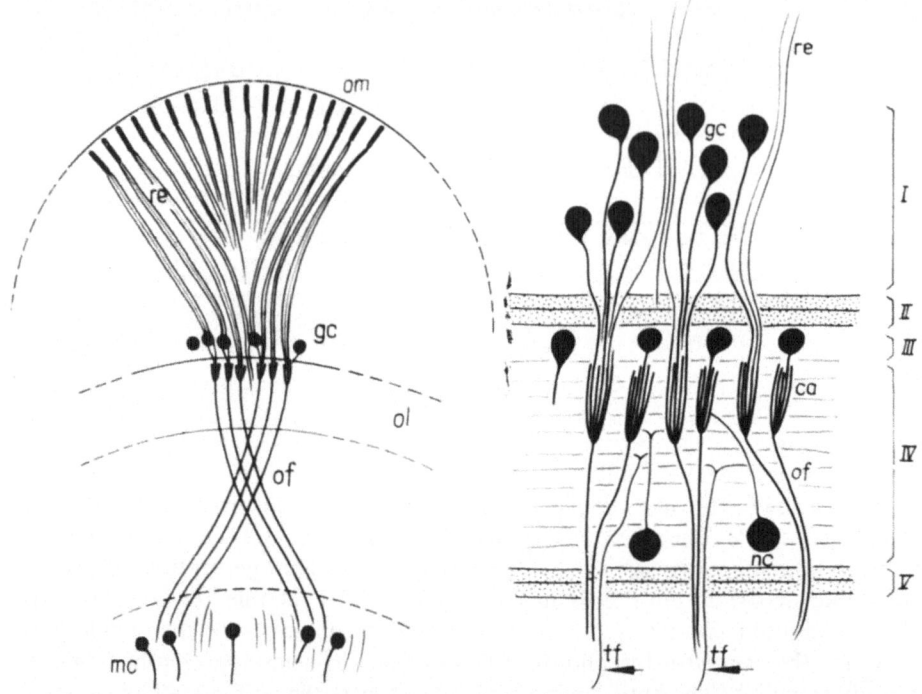

FIG. 1. Schematic view of the retina and the peripheral part of the optic lobe in longitudinal section. Retinula fibers (re) originating from retinula cells in the ommatidia (om) enter the optic lamina (ol), where they terminate on the axons of ganglion cells (gc). The latter leave the lamina as optic fibers (of) and, after a decussation in the optic chiasma, terminate below the medulla nerve cells (mc)

FIG. 2. Schematic view showing the five-layered optic lamina. I — ganglion cell layer, II — distal glial sheet, III second ganglion cell layer, IV — columnar region, V — proximal glial sheet, re — retinula fibers, gc — ganglion cells, ca — cartridges, nc — neurosecretory cells, tf — transverse fibers from central neuropile regions, of — optic fibers

synaptic connections of at least four different fiber types are found. The most striking elements of this region are the optic cartridges which occupy the upper half. In tangential sections, the cartridges are arranged in a strictly hexagonal array — similar to the ommatidial pattern of the retina — thus giving a periodic, geometrical appearance to the whole lamina.

Below the columnar region, sections show many light, round profiles of horizontal fibers and optic fibers in bundles. There are also a few cell bodies which prove to be neurosecretory by the Gomori criterion. Some of the horizontal fibers originate from these neurosecretory cells. Previous work has suggested that the majority of the transverse fibers run between the lamina and more proximal neuropile regions, even as far as the brain.

THE OPTIC CARTRIDGES

The cartridges, the main synaptic units of the lamina (Fig. 3), consist of one or two ganglion cell axons surrounded by seven retinula cell axons, which correspond exactly to the number of retinula cells of the ommatidia. Together with their unbranched form, this shows that each retinula cell axon runs to only one cartridge. There is no clearly organized chiasma between retina and lamina. A few crossing fibers and evidence from degeneration experiments suggest that while the seven axons from a single ommatidia do not necessarily run to one cartridge, they certainly do not run to seven different cartridges. The true situation is as yet unknown.

The ultrastructure of a single cartidge is more complex here than it is in insects (Trujillo-Cenóz 1965); neither synapses nor glia are similar. The two centrally located ganglion-cell axons send many spines into the bag endings of the presynaptic retinula cells. The latter contain synaptic ribbons, randomly dispersed synaptic vesicles, and large, rather empty mitochondria. The synaptic ribbons are always situated in incisions between the primary and secondary spines of the ganglion cells. The whole cartridge, which also includes two neurosecretory transverse fibers, is ensheathed by glial processes. In reality, the picture is much more complicated:

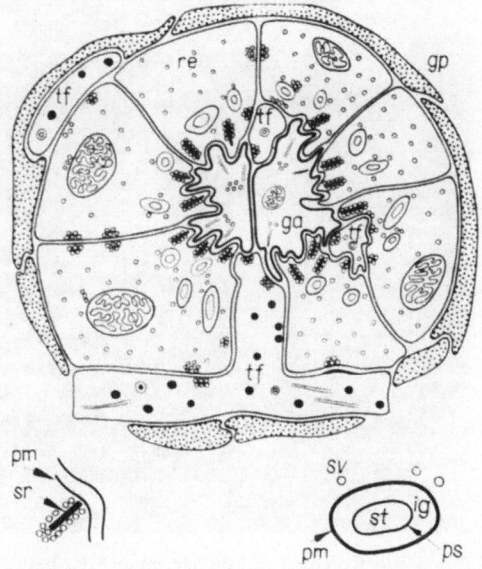

FIG. 3. Fine structure of a cartridge in cross section. Spines of the two central ganglion cell axons (ga) penetrate into the seven surrounding large retinula axon terminals (re). Synaptic ribbons (sr) lie in incisions between spines. An enlarged ribbon surrounded by synaptic vesicles is shown at the lower left. A cross section of an invaginated spine (st) is shown enlarged at the lower right. Note that presynaptic membrane (pm) is thicker than the postsynaptic (ps); ig — intracellular gap. Two larger and two smaller transverse fibers (tf) containing neurosecretory granules are also seen contacting retinula endings and ganglion cell axons. The whole cartridge is ensheathed by glial processes (gp). (Hámori and Horridge 1966b, slightly modified)

the primary, secondary, and, especially, the numerous tertiary spines of the ganglion cell axon (all colored yellow in Fig. 4) make numerous deep invaginations into the retinula axon bags, which results in an enormous enlargement of the contact area between the retinula axon and the axon of the lamina.

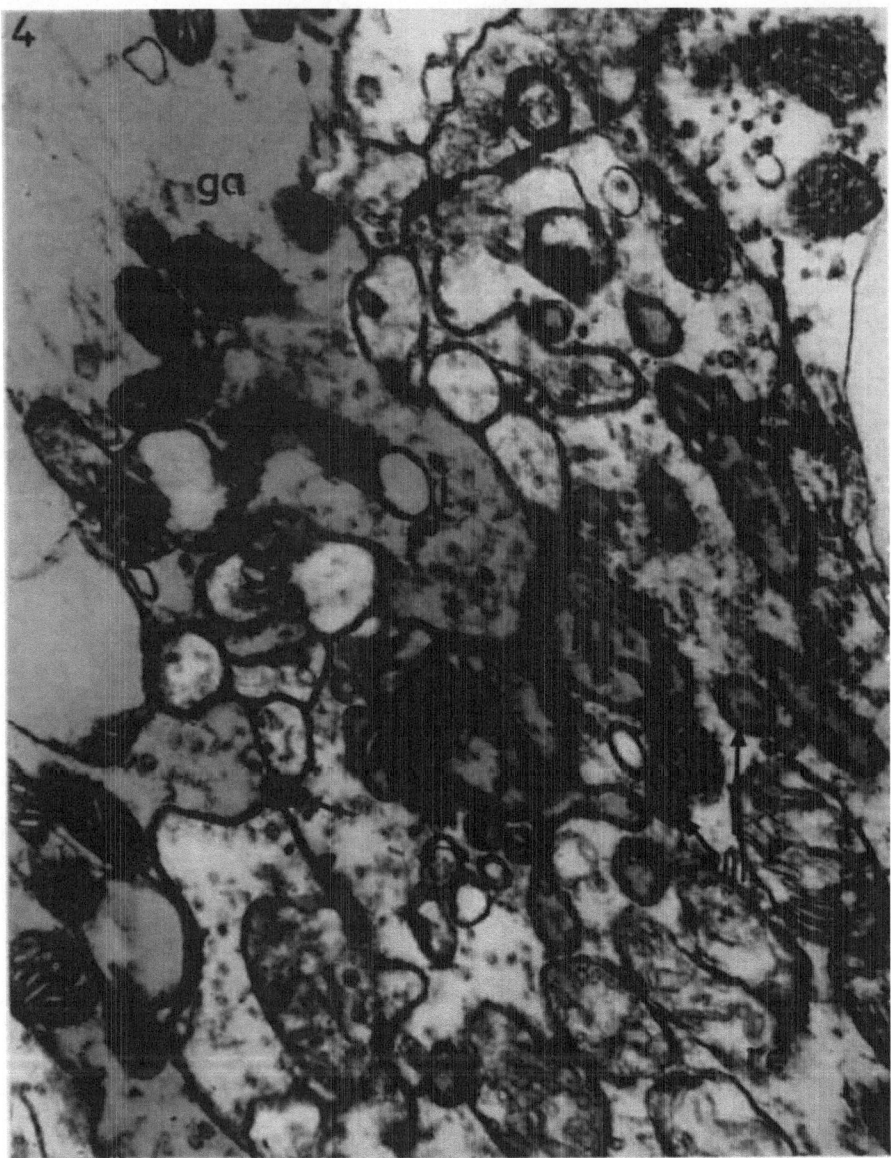

FIG. 4. Electron micrograph of contact area between retinula endings (re) and two ganglion cell axons (ga) in the cartridge region. For the sake of clarity, processes of the ganglion cell axons are colored yellow. I., II., III. — primary, secondary, and tertiary spines of ganglion axons, respectively. Retinula bags contain empty synaptic vesicles and large, pale mitochondria; compare with smaller, dark itochondria of the ganglion cell axons. Arrows show presynaptic ribbons

Two unusual features of the synaptic contacts, especially of the tertiary spines, can be observed in Fig. 4: (1) Although the synaptic membranes are not fused as in electrotonic synapses, the synaptic cleft of 7–13 μm is very narrow. (2) It is the presynaptic membrane which is thickened and not the postsynaptic membrane. As in the vertebrates, the retinula terminal bag of the lobster is crowded with synaptic vesicles of normal size, but, except around the synaptic ribbons, the vesicles are not aggregated as in synapses. This, and the narrowness of the cleft, suggest that the transmission may be electrical, in contrast to the case for locusts and flies, where the latency suggests chemical transmission at this synapse. It is curious that in vertebrates, octopus, and the lobster there is a comparable structure of the synapses of the primary visual fibers, with postsynaptic spines penetrating bag-like presynaptic terminals.

A comparison can be made with the chick ciliary ganglion, where Martin and Pilar (1963a, 1963b) found a dual electric and chemical transmission mechanism. There, the synapse develops into a labyrinthine structure similar to the spines within a bag, described here, but having presynaptic vesicles, membrane thickening, and close membrane appositions at various places. The whole synaptic complex is isolated and surrounded by a loose myelin sheath which is not effective except in the adult hen. Any final statement, however, can only be made after a careful electrophysiological investigation of the cartridges, which, in view of their complexity, will be a formidable task.

THE HORIZONTAL FIBERS

There are two types of horizontal fibers (Figs. 5 and 6).

(a) Fibers containing different kinds of granules, which resemble neurosecretory fibers. These are found inside the cartridge, where they are postsynaptic to retinula axon terminals and presynaptic to ganglion cell axons. The synaptic vesicles are normal and distinct from the secretory granules.

(b) Nonsecretory transverse fibers with normal synaptic vesicles at places of synaptic contact. Though a few may reach the cartridge region, the majority of these fibers are found between the cartridges and the proximal glial layer.

According to light-microscopic observations (Hámori and Horridge 1966a) the neurosecretory fibers are of local origin, with cell body located inside the columnar region (Fig. 2), whereas the empty fibers are the ones which come from distant regions. In addition to their synaptic contacts with retinula bags and to the ganglion cell axons in the cartridge, the neurosecretory fibers make synaptic contacts with non-neurosecretory horizontal fibers, mostly below and outside the cartridge region.

The main postsynaptic site for non-neurosecretory horizontal fibers is upon ganglion cell axons proximal to the cartridges, where the second-order axons have far fewer branching spines than in the cartridges (Fig. 5). The retinula axons make synapses upon transverse fibers of the secretory type. If retinula terminals excite the

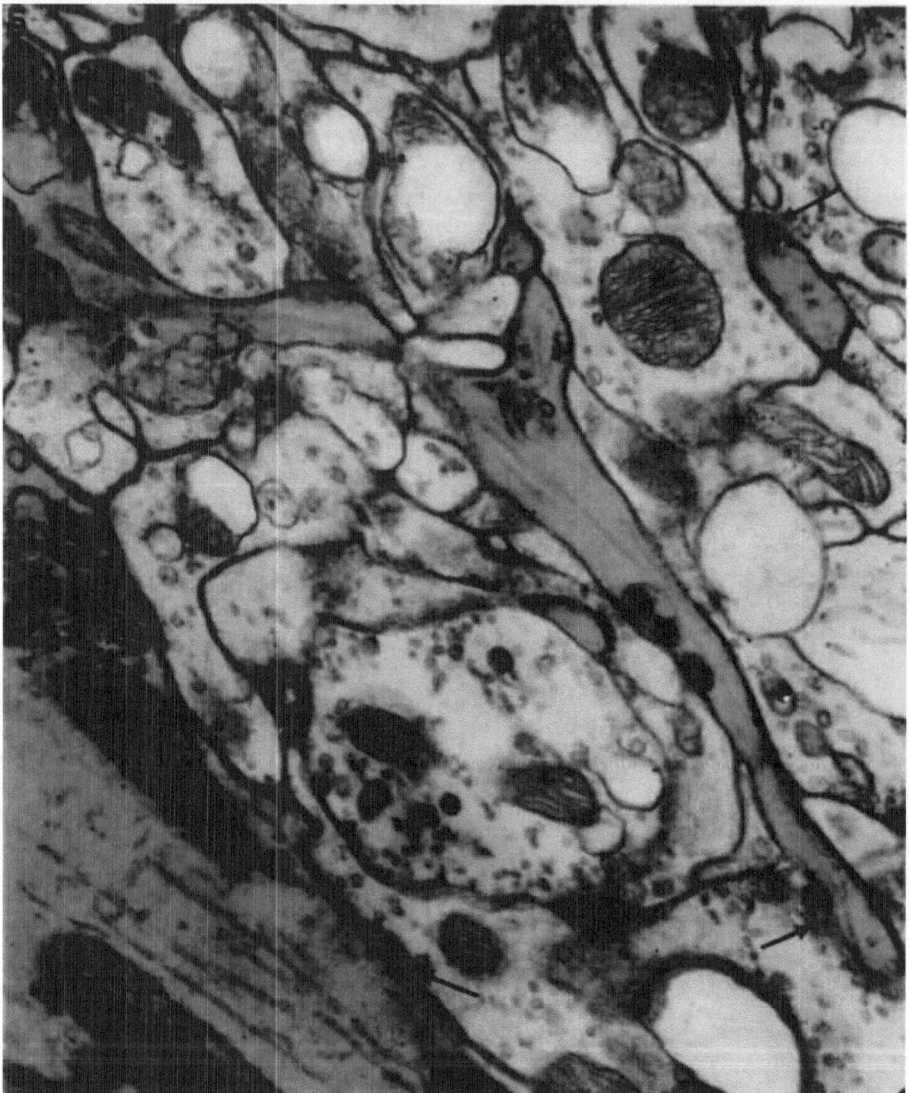

Fig. 5. Post-cartridge portion of ganglion cell axon (optic fiber), colored yellow, having only a few long spines. All other axonic profiles are transverse (uncolored) fibers, which at places marked by arrows make synaptic contacts with the smooth ganglion cell axon or its spines. Synaptic specializations between transverse fibers can be seen with accumulations of synaptic vesicles and thickening of synaptic membranes (at the arrows)

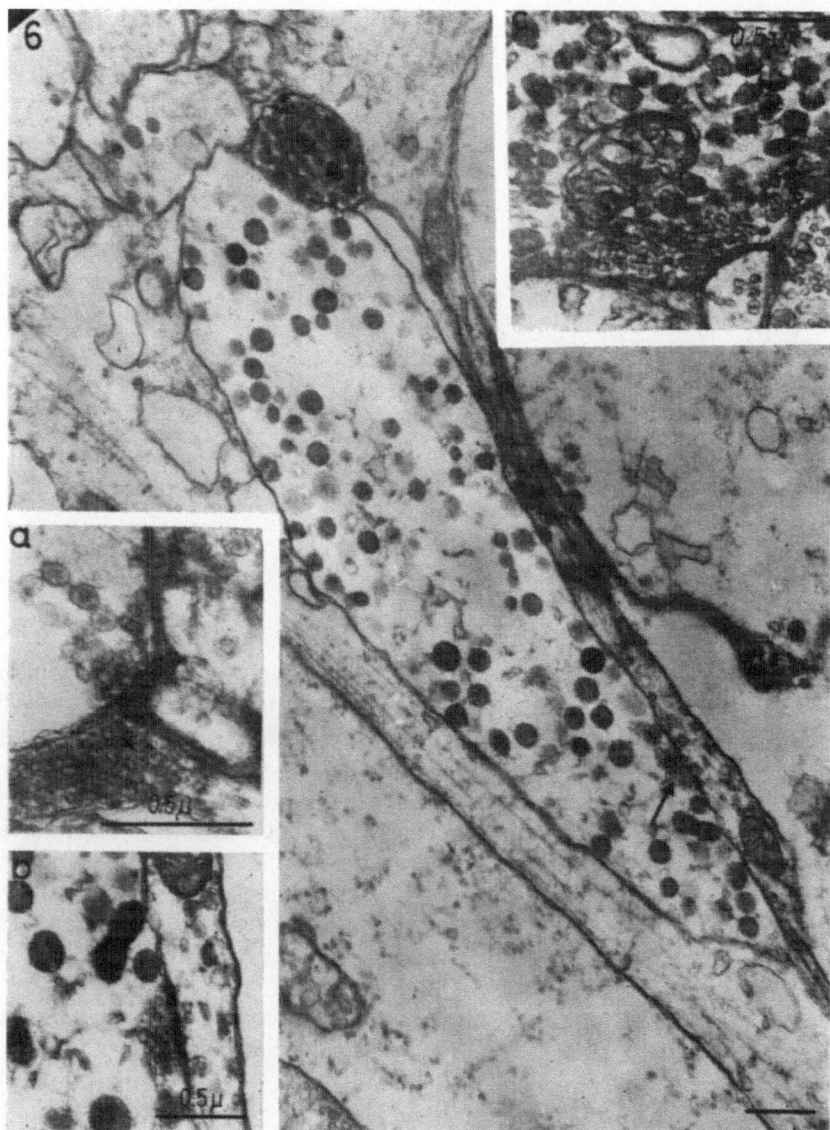

FIG. 6. Synaptic specialization of neurosecretory transverse fibers. The main figure shows two neurosecretory fibers which are in synaptic contact with each other (arrow). The synaptic area is shown somewhat enlarged in the inset (b), demonstrating thickened synaptic membranes and accumulations of presynaptic (empty) vesicles opposite the presynaptic membrane. Here, the neurosecretory granules are homogenous and are similar to the so-called elementary neurosecretory granules in vertebrates. Inset (a) shows a different type of neurosecretory fiber with smaller dense-core vesicles, whereas inset (c) demonstrates a transverse fiber with granules which can be considered as a form intermediate between those shown in inset (a) and inset (b). Note that in both endings [inset (a) and (c)], the empty, normal synaptic vesicles are concentrated opposite to the presynaptic membrane and that the postsynaptic membrane is somewhat thicker than the presynaptic membrane

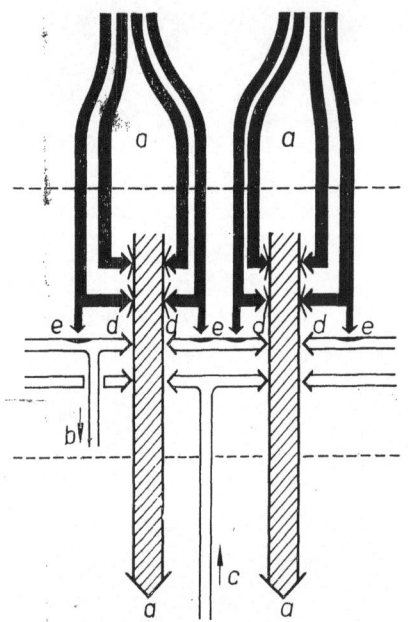

FIG. 7. Diagram of the probable information flow in the lobster optic lamina. Black arrows show endings of primary retinula fibers. Hatched arrows represent axons of the lamina ganglion cells, and open fibers demonstrate two kinds of transverse fibers which connect with the fibers of different cartridges. This scheme shows, a — preservation of information concerning angle of stimulus, b — a possible pathway sensitive to polarization plane, c — probable central control of synaptic transmission, d — an automatic gain control if synapse d is inhibitory, and e is excitatory

local horizontal fibers which in turn inhibit the ganglion cell axons, the combination would act as an automatic gain control of the main visual pathway (Fig. 7, e and d).

All transverse fiber synapses, irrespective of their neurosecretory or non-neurosecretory nature, have the appearance of typical chemical synapses. The pre- and especially the postsynaptic membranes (Fig. 6) are thickened, the synaptic cleft is moderately wide, and there is an accumulation of specific synaptic vesicles 30–50μm in diameter at the presynaptic membrane thickenings. There seems to be no way of distinguishing whether the secretory activity modifies the action of the lamina or whether it in some way measures the intensity of incident light and acts on deeper mechanisms of the hormonal system.

As seen in light-microscopic preparations, the horizontal fibers bridge very long distances and make contact with 40 or more neighboring cartridges without giving off side branches (Hanström 1924). If they are afferent, they collect from many similar sources (wide field units), and if efferent, they act on large areas of the lamina. It is tempting to compare this organization with similar examples in the central nervous system of the vertebrates. In the cerebellar cortex, for example, one 3-mm-long parallel fiber establishes synaptic contacts, without giving off any side branches, with about 300 Purkinje cells (divergence), whereas 250,000 parallel fibers synapse with each Purkinje cell (convergence) (Hámori and Szentágothai 1964). A similar synaptic system can be suggested for the transverse fiber-to-lamina ganglion cell contacts in the lobster, though the individual number of nerve fibers involved is much less.

COMPARISON WITH INSECTS

The nature of the fiber projection from the retina to the lamina is not yet known for a crustacean, but two types of projection are already known in insects. Both are electrotonic, without spikes in the primary axons.

In Diptera, the rhabdomeres of a single ommatidium are separated and the axes of the retinula cells point in different directions from one ommatidium. However, in neighboring ommatidia, there are different cells with axes which point in the same direction. These cells, six in number, have axons which converge anatomically at their first synapse in the lamina. This group of terminals forms, with two second-order neurons, an optic cartridge (Fig. 8A). There are as many optic cartridges as there are ommatidia, and each optic cartridge has the field of view of six retinula cells with superimposed axes (Braitenberg 1967). Unpublished recordings by Scholes show that there is linear summation of the excitation of the six terminals upon the second-order neurons, but the exact form of the synaptic interaction is unknown.

In the locust, as in most orders of insects, the rhabdomeres of one ommatidium are fused, and the 6–7 retinula cells of one ommatidium all have the same receptive field. In contrast to the dipterons, there is no chiasma of fibers between retina and lamina, and histological preparations suggest that all the retinula axons from an ommatidium go to one optic cartridge below. Basal retinula cells (eccentric cells) exist but are not investigated. Intracellular recordings from units of the lamina of the locust reveal that electrotonically conducted depolarization of the terminals of the retinula axons causes a *hyperpolarization* of units that are assumed to be postsynaptic neurons of the lamina. The postsynaptic response has a latency of > 2 msec, suggesting chemical transmission, and the following two important features: (1) the field of view of the second-order unit is as narrow as that of the retinula cells of one ommatidium, and (2) the low sensitivity to polarized light as compared to primary cell responses shows that some kind of summation of retinula cell responses occurs. At low light intensities, the postsynaptic response is a series of miniature potentials with an absolute sensitivity to photons as good as, or better than, that of a single retinula cell. This, and the multimodal distribution of the sizes of miniature potentials, again suggests summation of primary responses upon lamina neurons. At high light intensities, there is, however, depolarization of the same lamina neurons (Shaw 1967).

These results show that at the first synapse in the locust lamina, the angle of the incident illumination is preserved at the expense of combining the other attributes of the visual stimulus. Information about the plane of polarization is reduced, and differences in spectral sensitivity of retinula cells which look along a given axis are ignored. There is independent evidence that color vision is achieved by major regional differences in the compound eye, at least in bugs and dragonflies, and by particular retinula cells which pass right through the lamina in flies (*x* and *y* in Fig. 8A).

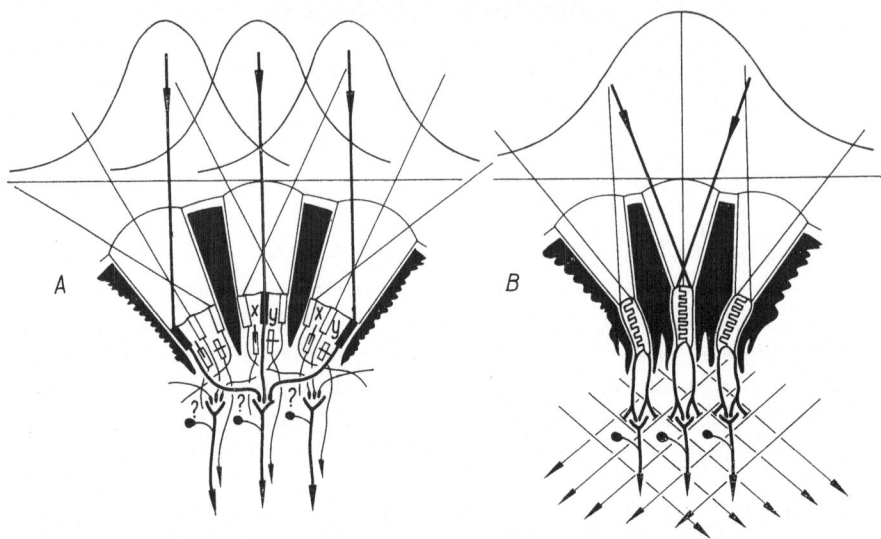

FIG. 8. Connections between primary retinula fibers and second-order neurons of the lamina. A — In Diptera, green-sensitive retinula cells from six different ommatidia converge upon one second-order neuron, thus bringing together excitation from one direction. Two blue-sensitive retinula cells x and y, with tubules at right angles to each other, have axons to different destinations from the other six. Angle, polarization plane, and color are thus coded in parallel channels. B — In arthropods with fused rhabdomeres, such as crustacea, the EM, physiological, and behavioral results are in accord with the convergence of different axons from one ommatidium, together with abstraction of color or polarization plane by separate horizontal fibers at lamina level

In the decapod Crustacea, we find the additional complication of two types of long horizontal fibers and a pattern of synapses which does not entirely agree with the picture found in insects. It is certain, however, that decapod Crustacea conserve very well the information about the angular position and movement of a visual stimulus in the horizontal plane (Horridge 1966). Since they have fused rhabdomeres, therefore, it can be predicted that in decapod Crustacea there is no chiasma between retina and lamina. This agrees with the finding that movement-perception fibers and the optokinetic response itself are hardly sensitive to the plane of polarization of the stimulating light (Horridge 1967), although individual retinula cell responses are extremely sensitive to the polarization plane (Shaw 1966). Three of the seven retinula cells have horizontally aligned tubules and the other four cells have vertical tubules. If all seven converge upon one optic cartridge, angle is conserved at the expense of polarization sensitivity. However, many Crustacea are sensitive to the net polarization plane over the whole eye. This suggests that some of the horizontal fibers of the lamina collect excitation from retinula cells sensitive to one plane, while other horizontal fibers collect from those sensitive to the plane at right angles. The proposal illustrated in Fig. 8B is a basis for further analyses of the connections.

SUMMARY

(1) In the main synaptic units of the lobster optic lamina — the optic cartridges — the terminals of the primary visual fibers are bag-like endings into which numerous spines of the ganglion cell axons penetrate to form a structure which is suggestive of electrical transmission.

(2) In contrast, the synapses of the transverse fibers which run between the cartridges appear to be typical chemical synapses.

(3) Both kinds of transverse fibers, irrespective of whether their origin is local (neurosecretory fibers) or extralaminar (empty fibers) are presynaptic to the ganglion cell axons (optic fibers).

(4) The fused rhabdom, the high optokinetic sensitivity, and the lack of sensitivity to polarized light in movement perception, taken together suggest that retinula cells from one ommatidium converge in one optic cartridge. If this is so, then polarization sensitivity may depend on horizontal fibers of the lamina which receive input from selected retinula cells.

REFERENCES

BRAITENBERG, V. (1967): Patterns of projection in the visual system of the fly. I. Retina-lamina projections. *Experimental Brain Research* **3** 271—298.

HÁMORI, J. and HORRIDGE, G. A. (1966a): The lobster optic lamina. I. General organization. *J. Cell Sci.* **I** 249—256.

HÁMORI, J. and HORRIDGE, G. A. (1966b): The lobster optic lamina. II. Types of synapse. *J. Cell Sci.* **I** 257—270.

HÁMORI, J. and SZENTÁGOTHAI, J. (1964): The "crossing-over" synapse: an electron microscope study of the molecular layer in the cerebellar cortex. *Acta biol. Acad. Sci. hung.* **15** 95—117.

HANSTRÖM, B. (1924): Untersuchungen über das Gehirn, insbesondere die Sehganglien der Crustaceen. *Arkiv. Zool.* **16** 101—119.

HORRIDGE, G. A. (1966): Study of a system, as illustrated by the optokinetic response. *Symp. Soc. exp. Biol.* **20** 179—198.

HORRIDGE, G. A. (1967): Perception of polarization plane, colour and movement in two dimensions by the crab, *Carcinus*. *Z. vergl. Physiol.* **55** 207—224.

MARTIN, A. R. and PILAR, G. (1963a): Dual mode of synaptic transmission in the avian ciliary ganglion. *J. Physiol.* **168** 443—463.

MARTIN, A. R. and PILAR, G. (1963b): Transmission through the ciliary ganglion of the chick. *J. Physiol.* **168** 484—475.

SHAW, S. R. (1966): Polarised light responses from crab retinula cells. *Nature* **211** 92—93.

SHAW, S. R. (1967): Organization of the locust retina. *Symp. Zool. Soc. London* **23** (in press).

TRUJILLO-CENÓZ, O. (1965): Some aspects of the structural organization of the intermediate retina of dipterans. *J. Ultrastruct. Res.* **13** 1—33.

DISCUSSION

B. Csillik: The possibility of an electrical coupling in retinula cell synapses is of great importance. So far, only in very few cases could an electrotonical transmission be proved unequivocally (club endings on Mauthner cells, ciliary ganglion of the chick). In the ciliary ganglion, Martin and Pilar described electrical coupling as early as 1961, yet, in spite of numerous studies on this topic, it was not until 1966 that De Lorenzo finally proved membrane fusion of pre- and post-synaptic membranes electron-microscopically. Therefore, the lack of such fused membranes in the present material does not by any means exclude the possibility of the actual existence of such fusions; it should be kept in mind that it is extremely difficult to find such tiny places in the relatively huge electron-microscopic specimens. However, the final proof for electrical transmission would be an electrophysiological demonstration of the absence of a synaptic delay, and I wonder if anybody has done such studies on retinula cells?

J. Hámori: I agree that the improvement of the fixation technique and a further set of observations may lead to the discovery of real electrotonic contacts (fused membranes) between the retinula terminal and the postsynaptic ganglion cell axon. As far there is no electrophysiological evidence for electric transmission in the lobster.

P. Röhlich: According to the observations of Trujillo-Cenóz and Melamed, presynaptic structures (e.g., synaptic ribbons) may occasionally be found in the process of the unipolar cell, i.e., at the postsynaptic side. Did you see such structures?

J. Hámori: We have never seen synaptic ribbons in the lobster ganglion cell axons in the cartridge region.

Symposium on Neurobiology of Invertebrates 1967 (123—133)

THE MODE OF RELEASE OF NEUROSECRETORY MATERIAL IN THE FRESHWATER PULMONATE *LYMNAEA STAGNALIS* L. (GASTROPODA)

A. NOLTE

Department of Zoology, University of Münster
Münster, GFR

Among gastropods, a morphologically established neurosecretory system has been found in the cerebral ganglia of some Basommatophora and Stylommatophora (Van Mol 1960; Kuhlmann 1963; Röhnisch 1964; Joosse 1964; Nolte 1964, 1965; Simpson *et al.* 1966a; Lever *et al.* 1965; Cook 1966). There are groups of neurosecretory cells at constant location forming typical neurosecretory tracts by which the secretory products are transported into the neurohemal areas. The various criteria for identification of neurosecretory cells by light- and electron-microscopy are the same as used for vertebrates and higher invertebrates (staining properties, elementary granules, etc.). Cautious interpretation may be necessary, as no definitive functions of neurosecretion have as yet been assigned. Nevertheless, the arrangement of the cells and their processes, the enlargement of the storage space by branching out within the neurohemal areas (Nolte 1965), and the termination in proximity to blood vessels or blood sinuses are strongly indicative of the existence of 'definite' neurosecretion in these gastropod species.

Two types of neurohemal areas have been described (Fig. 1): (1) The *lateral* ones, represented by the nervus arteriae cerebralis (mainly Stylommatophora) or the periphery of the nervus labialis medius (mainly Basommatophora), are simple and contain only glial cells in addition to the neurosecretory fibers. (2) The *dorsal* neurohemal areas, represented by the intercerebral commissure, are more complicated as result of the extension of the neurosecretory fibers into the mediodorsal bodies. Electron-microscope investigations of these areas have been reported only for Basommatophora (*Helisoma tenue*, dorsal neurohemal area – Simpson *et al.* 1966b; *Planorbarius corneus*, dorsal and lateral areas – Nolte 1964, 1965).

The present study concerning the lateral neurohemal area of *Lymnaea stagnalis* was undertaken to obtain more information about the storage-release area of the median lip nerve.

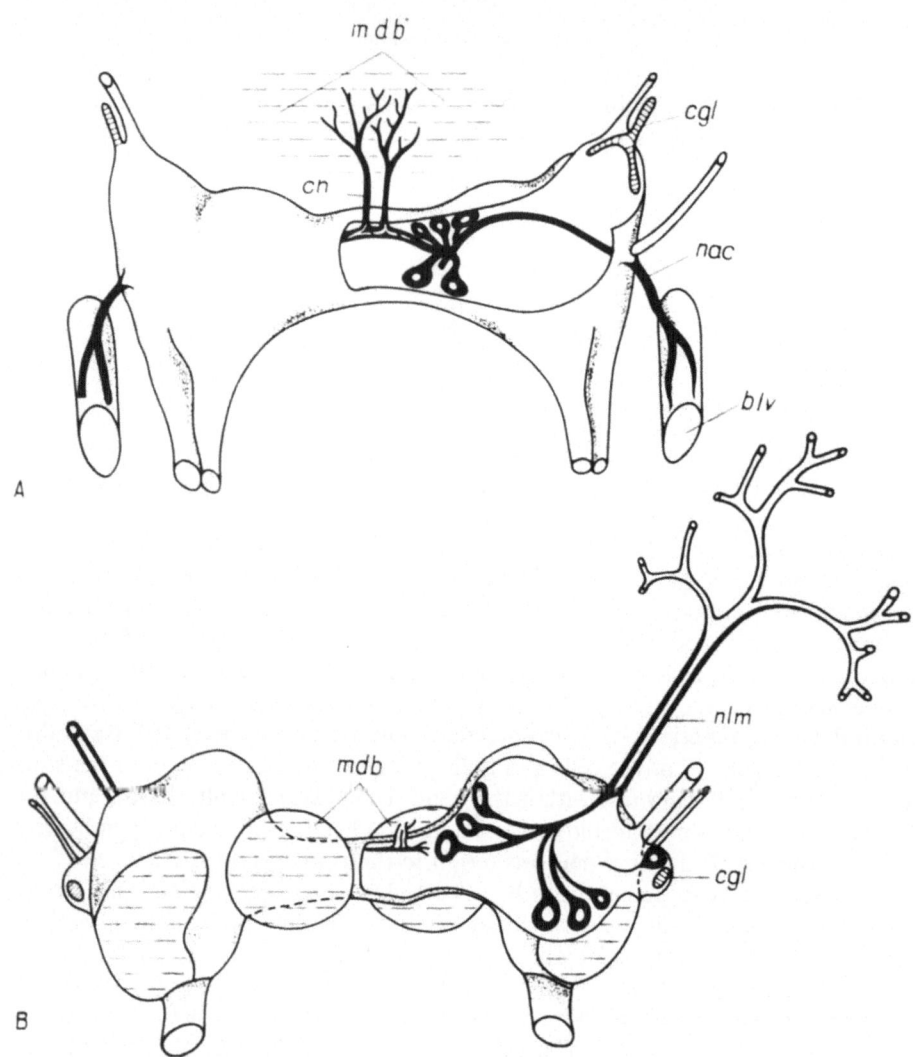

FIG. 1. The neurosecretory system. A — stylommatophoran typ; B — basommatophoran typ; blv — blood vessel; cgl — cerebral gland; cn — commissural nerve; mdb — mediodorsal body; nac — nervus arteriae cerebralis; nlm — nervus labialis medius

MATERIAL AND METHODS

The proximal and distal parts of the nervus labialis medius of 17 specimens were removed without anaesthetizing the snails, fixed according to the method of Wohlfarth-Bottermann (1957), and embedded in methacrylate or Epon (Luft

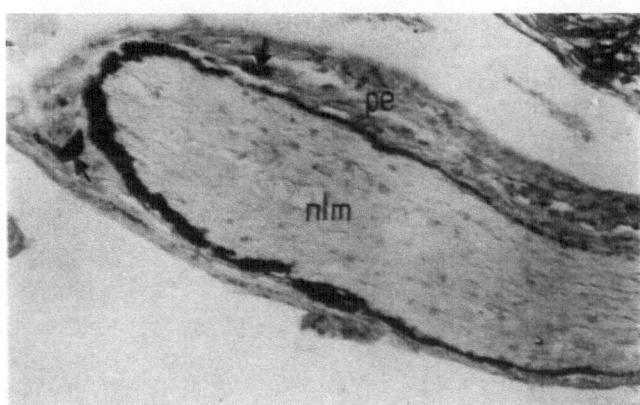

FIG. 2. Nervus labialis medius (nlm) of *Lymnaea stagnalis*. Note intense staining of neurosecretory fibers with paraldehyde-fuchsin. Two small branches (↗) in the perineurium (pe). × 600

1961). Sections were cut on the Leitz ultramicrotome and examined in a Siemens Elmiskop I a. Primary magnification: 2600, 8000, 20,000, or 40,000 times.

RESULTS

Under the perineurium, neurosecretory fibers were found along the whole nerve. The layer of neurosecretory fibers is thicker in the proximal part of the nerve near the ganglion, especially at one side of the nerve (Fig. 2). The light-microscopic aspect of bulb-like ends (6–10 μm) situated perpendicularly to the perineurium has been described for this species by Joosse (1964), suggesting that these are the widened ends of the axons where storage occurs. In electron micrographs, these endings seem to be far more numerous, sometimes having a diameter of only 1 μm (Fig. 3). In its course along the nerve, each neurosecretory fiber branches off into many protrusions directed toward the perineurium. Thus, the storage space for the neurosecretion products will be greatly enlarged and the number of terminals belonging to one of the hundred neurosecretory cells becomes very high.

Few *glial cells* are distributed among the neurosecretory fibers (Fig. 3). Their long, dense processes invest groups of neurosecretory fibers and some reach the perineurium. Between the terminals and the perineurium, glial processes are very rare, whereas, in peripheral nerves, glial cells are normally situated under the perineurium. A basement membrane and a nonfibrous mucopolysaccharide layer surround the nerve.

The perineurium (nerve capsule) consists of a more or less thick layer of fibrous connective tissue containing collagen fibrils (Figs. 3 and 4) and few fibrocytes. Of special interest in respect to our problem is the blood supply of this nerve. Embedded in the nerve capsule, small vessels lined by low myoendothelial cells send offshoots in all directions (Figs. 3 and 4). Thus, the distance between the ter-

A. NOLTE

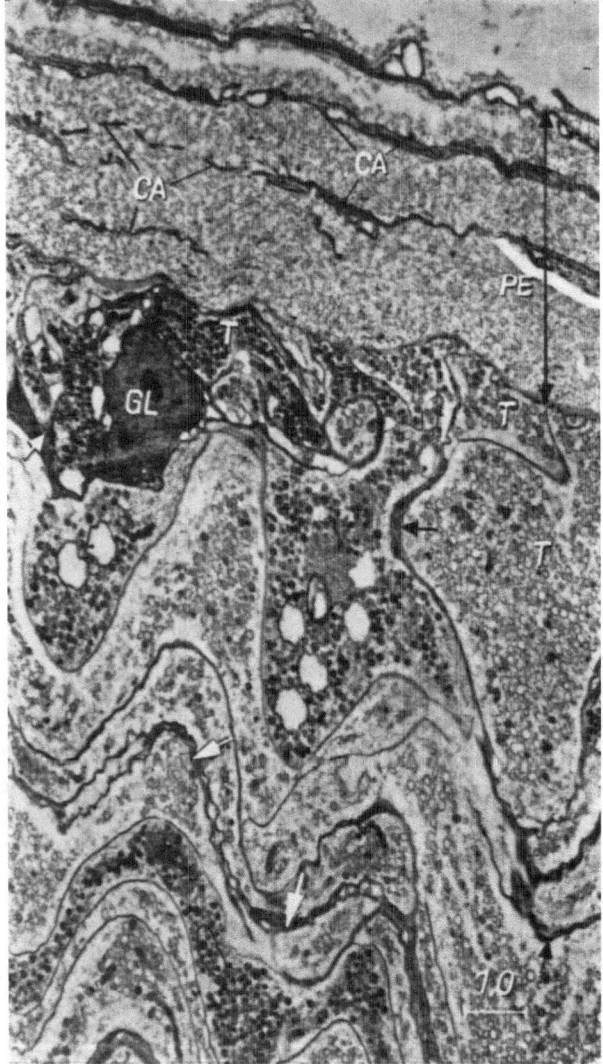

FIG. 3. Periphery of nervus labialis medius of *Lymnaea stagnalis* with neurosecretory fibers and terminals (T). CA — capillaries; GL — glial cell and glia cell processes (↗); PE — perineurium. Methacrylate

minals of neurosecretory fibers and the circulating blood is relatively short, and by contraction of the myoendothelial cells the blood is poured through the perineurium. There are also small bundles of neurosecretory fibers leaving the periphery of the lip nerve which contact the capillaries of the perineurium (Fig. 4).

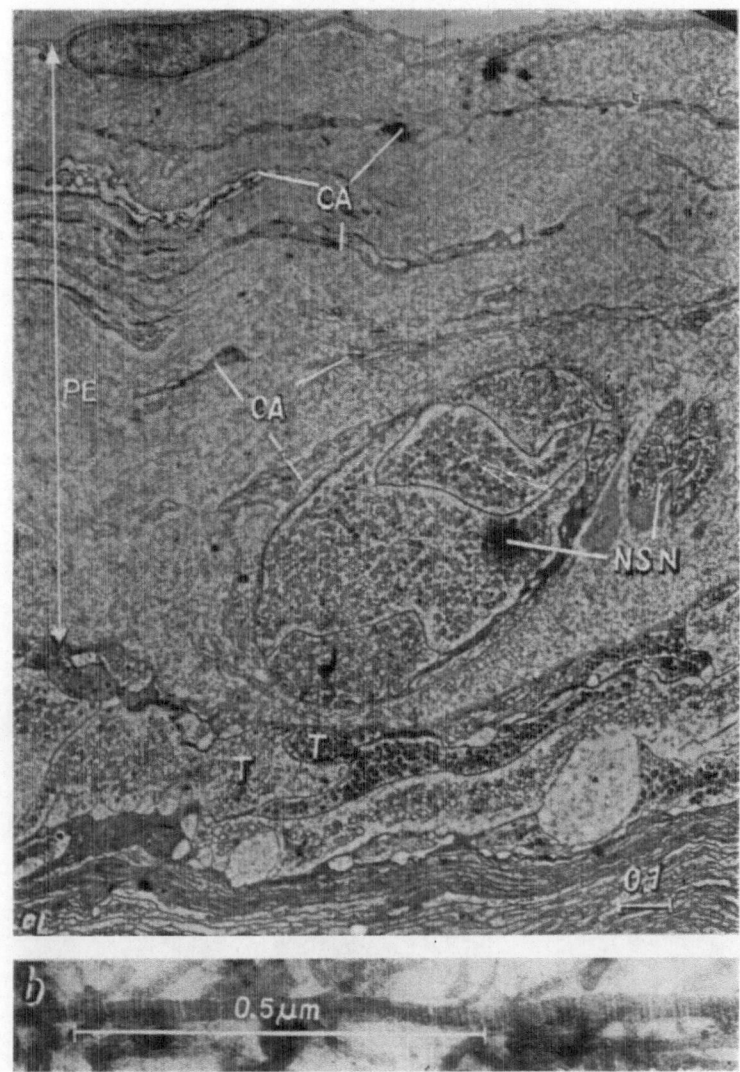

FIG. 4. Periphery of the nervus labialis medius of *Lymnaea stagnalis*. a — with two small neurosecretory nerves (NSN) in the perineurium (PE). CA — capillaries; T — terminals. Methacrylate. b — Collagen fibril of the perineurium

NEUROSECRETORY GRANULES

Apart from scarce mitochondria, neurotubules, and lamellated bodies, the neurosecretory fibers and their terminals contain different amounts of granules (diameters of 1200–2200 Å) and few small clear vesicles (diameter of 300 Å). Neuro-

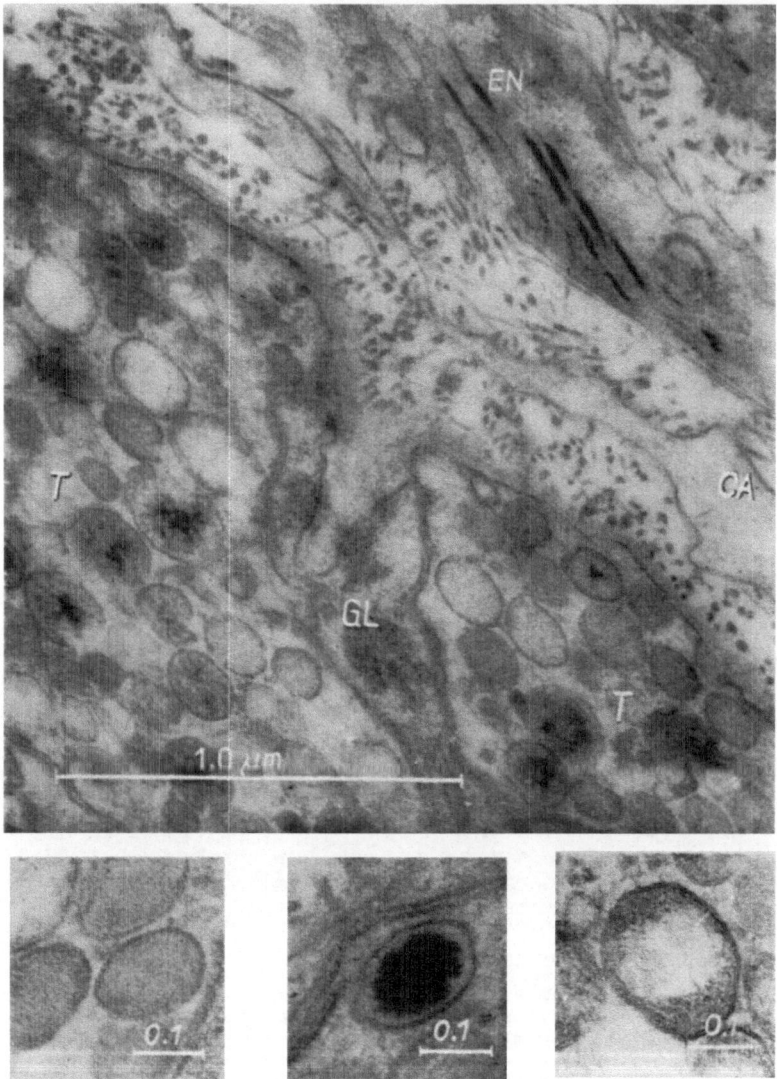

FIG. 5. Terminals (T) with neurosecretory granules of different characteristics. CA — capillary; EN — endothelial cell with myofibrils; GL — glial cell process. Epon

filaments are visible only in preterminal fibers with a small granule population. In Epon-embedded material, the granules are a bit more voluminous than in methacrylate-embedded ones.

The ultrastructural aspect of granules is highly variable. Generally, different aspects occur within one fiber or terminal (Fig. 5). Sections with uniform granules

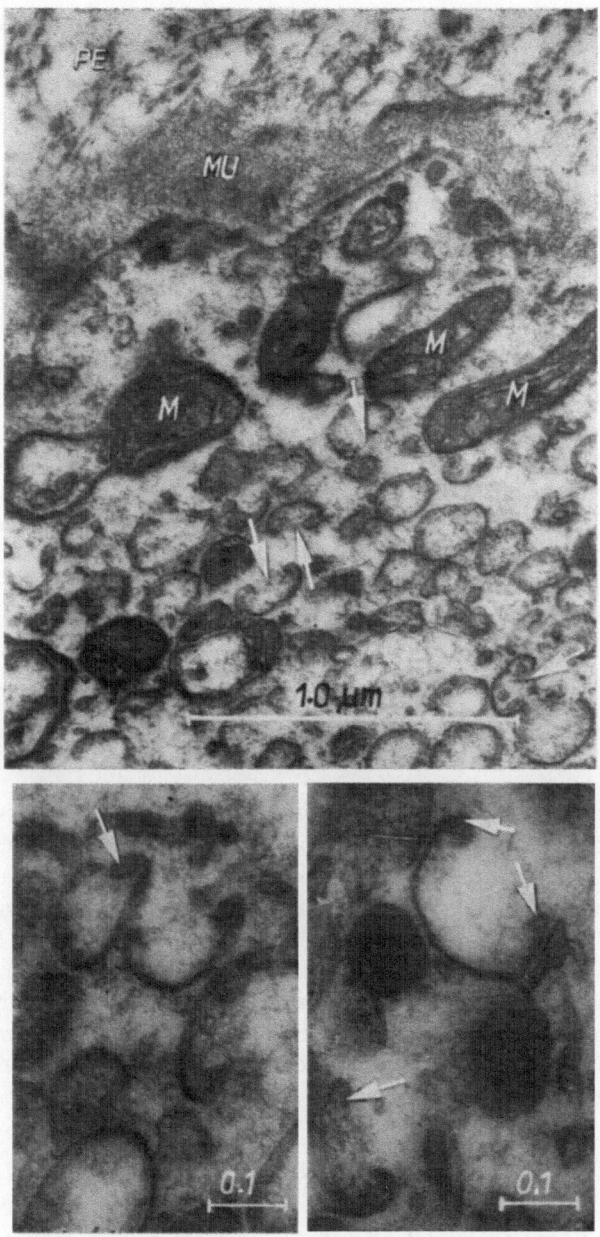

FIG. 6. Ruptured granules (↗) and small vesicles. M —
mitochondria; MU — mucopolysaccharide layer; PE —
perineurium. Methacrylate

are occasionally observed. Three main types of granules were visualized (Fig. 5): (1) granules with dense cores of different sizes, consisting of irregular particles, (2) dense granules with a homogenous distribution of fine granular contents showing different degrees of electron density and close application of membrane to the contents, and (3) electron-lucent granules with a more or less distinct accumulation of granular or fibrous material at the inner surface of the membrane. As there are all kinds of transition stages between these aspects, it seems questionable to interpret them as different types of granules. Disaggregation of the dense core, and a decrease of electron density, sometimes attended by membrane separation and disappearence of the granule contents from the center, may be successive stages of one granule type.

In the electron micrographs (1200 taken of 17 specimens), I could find neither granules outside of the fibers within the perineurium nor fusion between the membranes of granules and the plasmalemma. Therefore, I assume that the secretory material (hormones and/or carrier protein) has to diffuse out through the boundary membrane of the granules. The gradual loss of density of the granule contents may be the result of this discharge. It is of interest that granules of low electron density or electron-lucent aspect are slightly larger than those with dense contents. Changes of membrane permeability could be the reason for this. Further indication that the given interpretation of release may be correct is the frequent occurrence of ruptured granules (Fig. 6) among the electron-lucent ones. The scrolled edges of their membranes are curled inwardly.

SMALL VESICLES

Vesicles corresponding in size and aspect to synaptic vesicles are often described in neurosecretory terminals (De Robertis 1962). In *Lymnaea stagnalis*, they have a mean diameter of 300 Å and are regularly encountered in fibers or terminals with pale elementary granules in greater amount (Fig. 6) or in those with few granules. They sometimes lie associated with the membrane of the granules or 'inside' ruptured granules. If these vesicles carry transmitter substances involved in the process of releasing neurosecretory products, it is difficult to understand why they are more numerous in depleted fibers than in those with closely packed granules. Thus, it may be possible that at least some of them are products of the breakdown of granule membranes (Fridberg *et al.* 1966).

CONCLUSION

At the periphery of the nervus labialis medius (lateral neurohemal area) of *Lymnaea stagnalis*, the processes containing neurosecretory elementary granules make numerous contacts with the perineurium by means of perpendicular protrusions along the whole nerve. In order to release the neurosecretory material into the blood stream, the direct anatomical pathway would be the discharge from the terminals into the perineurium passing the nonfibrous mucopolysaccharide layer, the fibrous connective tissue layer, and the endothelial cell of the capillary. Neurosecretory granules outside of the fibers and their terminals could not be encountered. This agrees with the results of most investigations about vertebrate and invertebrate neurohemal organs. Therefore, the release of the neurosecretion material through the membrane of the granules seems to be the rule. The varying characteristics of the granule contents, especially the loss of electron density, may correspond to the releasing process.

As already mentioned, the varying characteristics of the granules appear not only in the terminals but in neurosecretory fibers far away from the perineurium as well. If the lowering of density and the occurrence of pale or 'empty' granules are the result of the release, we must consider that the release of neurosecretory substances is not restricted to the terminals in our subject. There are two possibilities for explaining how the released substances may reach the blood stream: Either they take the path to the terminals and diffuse there into the perineurium, or they pass into the intercellular spaces between the fibers or glial processes and from there into the perineurium.

Solitary axons or small bundles of them containing granules similar in size and characteristics to neurosecretory elementary granules have been encountered in almost all ganglia and many nerves of gastropods (Gerschenfeld 1963; Schlote and Hanneforth 1963; Simpson et al. 1966). Therefore, one must be cautious in interpreting granules as being neurosecretory. They may in some cases contain neurotransmitters. However, the great accumulation of these granules within the peripheral fibers of the nervus labialis medius, the formation of numerous terminals along each fiber, the lack of glial processes between terminals and perineurium, and the varying characteristics of granules within one fiber, taken together seem to suggest an interpretation of these granules as 'definite' neurosecretory granules and the periphery of the median lip nerve in some Basommatophora (Nolte 1964, 1965) as a 'definite' neurohemal area.

SUMMARY

The periphery of the nervus labialis medius of *Lymnaea stagnalis* has been investigated by electron microscopy.

1. The fibers from the neurosecretory cells of the cerebral ganglia occupy the periphery of the nerve under the perineurium. Along the nerve, each neurosecretory fiber branches into many terminals which lie in contact with the perineurium.

2. The neurosecretory granules show various characteristics within one fiber or termination. The changes of these characteristics have been discussed in relation to the release of the contents through the boundary membrane of the granules.

3. Neither neurosecretory granules were observed outside of the terminals or fibers, nor a fusion of the granule membrane with the cell membrane.

REFERENCES

COOK, H. (1966): Morphology and histology of the central nervous system of *Succinea putris* (L.). *Arch. néerl. Zool.* **17** 1—72.

DE ROBERTIS, E. (1962): Ultrastructure and function in some neurosecretory systems. *Mem. Soc. Endocrinol.* **12** 3—17.

FRIDBERG, G., BERN, H. A. and NISHIOKA, R. S. (1966): The caudal neurosecretory system of the isospondylous teleost, *Albula vulpes*, from different habitats. *Gen. and comp. Endocrinol.* **6** 195—212.

GERSCHENFELD, H. M. (1963): Observations on the ultrastructure of synapses in some pulmonate molluscs. *Z. Zellforsch.* **60** 258—275.

JOOSSE, J. (1964): Dorsal bodies and dorsal neurosecretory cells of the cerebral ganglia of *Lymnaea stagnalis* L. *Arch. néerl. Zool.* **16** 1—103.

KUHLMANN, D. (1963): Neurosekretion bei Heliciden (Gastropoden). *Z. Zellforsch.* **60** 909—932.

LEVER, J., DE VRIES, C. M. and JAGER, J. C. (1965): On the anatomy of the central nervous system and on the location of neurosecretory cells in *Australorbis glabratus*. *Malacologia* **2** 219—230.

LUFT, J. H. (1961): Improvements in epoxy resin embedding methods. *J. biophys. biochem. Cytol.* **9** 409—414.

NOLTE, A. (1964): Ultrastruktur des "Neurosekretmantels" des Nervus labialis medius von *Planorbarius corneus* L. (Basommatophora) *Naturwissenschaften* **51** 148.

NOLTE, A. (1965): Neurohämal-"Organe" bei Pulmonaten (Gastropoda). *Zool. Jb. Anat.* **82** 365—380.

RÖHNISH, S. (1964): Untersuchungen zur Neurosekretion bei *Planorbarius corneus* L. (Basommatophora). *Z. Zellforsch.* **63** 767—798.

SCHLOTE, F.-W. and HANNEFORTH, W. (1963): Endoplasmatische Membransysteme und Granatypen in Neuronen und Gliazellen von Gastropodennerven. *Z. Zellforsch.* **60** 872—892.

SIMPSON, L., BERN, H. A. and NISHIOKA, R. S. (1966a): Survey of evidence for neurosecretion in Gastropod molluscs, *Am. Zoologist,* **6** 123—138.

SIMPSON, L., BERN, H. A. and NISHIOKA, R. S. (1966b): Examination of the evidence for neurosecretion in the nervous system of *Helisoma tenue* (Gastropoda Pulmonata). *Gen. and. comp. Endocrinol.* **7** 525—548.

VAN MOL, J. J. (1960): Phénomènes neurosecrétoires dans les ganglions cérébroides d'*Arion rufus*. *C. R. Acad. Sci. (Paris)* **250** 2280—2281.

WOHLFARTH-BOTTERMANN, K. E. (1957): Die Kontrastierung tierischer Zellen und Gewebe im Rahmen ihrer elektronenmikroskopischen Untersuchung an ultradünnen Schnitten. *Naturwissenschaften* **44** 287—288.

DISCUSSION

M. Mirolli: Have you seen large spaces between the plasmalemma of the neuron soma and the glial cell lamellae? And have you seen evidence of breakage of the membrane of the glia cells?

What I have in mind is the possibility that those broken granules with the broken membrane that you have seen were an osmotic artifact. If this were true, one would also expect to find broken glia cell and very elongated intracellular spaces in the cortex of the ganglion.

A. Nolte: We have not examined the ganglia but only the lip nerves. The membranes of the glial cells of the nerve are intact. The granules with a broken membrane are to be found among those with an intact membrane, and in different lip nerves fixed at the same time you may see different amounts of such granules. It is possible that the membrane of the granules is more unstable after releasing the secretory material and so the breakage — caused by the fixation process or not — may somehow be related to the functional stage of the granules.

G. A. Cottrell: The results of some of our fractionation experiments with *Mercenaria* ganglia suggest that granules of the type you have described (1250–2500 Å and uniformly granulated) contain an unidentified cardioexcitor, substance X.

A. Nolte: In bivalves, it may be difficult to decide whether these granules are neurosecretory or not, as neurosecretory cells are not lying in groups within the ganglia, and neurohemal areas comparable to those of Basommatophora do not exist. The similar sizes and characteristics of the granules in different groups of animals do not give evidence for a similar function.

J. Salánki: Did you inject reserpine into the animal, and if so, into what part of it? As the animals live in water, is it not possible that a big part of the reserpine dissolves out from the animal, and the effective concentration becomes very low?

A. Nolte: Yes, we did inject reserpine into the hemolymph of the foot (50 mg/kg in one injection or the same dose again after one day). We do not know whether the reserpine dissolves out from the animals or not, but nevertheless, the snails show an effect by moving around far less than the control animals. No decrease of neurosecretory material could be seen.

P. Röhlich: The accumulation of the boutons filled with vesicles having a dense content suggests that a release of substances takes place in this region. To approach the chemical nature of the substances and the mechanism of release it would be most interesting to find pharmaka which can evacuate the content of the vesicles. Did you use such substances?

A. Nolte: Yes, we did try to affect the release of neurosecretory substances by injecting different pharmaka (acetylcholine, 5–HT, atropine, pilocarpine). In light-microscopic studies no effect could be seen with respect to the neurosecretory system.

P. Röhlich: The curling membranes of the vesicles seem to be artifacts due to osmium fixation. Although there are many difficulties when using aldehydes for the fixation of invertebrate tissues, a control fixation with glutaraldehyde would be of much help in clarifying the situation. According to my experience on invertebrate material, I would recommend a very short fixation in aldehydes and a rapid washing in buffer prior to osmium fixation.

A. Nolte: Thank you for the suggestion. We did not have good results with glutaraldehyde-fixation till now.

Symposium on Neurobiology of Invertebrates 1967 (135—142)

AUTORADIOGRAPHIC EXAMINATION OF THE DISTRIBUTION OF ^{35}S-CYSTEINE IN THE SPECIAL AND SECRETORY CELLS OF THE OPTIC TENTACLE AND TENTACULAR GANGLION OF PULMONATA

J. Bierbauer, J. Kiss and B. Vigh

Department of Histology and Embryology, University Medical School
Budapest, Hungary

In our earlier investigations, we examined the cytology of the optic tentacle in Pulmonata (Bierbauer and Török 1964; Bierbauer, Török and Teichmann 1965). Initially we studied the collar cells surrounding the hand-shaped ganglion, followed by a study of the light-microscopic cytomorphology and cytochemistry of the lateral processed *A* cells and the lateral processed metachromatic *B* cells which lie between the dermomuscular layer and the retractor muscles. More recently we conducted an electron-microscopic examination of the collar cells and the lateral *A* and *B* cells (Röhlich and Bierbauer 1966). In addition, the tentacular ganglion was studied in connection with the collar cells (Bierbauer 1967).

In the present paper, we use autoradiographic methods for the study of the collar cells, the lateral processed *A* and lateral processed *B* cells, and the tentacular ganglia.

THE COLLAR CELLS

The collar cells have a large, presumably polyploid nucleus. The longitudinally sectioned cell is pear-shaped, and its process is similar to the axon of the typical nerve cell, as it clearly can be observed in some cases by silver impregnation when it runs among the fibers of the tentacular ganglion. The collar cells have a strongly basophil cytoplasm in which numerous Gomori-positive granules (granulum II) and a few Gomori-negative granules (granulum I) are seen. Under the electron microscope, the cytoplasm shows high density. A correspondence can often be observed between the elements of the ergastoplasm and the large Golgi areas. In some cases, vesicles separate themselves from the surface of the neighboring ergastoplasmic sacs; however, it can also be observed that vesicles unite with the Golgi cysternae. Often developing in the concavity of the Golgi cysternae, are secretion vacuoles containing a thin material of modest density. More compact vesicles, some hundred Å in diameter, can regularly be found around the secretion vacuoles in the cytoplasm as well. They remind one of the elementary neurosecretory granules. Rarely, ultracytoplasmatic filaments can also be observed in the cells. Lane (1964) described microtubuli in the cytoplasm of the cells. There are also dense bodies greater than 1 μ scattered in the cytoplasm. It is possible that these dense bodies correspond to the Gomori-negative granules (granulum I) and they presumably contain lipofuscine. In the periphery of the cells, some areas are found which have a diameter of 1–2 μ and have the characteristics of nerve fibers; their cytoplasm contains scattered dense-core vesicles. The collar cells were thought by many authors (Flemming 1872; Retzius 1892) to be nerve cells and by other authors (Demal 1955; Flemming 1870) to be secretory cells. Synthesizing this dual character, Lane (1962) described the collar cells as neurosecretory cells. On the basis of their Gomori and DDD positivity, we also thought for a time that this could be possible. On the basis of further investigations, the collar cells could be interpreted as special neurosecretory cells, i.e., as specially modified neurons differing from the neurosecretory cells in the classical sense. We hope that we could further clarify the problems concerning the collar cells by means of our present embryological and regeneration investigations. Our latest studies allow us to suppose that the collar cells have an important role in the regulation of the glandula hermaphroditica. This fact seems to be confirmed by the works of Peluet and Lane (1961) as well as Gottfried (1967).

THE LATERAL PROCESSED *A* AND LATERAL PROCESSED METACHROMATIC *B* CELLS

The lateral processed *A* and *B* cells lie in rows under the dermomuscular layer of the optic tentacle. Many authors described these cells as nerve cells (Hanström 1928; Veratti 1900). Later, Lane (1962), on the basis of their Gomori-positivity, supposed that they could be neurosecretory cells. After the most recent light- and electron-microscopic studies we can surely declare that they are secretory cells (Röhlich and Bierbauer 1966). The lateral processed *A* cells contain lipoproteides. Their nuclei are large, presumably of the polyploid type. Under the electron microscope the cytoplasm seems to be compact. The secretory vacuoles have a diameter of 1.45 μ and they contain a light material with a homogeneous structure. Complex bodies are also seen in the cytoplasm of the cells. It is possible that these bodies correspond to lipofuscine. The processed *B* metachromatic cells differ from the *A* cells — namely, their nuclei are small, they contain mucoproteides, and show metachromasia. Electron-microscopically, the content of the ergasto-plasmic cysternae and the secretory vacuoles is much more dense than could be observed in the lateral processed *A* cells.

THE MUCOUS SECRETORY CELLS OF THE DERMOMUSCULAR LAYER (*D* CELLS)

The mucous secretory cells of the dermomuscular layer stain positively by muco-polysaccharide staining processes and show metachromasia. They are very similar to the lateral processed metachromatic *B* cells under the electron microscope, as the content of the ergastoplasmic cysternae and secretory vacuoles show a similar density.

THE TENTACULAR GANGLION

In the hand-shaped tentacular ganglion there are many little ganglion cells, although a few gigantic cells could be observed, having perhaps polyploid nuclei (Bierbauer 1967). Following cytomorphological criteria, we assume that these are motor cells; they show no secretory activity. We also observed that the lateral nerve fibers formerly described by us, fibrae tentaculares laterales, start from the finger-like processes of the tentacular ganglion (Bierbauer 1967). Little bipolar ganglion cells were discovered on the surface of the fibers not only by means of silver impregnation but also in stained preparations by the electron microscope (Röhlich and Bierbauer 1966). In the tentacular ganglion, Hanström (1925, 1928) described a reflex pathway which consists of sensory, associative, and cerebral ganglion cells. We suppose that there exists another pathway consisting of sensory, associative, and large ganglion cells. This supposition may be confirmed as follows: if we extirpated the optic tentaculum, i.e., disconnected the central ganglion, the contractor muscles would also contract in this case (Bierbauer 1967).

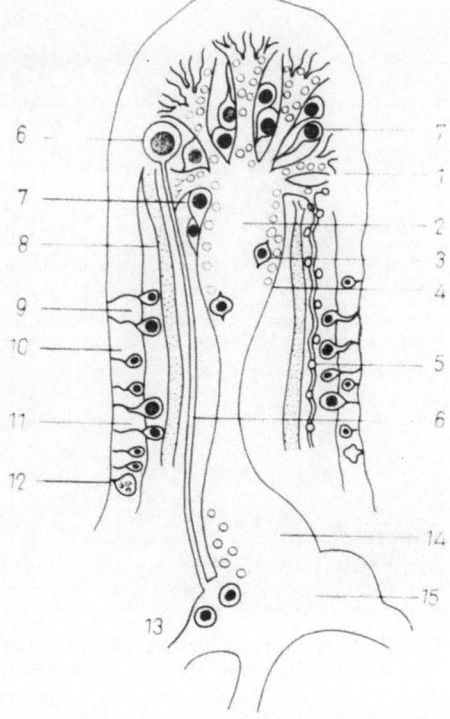

FIG. 1. Scheme of the tentacular ganglion. 1 — Dermomuscular layer; 2 — tentacular ganglion; 3 — large ganglion cells; 4 — small ganglion cells; 5 — fibrae tentaculares lat. and small ganglion cells; 6 — eye and optic nerve; 7 — collar cells; 8 — retractor muscles; 9 — lateral processed *A* cells; 10 — *D* mucous cells of the dermomuscular layer; 11 — lateral processed *B* cells; 12 — lime glands; 13 — postcerebrum; 14 — procerebrum; 15 — mesocerebrum

AUTORADIOGRAPHIC EXAMINATIONS

Material and method

In the present study, the species *Arianta arbustorum*, *Arion circumscriptus* and *Helicella obvia* (belonging to the suborder Stylommatophora) were used in specific groups. The animals received ^{35}S-cysteine by intracoelomal injection in an amount of 5 μC/g body weight. The specific activity of the isotope was 37.4 mC/mM. Forty-eight hours following the injection, the experimental materials were fixed in Bouin's fluid, embedded in paraffin, and cut to 5−6 μ thick sections. For autoradiography we used Kodak AR 10 stripping emulsion with an exposure time of 5 weeks. After developing in Kodak D 19b developer (5 min, 18–20 °C) fixation was made in Kodak acid-fixing powder. We stained the autoradiograms by hematoxalin according to Bartha and methyl-green-pyronin in the usual way. The appearance and density of the silver grains were evaluated qualitatively.

COLLAR CELLS

We observed that there is a great isotope incorporation in the cytoplasm of the collar cells after the administration of ^{35}S-cysteine. A significant density of the sil-

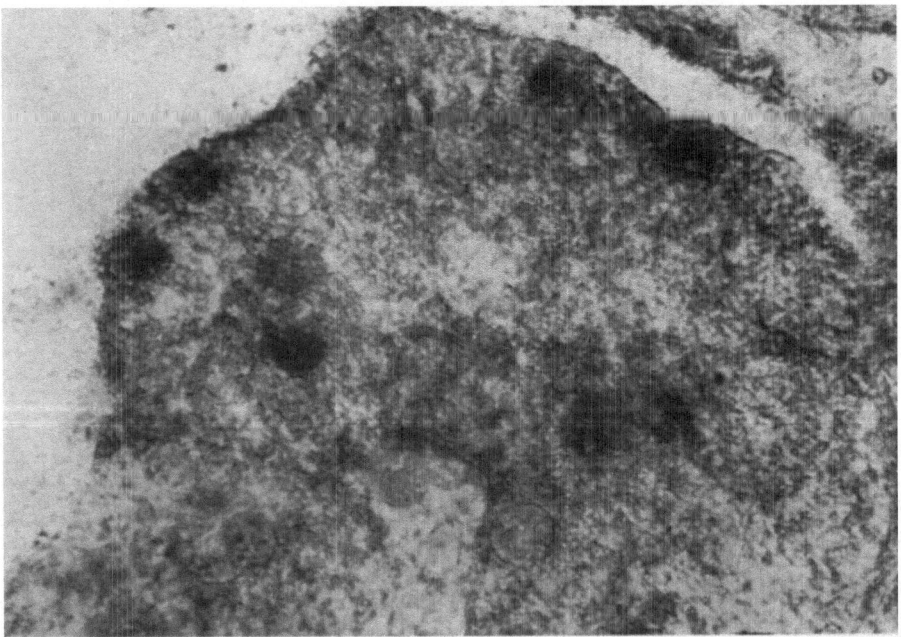

FIG. 2. A strong incorporation can be observed on the autoradiograms of the collar cells after the administration of ^{35}S-cysteine. The high density of the granules shows the degree of incorporation

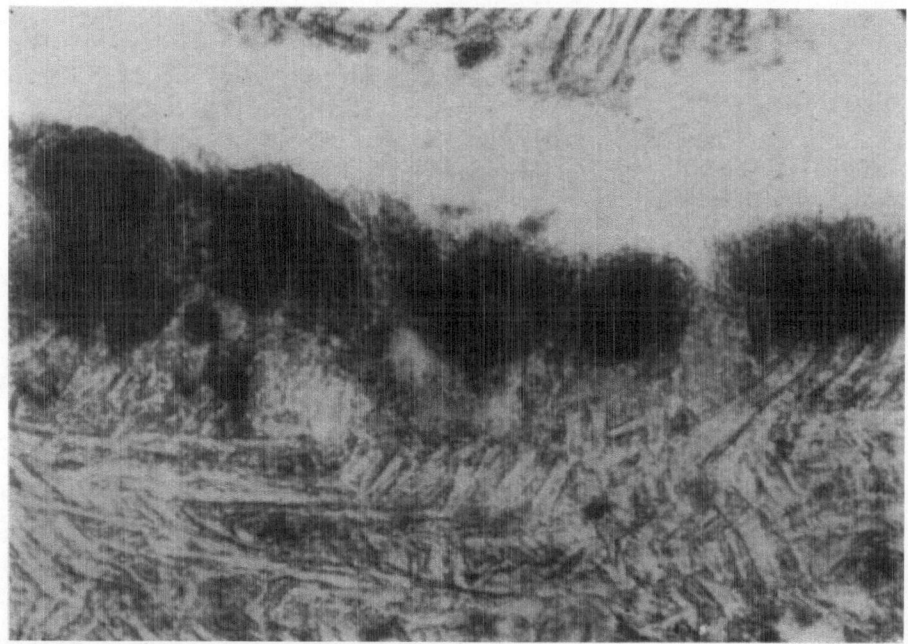

FIG. 3. The lateral processed *A* cells and *B* cells have a great activity, as shown by the density of the granules on the autoradiogram

ver grains demonstrates this result. Among different collar cells, we could not observe any differences concerning the density of the granules. This fact allows us to draw the conclusion that the isotope uptake of the collar cells is similar, or that the cells are in a similar phase of function. It is possible, although we have no data, that the degree of uptake correlates with biological cycles. There are quantitative autoradiographic studies in progress concerning this hypothesis.

THE LATERAL PROCESSED *A* CELLS

As our autoradiographic examinations showed, there is a lower density of the granules in the lateral processed *A* cells than in the collar cells, a fact which means a lower isotope incorporation. We have to remark that there is a difference between the density of granules of different secretory *A* cells. It is possible that the uptake changes correlate with the phase of the secretory cells.

THE LATERAL PROCESSED *B* CELLS

There is a large uptake of ^{35}S-cysteine in the lateral processed *B* cells. The density of the granules is different in certain cells or cell groups. It could be supposed

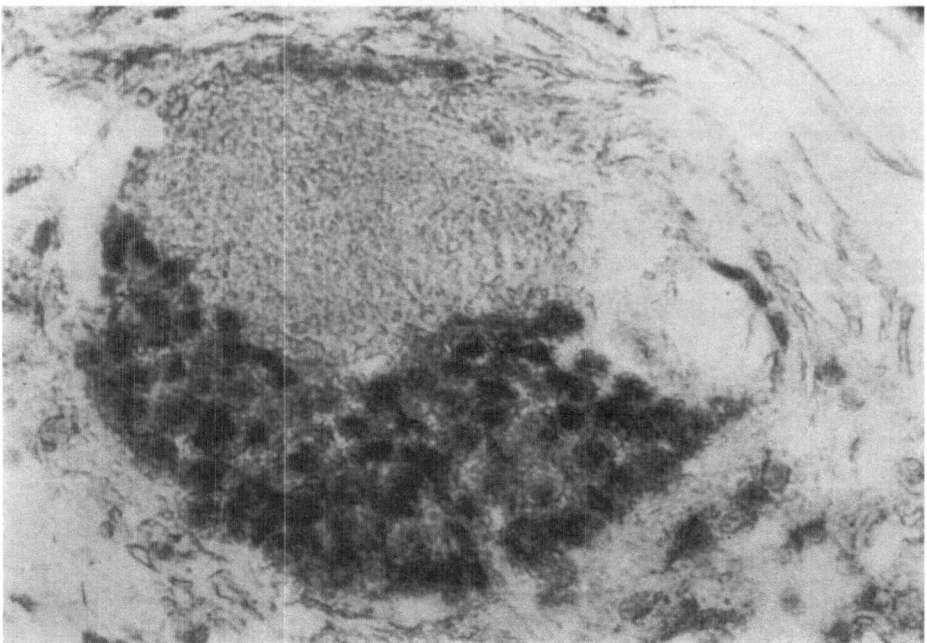

FIG. 4. There is a modest incorporation of ^{35}S-cysteine in the tentacular ganglion. The ganglion cells here have a similar isotope accumulation

that the quantity of the uptake changes depend upon the phase of the mucoid secretory cells, similar to the case for the lateral processed *A* cells. In addition, our cytochemical investigations allow us to assume functional differences between the different cells.

THE TENTACULAR GANGLION

In comparison with the former cell types, there is a lower uptake in the small and the large ganglion cells as well as in the fibers. This is in good agreement with our former investigations, i.e., that we could not observe neurosecretory phenomena in the field of the ganglion.

THE DERMOMUSCULAR LAYER AND THE HEMOLYMPH

In the connective tissue and smooth muscle cells of the dermomuscular layer there is a completely negative picture. The hemolymph is also negative, but the amoeboid cells in it show dense granulation. Here and there similar cells can also be observed in the connective tissue.

The degree of incorporation of ^{35}S-labeled cysteine into the different cell types was not determined quantitatively in the present study, although there is a good qualitative difference between the special and secretory as well as nonsecretory cells. The change of the density in the secretory cells of different type allows us to conclude that there are species differences among the cells. We hope to further clarify these problems in the course of our present examinations.

SUMMARY

On the basis of our former light- and electron-microscopic studies it was suggested that the collar cells are special neurosecretory cells differing from the classical neurosecretory cells. The optic tentacles have an important role in the regulation of the gametogenesis of the hermaphroditic gland. It is possible that the collar cells have special functions inside the tentacle.

In our present autoradiographic studies we could observe that there is a great isotope incorporation in the cytoplasm of the collar cells after the administration of ^{35}S-cysteine. A significant density of silver granules demonstrates this result. On the basis of our light- and electron-microscopic studies, the lateral processed A cells were considered to be secretory cells, even though they do not show mucous secretion. These cells also show high activity in our autoradiographic pictures, but the density of the granules is less than that in the collar cells.

The lateral processed metachromatic B cells produce mucous secretion. There is a significant incorporation of ^{35}S-cysteine in their cytoplasm.

We also found incorporation in the mucous secretory cells of the dermomuscular layer. The density of granules here is similar to that in the lateral processed metachromatic B cells. The incorporation is lower for both small and large ganglion cells as well as in the fibers and is less than for the former cell groups.

There is no activity in the connective tissue and smooth muscle cells of the dermomuscular layer; however, the amoeboid cells in the hemolymph show dense granulation on the autoradiograms.

REFERENCES

Bierbauer, J. (1967): Examination on the tentacular ganglion system of Pulmonates. *Gen. Comp. End ocrinol.*

Bierbauer, J. and Török, L. J. (1964): Cytological and neurosecretory investigations on the optic tentacle of pulmonata. *Acta biol. Acad. Sci. hung.* **15** Suppl. 6 39—40.

Bierbauer, J, Török, L. J. and Teichmann, I. (1965): Cytologische und histochemische Untersuchungen am neurosekretorischen System der Augententakel von Pulmonaten. *Zool. Jb. Physiol.* **71** 545—551.

Demal, J. (1955): Essai d'histologie comparés des organes chemorecepteurs des gastropodes. *Acad. Roy. de Belgique,* **29** 1—83.

Flemming, W. (1870): Untersuchungen über Sinnesepithelien der Mollusken. *Arch. mikr. Anat.* **6** 439—471.

FLEMMING, W. (1872): Zur Anatomie der Landschnekkenfühler und zur Neurozoologie der Mollusken. *Z. wiss. Zool.* **22** 365—378.

GOTTFRIED, H. (1967): Aspects of the reproductive endocrinology of the giant lang slug *Ariolimax californicus* (Stylommatophora: Gastropoda) *Gen. Comp. Endocrinol.*

HANSTRÖM, B. (1925): Über die sogenannten Intelligenzsphären des Molluskengehirns und die Innervation des Tentakels von *Helix. Acta Zool. VI* 183—215.

HANSTRÖM, B. (1928): *Vergleichende Anatomie des Nervensystems der wirbellosen Tiere.* Berlin.

LANE, N. J. (1962): Neurosecretory cells in the optic tentacles of certain pulmonates. *Quart. J. micr. Sci.* **103** 211—226.

LANE, N. J. (1964): The fine structure of certain secretory cells in the optic tentacles of the snail, *Helix aspera. Quart. J. micr. Sci.* **105** 35—47.

PELLUET, D. and LANE, N. J. (1961): The relation between neurosecretion and cell differentiation in the ovotestis of slugs (Gastropods: Pulmonata). *Canad. Zool.* **39** 691—805.

RETZIUS, G. (1892): Das sensibile Nervensystem der Mollusken. *Biol. Unters. N. F.* **4** 11—29.

RÖHLICH, P. and BIERBAUER, J. (1966): Electron-microscopic observations on the special cells of the optic tentacle of *Helicella obvia* (Pulmonata). *Acta biol. Acad. Sci. hung.* **17** 359—373.

VERATTI (1900): *Ricerche sui sistema nervoso dei Limax.* Reale Instituto Lombardo di Science e Lettere. **18** 160—175.

PROCESSES ON CELLULAR LEVEL

Symposium on Neurobiology of Invertebrates 1967 (145—167)

IONIC BACKGROUND OF ACTIVITY IN GIANT NEURONS
OF MOLLUSCS

P. G. KOSTYUK

A. A. Bogomolets Institute of Physiology, Academy of Sciences of the USSR
Kiev, USSR

The giant neurons of gastropodal molluscs are very convenient not only for investigating the mechanism of nervous activity of these animals, but also for solving various problems of general interest (particularly in the field of action potential generation, synaptic transmission, etc.). These cells are convenient because of their large dimensions and their superficial location, which makes them suitable for extensive microelectrode analysis. Giant neurons have been used in our department for several years for the investigation of ionic relations between the extra- and intracellular medium in the resting state and of ionic processes during spike generation; this paper summarizes the results of these investigations.

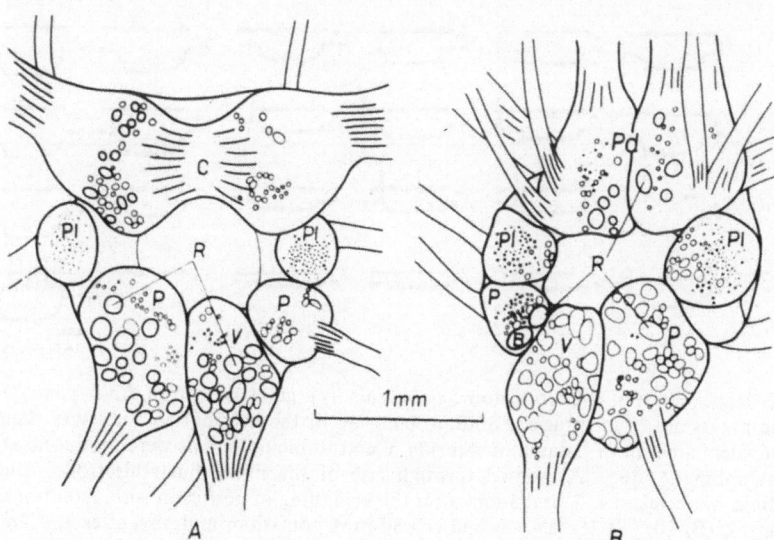

FIG. 1. Central nervous system of *Planorbis corneus*. C — cerebral ganglia, Pl — pleural ganglia, P — parietal ganglia, V — visceral ganglia, Pd — pedal ganglia, R — giant neurons two of which (A and B) are located in the smaller parietal ganglion

Experiments were performed on giant neurons of the terrestrial snail *Helix pomatia* and the freshwater mollusc *Planorbis corneus*; several measurements were also made on the freshwater mollusc *Lymnaea stagnalis* and the deep-sea mollusc *Tritonia diomedia* from the Pacific. The structure of the *Helix* nervous system is well known; the nervous system of *Planorbis* is less well known, and therefore its structure is shown in Fig. 1. As in *Helix*, different ganglia in this nervous system do not fuse together, and can be easily differentiated.

IONIC COMPOSITION OF THE HEMOLYMPH IN MOLLUSCS

The ionic composition of the internal medium in molluscs is rather variable, especially in ecologically different species, and is also subjected to large seasonal variations. Therefore, Sorokina and Zelenskaya (1967) made systematic studies of the ionic composition of the hemolymph from *Helix* and *Planorbis* during different seasonal periods.

The pH of the hemolymph changes from 7.0 to 8.2. Minimal values of pH were obtained in spring and maximal values during autumn and winter. The mean value of pH in hemolymph is 7.76 ± 0.08 in *Helix* and 7.40 ± 0.10 in *Planorbis*. As in the blood of higher animals, the main buffer systems are bicarbonate and

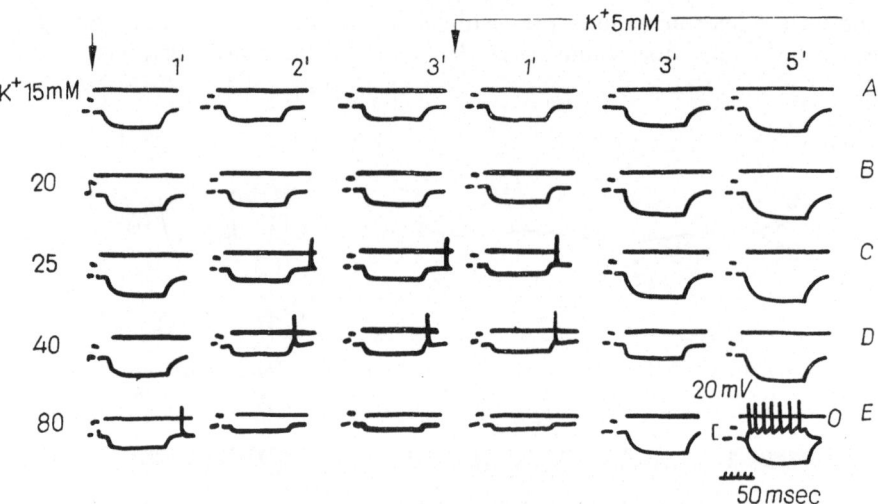

FIG. 2. Dependence of the electrotonic potentials in a giant neuron of *Helix pomatia* on the outside potassium concentration. Sodium chloride in the external solution was replaced by an equivalent amount of potassium chloride. Electrotonic potentials were produced by inward current pulses $(2 \cdot 10^{-8}$ A) passed through one of the intracellular electrodes. Successive recordings were taken 1, 2, and 3 min after the beginning of perfusion with a solution containing (A) 15, (B) 20, (C) 25, (D) 40, and (E) 80 mM potassium and also after 1, 3 and 5 min washing with normal saline solution. A standard calibration pulse is shown at the beginning of each recording on this and other figures. Response to a depolarizing pulse is shown in the last recording

TABLE 1

Electrolyte composition of the hemolymph (mM)

Ion	Helix	Planorbis
sodium	107.21±0.64	46.00±0.38
potassium	4.92±0.08	1.34±0.20
calcium	9.20±0.23	4.56±0.12
magnesium	8.83±0.49	1.49±0.02
phosphate		
total	0.65±0.02	0.48±0.03
acid-soluble	0.28±0.004	0.21±0.001
protein	0.20±0.006	0.09±0.003
inorganic	0.05±0.002	0.09±0.001
lipid	0.09±0.001	0.08±0.004
ammonium	0.03±0.002	0.02±0.006
chloride	85.14±2.14	32.87±1.89
sulphate	0.59±0.04	0.22±0.003
bicarbonate	5.71±0.30	6.93±0.42
nitrate	0.02±0.003	0.01±0.004

proteins, especially hemoglobin and hemocyanine. However, contrary to the case for higher animals, the concentration of monovalent cations is low. At the same time, the concentration of divalent cations (calcium and magnesium) is rather high (see Table 1). Phosphates are present in the hemolymph in comparatively low concentrations.

It can be seen from Table 1 that the differences in the ionic compositions of the hemolymph from the terrestrial snail *Helix* and freshwater snail *Planorbis* are extremely large; their seasonal variations are also quite different. If *Planorbis* are kept in steady living conditions in an aquarium, the variations in the concentrations of different ions are small. They become much larger in animals living under natural conditions. During summer and autumn, the concentration in the hemolymph of potassium, sodium, calcium, and chloride ions is maximal; later, it starts to decrease, and falls to a minimal level in spring, which probably coincides with the period of high water and the decrease in salt concentration in rivers (see Table 2). In *Helix*, the seasonal variations of inorganic ion concentrations (especially of sodium and calcium) are much higher. They have two maxima: one at the end of summer and one in winter and early spring. In summer, there is an increase mainly in concentration of potassium and sodium ions, but in winter and spring there is a special increase in the amount of calcium (up to 25% above the summer concentration). At the same time, the potassium concentration is not changed, and at the end of this period it even decreases.

In those species of *Helix* which were kept during winter in warm, moist conditions and which remained active, the amount of inorganic ions in the hemolymph stays approximately at the same level.

TABLE 2

Seasonal changes in the electrolyte composition of the hemolymph

Month	Ion concentration in hemolymph (mM/l)			
	K$^+$	Na$^+$	Ca^{++}	Cl$^-$
Helix				
January	4.26±0.04	98.20±0.36	9.15±0.07	79.56±1.28
February	4.02±0.06	105.64±0.97	9.98±0.08	89.78±2.34
March	4.03±0.09	110.76±0.57	11.25±0.15	92.48±2.13
April	3.98±0.08	108.30±0.59	12.06±0.29	92.69±1.98
May	4.54±0.03	91.26±0.48	6.67±0.13	74.66±1.94
June	5.51±0.01	109.23±0.54	8.74±0.08	85.26±1.15
July	5.98±0.02	125.68±0.09	9.23±0.08	88.23±2.31
August	5.64±0.03	116.62±0.36	9.06±0.11	88.92±2.38
September	6.61±0.05	115.00±0.08	8.96±0.09	86.80±1.34
October	5.63±0.04	102.70±0.09	8.70±0.11	82.50±1.58
November	5.03±0.04	100.78±0.15	8.76±0.09	80.80±1.24
December	4.08±0.05	98.36±0.39	8.67±0.09	79.50±1.34
Helix kept in laboratory conditions				
January	4.28±0.06	98.34±0.62	9.27±0.08	79.80±1.17
February	4.13±0.04	97.68±0.36	9.64±0.08	79.60±1.32
March	4.01±0.09	92.73±0.67	9.68±0.07	75.32±1.28
April	3.67±0.11	84.32±0.89	8.17±0.09	74.68±1.87
Planorbis				
January	1.28±0.10	46.50±0.30	4.19±0.04	31.82±1.03
February	1.11±0.24	43.82±0.46	3.85±0.26	31.64±1.90
March	0.98±0.26	42.34±0.87	3.28±0.19	31.70±0.98
April	0.83±0.18	40.97±0.98	3.33±0.15	30.68±1.90
May	0.91±0.01	40.21±0.20	3.92±0.09	29.82±2.30
June	1.23±0.01	44.31±0.09	5.07±0.03	32.81±1.64
July	1.31±0.04	44.63±0.40	4.43±0.01	32.43±1.23
August	1.50±0.03	49.61±0.45	4.74±0.04	34.00±1.53
September	1.65±0.05	50.33±0.10	5.42±0.02	35.00±0.95
October	1.80±0.04	50.43±0.06	5.36±0.02	34.80±1.96
November	2.00±0.09	51.30±0.09	5.68±0.01	32.42±0.75
December	1.52±0.04	47.46±0.10	4.81±0.04	32.27±1.75

The total ionic concentrations that mainly determine the osmotic pressure of the hemolymph are 242.77 and 93.53 mM/l for *Helix* and *Planorbis*, respectively.

THE IONIC COMPOSITION OF THE NERVOUS SYSTEM IN MOLLUSCS

The main difficulty in the photometric determination of the concentration of inorganic ions in the neurons of the nervous system is the problem of the value of the intercellular space in the ganglia. The determination is also approximate because by this method we cannot differentiate ions contained in different types of neurons and in glial cells.

Zelenskaya, Oleinikova and Sorokina (1967) have made measurements of intercellular space using several methods (such as determination of the rates of inflow and outflow of tracer sodium and sulphate and measurements of chloride, inuline and sucrose distribution). Such methods gave the following values:

	Planorbis	Helix
chloride	32.12%	32.23%
sucrose	33.22%	38.20%
sulphate	25.70%	30.96%
inuline	4.90%	4.50%
sodium	42.85%	41.82%

The inuline method is obviously incorrect because it gives very low values which are quite different from those of the other methods. On the basis of data given by the other methods, the mean values of the intercellular spaces for the ganglia of *Helix* and *Planorbis* can be calculated as 37.4% and 34.1%, respectively. Taking these values, the mean concentrations of inorganic cations in cells of the nervous system of the mollusc are shown in Table 3 (Sorokina and Kholodova 1967). Obviously, there is clear asymmetry in the distribution of sodium and potassium ions between the inside and outside of the cells; the differences in transmembrane concentration gradients (especially for potassium) between both species are quite marked. There is also a very high concentration of calcium and magnesium in the central nervous system; their total amount is almost equal to the amount of potassium. It is highly improbable that the neuroplasm contains such large amounts of divalent cations even under bound conditions. A certain

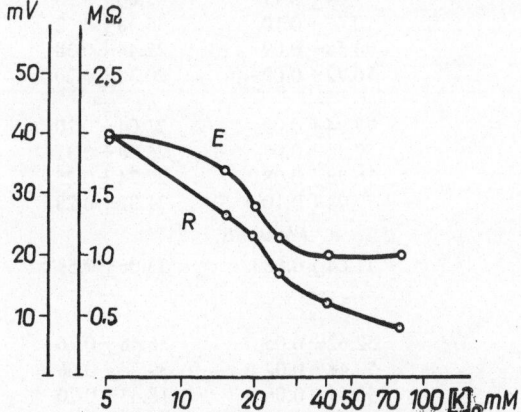

FIG. 3. Relations between external potassium concentration and resting potential (E) and membrane resistance (R) in a giant neuron of *Helix pomatia*. Potassium concentration was increased by equimolar substitution of sodium

amount of such ions is probably deposited outside the cells. It was found in several freshwater invertebrates that the connective tissue contains calcium granules which can act as calcium reservoirs.

It is important to stress that the amount of cations indicated above is only to a small extent compensated in the neurons by inorganic anions (chloride, sulphate, and phosphate). The deficit of inorganic anions in *Helix* is about 156 mequiv, and in *Planorbis* about 90 mequiv/kg cell water. This deficit can partly be attributed to phospho-organic compounds. The deficit in free anions can be also covered by fixed negative charges of proteins.

Seasonal variations of intracellular ionic composition were not studied in detail. Potassium concentration remains at approximately the same level throughout the year. Sodium concentration in neurones is minimal in summer. In winter and especially in spring the sodium concentration increases relative to the summertime value by a factor of 2–3, and the sodium transmembrane gradient decreases. This is obviously compensated by an increased calcium gradient, but it is impossible to calculate the latter because of uncertainty about the status of calcium in tissue.

TABLE 3

Ion concentration in neurons calculated from flame-photometry data

Month	Ion concentration in neurons (mM/kg w. w.)		
	K^+	Na^+	Ca^{++}
Helix			
January	—	—	—
February	78.06±0.12	29.74±0.38	8.94±0.09
March	82.36±0.17	38.76±0.52	12.62±0.15
April	75.64±0.17	43.80±0.58	10.38±0.23
May	70.46±0.10	36.38±0.62	3.56±0.08
June	74.58±0.09	22.44±0.38	3.38±0.05
July	76.72±0.09	20.38±0.20	3.98±0.04
August	—	—	—
September	77.94±0.09	21.04±0.20	5.63±0.02
October	77.30±0.05	25.38±0.31	10.38±0.11
November	76.78±0.08	27.84±0.25	4.43±0.01
December	79.74±0.10	31.82±0.28	5.36±0.04
Planorbis			
January	71.64±0.08	33.06±0.59	9.00±0.08
February	—	—	—
March	—	—	—
April	52.62±0.05	36.48±0.76	3.64±0.07
May	50.48±0.07	39.74±0.68	4.38±0.05
June	53.81±0.06	18.41±0.70	8.72±0.03
July	54.60±0.03	13.64±0.32	10.86±0.03
August	57.20±0.08	11.40±0.15	9.92±0.04
September	62.10±0.06	10.94±0.10	9.73±0.01
October	65.54±0.04	11.82±0.18	4.38±0.08
November	68.78±0.03	12.96±0.20	6.58±0.11
December	68.41±0.09	22.73±0.32	6.89±0.11

ACTIVITY OF IONS IN THE NEUROPLASM

The values of intracellular ionic concentrations can give us only a preliminary idea about the ionic forces that are causing active processes in a nerve cell. It would be much more important to measure directly the activity of ions in the protoplasm. For such measurements, intracellular microelectrodes were prepared from hydrogen, sodium, and potassium glass; the outside tip diameters of these electrodes varied from 0.5 to 1.0 μ. The microelectrodes were covered with glass insulation (except for the tip several microns in length). The selectivity constant of the hydrogen glass NCS 22–6 is so high ($1.2 \cdot 10^{-7}$) that such electrodes have no sodium or potassium function over the pH range from 1 to 10. The measurements with such electrodes were therefore quite simple. The hydrogen-selective electrodes were calibrated in standard buffer solutions and could be used for measurements several times. The selectivity constant of the sodium glass NAS 11–18 is lower: in the pH range from 6 to 8 it varies from 0.01 to 0.015, but it is still sufficiently high to avoid considerable errors from the presence of other cations in the medium. The potassium glass KABS 20–9–5 has the lowest selectivity (0.11–0.23). Therefore, potassium electrodes cannot be used, for example, for measurements of the activity of potassium ions in the hemolymph because of the presence of large amounts of sodium. Sodium and potassium electrodes were calibrated in mixed salt solutions for the determination of their action region.

For work with intracellular cation-selective microelectrodes, the location of the reference electrode is important. Usual open-tip microelectrodes filled with a potassium chloride solution and connected to a silver-silver chloride half-cell were used as reference electrodes. The most convenient position for such electrodes was inside the cell under investigation; however, a regular introduction of two such electrodes into the same cell was quite difficult. Therefore, the reference electrode was usually placed extracellularly, and in a special series of measurements the mean value of the resting potential of the investigated cells was determined. The values of the resting potential were then subtracted from the value of the potential difference between the cation-selective and reference electrodes.

Measurements made by Sorokina (1966) have shown that the mean value of potassium ion activity (a_k) is 73 $\pm$ 0.4 mM in *Helix* and 34$\pm$0.2 mM in *Planorbis* (data from measurements on 72 and 90 cells, respectively, during summer-autumn). The activity of sodium ions is, respectively, 13.0$\pm$0.01 mM and 8.0$\pm$0.02 mM (data from 24 and 18 measurements). The results of such measurements are presented in more detail in Table 4. The ratio between the activity of potassium or sodium ions and their analytical concentrations can be considered as activity coefficients of these ions; they are 0.78–0.73 for potassium and 0.53 for sodium. Such coefficients are only approximate, because we cannot be quite sure that the decrease of activity compared to the activity in a water solution of the same concentration found by a glass microelectrode indicates a real decrease of the energy of the ions. The decrease in such coefficients can also be a result of the isolation of a certain amount of ions in a closed intracellular space inaccessible to the

TABLE 4

Ion activity in neurons measured with intracellular cation-selective glass microelectrodes

Helix pomatia

Expt. No	N.M.	MP (mV)	a_k	γ_k	C_k (M/kg. f.w)
1	10	53.6	0.073	0.78	
2	5	54.6	0.073	0.78	
3	12	52.4	0.072	0.77	
4	10	60.7	0.074	0.79	
5	5	50.9	0.073	0.78	
6	12	58.6	0.074	0.79	
7	10	54.3	0.072	0.77	
8	5	52.4	0.073	0.78	
Mean ± S.D.		54.7±0.8	0.073±0.0004	0.78	0.0933±0.0013

Expt. No	N.M.	MP (mV)	a_{Na}	γ_{Na}	C_{Na} (M/kg. f.w.)
1	3	65.2	0.013	0.42	
2	8	50.4	0.012	0.40	
3	2	58.7	0.012	0.40	
4	3	53.4	0.012	0.40	
5	3	52.6	0.013	0.42	
6	2	55.8	0.013	0.42	
7	3	52.9	0.013	0.42	
Mean ± S.D.		54.2±0.6	0.013±0.00001	0.42	0.03106±0.00109

Planorbis corneux

Expt. No	N.M.	R.P. (mV)	a_k	γ_h	C_k (M/kg. f.w.)
1	10	51.3	0.039	0.73	
2	15	50.7	0.039	0.73	
3	25	55.8	0.038	0.72	
4	5	53.3	0.039	0.73	
5	10	56.7	0.038	0.72	
6	10	53.4	0.039	0.73	
7	10	55.2	0.037	0.73	
8	5	50.1	0.039	0.73	
Mean ± S.D.		53.3±0.6	0.039±0.0002	0.73	0.0534±0.0010

Expt. No	N.M.	MP (mV)	a_{Na}	γ_{Na}	C_{Na} (M/kg. f.w.)
1	3	51.7	0.007	0.42	
2	2	50.8	0.008	0.48	
3	3	49.4	0.008	0.48	
4	3	55.6	0.008	0.48	
5	3	53.6	0.008	0.48	
6	4	55.0	0.007	0.42	
Mean ± S.D.		52.7±0.8	0.008±0.00002	0.46	

measuring selective electrode, or of partial chemical binding. But at least it is certain that the status of potassium and sodium ions inside giant neurons is different. Such a difference is now also established on several other objects — for example, striated muscle fibers and giant axons. The potassium ions in the protoplasm, in contrast to the sodium ions, are electrochemically free and can participate completely in the generation of a resting potential. Mean equilibrium potentials calculated on the basis of such measurements are −74 mV for *Helix* and −87 mV for *Planorbis* neurons.

Resting potential measurements of giant neurons carried out in several groups gave a much lower value: about −50 mV. This value changes from about −60 to −40 mV in different seasons, being minimal in spring (see Fig. 4). Thus, the membrane of a giant neuron, as the membranes of most other excitable cells, never has the properties of an ideal potassium electrode; even in the resting state it obviously has a considerable sodium permeability, and sodium ions passing through the membrane are polarizing it in the opposite direction. Using the Hodgkin–Horowicz equation for calculation of the relative sodium permeability of the membrane.

$$E_m^{\cdot} = \frac{RT}{F} \, ln \, \frac{a\,K_{in} + \alpha a\,Na_{out}}{a\,K_{out} + \alpha a\,Na_{in}}$$

we obtain mean values for α of 0.042 for *Helix* and 0.071 for *Planorbis*. They increase in spring and decrease later in summer. The increase in intracellular sodium concentration is probably a reflection of this change in relative membrane permeability. Mean equilibrium potentials for sodium ions calculated from activity-measurement data are +37 mV and +39 mV for *Helix* and *Planorbis*, respectively; at the present time, we have no data about their seasonal variations.

Measurements of chloride ion activities in *Helix* cells were made by Kerkut and Meech (1966). Their measurements indicated very interesting differences of chloride activity in *H* and *D* cells. In the first, activity was 8.7±0.4 mM, and in the second, 27.5±1.5 mM. Correspondingly, there must be a big difference in chloride equilibrium potentials; in *H* neurons, −59.4±1.4 mV and in *D* neurons −39.4±0.8 mV.

It would be very important to determine the relative permeability of the resting membrane to potassium and chloride ions, as was done for striated muscle fibers. On the basis of experiments in which chloride ions in the outside solution were replaced by ions with larger diameters (Gerasimov, Kostyuk and Maisky 1965c), the conclusion can be made that the role of chloride conductance in the whole resting conductance is not high; such replacement brings about a small increase in membrane resistance (mean increase 16%) and insignificant hyperpolarization.

The behavior of the cell membrane is much closer to the behavior of an ideal potassium electrode when the potassium concentration in the outside solution is increased. Such an increase produces a large decrease in membrane resistance (Figs. 2 and 3) obviously due to an increase in potassium conductance. In such conditions, the relation between the logarithm of the external potassium concentration and the resting potential becomes linear with a slope close to the theoretical for a potassium electrode (Kostyuk 1964; Gerasimov, Kostyuk and Maisky 1965d). Identical data were recently obtained by Kerkut and Meech (1967).

Finally, measurements with a hydrogen microelectrode of hydrogen ion activity in the protoplasm of giant neurons gave pH value of 7.26 ± 0.08 for *Helix* and 7.26 ± 0.10 for *Planorbis* (Sorokina 1965). The pH values are very stable and are subjected to only small seasonal variations. Such variations are much less than the seasonal variations of the resting potential (see Fig. 4). Only in animals under poor conditions (for instance, those kept in the laboratory throughout the winter) did the value of intracellular pH decrease, sometimes reaching 6.9–6.8.

Several measurements were made on neurons with two microelectrodes (hydrogen-selective and usual microelectrodes) remaining inside for a long time (from 2 to 5 h). In such experiments there was an especially clear absence of relation between resting potential and intracellular pH. Despite a slow decrease in resting potential, the value of intracellular pH remained almost constant (Fig. 5). This indicates that in giant neurons (as in other cells) the value of intracellular pH is not regulated by passive redistribution of hydrogen and hydroxyl ion through the membrane and is governed by some other mechanism, possibly by active transport.

PHYSICAL PROPERTIES OF THE CELL MEMBRANE
IN GIANT NEURONS

The surface membrane of giant neurons has specific physical properties, manifested by a very large time-constant. The input resistance of such neurons varies from 1 to 34 MΩ. If we approximate the soma as a sphere with one axon and measure its diameter, we can calculate the specific resistance and capacity of the membrane; for *Helix* neurons they are $987\pm260\ \Omega\cdot cm^2$ and $29.8\pm7.2\ \mu F/cm^2$

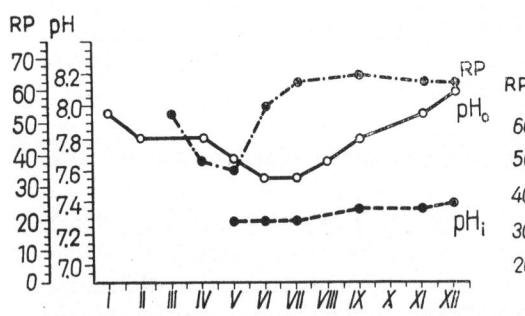

FIG. 4. Seasonal changes of resting potential (RP), pH of hemolymph (pH$_o$), and intracellular pH (pH$_i$) in ganglia of *Helix pomatia*

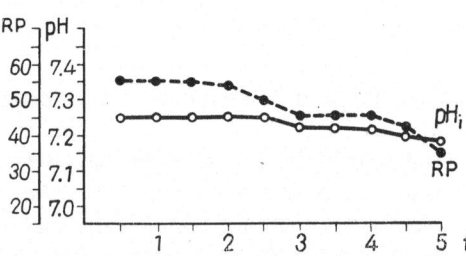

FIG. 5. Changes of resting potential (RP) and intracellular pH (pH$_i$) of a single cell of *Helix pomatia*; t is the time in hours after the introduction of electrodes into the cell

respectively (Maisky 1963). These values are close to the values obtained by Fessard and Tauc (1956) for *Aplysia* neurons — 2200 $\Omega \cdot cm^2$ and 23 $\mu F/cm^2$. Thus, the possible reason for the large time constant of the membrane is a high specific capacity, connected with some unknown properties of its molecular structure.

It is important to stress that such calculations are very approximate. The surface of the neuronal soma is not a simple sphere. Electron-microscopic pictures reveal that the membrane is highly folded, and, consequently, the real value of the cell surface cannot be calculated by a simple geometrical method (Maisky and Khomutovsky 1965).

Normally, the surface membrane of giant neurons has no marked rectifying properties. The membrane depolarized by high external potassium also does not show any rectification; the presence or absence of chloride ions does not play any role here. Thus, this membrane is quite different from the membrane of striated muscle fibers, which reveals strong anomalous rectification in solutions with high potassium content. Marked delayed rectification appears in the membranes of neurons in sodium-free solutions with blocked spike-generation and of neurons narcotized with nembutal (Gerasimov, Janishevsky and Skubalianka 1967).

ACTION POTENTIAL PRODUCTION IN GIANT NEURONS

Both the synaptic and the direct stimulation of giant neurons are well known and do not require special description. The spike appears when the depolarization reaches the critical level which is the same for synaptic and direct stimulation (see Gerasimov, Kostyuk and Maisky 1964a). The mean spike amplitude in *Helix* neurons is 94.4$\pm$8.6 mV (Gerasimov and Maisky 1963), in *Planorbis*,

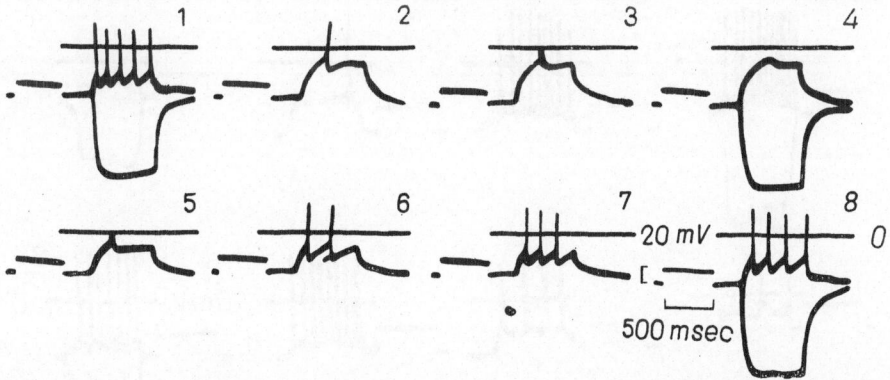

FIG. 6. Responses of giant neuron of *Lymnaea stagnalis* in an isotonic calcium chloride solution. On each record, responses to a depolarizing (0.3 · 10⁻⁸ A) and hyperpolarizing (1 · 10⁻⁸) current are superimposed. Currents were passed through one of the intracellular microelectrodes. Osc. 1 — response in normal saline, osc. 2–4 — in isotonic calcium chloride after 1, 2 and 3 min of perfusion, osc. 5–8 — during washing with normal saline for 1, 2, 3 and 4 min

86.9 ± 5.8 mV (Maisky and Gerasimov 1964), and in *Tritonia*, about 80 mV (Gera-simov and Maisky 1965). The maximal value of the ionic current during the rising phase of the action potential in *Planorbis* neurons reaches $2.0 \cdot 10^{-7} - 2.8 \cdot 10^{-7}$ A (Magura 1966).

However, there is an important difference in spike generation by cells of different species, which can be revealed if the cells are placed in solutions with altered ionic composition.

The production of action potentials in neurons of *Planorbis* and *Lymnaea* is connected in the usual way to the presence of sodium ions in the external medium. If the ganglion is immersed in a sodium-free solution, the spike amplitude rapidly diminishes, the number of spikes in a repetitive response decreases, and their duration increases. After 3–5 min, the spikes disappear completely, and only small local responses remain (Fig. 6). At the same time, the membrane becomes hyperpolarized. Finally, the cells lose excitability completely, despite the high resting potential and normal membrane resistance. Reimmersion in normal saline solution produces a transient depolarization of the membrane and complete restitution of excitability (Gerasimov, Kostyuk and Maisky 1964b).

Sometimes the cells of these molluscs cannot tolerate the immersion in a so-dium-free solution − after a transient hyperpolarization they depolarize, and the resistance of the membrane decreases (see Fig. 7). The generation of action potentials also stops. However, these changes are completely reversible, and the normal functional status of the neurons is recovered after perfusion with normal saline solution.

The same is characteristic of the action potentials of the deep-sea molluscs *Tritonia*, although the excitability in this case was lost with longer delay after

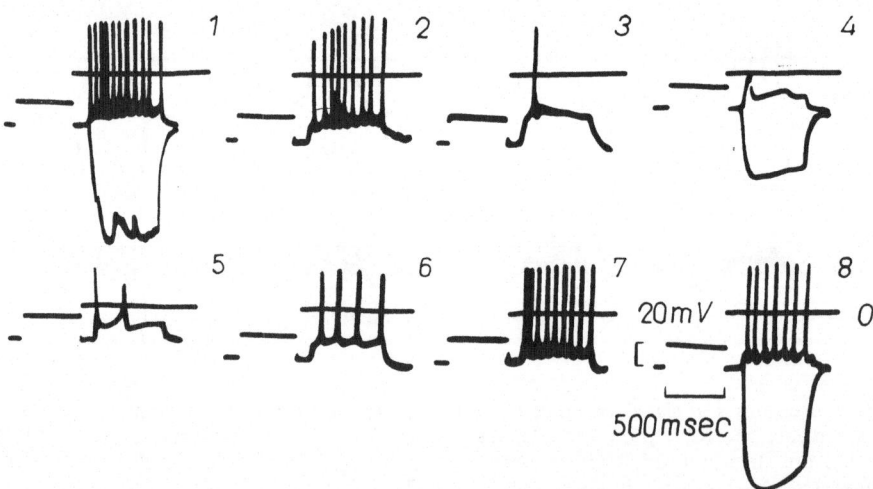

FIG. 7. Response of a giant neuron of *Planorbis corneus* in isotonic calcium chloride solution.
(Explanation the same as for Fig. 6)

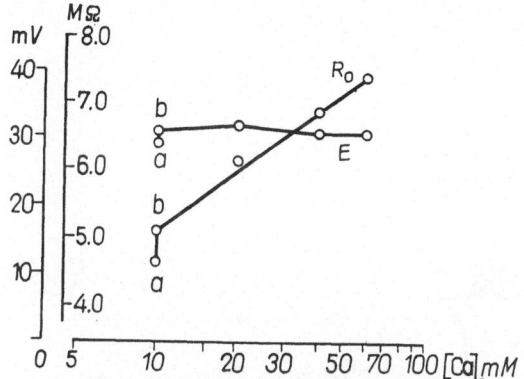

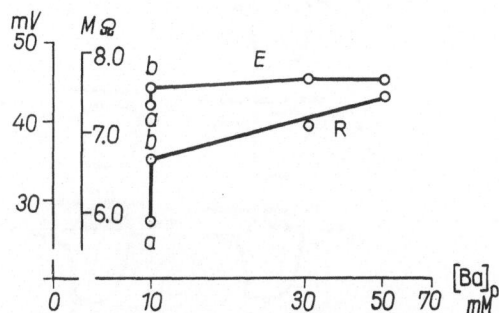

FIG. 8. Relation between external calcium concentration and resting potential (E) or membrane resistance (R_o) in a giant neuron of *Helix pomatia*. The change from 'a' to 'b' occurred during the substitution of sucrose for sodium chloride in the external solution; the necessary amount of $CaCl_2$ was then substituted for sucrose

FIG. 9. Relation between external barium concentration and resting potential (E) or membrane resistance (R) in a neuron of *Helix pomatia*. (The same procedure was followed as in Fig. 8)

immersion in sodium-free solution (Magura and Gerasimov 1966; Veprintsev *et al.* 1967).

The processes in *Helix* neurons are quite different. In this case, the neurons continue to produce normal action potentials many hours after the replacement in the external solution of sodium ions with choline-chloride or sucrose (Gerasimov 1964). Such action potentials are definitely dependent on the presence of calcium ions in the external solution, and therefore we have studied the effect of calcium ions on *Helix* neurons in more detail (Kostyuk 1964; Gerasimov, Kostyuk and Maisky 1965e). As was indicated above, the replacement of normal saline solution with a sodium-free solution containing usual amounts of calcium and potassium ions produces a hyperpolarization of the membrane (Fig. 8), and the new level of the resting potential is reached in 2–3 min. A subsequent increase in calcium concentration to 20 mM brings a small additional hyperpolarization (mean value 4.1 ± 0.8 mV). If the calcium concentration is further increased (to 60 mM) there is no parallel increase in resting potential, and the membrane may start to depolarize. The same effect can be produced if barium ions are added to the solution instead of calcium (Fig. 9). In contrast to the small changes in resting potential, the resting resistance changes greatly if calcium concentration is increased. In the concentration range studied (up to isotonic), the membrane resistance increases proportionally to the logarithm of calcium concentration (see Fig. 8). The slope is different in different cells and reaches several megohms per logarithmic unit of concentration. An increase in barium concentration produces similar changes in membrane resistance. All changes are completely reversible when the calcium concentration is returned to normal.

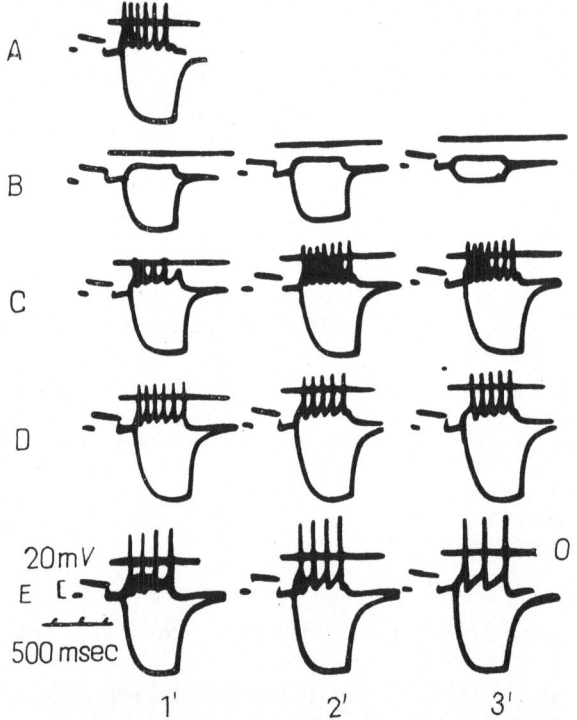

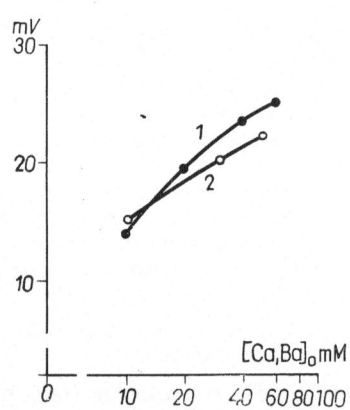

Fig. 10. Response of a giant neuron of *Helix pomatia* in solutions with different calcium chloride concentrations. Recordings were taken as in Fig. 6. A — in normal saline solution; B — 1, 2 and 3 min after immersion in calcium-free solution; C — again in normal saline solution; D — in sodium-free solution with 10 mM CaCl₂; E — in isotonic calcium chloride solution

Fig. 11. Relation between external calcium (curve 1) or barium (curve 2) concentration and the spike overshoot. A giant neuron of *Helix pomatia*

Against the background of such changes in the properties of the resting membrane, characteristic changes in the action potentials appear. If calcium is removed from the external medium, the membrane depolarizes and its resistance decreases; at the same time, there is a rapid decrease in spike amplitude. In a few minutes, it becomes impossible to produce spikes by direct depolarization of the cell (Fig. 10B). In contrast to this, an increase in calcium concentration in the external solution brings an increase in the spike amplitude. The overshoot of the spike in calcium solutions is almost linearly dependent on the logarithm of the external calcium concentration (Fig. 11). The slope of this relation is different in different cells, but in several cells it reaches the theoretical level for a calcium electrode (29 mV per logarithmic unit of calcium concentration). We observed no changes in the overshoot that exceeded the value theoretically possible for a calcium electrode (the change in the whole amplitude of the spike could be larger because of membrane hyperpolarization under the influence of calcium ions).

Of course, critical conditions for spike generation change in such solutions because of an increase in resting potential and especially in the resistance of the membrane (for instance, the threshold level of depolarization increased). There is also a specific change in subsequent spikes during a rhythmic discharge. Usually, a prolonged depolarizing pulse produces a rhythmic discharge with single spikes of equal amplitude. In a sodium-free solution, there is, with increased calcium concentration, a progressive increase in the amplitudes of subsequent spikes (see Fig. 9). During a depolarizing pulse 1-sec long, the amplitude of spikes can increase up to 10 mV.

Changes of action potentials in solutions with different calcium concentration are completely reversible. A change from a solution with a high calcium content to a solution with a low one also restored the normal shape of subsequent spikes in a rhythmic discharge.

Our findings were recently confirmed by Meves (1966). In his experiments with tetrodotoxin, it was shown that 'calcium' spikes, in contrast to 'sodium' spikes, are not blocked by this drug. 'Calcium' spikes are also much less sensitive to the narcotic action of nembutal (Gerasimov and Janishevsky 1967).

It would be very important to find out what ions produce the action potential in *Helix* neurons in the natural ionic medium (in hemolymph). The effect of calcium-free solutions seems to show that calcium ions are also very important for spike generation during the presence of sodium ions; however, we cannot exclude the possibility that calcium ions are really carrying charges only during spike generation in sodium-free solutions, and that the ionic mechanisms of action potential generation can to some extent vary if the cells are immersed in different saline solutions. Not all cells in the ganglion are producing action potentials in sodium-free solutions; some of them can lose excitability in such conditions and an increase in calcium concentration does not restore it. We have no data to indicate to which functional groups the cells with different sodium sensitivity belong.

In any case, it seems from such experiments that the membrane of *Helix* neurons has a less ion-specific ion-carrying mechanism for the depolarization during excitation than most other excitable cells, probably because of large seasonal variations of the ionic composition of the hemolymph and a corresponding variability of the transmembrane ionic gradient. Under different conditions, this mechanism can use both mono- and divalent cations of the extracellular medium as charge carriers.

Perhaps it is also possible to consider as an indication of certain reorganizations in the ionic mechanisms of action potential generation, the rapid changes in spike amplitude during rhythmic discharge in high-calcium solutions. The progressive increase in spike amplitude occurs without increase in their duration, which means it is due to an increase in the rate of rise of the action potentials. It seems that during a rhythmic discharge under such conditions a progressive increase

P. G. KOSTYUK

develops in the effectiveness of the divalent-ion-carrying mechanism, and the potential difference at the activated membrane has time before inactivation to come closer to the equilibrium potential.

Very special changes in action potential generation occur in giant neurons if barium ions are introduced into the extracellular medium (Gerasimov, Kostyuk and Maisky 1965b). In such conditions, the critical depolarization produces, instead of the usual action potential, so-called 'prolonged' action potentials. The cell membrane becomes fixed in an excited (depolarized) state for a long period of time (many seconds) and only after this delay returns to the resting state.

The addition of barium ions to the medium produces such changes in all investigated neurons of *Helix* and *Planorbis*, although the effectiveness of this action of barium ions is different in different species. In *Helix* neurons, the transition from normal to prolonged action potentials starts when the barium concentration reaches 30 mM, but in *Planorbis*, 4 mM is enough for the appearance of prolonged action potentials. Prolonged action potentials in *Helix* neurons can also occur for long periods in sodium-free solutions.

The transition from normal to prolonged action potentials takes 2–3 min, and during this period, several transitory forms of action potentials can be observed (see Fig. 12). The rising phase of the action potential is not changed, but on the descending phase, a plateau appears with slowly increasing duration. Initially this plateau disappears together with the break of the direct depolarizing current, but later on, the mechanism producing it becomes so stable that the plateau remains many seconds after cessation of the stimulation.

During the prolonged action potential, the potential difference at the membrane is fixed approximately at the zero level; after the end of the action potential, a prolonged hyperpolarization sometimes occurs (up to 1-sec long) with a maximal value of the undershoot of up to 20 mV.

For the estimation of changes in the ionic conductance of the membrane during a prolonged action potential, its resistance was measured by short pulses of inward (anelectrotonic) current passed through one of the intracellular microelectrodes. During the plateau, the resistance of the membrane is greatly decreased, to a level of 20–30% from the resting value. The greatest decrease in resistance is

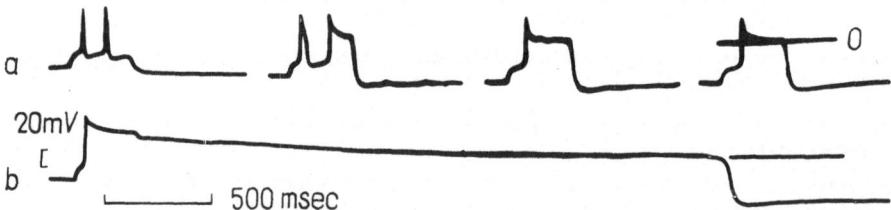

FIG. 12. Changes in action potentials of a *Planorbis* neuron 1 (osc. 2), 2 (osc. 3), 3 (osc. 4) and 4 (osc. 5) min after the beginning of perfusion with solution containing 20 mM barium chloride. Osc. 1 was taken before the action of barium. Responses were produced by direct depolarization of the cell membrane through one of the intracellular electrodes; the duration of the current pulse is identical in all cases

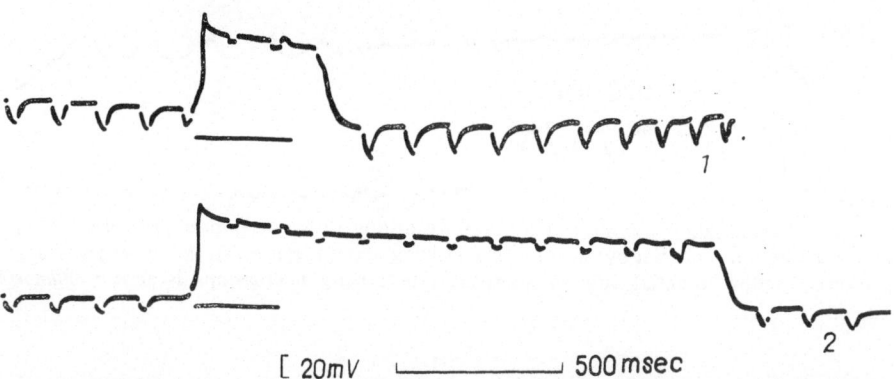

FIG. 13. Membrane resistance measurements in *Planorbis* neurons during prolonged action potentials. Strength of the measuring inward current pulses is 10^{-8} A. The prolonged action potential was produced by direct depolarization (duration of the depolarizing current indicated by the bar below each recording). Barium concentration 4 mM (upper recording) and 20 mM (lower recording)

observed at the beginning of the action potential; later, it slowly increases in parallel with slow repolarizing changes (Fig. 13). It is possible to abolish the prolonged action potential by passing hyperpolarizing pulses through the membrane during the plateau. Sometimes, such abolition can be produced even by those weak pulses (10^{-8} A) that are used for the resistance measurements. The hyperpolarizing pulses produce a rapidly-rising active hyperpolarization ('repolarizing response') which can be transient or steady, returning the membrane potential to the resting level. Examples of active repolarizing responses are shown in Fig. 14. In this figure, an usual prolonged action potential of the same cell is also shown; in this case, spontaneous repolarizing waves appeared at the end of the response, but hyperpolarizing pulses abolished the plateau almost three times as rapidly.

The active nature of the repolarizing response can also be proved by the fact that a refractory period develops after it − a second hyperpolarizing pulse is usually ineffective if applied after a short interval.

The large decrease of membrane resistance during the plateau of the action potential obviously indicates that barium ions are blocking the inactivation of the ion-carrying mechanism in the membrane. The activation and inactivation processes for the inward ionic current, usually closely coupled, become uncoupled during the prolonged action potential. Since such action potentials can be produced both in sodium-containing and sodium-free solutions, it can be concluded that this action of barium ions does not depend on the nature of the current-carrying ion.

The uncoupling of activation and inactivation processes for the inward current is not the only change produced by barium ions in the action potential mechanism. If the conduction changes were only of this kind then the membrane potential would be fixed during the plateau at the level of the maximum spike; in fact, it is

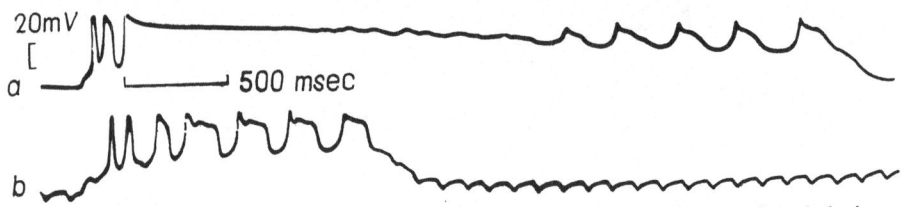

FIG. 14. Repolarizing responses. The lower recording shows responses produced during a prolonged action potential by short hyperpolarizing current pulses. In the upper recording, a prolonged action potential with spontaneous repolarizing fluctuations is shown. *Planorbis* neuron

fixed near the zero level. This is probably a result of steady parallel increases in the membrane permeability to some other ions — most likely to potassium — displacing the membrane potential in the repolarizing direction.

AFTER-POTENTIALS IN THE GIANT NEURONS

The development of the action potential in the giant neuron is always followed by a large after-hyperpolarization, the duration of which is about 200–400 msec, and during which the ionic conductance of the membrane is increased. Resistance measurements give a decrease of about 20% (Gerasimov, Kostyuk and Maisky 1964a). The membrane potential during the after-hyperpolarization is displaced toward the potassium equilibrium potential, indicating a predominant increase in potassium permeability. However, Chiarandini and Stefani (1967) have recently shown that the giant neurons differ in this respect: in some of them, the membrane potential comes very close to the potassium equilibrium potential during the after-potential, but in others it remains less by about 10 mV. It is suggested that giant neurons undergo at least two types of permeability changes during after-potentials: (1) an increase in potassium permeability only and (2) an increase in both potassium and chloride permeability.

In certain conditions, it is possible to produce a hyperpolarizing response connected with an increase in potassium permeability of the membrane without a preceding action potential (Gerasimov, Kostyuk and Maisky 1965a). A long-lasting hyperpolarizing response (about 250-msec long) occurs after the break of a strong hyperpolarizing current (Fig. 15). The active nature of this response can be seen from the fact that there exists a hyperpolarizing threshold necessary for its origin. Its development begins when the anelectrotonic potential falls to the level of 60–70 mV; this level corresponds to the maximal value of undershoot during the usual after-hyperpolarization. There is a strong decrease of membrane resistance during the hyperpolarizing response — up to 50% of its original value. If the membrane potential is artificially shifted by direct current in the hyperpolar-

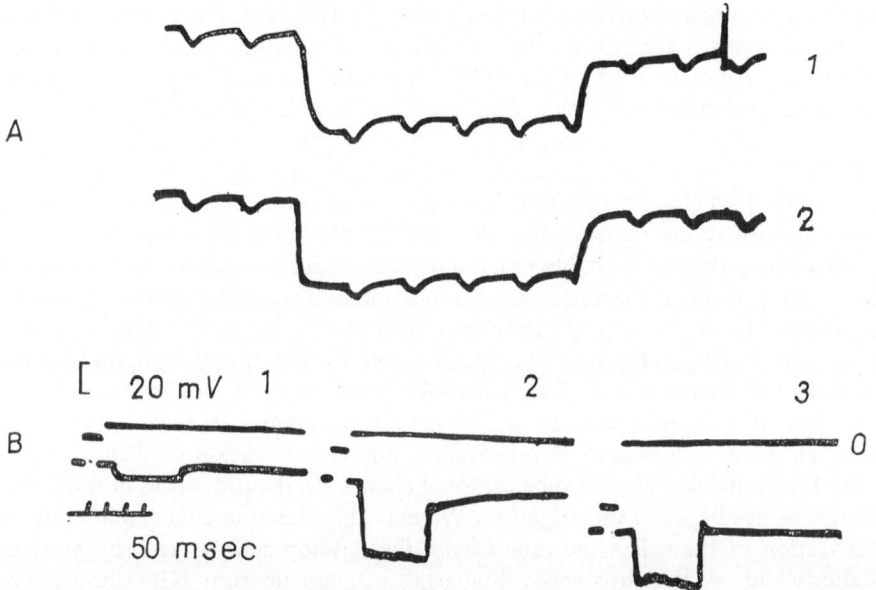

FIG. 15. Postanodal hyperpolarization in *Planorbis* neurons. A — resistance measurements by short inward current pulses in the presence of postanodal hyperpolarization (osc. 1) and without it (osc. 2) B — changes of postanodal hyperpolarization during steady shift in membrane potential. Osc. 1–2 — normal resting potential and different strength of the anodal pulse; osc. 3 — increase in resting potential by direct current passed through one of the intracellular electrodes

izing direction to the level of 60–70 mV, then the hyperpolarizing response disappears; with further steady hyperpolarization of the membrane, it changes into a depolarizing response. Thus, the equilibrium potential for this response is below the potassium equilibrium potential; the increase in ionic conductance during the response probably has a mixed nature.

DISCUSSION

Most important for the functioning of the nervous system, particularly in molluscs (and especially in terrestrial forms), is the instability of the ionic content of their internal medium (hemolymph). The variations of the ionic content are quite complex and can be manifested in separate increases of monovalent ions (during the summer-autumn period) or of divalent ions (during spring). The large increase in divalent ion concentration is obviously characteristic for molluscs and connected with the construction of the shell.

The functional properties of the central neurons in molluscs are also not constant and are subjected to seasonal variations. The mean value of the resting potential (and, consequently, of cell excitability) changes during the year, being smallest in spring and increasing toward summer. However, it is obvious that such functional changes are not directly connected with changes in external ionic concentrations; depolarization of the cell membrane is maximal in periods when the potassium concentration outside is minimal, and not vice versa. Our data does not seem to confirm the suggestion made by Kerkut and Meech (1967) on the basis of indirect evidence that the reasons for such changes in resting potential are for the most part changes of intracellular potassium concentration. At least photometric investigations indicate that the intracellular potassium concentration is approximately steady. Probably much more important are changes in the selective permeability of the membrane toward potassium and sodium. As far as is possible to conclude from the largely increased entrance of sodium into the neurons in spring, the P_K/P_{Na} ratio must be much lower during this period than during summer. This increase in sodium permeability of the resting membrane can be the immediate reason for the decrease in the resting potential level which is masking the small direct influence of changed-potassium concentration gradients.

The functional meaning of such seasonal changes in the properties of the resting membrane need special investigation. We can only mention that a moderate depolarization of the cell membrane favors the development of autorhythmic oscillations and spontaneous spike discharges in giant neurons (Gerasimov, Kostyuk and Maisky 1964a).

The properties of the activated membrane are also variable in giant neurons; this variability is probably based on a lower specificity of the ionic mechanisms of spike-generation compared to animals with more constant internal media. The ion-carrying mechanisms of the activated membrane can use for charge transfer not only mono- but also divalent ions, and this capability is well developed in molluscs with an especially large seasonal increase in concentration of the calcium transmembrane concentration gradient. It is quite possible that this property of ion-carrying mechanisms is important for the production of normal action potentials throughout the whole year.

The data also suggest several questions of general interest. Our experiments, as well as experiments of Kerkut and Meech (1967), have shown that the resting potential during each seasonal period is at a very stable value, and is not very dependent on wide-range changes in external potassium concentration. Only after a large increase in external potassium, enough to cause an enormous increase in the ionic conductance does the membrane function come close to the function of a potassium electrode. Membranes of other excitable cells are much more sensitive to changes in potassium concentration gradient. It would be very important to know the mechanism which secures such a high stability of membrane polarization. A possible explanation could invoke a predominant role for chloride ions in supporting the resting potential, but our data indicate that the chloride conductance in resting conditions is relatively low. The maintenance of a steady resting potential in solutions with different potassium concentrations also does not depend on the presence or absence of chloride ions in the solution. Thus, it is possible that the main factor here is also a changing relation between the resting potassium and sodium permeabilities of the membrane, with their balance keeping the resting potential at a steady level within certain limits.

No less important is the nature of the mechanism which allows the use of different ions as charge carriers during the generation of an action potential. It is possible that ions can be carried through the same channels, but arguing against this suggestion are data concerning the different actions of pharmacological agents on 'sodium' and 'calcium' spikes. Therefore, we cannot exclude the possibility that in the same electrically excitable membrane, different channels can exist for inward movements of ions, and that the quantitative relations between such channels are not fixed and can be regulated by the composition of the external ionic medium.

SUMMARY

1. The ionic composition of the hemolymph in molluscs shows considerable seasonal variation. Such variations are larger in the terrestrial *Helix* than in the freshwater *Planorbis*. In *Helix*, they are manifested by a substantial increase in calcium concentration during spring and monovalent cation concentration during summer and autumn.

2. Photometric determination of the intracellular ionic concentration also shows variability. There is an increase in sodium and calcium content during spring, but the potassium concentration shows little change. The intracellular activity of potassium ions, as determined by intracellular selective-glass micro-electrodes, is close to its analytical concentration, but the activity of sodium ions is considerably lower than its analytical concentration.

3. The mean resting potential in giant neurons of molluscs is substantially lower than the potassium equlibrium potential. It changes according to the Nernst equation for a potassium electrode only for the considerably high external concentrations of potassium which bring about a large increase in ionic conductance of the membrane. The mean value of the resting potential is also subjected to seasonal variations, being minimal in spring for *Helix*.

4. The generation of action potentials in *Planorbis* is dependent upon the presence of sodium ions in the external solution, but in *Helix* it goes on in a sodium-free medium containing calcium ions. The properties of the activated membrane during the generation of 'calcium' spikes are close to the properties of a calcium electrode.

5. The introduction of barium ions in the external solution produces dramatic changes in the action potential of a giant neuron: the blockage of the inactivation of the increased ionic conductance of the membran and the production of prolonged spikes lasting for many seconds.

6. The generation of an action potential in a giant neuron is always followed by a large after-hyperpolarization; this conclusion is based on marked increases in the ionic conductance of the membrane. Similar changes in the membrane can be produced under some conditions without a preceding spike.

7. Possible mechanisms of seasonal variation in the properties of the resting membrane of giant neurons and the nature of differences in spike generation in neurons of different species of Mollusca are discussed.

REFERENCES

CHIARANDINI, D. J. and STEFANI, E. (1967): Two different ionic mechanisms generating the spike "positive" after-potential in molluscan neurones. *J. gen. Physiol.* **50** 1183—1200.

FESSARD, A. and TAUC, L. (1956): Capacité, résistance et variations actives d'impédance d'un soma neuronique. *J. Physiol. (Paris)* **48** 541—544.

GERASIMOV, V. D. (1964): Influence of changes in the ionic composition of the medium on excitatory processes in giant neurones of Mollusca. *Fiziol. Zh. USSR* **50** 457—463.

GERASIMOV, V. D. and YANISHEVSKY, L. (1967): The influence of urethane on the electrical activity and the voltage-current relation of the membrane of different nerve cells of molluscs. In *"Evolutionary Neurophysiology and Neurochemistry"*. Leningrad 12—15.

GERASIMOV, V. D. and MAGURA, I. S. (1965): Electrical activity of giant neurons of the nudibranchiata molluscs *Tritonia diomedia*. *J. evol. biochem. physiol.* **1** 360—363.

GERASIMOV, V. D. and MAISKY, V. A. (1963): Electrical activity of giant nerve cells of the garden snail. *Fiziol. Zh. USSR* **49** 1099—1104.

GERASIMOV, V. D., JANISHEVSKY, L. and SKUBALIANKA, E. (1967): Rectifying properties of giant nerve cell membranes in solutions with different ionic composition. *Biofizika* **12** 97—103.

GERASIMOV, V. D., KOSTYUK, P. G. and MAISKY, V. A. (1964a): Peculiarities in generation of action potentials in the giant neurons of Mollusca. *Fiziol. Zh. USSR* **50** 1321—1328. (*Feder. Proc.* **24** T-763—T-767. 1965).

GERASIMOV, V. D., KOSTYUK, P. G. and MAISKY, V. A. (1964b): Excitability of giant nerve cells of various representatives of pulmonary molluscs in solutions free of sodium ions. *Bull. exsp. Biol. Med.* **58** 9 3—7. (*Feder. Proc.* 24, T 676, 1965).

GERASIMOV, V. D., KOSTYUK, P. G. and MAISKY, V. A. (1965a): Giant nerve cell responses to break of hyperpolarizing current. *Fiziol. Zh. USSR* **51** 703—710.

GERASIMOV, V. D., KOSTYUK, P. G. and MAISKY, V. A. (1965b): Prolonged action potentials of giant nerve cells. *Fiziol. Zh. USSR* **51** 1434—1441.

GERASIMOV, V. D., KOSTYUK, P. G. and MAISKY, V. A. (1965c): Ionic conductance of the membrane of a giant nerve cell of a snail. *Biofizika* **10** 82—89.

GERASIMOV, V. D., KOSTYUK, P. G. and MAISKY, V. A. (1965d): Changes of electrical characteristics of the giant neuron membrane during an increase in external potassium concentration. *Biofizika* **10** 272—280.

GERASIMOV, V. D., KOSTYUK, P. G. and MAISKY, V. A. (1965e): The influence of divalent cations on the electric characteristics of the membrane of giant neurons. *Biofizika* **10** 447—453.

KERKUT, G. A. and MEECH, R. W. (1966): The internal chloride concentration of H and D cells in the snail brain. *Comp. Bioch. Physiol.* **19** 819—832.

KERKUT, G. A. and MEECH, R. W. (1967): The effect of ions on the membrane potential of snail neurones. *Comp. Biochem. Physiol.* **20** 411—429.

KOSTYUK, P. G. (1964): Grundvorgänge in Riesen–Nervenzellen. *Nova Acta Leopoldina (Leipzig)* **28** M. 169, 65—72.

MAGURA, I. S. (1966): Quantitative characteristics of action potentials of the soma of giant neurons in *Planorbis corneus* during spontaneous rhythmic discharge. *Fiziol. Zh. Kiev* **12** 770—775.

MAGURA, I. S. and GERASIMOV, V. D. (1966): The action of calcium on the electrical activity of giant neurons of the molluscs *Tritonida diomedia*. *J. evol. biochem. physiol.* **2** 347—352.

MAISKY, V. A. (1963): Electrical characteristics of the surface membrane of giant nerve cells of *Helix pomatia. Fiziol. Zh. USSR* **49** 1468—1474.

MAISKY, V. A. and GERASIMOV, V. D. (1964): An electrophysiological investigation of giant neurones of some representatives of pulmonary Molluscs. *Bull. exp. Biol. Med.* **58** *12* 21—26.

MAISKY, V. A. and KHOMUTOVSKY, O. A. (1965): Certain features of the electrical response and submicroscopic structure in unipolar giant neurons of the mollusc *Planorbis corneus. J. evol. biochem. physiol.* **1** 351—359.

MEVES, H. (1966): Das Aktionspotential der Riesen-Nervenzelle der Winbergschnecke *Helix pomatia. Pflügers Arch. ges. Physiol.* **289** R–10.

SOROKINA, Z. A. (1965): Measurement of hydrogen-ion activity inside and outside nerve cells of mollusc ganglia. *J. evol. biochem. physiol.* **1** 343—350.

SOROKINA, Z. A. (1966): Activity of potassium and sodium ions in giant neurons of molluscs. *Fiziol. Zh. Kiev* **12** 776—780.

SOROKINA, Z. A. and KHOLODOVA, YU. D. (1967): Ionic composition of nervous ganglia in molluscs. *J. evol. biochem. physiol.* (in press).

SOROKINA, Z. A. and ZELENSKAYA, V. S. (1967): Peculiarities of electrolyte composition of molluscan hemolymph. *J. evol. biochem. physiol.* **3** 25—30.

VEPRINTSEV, B. N., GERASIMOV, V. D., KRASTS, I. V. and MAGURA, I. S. (1966): Influence of the ionic composition of a medium on the action potential of giant neurones of the nudibranchiate mollusc *Tritonia. Biofizika* **11** 1000—1007.

ZELENSKAYA, V. S., OLEINIKOVA, T. N. and SOROKINA, Z. A. (1967): The extracellular space in nervous ganglia of molluscs. *J. evol. biochem. physiol.* (in press).

DISCUSSION

J. Salánki: As your cells were not isolated, the changes in the ionic concentration of the bath fluid affected the other cells as well. Furthermore, this effect may, through synaptic transmission, influence the excitability of the investigated cells. What is your opinion about this?

P. G. Kostyuk: Since the recording from single cells was intracellular, we could easily observe synaptic influences from other cells, in the form of spontaneous postsynaptic potentials. All changes of resting potential described above were certainly produced by direct influences on the membrane.

Symposium on Neurobiology of Invertebrates 1967 (169—199)

ELECTRICAL PROPERTIES AND TEMPORAL ORGANIZATION IN OSCILLATORY NEURONS*

(APLYSIA)

A. Arvanitaki and N. Chalazonitis

Département de Neurophysiologie Cellulaire
Institut de Neurophysiologie et de Psychophysiologie
Centre National de la Recherche Scientifique.
et Laboratoire d'Electrobiologie, E.P.H.E.
Marseille, France

The most fascinating aspects of brain activity are periodic ones, recorded under various conditions: e.g., as recurring electrical patterns in 'spontaneous' activities, and as after-discharges following peripheral stimulation or ensuing from direct stimulation of the brain.

Although slow potential changes associated with bursts of spikes as recorded from the cortex were known for some time (Adrian and Matthews 1934; Adrian and Moruzzi 1939; Grey Walter 1959), the mechanism and the site of generation of such waves and their precise relationship to the spike discharge remained poorly defined.

Very early we were particularly interested in studies of single-cell rhythmicities. At the time, brain investigators and neuroelectrophysiologists were suggesting loop-like circuits and 'reverberating' activities to explain the periodic appearance of electrical signals. However, in 1936, it was clearly shown, in a Cephalopod isolated axon, the *Sepia* giant axon, how, under appropriate conditions, a local, graded potential may develop into spontaneous oscillatory activity (Fig. 1) (Arvanitaki 1936, 1938, 1939; Cole 1941; Cole and Curtis 1941; Hodgkin 1948). This was the first step towards the general concept of the dynamic behavior of the neuronal membrane.

Displayed in a nerve fiber such behavior was only a 'model'. However, from the very beginning, it was realized that assuming it could be proved operative in central nerve cells under normal conditions, the neuromembrane dynamics concept could be far-reaching in functional applications.

The disclosure in the visceral ganglia of *Aplysia* of a number of nerve cells, identifiable on the basis of morphological and functional criteria (Arvanitaki and Chalazonitis 1955, 1958) and the systematic exploration of such units through either extracellular or intracellular microelectrodes, revealed that central neurons do indeed normally work as oscillators (Fig. 2). Two main functional types of oscillators are formally distinguished in *Aplysia* and in *Helix* ganglia under normal

* The work reported in this paper was supported in part by the Centre National de la Recherche Scientifique, France, and in part by Research grants form the National Institutes of Health under grant NB 03337.

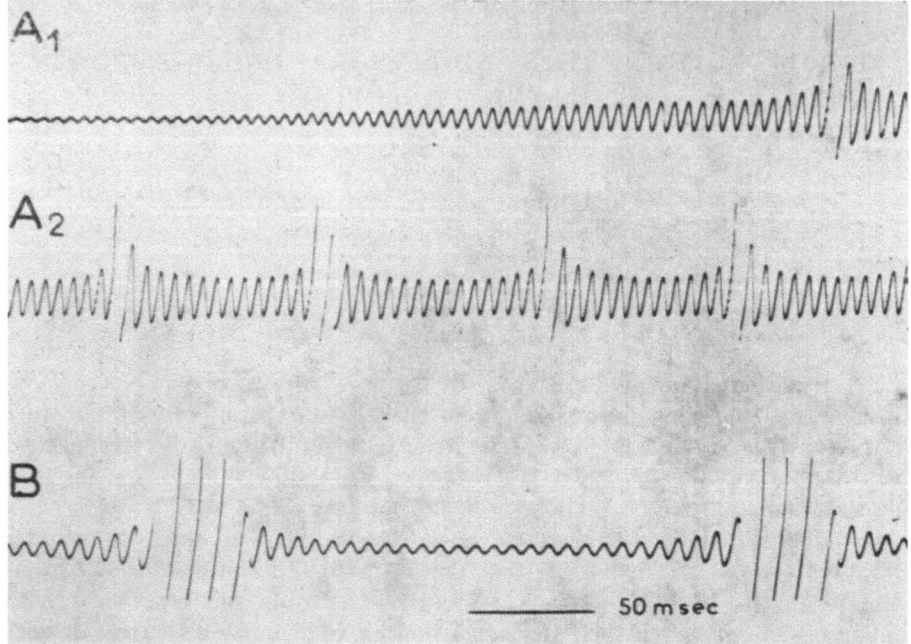

Fig. 1. Spontaneous graded oscillatory activity and recurring spikes in the isolated *Sepia* giant axon. Continued recording from A_1 to A_2. Some seconds after the local application of 10 % sodium citrate in physiological solution, initiation of oscillatory activity of increasing amplitude until, the threshold being reached, a spike is generated. Thereafter – spaced by a number of local oscillatory potentials – regularly recurring spikes are recorded. B — In another axon, bursts of spikes, spaced by a number of local oscillations, recur regularly. All recordings are through extracellular microelectrodes. (Arvanitaki 1939)

homeostatic conditions and apparently in the absence of any extrinsic synaptic input.

A number of identifiable neurons are found to be regularly autoactive, i.e. spiking at constant frequencies, each neuron exhibiting its own frequency. Extending from 0.05 to 15 Hz, a large spectrum of frequencies is thus operating among the cells of the *Aplysia* centers (Arvanitaki and Chalazonitis 1958). In addition to that kind of regular periodic activity — denominated monotonic — the behavior of identifiable nerve cells, initially designated the *Br*-type cells,* appeared to be of particular interest. These cells were disclosed in the *Aplysia* visceral ganglia, and their activity was first studied through extracellular microelectrodes and later intracellularly (Arvanitaki and Chalazonitis 1949b, 1955a, 1955b, 1955c, 1958, 1961a; Arvanitaki 1962a).

* Because of the topological criterion for their recognition was their proximity to the branchial nerve efference from the visceral ganglion (as viewed on its dorsal face, Fig. 3).

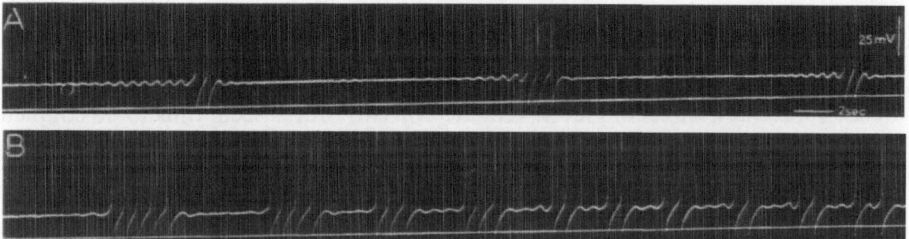

FIG. 2. Graded local oscillatory activity and intermittent spiking in the giant neuron of *Helix pomatia*. A — Impaled with two microelectrodes, the neuron was injected with a linearly increasing outward current and simultaneously its activity was recorded. Local graded oscillatory potentials and intermittent spiking are shown. B — In another neuron, under the action of the linearly increasing outward current, intermittent bursts spaced by local oscillatory potentials are found. (From Arvanitaki and Takeuchi unpublished)

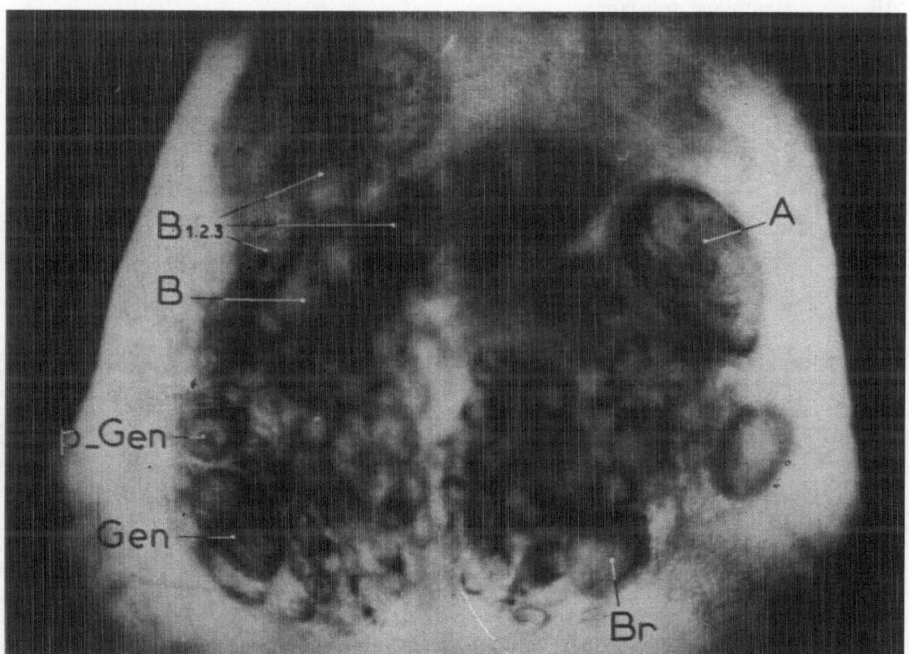

FIG. 3. Identifiable neurons as viewed at the dorsal surface of the *Aplysia* visceral ganglia. Among other identifiable neurons such as the *A* neuron, the four *B* neurons near the afference of the pleurogenital connective, the *Gen*, the *para-Gen*, etc., most remarkable is the one in the vicinity of the emerging branchial nerve, called the 'branchial' neuron (the *Br* cell). Such neurons are identifiable from morphological, topological and bioelectrical criteria. (Arvanitaki and Chalazonitis 1958)

Such cells exhibit 'slow-waves' of their membrane potential recurring at constant low frequency–under normal homeostatic conditions, in the absence of any synaptic input–and concomitantly, a high-frequency burst of spikes elicited on each wave at a threshold level. Kinetics of two orders seem thus to be coupled in the processings of such cell activity, hence its designation as 'bimodal'.

Our aim here is to define the bioelectrical characteristics of the 'waving neuron' membrane; to this end, the effects determined by transmembrane currents and by transitions in environmental factors will be analyzed. The synaptic correlations will be proven to be of particular interest.

The functional significance of this cell behavior, as implied in the nerve center temporal organization and information processing, will be considered.

Abbreviations used in this work:

MP	Membrane potential
MR	Membrane resistance
S–W	Slow-wave
W–N	Waving neuron
F_W	Frequency of slow-wave
F_{sp}	Frequency of spikes
EvkP	Evoked potentials
EPSP	Excitatory postsynaptic potential
IPSP	Inhibitory postsynaptic potential
$I.pO_2$	Intracellular partial pressure of oxygen
$E.pO_2$	Extracellular partial pressure of oxygen

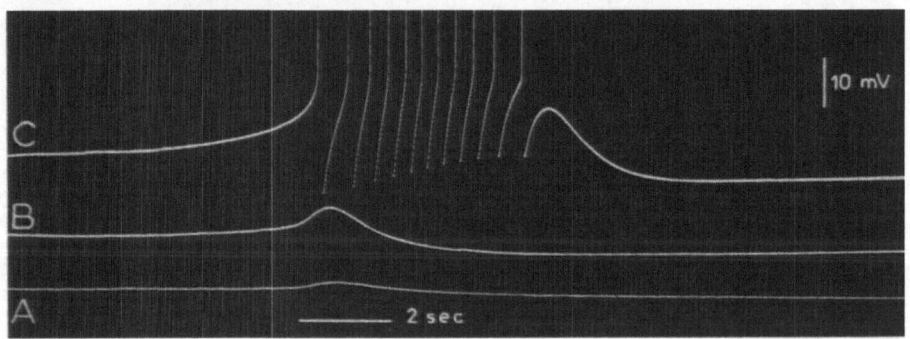

FIG. 4. Subthreshold S–W's and threshold MP requirements for the bimodal pattern of the *Br*-type cell activity. A and B — The neuron, at its hyperpolarized state, only exhibits subthreshold biphasic waves of many seconds duration. C — Slightly depolarized, the same cell spontaneously displays a large depolarizing wave bearing a burst of spikes with a characteristic evolution of their frequency. The development of a repolarization phase ends the cycle

CHARACTERISTICS OF THE SPONTANEOUSLY W–N

Control Effected by the MP Level

Recorded at high magnification, under homeostatic conditions, the developing S–W is frequently devoid of any synaptic sign (see Fig. 4). This fact is taken as a simple criterion for considering the S–W genesis as a 'spontaneous' event, i.e. as one proceeding from energy 'turnovers' and from specific cytostructural organization proper to the nerve cell unit being considered.

In the same context, the requirements of a 'threshold' MP level for the genesis of the S–W pattern are significant: In the hyperpolarized *Br* neuron, only low-rate spontaneous S–W's, subliminal for a burst generation, may be elicited (Fig. 4A). A long-lasting repolarizing phase following the depolarizing one is notice-

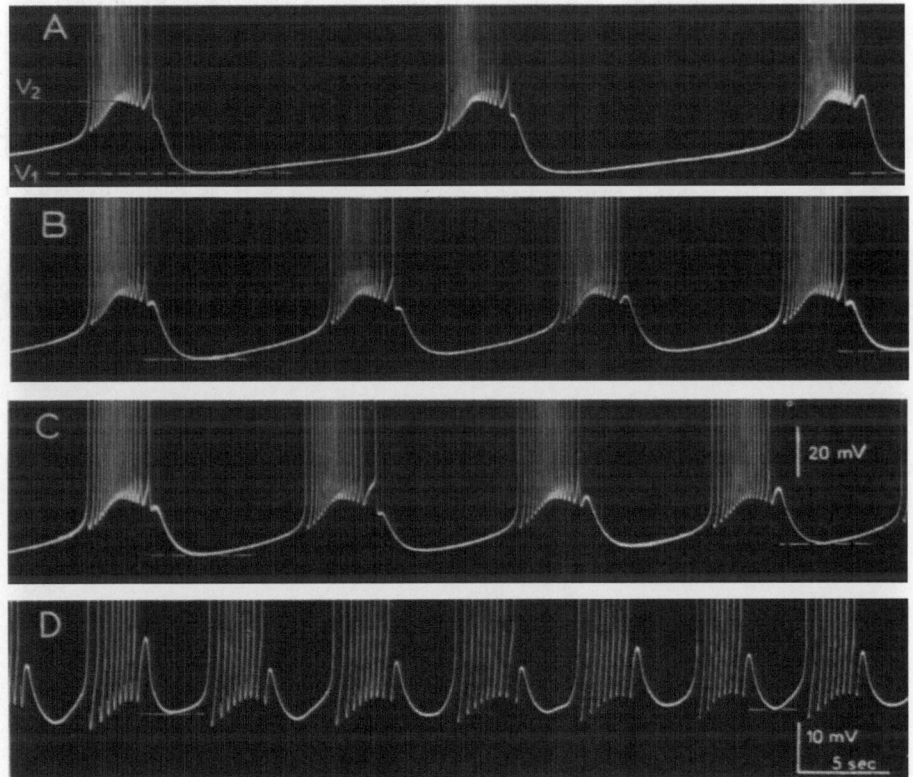

Fig. 5. The evolution of the S–W pattern characteristics in relation to the MP level. A–C — Continuous recording of spontaneous activity from a *Br*-type cell evolving under the continuous variation of an environmental factor (in this case, intracellular pO_2 continuously decreasing). The highest value of the MP level (V_1) reached during the S–W hyperpolarization phase (and indicated here by horizontal broken lines) is taken as a reference for defining the evolving state of the cell. D — Sixty seconds after C

able (Fig. 4B). In the same neuron, a further depolarization of the membrane by approximately 5–10 mV permits the autoregenerative development of the S–W and the discharge of spikes (Fig. 4, C).

Threshold level requirements for the S–W and its possible isolation as a 'naked' diphasic slow action potential (Fig. 4, A and B) are appropriate signs settling the S–W generation as a 'spontaneous' process.

A most significant bioelectrical parameter, determinant of the S–W pattern rates proves to be the MP value (V_1) at the highest level reached by the repolarization phase of the S–W (Fig. 5). The V_1–V_2 'span' where V_2 is the highest MP value reached by the afterpotential of the spikes in the burst — is also significant.

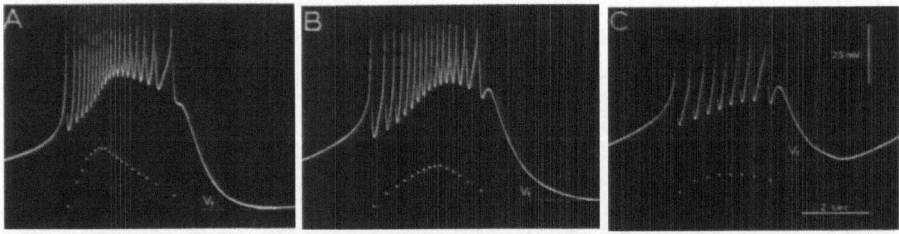

FIG. 6. The spike frequency characteristic evolution within the S–W as related to the (V_1) level. From a continuous recording where V_1 decreased from 68 mV to 64 mV, five S–W patterns were omitted between A and B and five between B and C. The spike instantaneous frequency (indicated by dots), discloses the kinetics of the process responsible for the S–W pattern. Its initial rate and the maximum frequency reached within the burst increase with the V_1 value

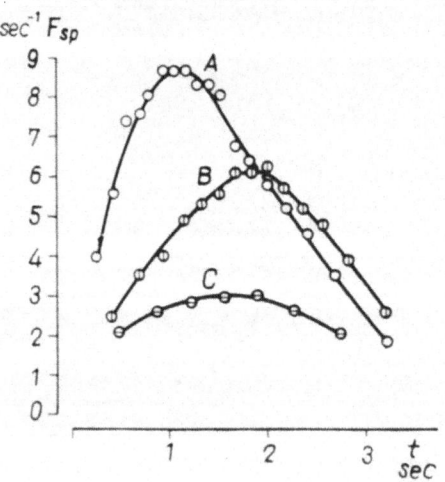

FIG. 7. Diagram of the spike frequency (F_{sp}) evolutions within three S–W patterns in which the V_1 characteristic values decrease: respectively 54, 50 and 45 mV

This varies with V_1 and seems to control the highest frequency reached by the spikes in the burst and the S–W frequency. The larger the $V_1 - V_2$ 'span' the higher the frequency of the spikes in the burst, and the longer the ensuing S–W period.

The time distribution of the spikes frequency within the burst (Figs. 6 and 7) suggests the display of an intrinsic, alternately increasing and decreasing outward current underlying the S–W pattern generation. The kinetics of this process $i(t)$, of which both rate and amplitude appear as V_1-dependant, should determine the spike frequency function $F_{sp}(i, t)$ within the pattern.

Input of EPSP's and Responsiveness of the Br cell in Relation to its MP Level

As a consequence of the above relation, and insofar as invasion of a cell by EPSP's determines membrane currents and displacements of the MP level, it was expected that a summation of EPSP's would contribute to the control of the W–N activity. In a highly hyperpolarized neuron (Fig. 8A), a high frequency of EPSP's is determinant of only a linearly increasing membrane depolarization soon followed by a repolarization, in spite of the still continuing EPSP input. In the same cell, but slightly depolarized (by ~ 5 mV) the same series of EPSP's now determines a depolarization at an exponentially increasing rate, suitable for promoting the MP to a threshold level for a S–W initiation (Fig. 8B). In the same cell, less hyperpolarized, but still silent, a single EPSP may elicit, although with a long latency, a graded subthreshold S–W (Fig. 8C). Finally, in the same neuron, a further decrease of the MP level allows the spontaneous generation of the S–W pattern (Fig. 8D).

Periodic Changes of the Membrane Resistance Concomitant to the Recurring S–W Patterns

Besides the basic MP change bearing the bimodal patterned activity, concomitant periodic changes of the membrane resistance are worthy of consideration.

These changes are evinced by the MP displacements due to brief square pulses of inward – or outward – current (Figs. 9 and 10). Parallel with the spontaneously developing slow depolarization, the MR progressively decreases to reach a minimum when the S–W depolarization is at its maximum. At that level, the MR is so greatly reduced that the inward pulses of current may be unable – as is in the case in Fig. 9 – to determine any diminution of the spike frequency within the pattern.

If, as already suggested, an intrinsic outward generator current alternately increasing and decreasing underlies the changes of the S–W MP, it should indeed also be responsible for the changes in the MR. In fact, the whole pattern of activ-

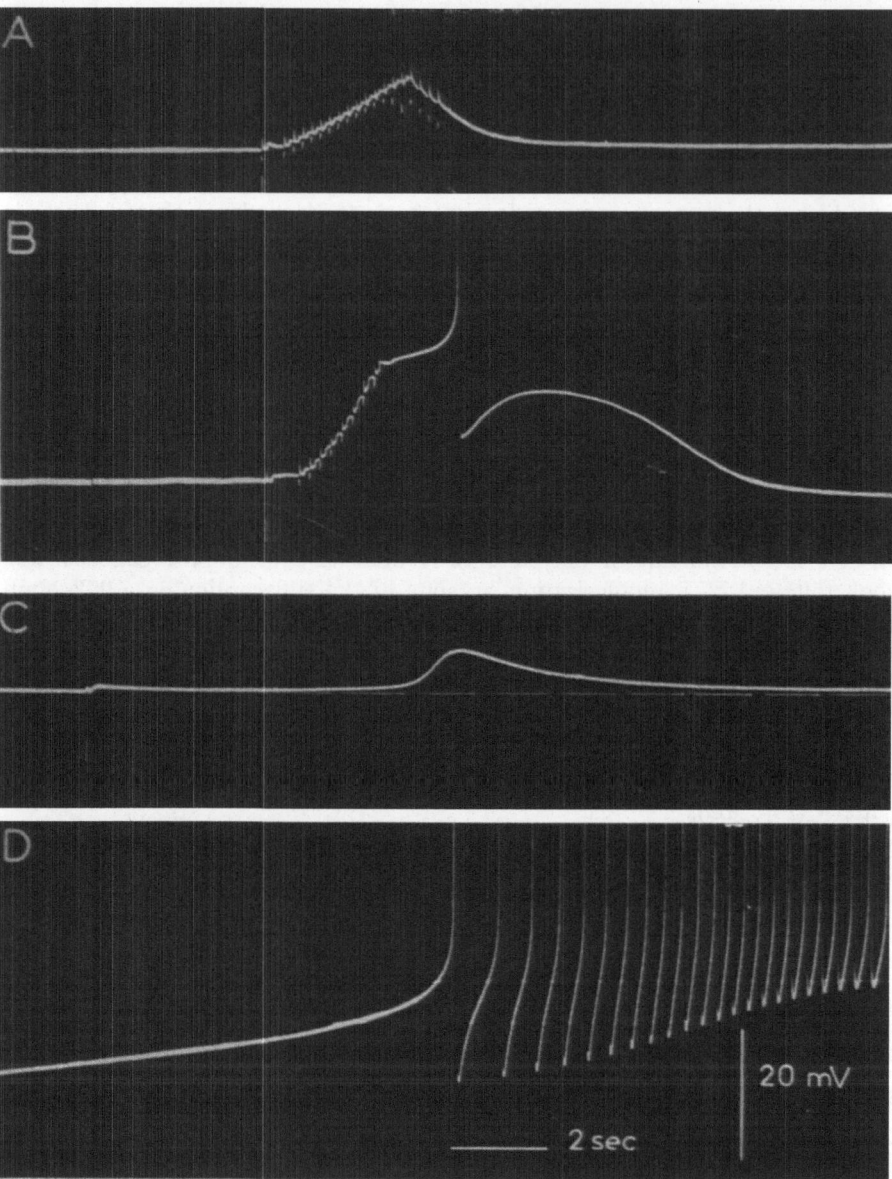

FIG. 8. The *Br*-cell behavior to an input of EPSP's as related to the membrane potential leve·
A — Cell highly hyperpolarized. Stimulation of the pleurogenital connective (connective B)
a series of EPSP's evokes a linear depolarization which turns suddenly into repolarization
(arrow). B — Slightly depolarized (by 5 mV) under the action of the same input as in (A) the
same cell develops here an exponentially increasing depolarization which leads to the generation of a short S–W (with emission of only one spike). C — The cell being further depolarized
(by 15 mV) a single EPSP elicits (after a long latency) an abortive S-W. D — Some 30 sec
later, spontaneous S–W activity is fully established

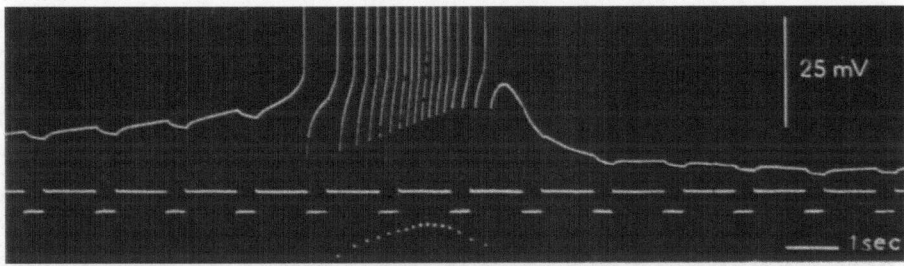

FIG. 9. Membrane resistance changes concomitant to the S–W pattern. *Br*-type neuron of a high V_1 value. Increase in membrane conductance at both the depolarized and the hyperpolarized level

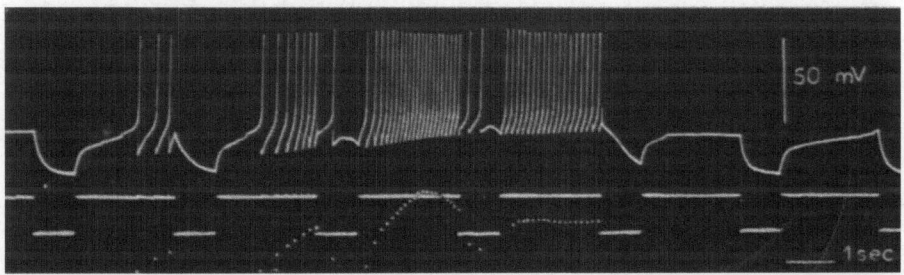

FIG. 10. Transient compensations of the intrinsic outward current by injecting inward current pulses. Simultaneous recordings of injected square current pulses (*i*) of the MP displacements (*V*) and of the spike frequency (F_{sp}). The intensity of injected current pulses is higher here than in Fig. 9. It is just sufficient to abolish the spiking in *m* and *n*. This fact would indicate that the intensity of the inward current pulse just balances the intrinsic outward current underlying the S–W pattern. Notice the concomitant changes of the spike frequency

ity characteristic of the *Br*-cell behaviour may satisfactorily be duplicated in a silent neuron by injecting a suitably shaped increasing then decreasing outward current. Conversely, an inward current pulse of appropriate intensity, whenever just able to balance the intrinsic outward current — and hence to abolish the spiking — would provide a measure of the instantaneous value of the underlying intrinsic outward generator current. In the case of Fig. 10, for example, the outward generator current would reach a maximum value of approximately 8 nA.

The Current/Voltage Relationship of the S–W Neuron Membrane

To study the *I/V* relation, a linearly varying outward current $i = at$ alternately increasing to a maximum and then decreasing was injected into a *Br*-type cell through an intracellular microelectrode and recorded together with the MP displacements (Fig. 11). Changes of the instantaneous value of the membrane

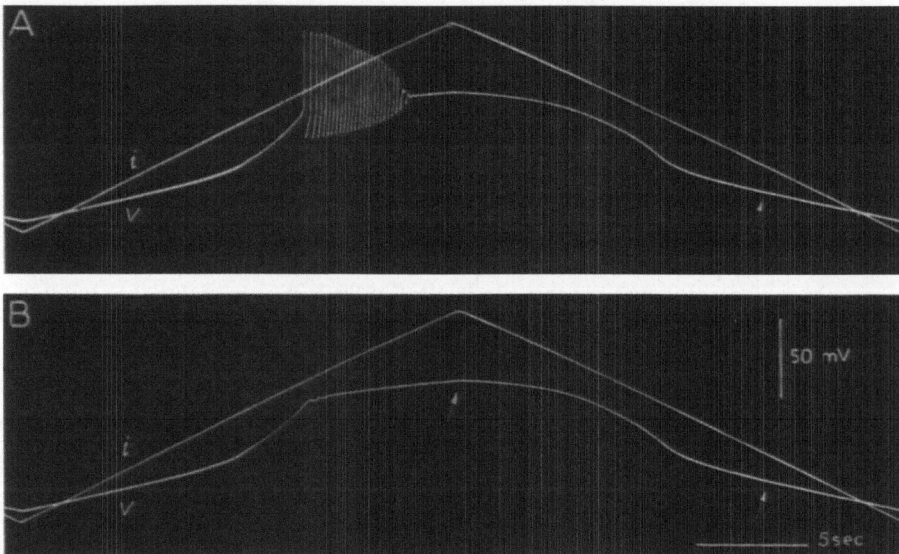

FIG. 11. A linearly varying outward current, alternately increasing and decreasing injected into a Br-type cell. The MP changes exhibit two domains (arrows) of low rate, i.e. of high conductance at the hyperpolarized and the depolarized levels, respectively and, interposed, two domains of high rate, i.e. of high resistance. In (B) the burst of spikes, happening abortively, reveals the undisturbed S–W potential course as is determined by the current. Note the flattened configuration of the wave due to reduced MR at both the depolarized and the hyper-polarized level (Arvanitaki and Romey 1966, 1968)

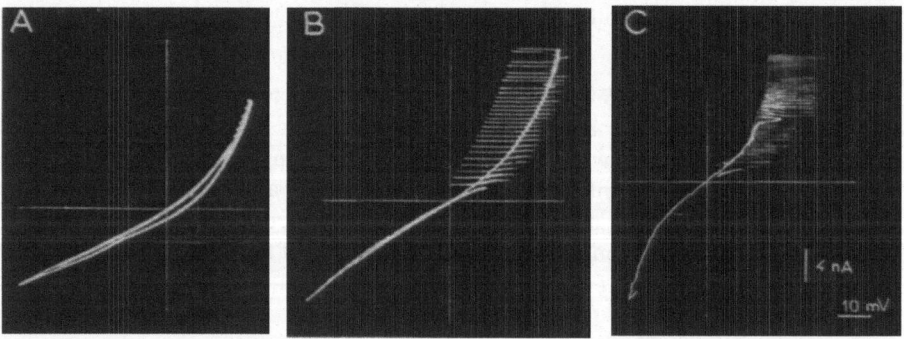

FIG. 12. Current/voltage relations in three identifiable neurons of *Aplysia* respectively. A — in the giant cell of the *A* type; *B* — the *B*-type and the *Gen*-type cells; C — the *Br*-type cell. Note high conductance in both the depolarized and the hyperpolarized state for the *Br*-type cell (Arvanitaki, Romey and Takeuchi 1967)

resistance dV/di then become immediately evident through the dV/dt changes. Two domains of minimum MR, at the depolarized and at the hyperpolarized states, respectively, are thus shown, the former being most obvious when the development of the spikes is aborted, as happens in Fig. 11B, for example.

Simultaneously, the current/voltage relationship was directly recorded on the oscilloscope $X–Y$ plane (Fig. 12). Changes of the current were recorded along the Y axis of the oscilloscope screen, while the concomitantly-elicited MP changes were recorded along the X axis (Arvanitaki and Romey 1966; Arvanitaki, Romey and Takeuchi 1967). Thus, for each given voltage, the corresponding slope on the I/V trace is indicative of the membrane conductance value.

In this way, Fig. 12 illustrates the I/V relations comparatively for three identifiable neurons of *Aplysia*. While the A-type nerve cell (see Fig. 12A) yields an I/V characteristic relation similar to that of the squid giant axon (Cole and Curtis 1941), in contrast, the Br-type cell exhibits a two-domain, large increase of the membrane conductance, i.e. in both the depolarized and the hyperpolarized states (Arvanitaki, Romey and Takeuchi 1967). A similar I/V relation was reported in the metacerebral neuron of *Helix* (Kandel and Tauc 1966). As a matter of fact, the Br-type cell, which is devoid of the common inhibitory synaptic regulation (IPSP's) may display − upon stimulation either of the pleurobranchial or the pleurogenital connective − series of postsynaptic potentials of the so-called 'amphoteric' or 'biphasic' BPSP type − a profound long-lasting hyperpolarization. The concomitant high increase of the membrane conductance renders ineffective any further synaptic action or current action (Arvanitaki and Chalazonitis 1964a, 1964b; Chalazonitis 1967b).

A cell exhibiting such a breakdown of the MR when alternately reaching the depolarized or the hyperpolarized domain thus behaves as a sort of 'bistable' system.

BEHAVIOR OF THE W–N UNDER INJECTED CURRENTS

The high degree of control proven to be exerted by the MP changes on the W–N activity drew attention to its behavior under injected currents.

Two microelectrodes were introduced into the cell, one for the injection of the current and the other for the recording of potentials.

It seemed opportune to examine first the possible 'adaptation' or 'accommodation' of the frequency of the S–W's to the constant current, allowance being made for the well-known accommodation of spike frequency to the same modifier. The analysis was directed to the study of the changes of the S–W frequencies and those of the spikes under outward and inward currents either of constant intensities, $i = k$, or of intensities linearly varying with time, $i = \alpha t$.

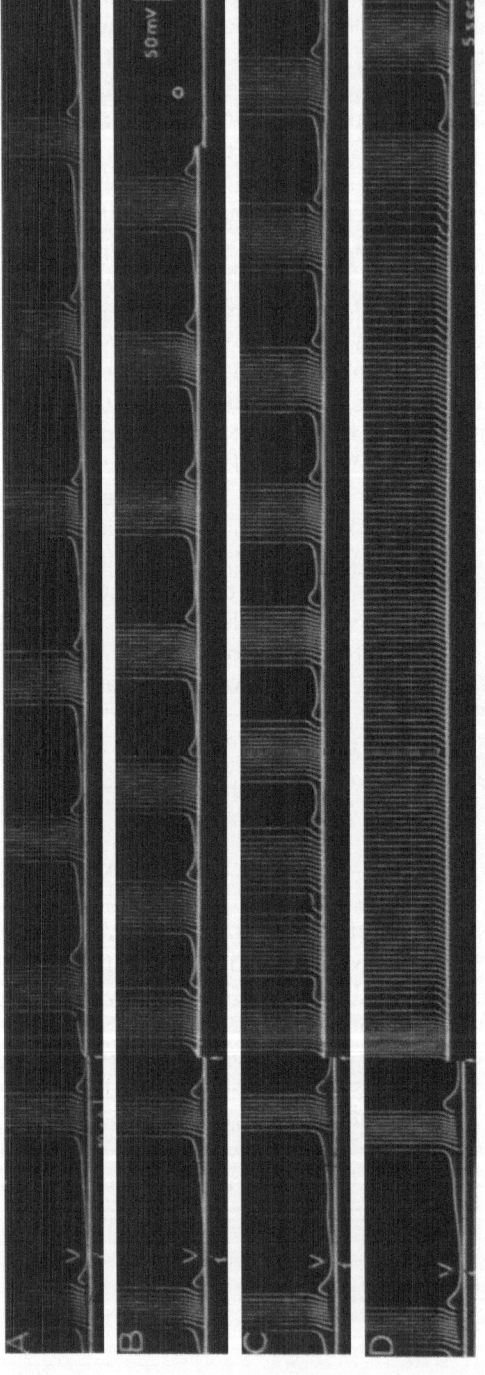

Fig. 13. Direct electrical stimulation of an *Aplysia Br* cell. *Accommodation*. *Br* cell, autoactive under normal conditions, exhibiting S–W patterns at a fairly constant frequency 0.05 sec^{-1}. Note at left the constancy of the normal periods (before injection of the current). Onsets of the constant outward currents through a second microelectrode are noted by arrows. The intensity of the current increases, to bottom from top respectively, 0.6, 1.7, 2.8, 3.9 nA. Note immediate increase of the S–W frequency at 'on' and 'accommodation' under the continuing dc current. At the highest intensity, fusion of the waves occurs, modulation of the spike frequency only is observable. Note at 'on' the depolarizing displacement of the S–W V_1 level, increasing with the current intensity. Slow repolarization as time proceeds, although current is still maintained

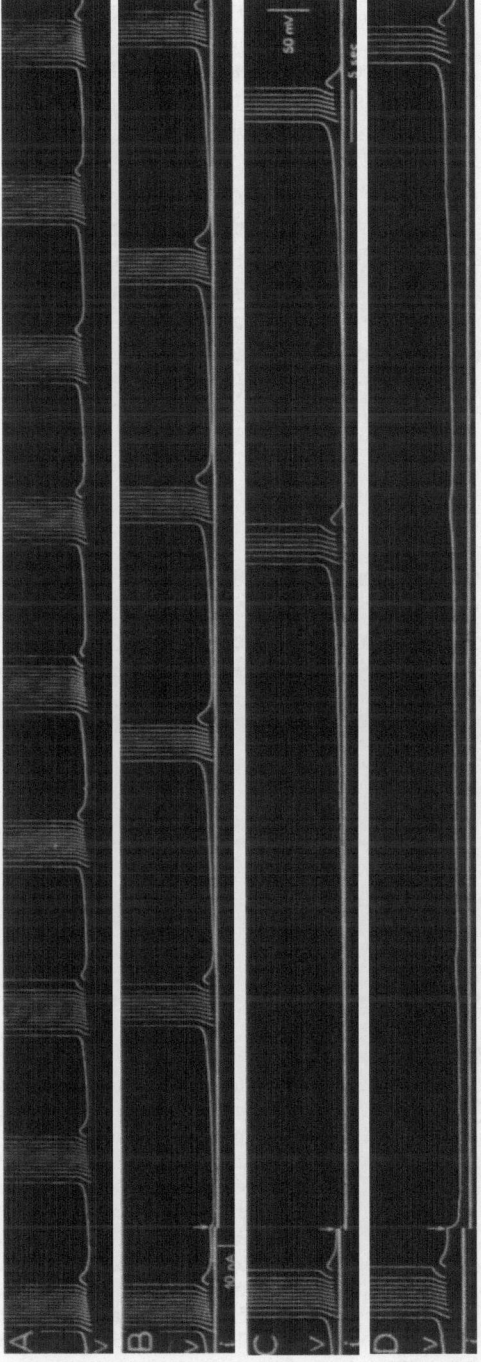

FIG. 14. The S–W frequency of the Br cell under constant inward currents. Accommodation. A — Br cell autoactive under normal conditions at a fairly constant frequency. B, C, D — under the action of constant inward current of increasing intensity, from top to bottom, respectively, 0.6, 1.7 and 2.8 nA. At 'on' of the current, note the immediate decrease in frequency and the subsequent slow recovery (accommodation) while the current persists

UNDER OUTWARD OR INWARD, CONSTANT CURRENTS: i = const

While the current is 'on', the frequency of the S–W as a function of time and intensity, $F_w(i, t)$, showed the following aspects. Immediately upon the establishment of an outward current, F_w reaches a maximum from which it starts decreasing exponentially with time (Fig. 13).

Conversely, upon establishment of an inward current, F_w immediately falls to a minimum; if the current is prolonged, the patterns reappear with a gradually increasing F_w (Fig. 14) (Chalazonitis, Takeuchi and Arvanitaki 1967).

Diagrams describing F_w as a function of time ($t = 0$ being the time the current was turned 'on') were established for the different values of i. For each neuron, a family of diagrams then becomes available (Fig. 15).

For each considered time t_n, the frequency $F_w(t_n)$ was computed, and was found to vary linearly as a function of i:

$$F_{w(t_n)} = \beta_n i$$

the constant β_n decreasing when t_n increases. The efficiency (or the yielding) of a constant outward current is thus shown to decrease as a function of both its intensity and its duration. This property indicates the 'accommodation' to the current of the mechanism responsible for the S–W pattern genesis. In the same way, 'accommodation' to an inward current has been shown as well.

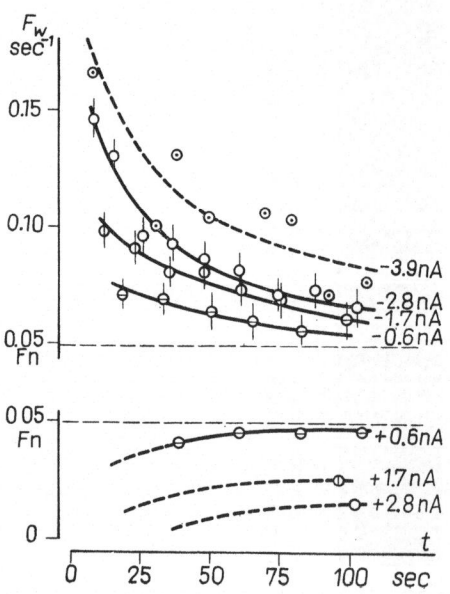

FIG. 15. The S–W frequency F_w of the Br cell as a function of outward constant currents (upper diagrams) and of inward constant currents (lower diagrams). Accommodation. F_n: initial steady frequency of S–W under normal conditions. As soon as the current is 'on', F_w reaches a maximum. This is followed by an exponential fall. Such a decrease of the current effectiveness as a function of time is the accommodation process. The latter may be described by the time constant of the F_w decay

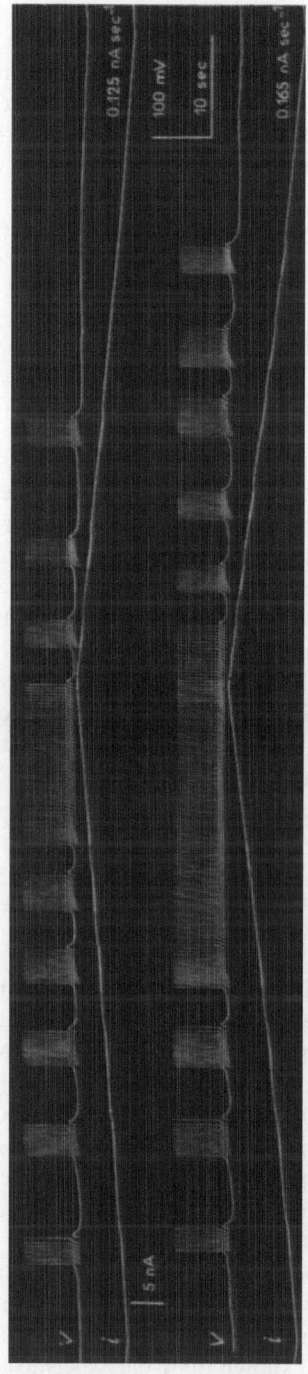

Fig. 16. The S–W frequency of the *Br* cell under linearly increasing then decreasing outward current. Through two intracellular microelectrodes, simultaneous recording of injected current (*i*) and of potentials (*V*). In two consecutive assays outward current linearly increasing with time at two rates, respectively, 0.125 nA sec.$^{-1}$ and 0.165 nA sec.$^{-1}$. The membrane depolarizes, the frequency of the slow-wave pattern, the frequency and the number of the spikes inside the pattern all increase as a function of the current increment. After a maximum, the current intensity decreases at the same rate (Arvanitaki and Romey, unpublished)

Under Linearly Increasing,   then Decreasing Outward Currents

In another series of experiments, injected outward currents were linearly increased with time, $i = \alpha t$, until a maximum, and then decreased at an equal rate (Fig. 16). It was shown that the frequency of the S–W's and the higher frequency of the spikes reached inside the waves increase linearly with log i (Arvanitaki and Romey 1968).

A general relation between current intensity and its derivative and F_w and its derivative may be tentatively presented:

$$[\beta(i) + \mu \frac{d\beta(i)}{dt}] = \lambda [F_w + \phi \frac{dF_w}{dt}].$$

β, being a log-type function.

A possible interpretation of accomodation to the current may be one associating it with an alteration of the I/V characteristics of the somatic membrane. Such an effect should result from a gradual transformation under the current of the membrane molecular configuration. Whatever the interpretation, 'accommodation' or 'adaptation' to the current, the notion of the S–W genesis as being an active autoregenerative membrane process — as is that of the spikes in any regularly autoactive cell is — strengthened (Hill 1936; Skoglund 1942; Kolmodin and Skoglund 1958; Frank and Fuortes 1960; Bradley and Somjen 1961; Granit et al. 1963).

In connection with this, it is worth noting that in regularly autoactive, or monotonous, neurons (i.e. spiking at constant frequency under homeostatic conditions), the above relations — which here concern S–W frequencies — hold indeed for spike frequencies (Arvanitaki, Costa and Takeuchi 1965).

BEHAVIOR OF THE W–N UNDER CHANGES IN ENVIRONMENTAL FACTORS: THE PARTIAL PRESSURE OF OXYGEN

Among different environmental factors which are worthy of consideration, changes in oxygen partial pressure (pO_2) deserve particular attention, because of the functional significance of such changes and the quantitative correlations that may thereby be established. Moreover, among different identifiable neurons, the *Br* neuron proved to be the most sensitive to pO_2 changes (Arvanitaki and Chalazonitis 1965b). As can be seen in the continuous recording of Fig. 5, for instance, the transition from the oxygenated state to hypoxia determined in 180 sec spectacular changes in the V_1 level of the membrane potential, the S–W frequency and the number and frequency of the spikes inside the patterns.

The direct observation and the measurement of the oxygen partial pressure inside a neuron *in vivo* have been achieved owing to the discovery of an intracellular hemoprotein in the *Aplysia* neurons (Arvanitaki and Chalazonitis 1949a). This was identified as a hemoglobin-like pigment shown to display specific absorbancy

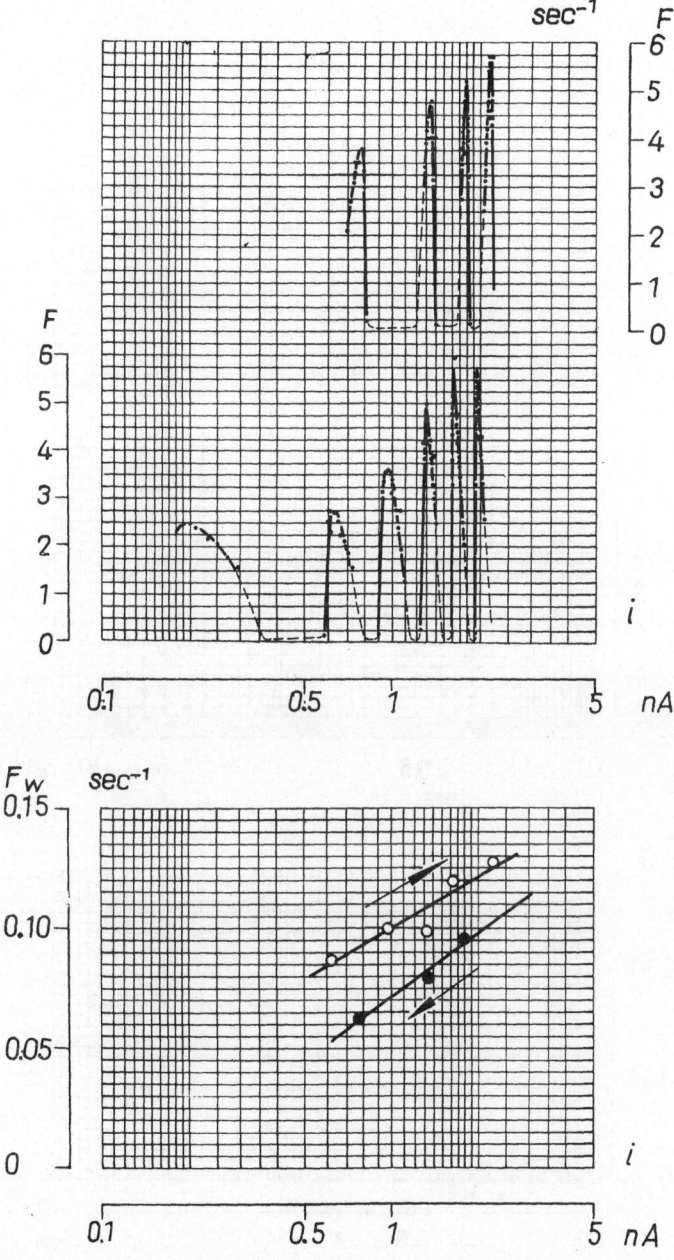

Fig. 17. Frequency of the S–W pattern (F_w) and of the spike frequency (F) inside the S–W pattern, as a function of the logarithm of the intensity of the outward current linearly varying with time (increasing, hence depolarizing; then decreasing at the same rate, hence repolarizing). The frequency of the S–W pattern (lower graph) and the higher frequency (upper graph) of the spikes inside the pattern vary linearly with log *i*

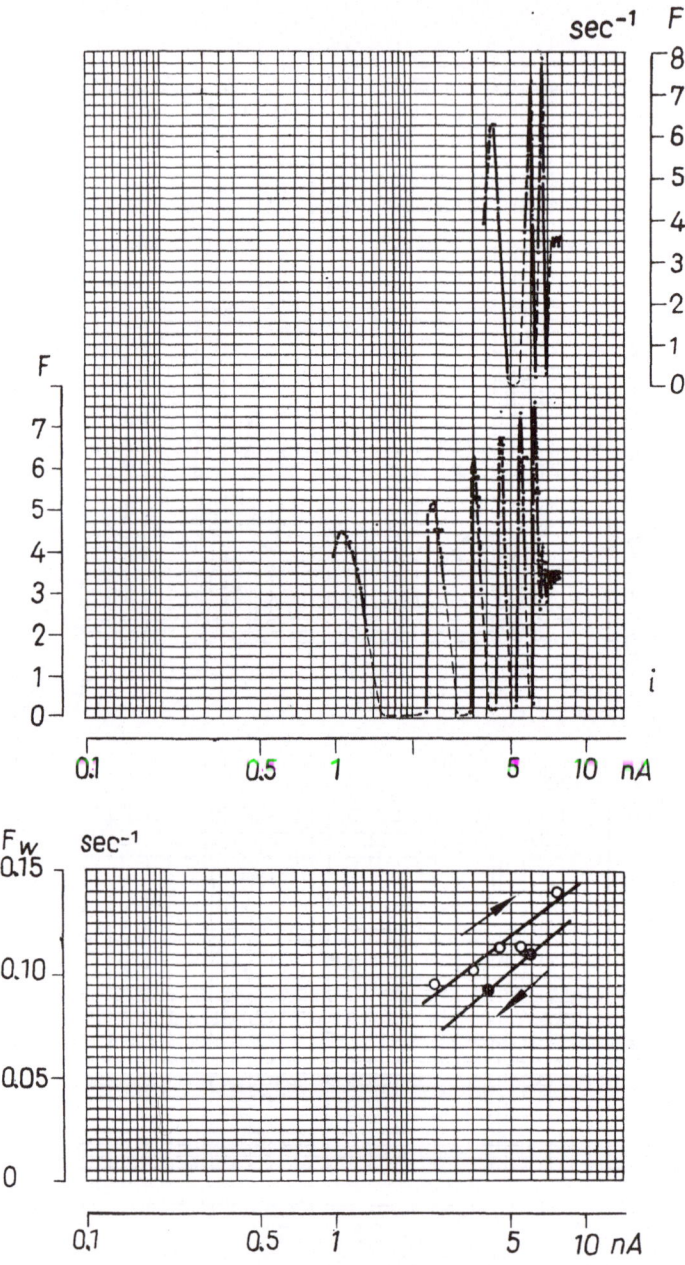

FIG. 18. Same experiment as in Fig. 17, but the rate of the current increase as a function of time is higher

changes during the reaction Hb + O_2 ⇌ HbO_2, and shown to behave as an intracellular oxygen storer (Chalazonitis and Arvanitaki 1951, 1956; Arvanitaki and Chalazonitis 1960; Chalazonitis and Gola 1965; Wittenberg et al. 1965, Chalazonitis 1967a).

The differential transmittancy changes between 580 mµ and 565 mµ in the *Aplysia* neuron, specific to the Hb ⇌ HbO_2 transitions, were recorded as a function of time simultaneously with the bioelectrical changes. Such transmittancy changes are approximately equal to only 5% of the total transmitted light, and are equivalent to absorbancy changes. On the other hand, absorbancy changes of this hemoglobin-like pigment are proportional to its percent saturation with oxygen. Therefore, the small transmittancy changes recorded *in situ, in vivo*, are linearly related to the changes in the percent of oxygen saturation of the intracellular hemoglobin (Chalazonitis and Gola 1965; Chalazonitis, Gola and Arvanitaki 1965a, 1965b). The intracellular pO_2 (I.pO_2) values were computed by reference to a saturation curve of the HbO_2 established on extracted pigment solutions.

The determination of pO_2 inside the cell was achieved by differential cine-spectrophotometry, by which the absorbancy changes of the intraneuronal hemo-protein during the transitions Hb + O_2 ⇌ HbO_2 were measured.

Alternating O_2 and N_2 admission in the chamber allowed the intracellular Hb + O_2 ⇌ HbO_2 changes to be determined. Such transitions were correlated to the concomitant bioelectrical events as recorded by intracellular microelectrodes (Fig. 19).

The Hb + O_2 ⇌ HbO_2 transitions were accomplished in approximately 100 sec. The external pO_2 values measured during the Hb + O_2 ⇌ HbO_2 changes varied from about 40–400 mmHg.

According to the oxygen saturation curve of the extracted pigment (at pH 7.5 in the absence of CO_2 in artificial seawater at 25 °C), the point of half saturation was at 13 ± 1 mmHg and the full saturation was at about 100 mmHg. On the

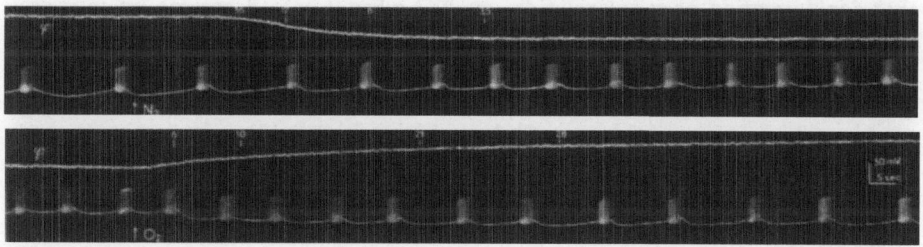

FIG. 19. Simultaneous recording of the activity of a *Br*-type cell and of the changes of the intracellular pO_2. Upper double-recordings: *Br* cell autoactive initially at a low frequency in the presence of oxygen. At the arrow, nitrogen is allowed to flow in the preparation chamber. The photometric trace (φ) signals (downward displacement) progressive desaturation of the intracellular hemoglobin. Simultaneously the membrane depolarizes and concomitantly the frequency of the patterns increase. Lower double-recordings: Oxygen is reintroduced in the chamber. The photometric trace signals (upward displacement) the reoxygenation of the intra-cellular hemoglobin. Simultaneously, the membrane repolarizes and F_w increases

other hand, spectroscopic examination of the ganglion just after the excision of the mantle muscle shows that the HbO_2 saturation in the cells *in situ* is close to 25 %; therefore the physiological intraneuronal pO_2 must be lower than 13 mmHg. A reasonable value to be considered provisionally as the normal intraneuronal pO_2 is 5–7 mmHg. Such a value is indicative of a 20–25 % saturation, in agreement with the spectroscopic observations.

Therefore, if the description of changes in a given membrane parameter is connected with this value as a reference point, the desaturation transition from 5 to 0 mmHg of $I.pO_2$ would be considered as a hypoxic transition, and the saturation from 7 to greater values as a hyperoxic $I.pO_2$ transition.

It was then tempting to evaluate in the *Br*-type cell the 'sensitivity' to pO_2 changes as manifested through changes of the membrane potential V_1 level and of the S–W pattern frequency, i.e. to determine the least pO_2 change able to produce a measurable change in the above functions.

From observations in 10 complete experiments conducted on the *Br*-type neuron, two intervals of $I.pO_2$ appeared to be of interest in this respect. The first is the 'hypoxic interval' lying between 6 and 2 mmHg ($I.pO_2$); within this, an increment of only 3 mmHg in the extracellular $E.pO_2$ is sufficient to determine a 5-mV repolarization and a corresponding decrease of the S–W frequency. The second is the 'normoxic' $I.pO_2$ interval lying around 7–5 mmHg (intraneuronal hemoglobin half-saturated). Here, an increment or a decrement of the $E.pO_2$ of 5 mmHg elicits a significant increase or a decrease of the MP.

In conclusion, during strictly aerobic conditions, small changes in the $E.pO_2$ (of less than 5 mmHg) are highly significant for the *Br* membrane activability. Therefore, if small fluctuations in pO_2 of the same order occur physiologically *in vivo* in the vicinity of the neuron (e.g. from vascular changes), pO_2 must be considered as a physiological environmental regulator of the neuron function.

In an attempt to understand the mechanisms for such a regulation of the S–W neuron modal behavior under varying $I.pO_2$, one can start with the hypothesis that depolarization or repolarization evoked when $I.pO_2$ decreases or increases would be brought about through changes of an intrinsic transmembrane outward current increasing (when $I.pO_2$ decreases) and decreasing (when $I.pO_2$ increases). For further information, the frequency of the S–W pattern under $I.pO_2$ decreasing (desaturation of the intracellular haemoglobin) and then increasing (resaturation of the intracellular haemoglobin) was plotted against the logarithm of

$$\frac{1}{I.pO_2} \text{ (Fig. 20).}$$

It is found that the frequency of the S–W varies linearly with the logarithm

of $\dfrac{1}{I.pO_2}$. Comparison with the relation giving F_w as a function of log i suggests that effects of outward current changes might be taken as an analogue of those of $I.pO_2$ changes. It is therefore suggested that along the sequence of mechanisms

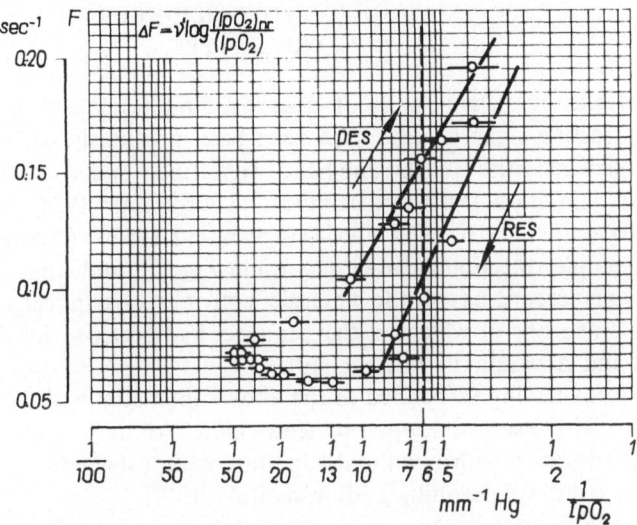

FIG. 20. Frequency of the S–W patterns F_w as a function of the logarithm of the ratio of intracellular pO_2 in normoxia $(I.pO_2)_{nr}$ to instantaneous value of $I.pO_2$. Decrease of intracellular pO_2 (desaturation of the intracellular haemoglobin) determines depolarization and increase of F_w. The reverse holds when $I.pO_2$ increases. Parallelism between increment of $(I.pO_2)_{nr}$ and increment of outward current. Thus, an equivalence tentatively is put forward of outward membrane current changes during desaturation of intracellular haemoglobin: a transduction of haemoglobin desaturation to a generator membrane current (Arvanitaki, Gola and Romey 1968)

controlled by $I.pO_2$ changes, an early process would involve the production across the neuromembrane of an outward generator current varying linearly with $\dfrac{1}{I.pO_2}$ and functionally interfering with preexisting 'intrinsic currents'. Consequently, diagrams such as those of Figs. 18 and 20 established from data in the same neuron, would allow for a given change in F_w, computation of the Δi change in the intrinsic outward current which should be determined by a given $I.pO_2$ change.

High sensitivity of the Br-type neuron extends to other environmental factors especially to temperature (Chalazonitis 1962).

It appears from the foregoing that the Br-type cells behave inside the nervous center as due sensors of environmental changes, at least for pO_2 and for temperature, a far-reaching property owing to the role imparted to those cells in the coordination of multiunit activities.

A MAJOR TRIAD OF IDENTIFIABLE NEURONS
INCLUDING THE *Br* TYPE

Activities of one or several pairs of neurons, simultaneously displaying respectively contrasting signs of activation have been previously defined as reciprocal or antergic* activities: on the one hand, excitation – spike or EPSP – and on the other hand, inhibition, IPSP (Chalazonitis and Arvanitaki 1958, 1961; Arvanitaki and Chalazonitis 1961b, 1963, 1965a; Arvanitaki 1962b; Chalazonitis 1963, 1967). For example, the double trace $(1_i, 1_e)$ in Fig. 21 and cont. reproduces the simultaneous recording of the antergic activities, synaptically elicited in two identifiable neurons of which another sequence had been previously published (Chalazonitis and Arvanitaki 1961, Fig. 10).

Pairs of neurons having responded as the above upon stimulation of the nerve path developed 20–30 sec later, opposite slow changes of up to 20-mV amplitude and up to 15-sec duration with or without superimposed postsynaptic potentials or spikes (Fig. 21 cont. Chalazonitis and Arvanitaki 1961).

The aim here is to disclose the various aspects of concerted actions that groups

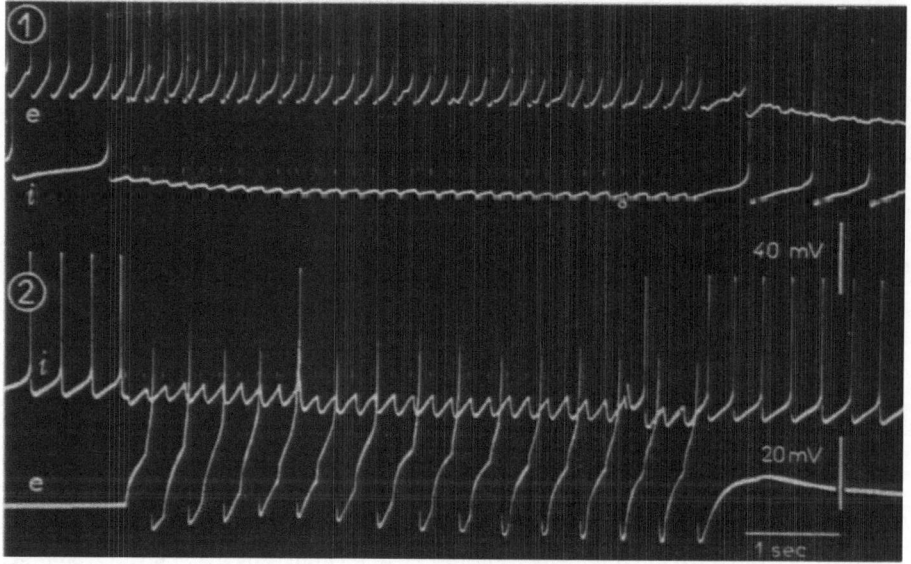

FIG. 21. Synchronous reciprocal synaptic activations: EPSP–IPSP on two pairs of identifiable neurons. The neurons (1_i) and (2_i) were initially autoactive at respective constant frequencies, with apparently no synaptic incidence: neuron (1_e) initially displayed IPSPs at a low frequency; neuron (2_e) was initially inactive. Just threshold stimulation at a frequency of 3 sec^{-1} to the pleurogenital connective. At the same frequency, neurons (1_i) and (2_i) display unitary IPSPs. Simultaneously, neurons (1_e) and (2_e) are activated; neuron (1_e) emitting one spike for one EPSP; neuron (2_e) emitting one spike for two EPSP's

---

* Proposed as an antonym to synergic.

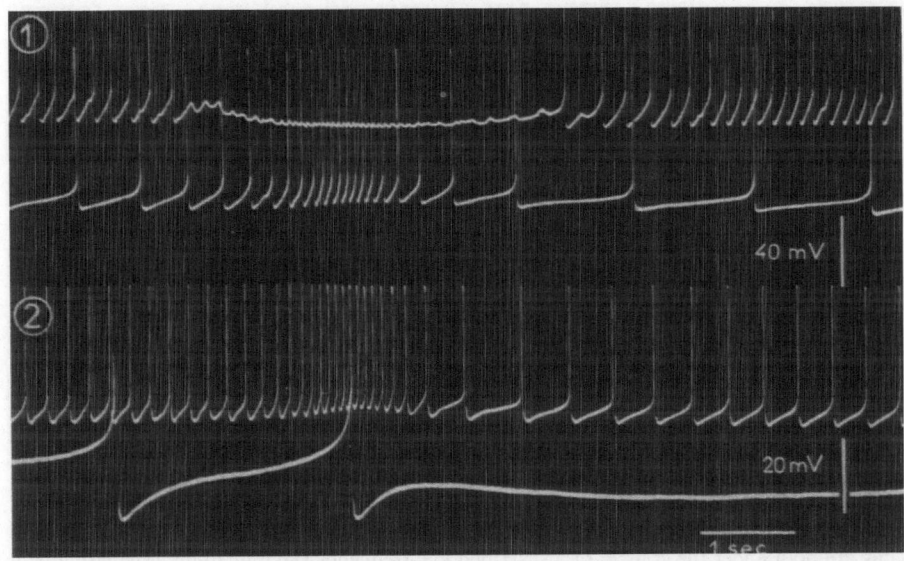

FIG. 21 cont. Recurrent patterns of reciprocal activities induced after the cessation of the stimulation of the nerve path. Continuation of Fig. 21 on the same two pairs of neurons. Stimulation was over for 35 sec, when neurons (1_i) and (2_i) which had been inhibited through the afferent nerve stimulation, now show a burst of high-frequency spikes, while the paired neurons undergo an 'avalanche' of IPSP's at high frequency (1_e) or hyperpolarization and decrease of the spike frequency (2_e). Such a pattern of activities repeated later three or four times, vanishingly

of identifiable neurons including *Br*-type neurons, can bring forth in an 'elementary brain' such as an *Aplysia* ganglion.

The research was conducted by examining successively through intracellular recordings various pairs of identifiable neurons of the visceral ganglion of *Aplysia depilans* and *californica*.

The study proceeded: (a) in the course of 'spontaneous' activities, i.e. in the absence of any deliberate stimulus; (b) by directly stimulating successively one by one the neurons in the group; and (c) by conveying to the units just-threshold rhythmic stimulations through different afferent nerve paths.

The neurons examined simultaneously with the *Br* cell were, in particular: the *B* neurons and the *Gen* neurons, both displaying IPSP's; the *para-Gen* neuron, which exhibits EPSP's and IPSP's concurrently (Arvanitaki and Chalazonitis 1958).

In several cases the simultaneous exploration of two pairs of neurons was conducted by using four independent microelectrodes. The systematic study of interneuronal anatomic relations was conducted paralelly by serial sections 1 or 2 μ thick of the ganglion.

Numerous attempts had already resulted in evidence of the existence for several pairs of identifiable neurons displaying regular reciprocal activation: EPSP versus IPSP.

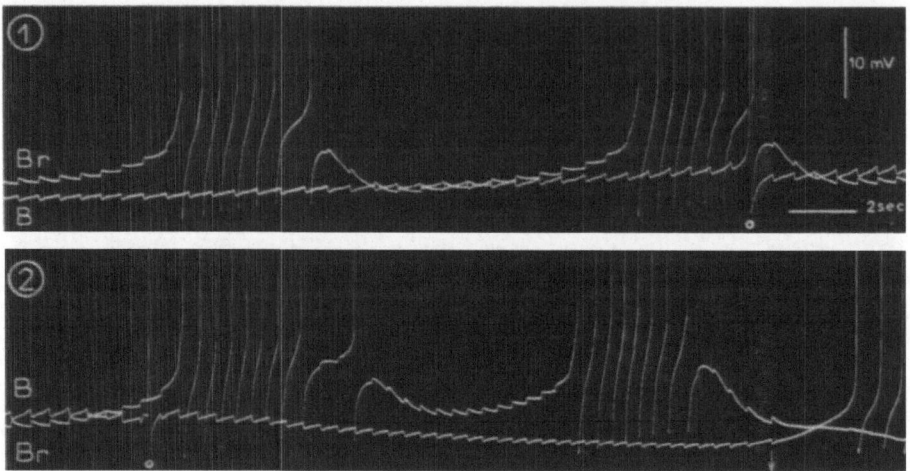

FIG. 22. Reciprocal synaptic activation under the action of a common monitor cell (M). Simultaneous recording of the activity from two neurons: the *Br* neuron and a *B* neuron. They are both under the control of a sustained low-frequency synaptic afflux originating from a third autoactive neuron, determinant of an antergic synaptic 'doublet': EPSP in *Br* synchronous with the IPSP in the *B* neuron. Traces (2) immediately follow traces (1). On the *Br* neuron, the frequency of the slow-waves and that of the discharges of spikes controlled by the EPSP—and therefore by the monitor cell, the activity of which they signalize – augment with the increase in the frequency of the latter (compare the *Br* groups in [1] and [2]). Regarding *B*, the activity is suppressed under the IPSP's: only a slow depressed oscillation of the membrane potential persists, allowing two single spikes (indicated by dots under the recording). The cessation of the synaptic afflux (arrow ↓) immediately sets off the initiation of the wave discharge from *B*, while a tendency for hyperpolarization is shown in the *Br* cell

The simultaneous recordings had shown that the genesis of the opposite postsynaptic potentials is synchronous in the two antergic neurons, within the limits of error of the method (Figs. 22 and 23). The statement implies that the number of synaptic delays interposed between the source of the information, i.e. of the stimulus whatever its nature, and the recorded signals would be equal in the paired neurons, i.e. that the origin of the signal is common.

Recent results have shown that antergic postsynaptic potentials are indeed conveyed from a common neuron (Kandel *et al.* 1967; Arvanitaki, Romey and Watanabe 1967). The recording at high speed of the simultaneous antergic postsynaptic potentials has clearly shown their common origin and their synchronous elicitation (Figs. 23 and 24).

Timing of the antergic EPSP–IPSP in the two follower cells reveals the typical mode of the frequency of the monitor cell spikes to be a function of time. The monitor cell displays rhythmic bursts of spikes up to 70–120 at frequency, exponentially rising with time and reaching up to 8 sec^{-1} (Fig. 23). Clearly, the monitor cell is autoactive on large S–W's; i.e. the monitor cell is an autoactive cell of the *Br* type. The monitor is a medium-sized cell of 100–150-μ diameter. When impaled with two microelectrodes (in order to inject current), it often loses its modal behavior and – owing to depolarization – appears as continuously autoactive (Figs. 24, 25 and 26). However, under the injected outward current, the frequency of the ensuing spikes shows an exponential increase [Fig. 24 (2, 3, 4) and Fig. 27]. The higher the intensity of the injected current, the higher the exponentially rising frequency. Such a behavior suggests that the injected current would activate mechanisms of the intrinsic generator currents responsible for the spontaneous bursting proper to the cell.

A further characterization of the monitor cell M as being of the *Br* type resides in its poststimulative dynamic behavior: on the one hand, at the cessation of any outward depolarizing current slow hyperpolarization develops with an amplitude and duration dependent on the precessor outward current [Fig. 24 (2, 3, 4)]; on the other hand, after the completion of this first poststimulative reaction, 20–30 sec later, a further consecutive spectacular hyperpolarization

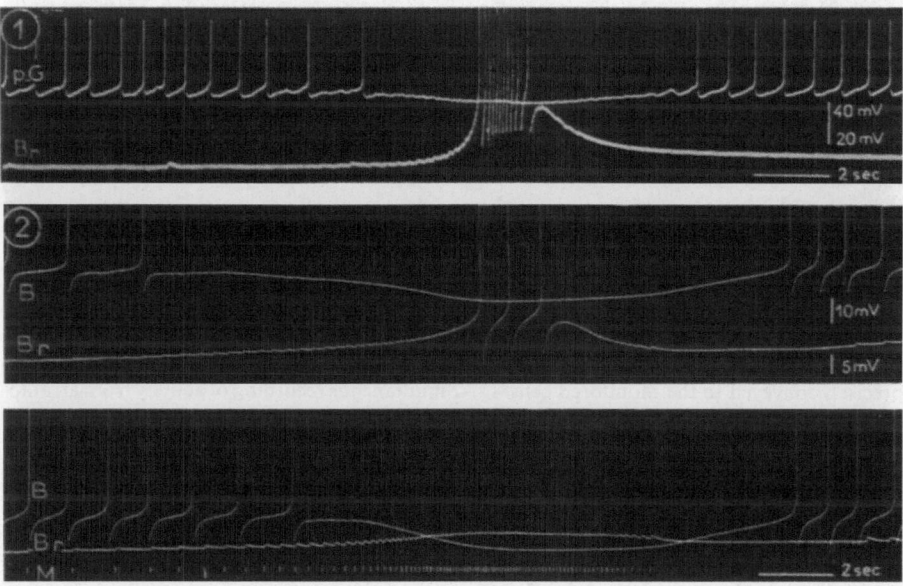

FIG. 23. Antergic patterns of paired neurons of the *Br* type under the synaptic control of a common monitor cell. 1 — Example of a recurring "reciprocal" pattern in a couple of identifiable neurons (of the *Br* type and *para-Gen* type). The burst of the *Br* neuron is under the control of a series of EPSP's of unitary origin, while in the *para-Gen* neuron, the IPSP's stem from at least two sources both contributing to its inhibition. The timing of the EPSP's discloses that of the monitor neuron, thus proving it to be of the *Br* type (from Chalazonitis 1963). 2 — Two sequences of activity simultaneously recorded from two paired neurons, *Br* and *B*, under the direct reciprocal control of a monitor cell (M). The spike frequency of the monitor cell is directly revealed from that of the EPSP's and of the IPSP's on the monitored cells, proving again that the (M) cell is of the *Br* type, i.e. normally bursting on S–W's. Note that in the lower paired recording, the *Br* cell was hyperpolarized. Thus, in spite of the high frequency of the summating EPSP's, no spiking results

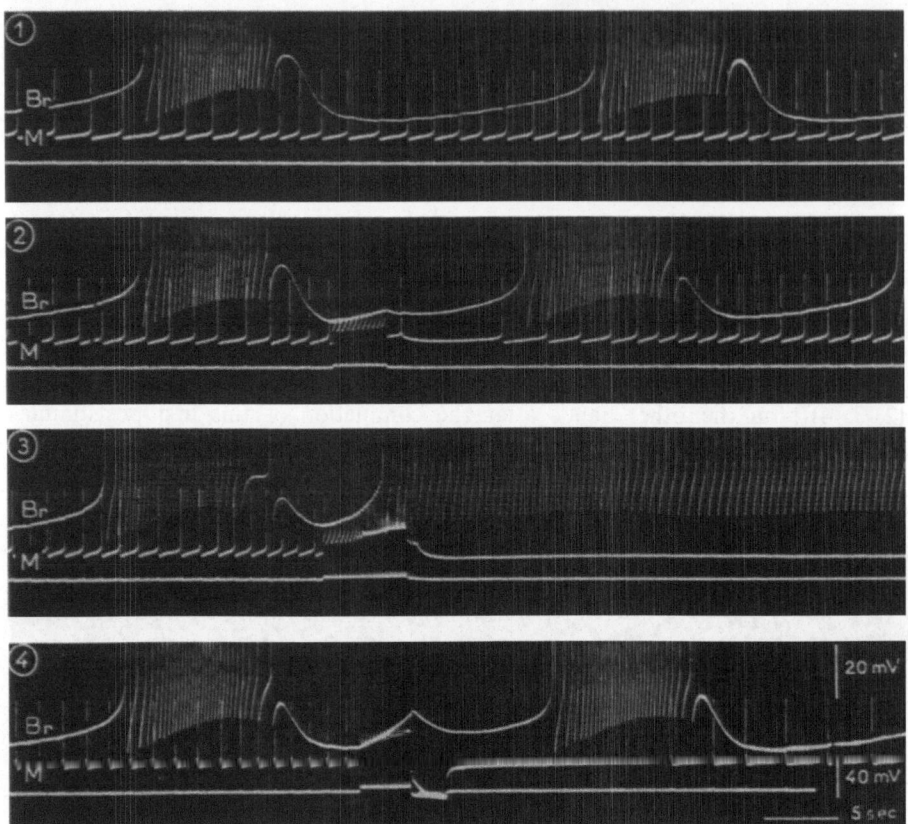

FIG. 24. Series of spikes determined by transmembrane current on this monitor cell are precisely conveyed to the monitored cells. 1 — Simultaneous recording of activity from an auto-active monitor cell (M) and a *Br*-type cell. Normal homeostatic conditions, in the absence of any deliberate extrinsic stimulus. Frequency of the S–W fairly constant. Each spike from (M) corresponds to an EPSP on *Br*. 2 — A pulse of depolarizing current injected into (M) elicits spikes of high frequency. These evoke in Br synchronous series of EPSP's depolarizing the membrane by summation. Notice, as a consequence, shortening of the period of the *Br* S–W pattern. Compare to the period seen in (1). 3 — A pulse of current injected into (M), of higher intensity and longer duration elicits high frequency of spikes conveying to Br a corresponding series of EPSP's. Their summation on the *Br* cell yields a powerful activation conducive to the premature initiation of a most prolonged burst of spikes. 4 — The pulse of outward current injected into (M), is of higher intensity than in (3). At off, with the abolition of the current there coincides a step of inward current. The depolarization developed on *Br* due to the summation of the EPSP's is abruptly cut off [compare to (3)]

of ~ 8 sec duration and ~ 20 mV amplitude is recorded [Fig. 25 (1)]. In most of the cases, such a bioelectric activity recurs spontaneously, and vanishingly, three to four times at 30–40 sec intervals. Moreover, it is observable not only subsequent to a direct current stimulation, but also following a synaptic stimulation [see Fig. 26 (1, 2) and Fig. 27 (2)].

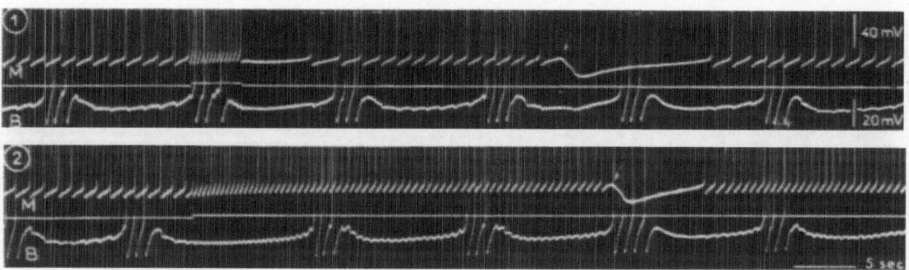

FIG. 25. Monitor cell evoking IPSP's to B-type cell. 1 — To each spike of the monitor cell is timed an IPSP on a B-type cell. A pulse of outward current injected into (M) determines a high-frequency spiking which evokes in B_2 a corresponding synchronous series of IPSP's. Notice on (M), 20 sec after the end of the current pulse a sinking hyperpolarization of the membrane of approximately 20mV amplitude which, intriguingly, is partly 'felt' in the B_2 cell. 2 — A long-lasting outward current injected into the monitor cell determines high frequency spikes which correspondingly evoke IPSP's in the B_2 cell. Notice on (M) the event of the hyperpolarizing sinking — although the outward current is still 'on'

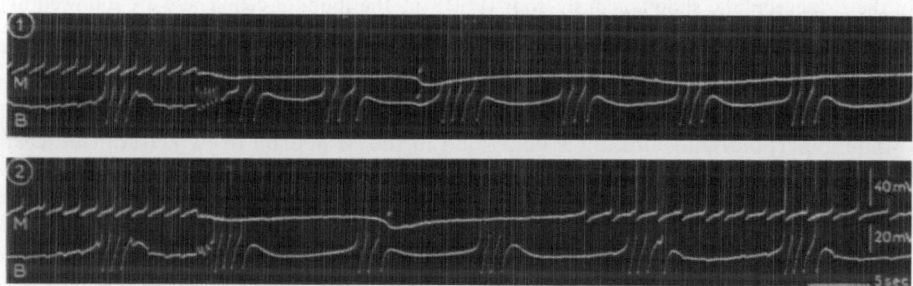

FIG. 26. Synaptic control to the monitor cell. Same pair of neurons as in Fig. 25. 1 — Seven short just-threshold stimuli were delivered to the pleurogenital connective (the so-called B connective). Simultaneously, a corresponding series of IPSP's are evoked on the monitor neuron while simultaneously the B neuron displays EPSP's. 2 — Seven short just-threshold stimuli delivered to the pleurobranchial connective (the so-called A connective). Synchronously, IPSP's are evoked on the monitor cell while, simultaneously the B neuron displays EPSP's. Notice, in both (1) and (2), development of the hyperpolarizing sinking approximately 25 sec after the synaptic stimulus

One more observation is of interest: in most of the recorded cases, the hyperpolarizing sinking is recorded simultaneously — although with reduced amplitude — in both of the monitored cells, thus achieving striking spatiotemporal patterns in concerted activities of such a 'triad' of neurons [Fig. 26 (1), Fig. 27 (1) and Fig. 27 (2)] (Arvanitaki, Romey and Watanabe 1968).

Two questions are then raised about: (1) the nature and the mechanism of the recurring hyperpolarizing sinking; (2) the mechanism of its propagation and its transmission to follower cells in the triad.

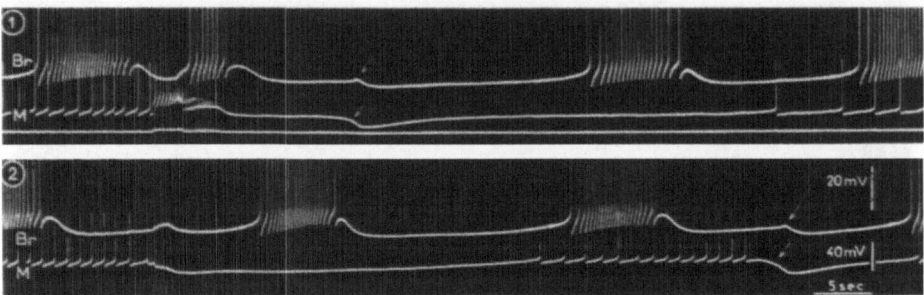

FIG. 27. Recurring hyperpolarizing sinks in the monitor neuron. Simultaneous recording in the monitor and the *Br* monitored neuron. 1 — Stimulation by direct outward current to the monitor neuron determines a high-frequency spiking, evoking in the *Br* cell synchronous EPSP's with summation and premature bursting. Following cessation of the current, very long poststimulative inhibition of the monitor neuron with, superimposed hyperpolarizing sings. Note on the *Br* abortion of the slow-wave pattern whose genesis was just coinciding with the hyperpolarizing sink in the monitor cell. 2 — Same of neurons as in (1). Four brief just-threshold stimuli delivered to the pleurogenital connective, evoke four EPSP's in the *Br* neuron and simultaneously four IPSP's in the monitor cell. These are followed by a hyperpolarizing sink, recurring 50 sec later (arrow). Note simultaneous abortive S–W in the *Br* neuron

The hyperpolarizing sinking is generated in the M neuron in two circumstances. It appears following: (a) a direct electrical activation of the M cell [Fig. 25 (1, 2) and Fig. 27 (1)]. It is then often shown as implying two steps: a first component of rather slow rate [signaled by the arrows in Fig. 25 (1, 2)], followed by a second rapid component; (b) a spontaneous burst of spikes [Fig. 27 (2)]; (c) an initial setting of hyperpolarization from synaptically evoked IPSP's [Fig. 26 (1, 2)].

In all the cases, the hyperpolarizing sinking appears not to be relatable directly to synaptic events, but rather to be an expression of the dynamic behavior of the neuron. The above remarks do not exclude classical ways for propagation and transmission of these 'slow' events but suggest electrotonic mechanisms possibly imparted to the synaptic events in the functional organization of the 'triad' performances.

Closing the Loop of the Triad

A further aim here is directed to providinge vidence for synaptic ways of controlling of the monitor cell itself in the 'triad'. Direct stimulation to various cells in the ganglion have not yet disclosed the connection, although it must exist. As a matter of fact, direct stimulation to either the pleurogenital or the pleurobranchial connective evoked in most of the cases IPSP's and long-lasting inhibition in the M cell, but evoked EPSP's in the neuron which in the triad was previously inhibited from M [Fig. 26 (1, 2)] and simultaneously, inhibiting postsynaptic potentials (of the 'biphasic' type, Chalazonitis, this Symposium) to the *Br*-type cell, which was previously activated by EPSP's from M. Thus, the 'triad' devic conducive to monitored paired contrasting patterns of activity, actually pertains to a 'loop' allowing the alternating reversal of the triad output.

Such a system is not unique in the nervous center studied here, the visceral ganglion of *Aplysia*. In fact, several parallel interneuronal systems of the same type exist, as do much more elaborated systems.

Acknowledgments: As quoted in the text, much recent work was accomplished with the skilfull collaboration of M. Gola, G. Romey, H. Takeuchi and Y. Watanabe, to whom acknowledgment is due.

We are greaty indebted to R. Chagneux and R. Fayolle for their expert technical assistance: to D. André and G. Cadenel who were responsible for the illustrations: to M. André and G. Blanchet for their constant care in the preparation of the manuscript.

REFERENCES

ADRIAN, E. D. and MATTHEWS, B. H. C. (1934): The interpretation of potential waves in the cortex. *J. Physiol.* **81** 440.

ADRIAN, E. D. and MORUZZI, G. (1939): Impulse in the pyramidal tract. *J. Physiol.* **97** 153—199.

ARVANITAKI, A. (1936): Variations lentes de potentiel associées au fonctionnement rhythmique des nerfs non myélinisés isolés. *J. Physiol. Path. gén.* **34** 1182—1197.

ARVANITAKI, A. (1938): *Les variations graduées de la polarisation des systèmes excitables. Signification fonctionnelle dans l'activité rythmique.* Ed.: Herman, Paris.

ARVANITAKI, A. (1939): Recherches sur la réponse oscillatoire locale de l'axone géant isolé de Sepia. *Arch. int. Physiol.* **49** 209—256.

ARVANITAKI, A. (1962a): Prolonged hyperpolarization outlasting excitatory and/or inhibitory afferent stimulation, in identifiable nerve cells. *Fed. Proceed.* **21** 348.

ARVANITAKI, A. (1962b): Information processing in elementary central nervous systems. Spatiotemporal bioelectric patterns. In *Information processing in the nervous system*, XXIIe Internat. Congr. IUPS., Leyden, 312.

ARVANITAKI, A. and CHALAZONITIS, N. (1949a): Réactions bioélectriques neuroniques à la photoactivation spécifique d'une hème-protéine et d'une carotène-protéine. *Arch. Sci. physiol.* **3** 37—44.

ARVANITAKI, A. and CHALAZONITIS, N. (1949b): Prototypes d'interactions neuroniques et transmissions synaptiques. Données bioélectriques de préparations cellulaires. *Arch. Sci. physiol.* **3** 547—566.

ARVANITAKI, A. and CHALAZONITIS, N. (1955a): Les potentiels bioélectriques endocytaires du neurone géant d'*Aplysia* en activité autorythmique. *C. R. Acad. Sci.* **240** 349—351.

ARVANITAKI, A. and CHALAZONITIS, N. (1955b): Variations lentes et périodiques du potentiel de membrane associées à des groupes de pointes (neurone géant d'*Aplysia*). *C. R. Acad. Sci.* **240** 462—464.

ARVANITAKI, A. and CHALAZONITIS, N. (1955c): Potentiels d'activité du soma neuronique géant (*Aplysia*). *Arch. Sci. physiol.* **9** 115—144.

ARVANITAKI, A. and CHALAZONITIS, N. (1958): Configurations modales de l'activité propres à différents neurones d'un même centre. *J. Physiol.* **50** 122—125.

ARVANITAKI, A. and CHALAZONITIS, N. (1960): Photopotentiels d'excitation et d'inhibition de différents somata identifiables (*Aplysia*). *Bull. Inst. océanogr. Monaco* **57** (1164) 1—83.

ARVANITAKI, A. and CHALAZONITIS, N. (1961a): Slow waves and associated spiking in nerve cells of *Aplysia. Bull. Inst. océanogr. Monaco* **58** (1224) 1—15.

ARVANITAKI, A. and CHALAZONITIS, N. (1961b): Motifs réitérés d'excitations-inhibitions réciproques, induites au niveau de cellules nerveuses contiguës (ganglion d'*Aplysia*). *C. R Acad. Sci.* **252** 192—193.

ARVANITAKI, A. and CHALAZONITIS, N. (1963): Patterned activities from identifiable "cold" and "warm" giant neurons (*Aplysia*). *1st internat. Symp. in "Olfaction and Taste"*, Stockholm, Ed.: Y. Zotterman, Pergamon Press, 377.

ARVANITAKI, A. and CHALAZONITIS, N. (1964a): Processus d'excitation d'un neurone autoactif sur ondes lentes. *C. R. Soc. Biol.* **158** 1119—1123.

ARVANITAKI, A. and CHALAZONITIS, N. (1964b): Processus inhibiteurs du^2 "courant générateur intrinséque" des neurones autorythmiques sur onde lente. *C. R. Soc. Biol.* **158** 1674—1677.

ARVANITAKI, A. and CHALAZONITIS, N. (1965a): Interactions entre onde lente du potentiel de la membrane neuronique et potentiels post-synaptiques d'excitation (neurone "branchial" d' *Aplysia*). *C. R. Soc. Biol.* **159** 1783—1787.

ARVANITAKI, A. and CHALAZONITIS, N. (1965b): Différenciations modales des activités électriques, par variation de la pression partielle de l'oxygène, sur trois neurones identifiables (*Aplysia depilans*). *C. R. Acad. Sci.* **261** 548—551.

ARVANITAKI, A., COSTA, H. and TAKEUCHI, H. (1965): Fréquences des potentiels oscillatoires et pointes de la membrane somatique par courants transmembranaires constants et linéairement croissants (neurone d'*Helix pomatia*). *C. R. Soc. Biol.* **159** 697—707.

ARVANITAKI, A. and ROMEY, G. (1966): Comportement de la membrane neuronique aux courants périodiques de basse fréquence. *J. Physiol. (Paris)* **58** 449—450.

ARVANITAKI, A. and ROMEY, G. (1968): In preparation.

ARVANITAKI, A., GOLA, M. and ROMEY, G. (1968): *C. R. Soc. Biol.* (In preparation).

ARVANITAKI, A., ROMEY, G. and TAKEUCHI, H. (1967): Variations de la relation caractéristique (V/i) de la neuromembrane en fonction de la pO_2. Implications fonctionnelles. *C. R. Soc. Biol.* **161**, 1629—1634.

ARVANITAKI, A., ROMEY, G. and WATANABE, Y. (1968): Acivités synaptiques synchrones et réciproques de neurones identifiables. *(Aplysia)*. Implications fonctionelles. *C. R. Soc. Biol.* (in press).

BRADLEY, K. and SOMJEN, G. G. (1961): Accomodation in motoneurones of the rat and the cat. *J. Physiol. (Lond)* **156** 75—92.

CHALAZONITIS, N. (1962): Inhibition thermique des ondes électriques lentes d'un neurone géant identifiable (neurone Br d'Aplysia fasciata). *C. R. Acad. Sci.* **255** 1652—1653.

CHALAZONITIS, N. (1963): Effects of changes in pCO_2 and pO_2 on rhythmic potentials from giant neurons. *Ann. N-Y Acad. Sci.* **109** 451—479.

CHALAZONITIS, N. (1967a): Intracellular pO_2 control on excitability and synaptic activability (in *Aplysia* and *Helix* idenfiable giant neurons). *Ann. N-Y Acad. Sci.* (in press).

CHALAZONITIS, N. (1967b): Synaptic properties of the oscillatory neurons (*Aplysia* and *Helix*). This Symposium.

CHALAZONITIS, N. and ARVANITAKI, A. (1951): Identification et localisation de quelques catalyseurs respiratoires dans le neurone d'Aplysia. *Bull. Inst. océanogr., Monaco* **48** (996) 1—20.

CHALAZONITIS, N. and ARVANITAKI, A. (1956): Chromoprotéides et succinoxydases dans divers grains isolables du protoplasme neuronique. *Arch. Sci. Physiol.* **10** 291—319.

CHALAZONITIS, N. and ARVANITAKI, A. (1958): Dérivation endocytaire simultanée de l'activité de différents neurones, *in situ*. *C. R. Acad. Sci.* **246** 161—163.

CHALAZONITIS, N. and ARVANITAKI, A. (1961): Slow changes during and following repetitive synaptic activation in ganglion nerve cells. *Bull. Inst. océanogr. Monaco*. **58** (1225) 1—23.

CHALAZONITIS, N. and GOLA, M. (1965): Enregistrements simultanés de la pO_2 intracellulaire et de l'autoactivité électrique du neurone géant (*Aplysia depilans*). *C. R. Soc. Biol.* **159** 1770—1776.

CHALAZONITIS, N., GOLA, M. and ARVANITAKI, A. (1965a): Microspectrométrie différentielle sur des neurones géants in vivo. (*Aplysia depilans*). — Mesure de la diffusibilité de l'oxygéne. *C. R. Soc. Biol.* **159** 2440—2445.

CHALAZONITIS, N., GOLA, M. and ARVANITAKI, A. (1965b): Oscillations lentes du potentiel de membrane neuronique, fonction de la pO_2 intracellulaire (neurones autoactifs d'*Aplysia depilans*). *C. R. Soc. Biol.* **159** 2451—2453.

CHALAZONITIS, N., TAKEUCHI, H. and ARVANITAKI, A. (1967): Accommodation des neurones à ondes lentes, au courant constant. *C. R. Soc. Biol.* (in press).

COLE, K. S. (1941): Rectification and inductance in the Squid giant axon. *J. gen. Physiol.* **25** 29–51.

COLE, K. S. and CURTIS, H. J. (1941): Membrane potential of the squid giant axon during current flow. *J. gen. Physiol.* **24** 551—563.

FRANK, K. and FUORTES, M. G. F. (1960): Accommodation of spinal motoneurones of cats. *Arch. ital. Biol.* **98** 165—170.

GRANIT, R., KERNELL, D. and SHORTESS, G. K. (1963): The behaviour of mammalian moto-neurones during longlasting orthodromic, antidromic and trans-membrane stimulation. *J. Physiol. (Lond).* **169** 743—754.

GREY WALTER, W. (1959): Intrinsic rhythms of the brain. In *"Handbook of Physiology"*. Neurophysiology 1, Ed.: J. Field, American Physiological Society, Washington, 279—298.

HILL, A. V. (1936): Excitation and accommodation in nerve. *Proc. roy. Soc.* B. **119** 305—355.

HODGKIN, H. L. (1948): THe local electric changes associated with repetitive action in a non-medullated axon. *J. Physiol.* **107** 165–181.

KANDEL, E. R. and TAUC, L. (1966): Anomalous rectification in the metacerebral giant cells and its consequences for synaptic transmission. *J. Physiol. (Lond.)* **183** 287—304.

KANDEL, E. R., FRAZIER, W. T. and COGGESHALL, R. E. (1967): Opposite synaptic actions mediated by different branches of an identifiable interneuron in *Aplysia*. *Science* **155** 346—349.

KOLMODIN, G. M. and SKOGLUND, C. R. (1958): Slow membrane potential changes in neurones. *Acta physiol. scand.* **44** 11—54.

SKOGLUND, C. R. (1942): The response to linearly rising currents in mammalian motor and sensory nerves. *Acta physiol. scand.* **4** *Suppl.* 12.

WITTENBERG, B. A., BRIEHL, R. W. and WITTENBERG, J. B. (1965): Haemoglobins of invertebrate tissues. *Biochem. J.* **96** 363—371.

Symposium on Neurobiology of Invertebrates 1967 (201—226)

SYNAPTIC PROPERTIES OF OSCILLATORY NEURONS
(*APLYSIA* AND *HELIX*)*

N. CHALAZONITIS

Département de Neurophysiologie Cellulaire
Institut de Neurophysiologie et de Psychophysiologie
Centre National de la Recherche Scientifique,
et Laboratorie d'Electrobiologie, E. P. H. E.
Marseille, France

INTRODUCTION

The electrical properties of the normally oscillating neuron (the neuron which, under normal conditions, is autoactive on slow-waves of the membrane potential) and its behavior with respect to current changes and to various environmental factors have well established its functional individuality (Arvanitaki and Chalazonitis, this Symposium). The main characteristics defining the functional entity of these neurons may be summarized as follows:

1. Evidence of spontaneous, low-frequency oscillations of the membrane potential (MP), very often devoid of any superimposed synaptic signal (Arvanitaki and Chalazonitis 1949; 1955a, 1955b, 1958).

2. Periodic changes in membrane resistance concomitant to the MP oscillations (Arvanitaki and Chalazonitis 1964a, 1964b).

3. Elicitation of the oscillations in resting neurons of this type by injecting a depolarizing current of constant intensity (Chalazonitis 1963; Arvanitaki and Chalazonitis 1965a).

4. Increase in the frequency of such oscillations by a depolarizing current of constant intensity.

5. Optimum oxygen (pO_2) requirements to elicit and maintain such oscillating activity (Chalazonitis *et al.* 1965; Arvanitaki and Chalazonitis 1965b).

Therefore, it seems appropriate to consider the physiological implications of such cells, i.e., to analyze: (a) the synaptic excitatory and inhibitory effects on such neurons, and (b) the way in which these cells assume the transfer and transformation of information, i.e., their synaptic interactions with other neurons.

Two types of oscillatory neurons will be considered here: — (1) neurons which normally oscillate on slow-waves (i.e., under normal — normoxic and normothermic — conditions); and (2) — neurons potentially oscillatory, i.e., neurons which fire regularly or are even silent under normal conditions, but which display oscillatory behavior following a synaptic afference or a brief current pulse input.

* The work reported in this paper was supported in part by the Centre National de la Recherche Scientifique, France, and in part by research grants from the National Institute of Health under grant NB 03337.

Abbreviations used in this work:

Br	type of neuron called the *branchial neuron*. Identifiable neuron visible on the dorsal surface of *Aplysia* abdominal ganglion near the emergence of the branchial nerve
O–N	oscillatory neuron
W–N	waving neuron
S–W	slow-wave
MP	membrane potential
P–O–N	potentially oscillatory neuron
LLH	long-lasting hyperpolarization
LLD	long-lasting depolarization
EPSP	excitatory postsynaptic potential
IPSP	inhibitory postsynaptic potential
G.IPSP	giant inhibitory postsynaptic potential
A.PSP	amphoteric postsynaptic potential
B.PSP	biphasic postsynaptic potential
Si–N	silent neuron
MR	membrane resistance

I. NORMALLY OSCILLATING NEURONS

(A) Excitatory synaptic effects

1. Subliminal actions

In oscillatory neurons (O–N) initially at rest because they are hyperpolarized, – e.g., under inward transmembrane current or under hyperoxia – the summation of EPSP's builds up a S–W. For an appropriate frequency and amplitude of the EPSP's, this S–W may only be subliminal, without any superimposed firing. In such a case, the amplitude of the EPSP's is continuously increasing all along the S–W. When the EPSP amplitude is compared at two isopotential points of the slow potential, the first EPSP is of lower amplitude than its conjugate one (Fig. 1). Thus, the iterative input of EPSP's on that neuron results in a kind of synaptic potentiation effect. Whatever the explanation, it is nevertheless remarkable that the summation of EPSP's, equivalent to a direct current, may produce a S–W. It is even possible for a S–W to be elicited by a single EPSP.

On the other hand, the effect of an EPSP may be simulated by the injection of a short depolarizing pulse of current (of 50-msec duration and of appropriate intensity), so that the summation of such short pulses similarly builds up a S–W. When the neuron is normally 'waving',* S–W's elicited by short pulses are real extra responses (Takeuchi and Chalazonitis, unpublished).

---

* The term 'waving' is proposed here by analogy with the term 'firing'; both mean repetitive responsiveness, the first by spikes, the second by waves.

2. *Liminal actions*

In an initially hyperpolarized, and hence, silent O–N, excitatory synaptic input of high frequency and amplitude may depolarize the membrane to a level suitable for a S–W elicitation (Fig. 2a). A sustained high-frequency synaptic input, by affording a continuous state of the membrane depolarization, may evoke a recurring sequence of S–W patterns which cease, as soon as the synaptic input is over (Fig. 2b).

If the O–N is initially autoactive on S–W's, a brief synaptic high-frequency excitatory drive will shortly elicit the whole S–W bimodal pattern of activity. It is interesting to note that following such an extra S–W, the spontaneous reiteration of the S–W pattern occurs after a longer period (Fig. 3).

It turns out that a synaptic excitatory input to the O–N is equivalent to a depolarizing direct current (Arvanitaki and Chalazonitis 1961b). For the same reason, repetitive injection of short (50-msec) pulses of depolarizing current may, as they imitate the EPSP effect, elicit extra waves (Fig. 3).

(B) Long-lasting inhibitory effects

1. *Gradually established hyperpolarization*

(a) HYPERPOLARIZATION BY EXCITATORY POSTSYNAPTIC POTENTIALS (EPSP's)

Inhibition in waving neurons differs from that of stable membrane neurons not only in the prolongation of the inhibitory response after the cessation of its afferent cause, but in the possibility of post-inhibitory depolarizing rebounds of the MP. This last property might better be qualified as a post-inhibitory 'oscillability'.*

Another remarkable property of the slow waving neuron is its peculiar way of effecting inhibitory transients: the synaptic input, apparently of the EPSP type, occurring on the depolarization plateau of the S–W elicits a gradual hyperpolarization. The net result is a shortening of the S–W plateau duration (compare Fig. 4a to Fig. 4b).

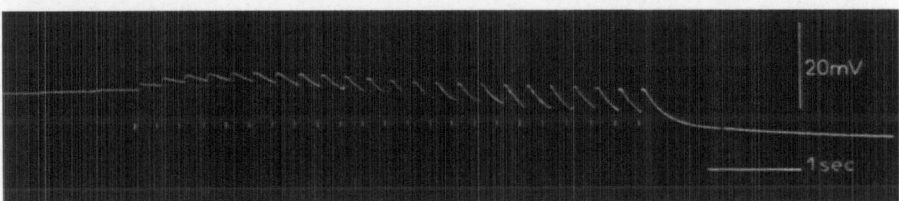

Fig. 1. Excitatory synaptic input (to a resting *Br* cell) eliciting a subliminal S–W. 5/sec EPSP's elicited by stimulation of the pleurogenital nerve (lower bars). Note the continuous increase in the amplitude of the EPSP's

* This word is coined to mean the ability to oscillate.

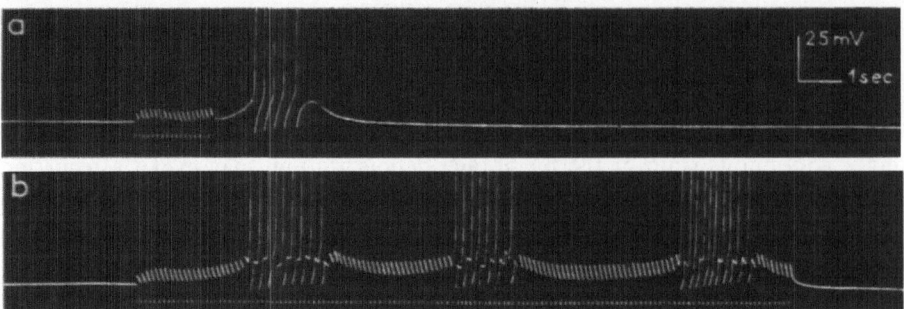

FIG. 2a. Excitatory synaptic input (lower bars) eliciting a liminal S–W. FIG. 2b. Long-lasting excitatory synaptic input repeatedly eliciting liminal S–W's

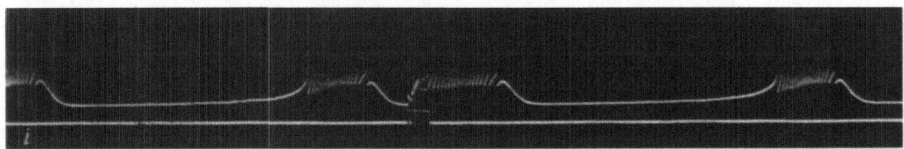

FIG. 3. Repetitive injection of eight short current pulses (lower trace) imitating the action of EPSP's and eliciting an extra S–W (Takeuchi and Chalazonitis, unpublished)

(b) HYPERPOLARIZATION BY ANTIDROMIC ACTIVATION

It is possible, by stimulating the pleurobranchial nerve in *Aplysia* visceral ganglion, to activate antidromically the *Br* cell through axonic branches or through its 'main neurite'.

When the antidromic potentials occur upon the depolarizing plateau of the S–W, the result is similar to the one reported above for EPSP-like potentials: the neuron is gradually hyperpolarized, and a long-lasting inhibition is established (Fig. 5b).

Hyperpolarization by axonal activation also occurs for oscillations elicited by dc: as is already known, a threshold depolarizing dc on *Br* cells initially at rest can elicit a long-lasting wave. Consequently, any supervening axonal activation is able to counteract the action of the depolarizing dc, and to establish a long-lasting hyperpolarization (LLH) (Figs. 6a, 6b and 7; Arvanitaki and Chalazonitis 1964b).

While the orthodromic stimulation enhances the firing frequency, an abrupt antidromic activation of the main neurite (or of the axon) not invading the soma is able to elicit a LLH not only on waving cells, but even on regularly firing cells (Fig. 8).

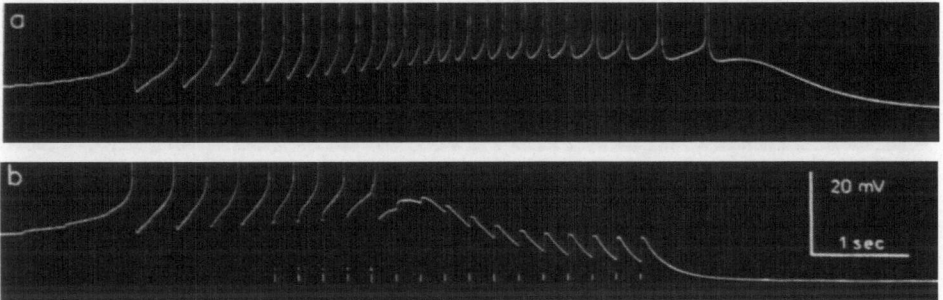

FIG. 4a. Normal S–W pattern

FIG. 4b. Same neuron as in Fig. 4a. Repetitive stimulation (lower bars) of the pleurogenital
nerve (starting on the ascending phase of the S–W) sets the pace for five spike generation
(compare to Fig. 4a pattern) followed by the elicitation of a LLH

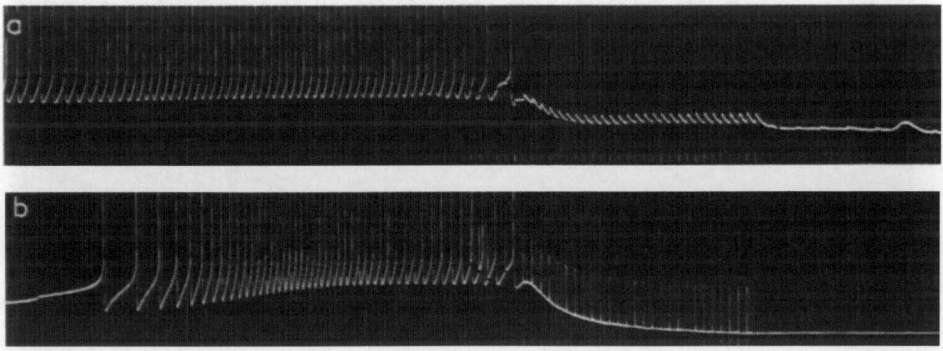

FIG. 5a. Same effect as in Fig. 4b, obtained with a higher frequency synaptic input (low
vertical bars)

FIG. 5b. The repetitive stimulation of the pleurobranchial nerve of *Aplysia*, starting on the
S–W plateau, elicits axonal branch potentials followed by hyperpolarization

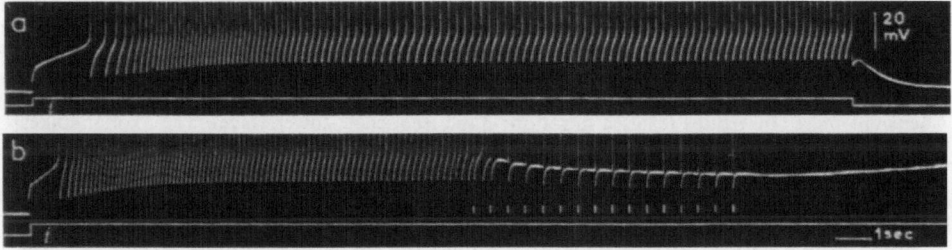

FIG. 6a. S–W discharge elicited by dc injection (*i*, lower trace) in an initially resting *Br* cell

FIG. 6b. Same neuron as in Fig. 6a. S–W discharge by dc of higher intensity (*i*). Repetitive
stimulation through the pleurobranchial nerve (lower bars) sets the pace for 16 antidromic
spikes interrupting the S–W discharge course. Note that as soon as the antidromic stimulation has ceased, the membrane starts redepolarizing under the continuing outward dc

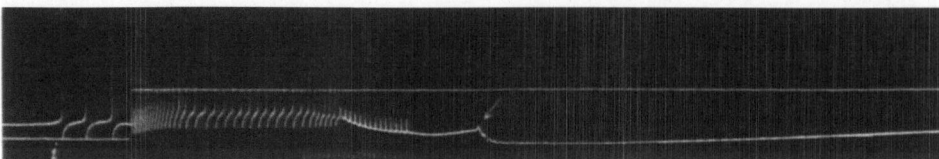

FIG. 7. Another *Br* waving cell. During outward, high-intensity dc injection (lower trace) the frequency of the waving is high. Superimposed repetitive axonal branch stimulation (lower vertical bars) interrupts the discharge course and the membrane repolarizes in spite of the continuing dc. Some seconds later, a spontaneous discharge (arrow) of amphoteric potentials elicits a secondary hyperpolarization, while dc is still continuing

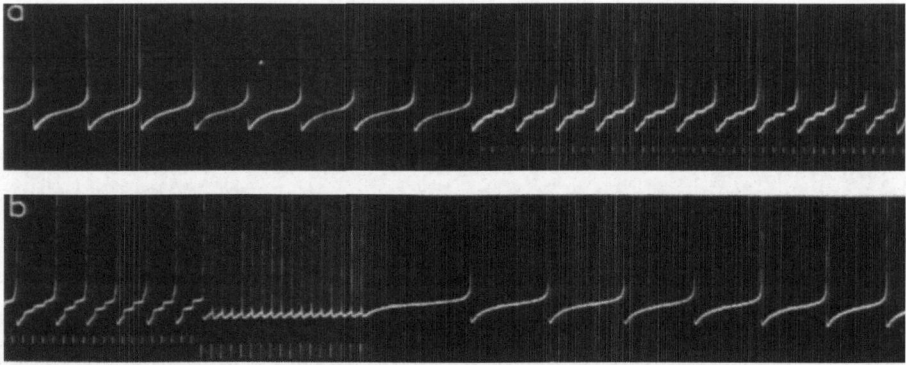

FIG. 8a. *Aplysia A* cell, initially autoactive (18 °C). When synaptically and repetitively stimulated (short vertical bars) its frequency increases

FIG. 8b. Continued from Fig. 8a. When the orthodromic synaptic stimulation is abruptly replaced by an antidromic stimulation (longer vertical bars), the cell spikes at the latter's pace. At 'off', the former low frequency autoactivity is gradually established

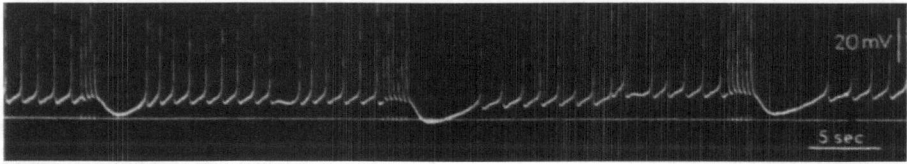

FIG. 9. A *Helix* waving neuron. Long hyperpolarizations by injection of brief depolarizing current pulses in groups (5-msec duration; 3/sec frequency, lower trace). The axonal spikes are followed by hyperpolarizing depressions (Takeuchi and Chalazonitis, unpublished)

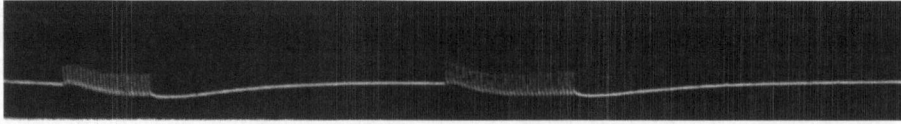

FIG. 10. Slow waving of the MP – on an initially resting *Br* cell – determined by repetitive stimulation of an axonal branch

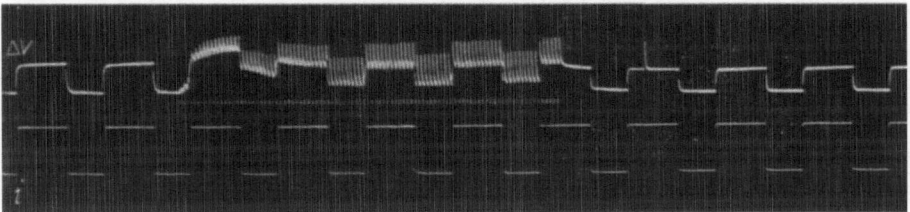

FIG. 11. Decrease in membrane resistance (MR) during and following an axon-branch stimulation. Upper trace, repetitive displacements (ΔV) of the MP of an *Aplysia* A cell due to short hyperpolarizing dc pulses (*i*). The repetitive axonal-branch activation (vertical bars) causes a decreased ΔV (due to a MR decrease). Following the cessation of the axon-branch activation, gradual recovery of MR

(c) HYPERPOLARIZATION BY DIRECT AXONAL BRANCH ACTIVATION

In oscillating and in regularly firing neurons of *Helix*, brief dc pulses (of $\approx$ 5-msec duration and of threshold intensity) preferentially activate the axonic membrane. Axonal potentials are elicited which are apparently devoid of any prepotential and are endowed with a small positive after-potential (Fig. 9). Such directly stimulated axonal responses are followed by a long-lasting inhibition. It must be specified that this inhibition is shorter than the one synaptically elicited. Axon branch potentials may also be elicited in the *Br* cell at rest by stimulating the appropriate nerve. Their summation leads to a LLH (Fig. 10). The membrane resistance decreases concomitantly with the summation of such axon branch potentials (Fig. 11).

2. *Abruptly established long-lasting hyperpolarization (LLH)*

(a) HYPERPOLARIZATION BY GIANT INHIBITORY POSTSYNAPTIC POTENTIALS (G.IPSP's)

The so-called giant inhibitory postsynaptic potentials (G.IPSP's) are of unitary origin and are remarkable for their amplitude (10–20 mV) and duration (several seconds). They were first described in S–W oscillating neurons of *Aplysia* inhibited under hyperthermia, and were called 'potentiels d'hyperpolarisation géants' (Chalazonitis 1962).

Such potentials are observable in both *Aplysia* and *Helix* on slowly oscillating neurons. They are either directly elicited by synaptic stimulation (Figs. 12 and 13) or occur after some delay from a preceding synaptic input of amphoteric potentials [see paragraph 2 (c) below]. These would activate indirect relays of neurons (a longer circuit of neurons) which in turn would inhibit, with a certain delay (3–4 sec), the S–W oscillating neuron (Fig. 14).

As to the mechanisms of the G.IPSP occurrence, two facts may be relevant for consideration: (1) a conspicuous decrease, almost annihilation, of the membrane resistance simultaneously with the G.IPSP course (Fig. 17), and (2) the possibility

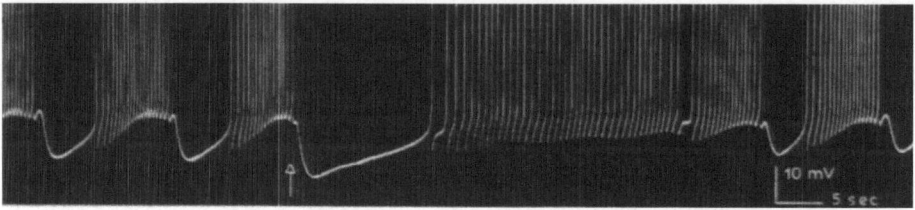

FIG. 12. *Helix* neuron displaying S–W's. At the arrow, a G.IPSP elicited by a single stimulus on the afferent nerve. Note post-inhibitory depolarizing rebound (Takeuchi and Chalazonitis, unpublished)

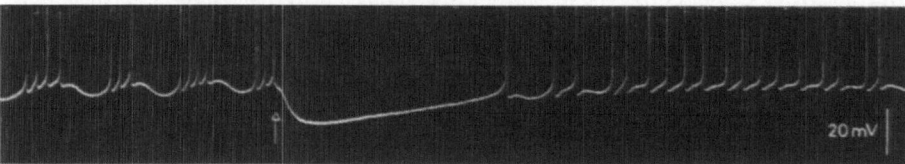

FIG. 13. Synaptically elicited G.IPSP (arrow) in a *Helix* waving neuron (Takeuchi and Chalazonitis, unpublished)

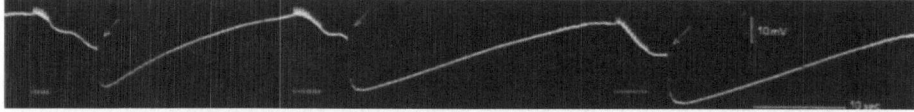

FIG. 14. Spontaneous G.IPSP's in *Aplysia depilans* neuron of the *Br* type elicited consecutively to graded inhibitions brought about synaptically and repetitively (lower bars)

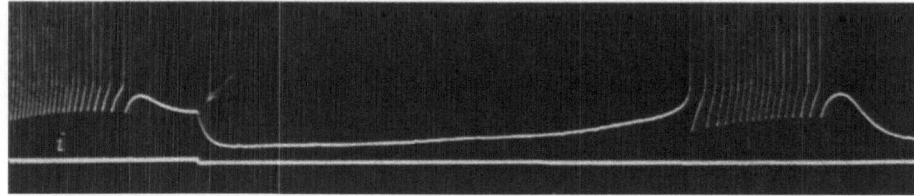

FIG. 15. Imitation of a G.IPSP (arrow) course by injection of a small hyperpolarizing dc during the falling phase of the S–W (this effect could also be produced by abrupt cessation of a depolarizing dc coinciding with the falling phase of the S–W) (Takeuchi and Chalazonitis, unpublished)

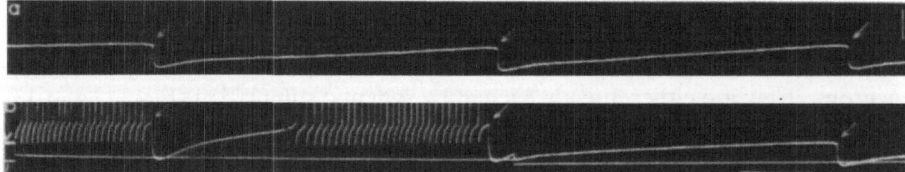

FIG. 16a. *Br*-type neuron of *Aplysia depilans* displaying 'spontaneously' (arrows) giant inhibitory postsynaptic potentials (G.IPSP's)

FIG. 16b. By injection of depolarizing dc the cell reaches its waving level. The S–W patterns are both cut off (arrows) by the G.IPSP's. Note: (1) the increase in amplitude of the G.IPSP's (arrows) when the cell is depolarized by dc; (2) the frequency of the G.IPSP's spontaneous recurrence, comparable in Fig. 16a and Fig. 16b (calibrations 20 mV and 10 sec)

of initiation of a G.IPSP by an abrupt decrease in the generator current intensity during the falling phase of a given S–W (Fig. 15; Takeuchi and Chalazonitis, unpublished).

(b) SPONTANEOUS PERIODIC ACTIVITY ELICITED BY PERIODIC OCCURRENCE OF G.IPSP

Recording from *Br* cells of *Aplysia depilans* specifically, it has been possible to reveal the so-called giant inhibitory potentials (G.IPSP) of 10–20 mV amplitude and 30–60 sec duration (Figs. 16a, 16b).

It is important to emphasize that a G.IPSP often gives rise to a conspicuous hyperpolarization, because at the instant of its incidence, the soma is subliminally depolarized (MP $\approx$ 45 mV). The incidence of the G.IPSP instantaneously brings the MP to more than 60 mV; thence, a LLD starts progressively. The depolarization probably proceeds from a growing generator current through a parasomatic area. Before the somatic depolarization can reach the firing level, a recurring G.IPSP again abruptly hyperpolarizes the membrane and the sequence is reproduced repeatedly. On the other hand, the abrupt large hyperpolarization is always accompanied by a conspicuous decrease of the membrane resistance (Fig. 17). On the ascending (redepolarizing) phase of the G.IPSP, the membrane resistance recovers progressively.

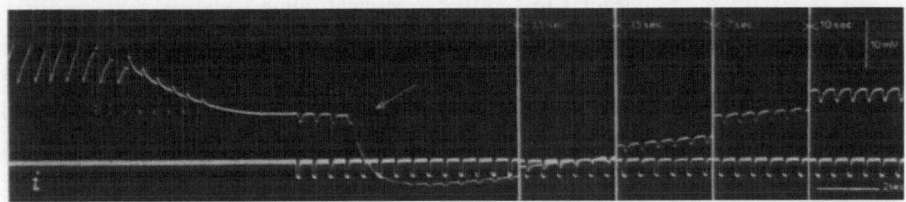

FIG. 17. *Aplysia Br* cell. Elicitation of a gradual hyperpolarization due to E.PSP input (vertical bars). The two first EPSP's (dotted) coincide with the two last spikes of the discharge. During the hyperpolarization course, injection of short inward dc pulses (lower trace). At the arrow – coincident with the fourth current pulse-generation of a G.IPSP. Note the annihilation of the MR and its gradual long-lasting recovery during the depolarizing phase of the G.IPSP

(c) HYPERPOLARIZATION BY AMPHOTERIC POSTSYNAPTIC POTENTIALS (A.PSP)

Negative-positive postsynaptic potentials, called amphoteric* have been described in the oscillating neurons of *Aplysia depilans* (Arvanitaki and Chalazonitis 1965c). As these potentials normally occur when evoked by a synaptic afference, they are qualified as postsynaptic events. They exhibit a minute initial depolarizing phase (0.5 mV amplitude) followed by an ample hyperpolarizing phase of long duration — i.e., a giant IPSP (Fig. 18).

* The term 'amphoteric' is used here to indicate the ability to displace the membrane potential in either direction, first depolarizing and then hyperpolarizing.

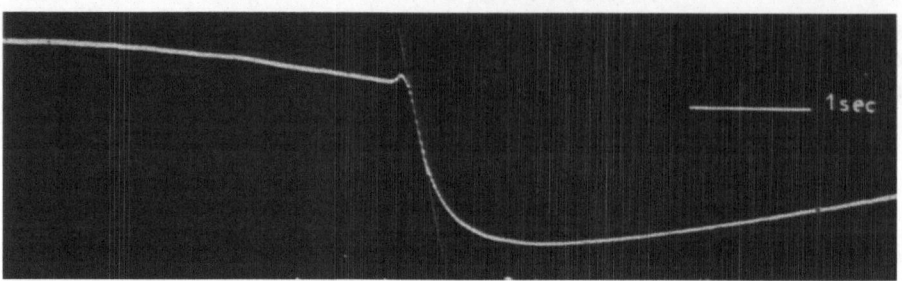

Fig. 18. In *Aplysia depilans Br*-type cell. Time course of the A.PSP. Note its minute depolarizing phase (0.5–1 mV) and its hyperpolarizing phase of several millivolts

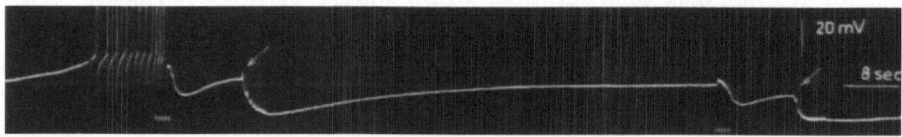

Fig. 19. *Aplysia depilans Br* cell, five EPSP's by nerve stimulation (vertical bars) introduced into the S–W discharge course. After some seconds, the synaptic hyperpolarization is followed by a burst of A.PSP's (arrow) and increased hyperpolarization. The same sequence is repeated later, as shown on the right side of the recording: five EPSP's followed by a burst of A.PSP's leading to a further hyperpolarization

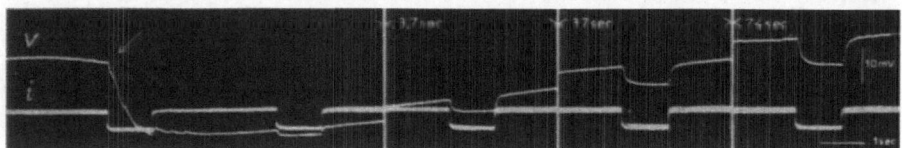

Fig. 20. *Aplysia depilans Br* cell. During the time course of a LLH injection of hyperpolarizing dc pulses (lower recording) probing MR changes. Strikingly coincident with the onset of a hyperpolarizing current pulse, a burst of six. A.PSP's superimposed on a 20-mV hyperpolarization of the membrane begins (arrow). Simultaneously, greatly reduced, the displacements of the membrane potential by the dc pulses indicate a substantial decrease in MR. The latter is very slowly reestablished to its initial value

Such potentials for S–W neurons, with their two asymmetric phases (an extremely small depolarizing phase preceding a large hyperpolarizing one) have been further analyzed in *Helix* preparations. Their shape has been studied as a function of the MP, the latter being displaced by a transmembrane current (Fig. 21; Chalazonitis and Takeuchi, unpublished). The hyperpolarizing phase of the A.PSP is reduced in a hyperpolarized membrane; conversely, it is increased when the neuron is depolarized.

The LLH raised by A.PSP is analogous to the one described above concerning the G.IPSP; it is also concomitant with a high local increase in the membrane conductance (Fig. 20). The first, minute depolarizing phase signalizes some local postsynaptic response evoked by some afference. The A.PSP has to be distinguished from the biphasic postsynaptic potential (B.PSP) previously described (Hughes and Tauc 1965), since in a B.PSP, the two phases are almost equivalent.

(d) SECONDARY LONG-LASTING HYPERPOLARIZATIONS

When LLH in *Aplysia Br* cells is established after a given summation of EPSP's, G.IPSP's, or A.PSP's, a second discharge of A.PSP's is often conspicuous 3–5 sec later (Figs. 14, 17 and 19). Such an interval seems too large to justify any lengthening of the synaptic processes after cessation of the nerve stimulation even in the case of a very long indirect pathway. However, the delayed long-lasting inhibition by A. PSP's could possibly be of synaptic origin if interneurons of the S–W type, delaying the information, were interposed.

(e) LONG-LASTING INHIBITION ESTABLISHED BY DC BIFOCAL SIMULTANEOUS ACTIVATION

With *Aplysia Br* cells it is sometimes possible by the injection of a suprathreshold depolarizing current to first elicit a somatic discharge which is then followed by a later discharge from a deeper area of the axonal or the neuritic type.

As soon as this axonal discharge starts, the somatic discharge is abolished and the cell hyperpolarizes (Figs. 22b and 23). This effect seems to be of the same type as the one elicited by stimulation of the pleurobranchial nerve, giving rise to axonal potentials (Figs. 5b and 6b). The interesting point here is that the direct stimulation of the axonal branches by a transmembrane dc occurs concurrently with the somatic stimulation. This is strikingly illustrated in Fig. 23, where, under the continuous action of an outward constant current, the S–W bursts of somatic spikes alternate with the regular perodic activity of a pair of axonal branches.

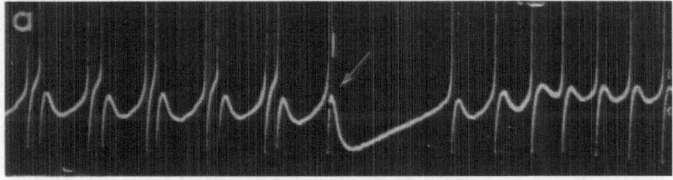

FIG. 21a. Normally W–N of *Helix pomatia*. Brief stimulation (upper vertical bar) of a nerve trunk eliciting an A.PSP. Note the supremacy of the hyperpolarizing phase; the slow recovery to the initial MP level followed by a depolarizing S–W

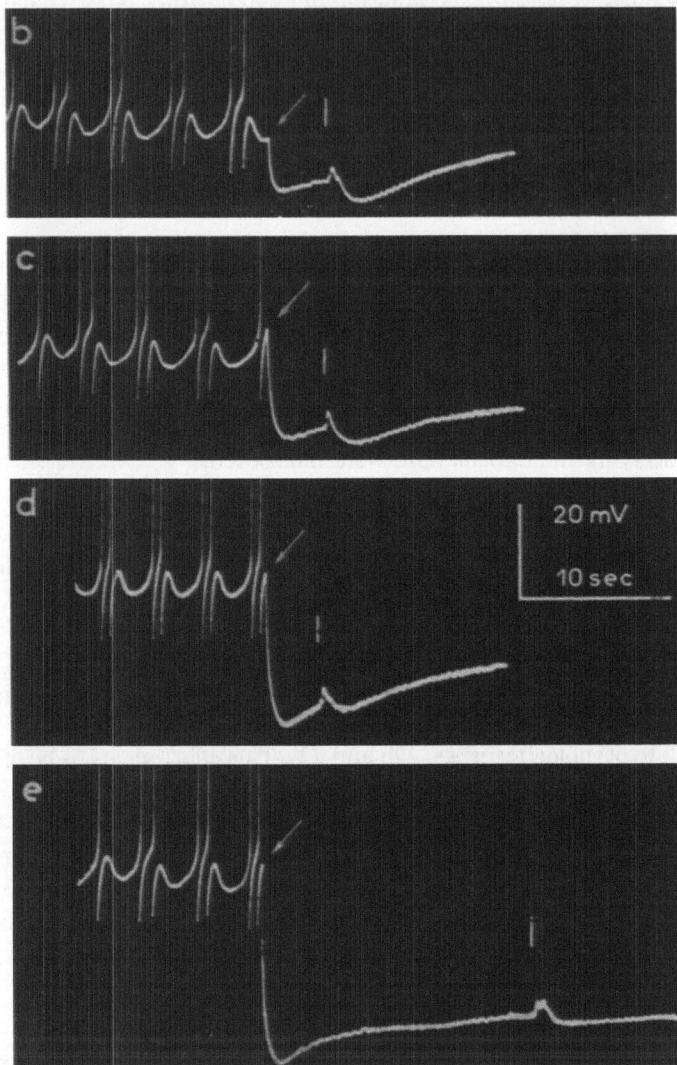

FIG. 21b–e. At the arrows, injection of inward dc of intensity increasing from b to e. The MP displacements increase accordingly. Note that as the MP increases, the A.PSP (signal at vertical bar) exhibits a conspicuous decrease of its hyperpolarizing phase with a slight increase of its depolarizing phase (Takeuchi and Chalazonitis, unpublished)

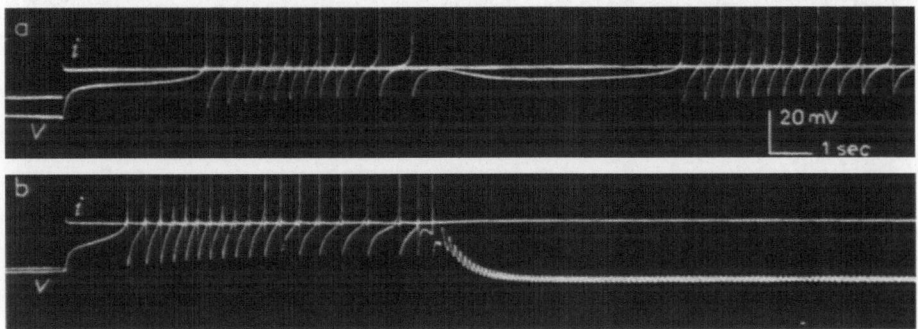

Fig. 22a. Initially resting *Br* cell; injection of an outward current (*i*) (upper trace); recurrent S–W's and bursts of spikes are displayed

Fig. 22b. The same cell injected with stronger dc (*i*) exhibits a somatic spike discharge which is interrupted by a discharge of axonal branch potentials (see lower vertical bars) hyperpolarizing the cell – in spite of the ongoing dc action – and determines a local increase of the membrane conductance

3. *Suggested mechanisms for long-lasting inhibitions*

A possible mechanism is based on the concept of a 'bifocal activation' leading to a diminution of the generator current (Fig. 24).

When for any reason the conductance of a remote axonal or neuritic area is increased, a leakage current escapes through that distant area and a diminution of the intrinsic generator current might ensue. Such a diminution would lead to a repolarization and hence to a decrease in the S–W frequency, or even to abolition. This would imply that as long as the leakage current persists, it would create a long-lasting somatic hyperpolarization.

But what could be the origin of the remote axonal increase in conductance? It might be due either to a synaptic activation or to a neuritic branch activation—whether direct or antidromic. Whatever the origin is, it has been observed that during any hyperpolarization obtained by any kind of activation of these remote areas, the membrane resistance always decreases. Therefore, it is suggested that a long-lasting hyperpolarization might be a postsynaptic event inherent to a bifocal activation in the same neuron (axon-hillock and remote axonal area), independent or not of any other presynaptic transmitter influence.

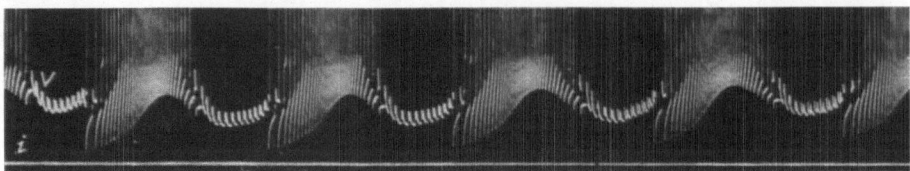

FIG. 23. *Aplysia depilans Br* cell submitted to a depolarizing dc of constant intensity activating alternatively a somatic area – which gives rise to a somatic spike discharge – and an axonal area which gives rise to axonal potentials. Somatic potentials are the ones on the rising phase of the S–W. Axonal and axon-branch potentials are conspicuous on the falling phase of the S–W

4. *Spontaneous oscillatory activity and post-inhibitory burst activity*

In various neurons of the higher centers of mammalians, exploration by micro-electrode reveals rhythmic activities not only of the regular monotonous type, but also those subsequent to synaptic activation (Li and Jasper 1953; Amassian and De Vito 1954; Buser and Albe-Fessard 1957; Andersen and Eccles 1962; Andersen and Sears 1964). Furthermore, rhythmic fluctuations of the membrane potential with a respiratory periodicity have been reported (R. Eccles *et al.* 1962; Sears 1964).

In *Helix* and *Aplysia* neurons autoactive at constant frequency, an afferent array of IPSP's gives rise at off to a post-inhibitory depolarizing rebound and a burst of spikes (Figs. 27 and 28). In *Aplysia* neurons, activity of this kind is commonly recorded (Tauc 1958, 1966; Fessard and Tauc 1960; Arvanitaki and Chalazonitis 1961a).

As a general rule, the maximum frequency of the post-inhibitory discharge increases as a direct function of the maximum level of the hyperpolarization reached during the inhibitory bombardment (Chalazonitis and Arvanitaki 1961).

In recent experiments, it has been established that not only a synaptic inhibitory input, but any cause of abrupt hyperpolarization of the somatic membrane, leads to a depolarizing rebound of the MP. Thus, a brief injection of hyperpolarizing current, through either an extracellular or an intracellular microelectrode, gives rise to an abrupt depolarizing rebound (Fig. 29; Takeuchi and Chalazonitis 1967). It is thus suggested that whenever a periodic inhibitory synaptic input is evident, the tendency to the S–W behavior of a neuron firing at constant frequency will be at the origin of the periodic bursts. It turns out that 'dynamic behavior' i.e., 'oscillability' is an intrinsic general property of firing neurons.

In mammalian centers, the interpretation of periodic activities of neurons among the most representative, such as the thalamic and the pyramidal cells, required spatial conditions. Andersen and Eccles (1962) postulated the possibility of feedback inhibitory systems and suggested the recurrent inhibition through collateral branching. The loop-like system would secure a regular timing in the generation of the inhibitory afferences.

Such a system is at least a two-cell system, securing a collateral inhibition

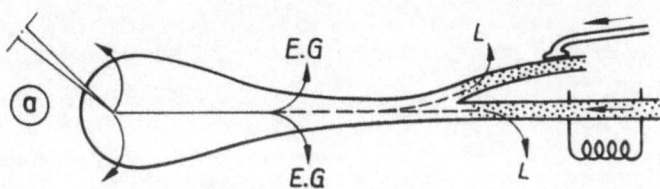

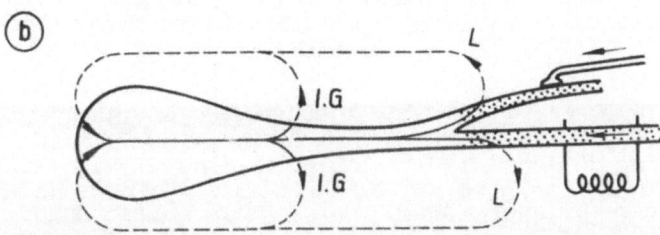

FIG. 24a. Hypothetical distribution of the generator current pathways. The anodal pole is the pipette inserted into the soma. (E.G) hypothetical outward lines of the 'extrinsic' generator current through the axon hillock area. Axon-branch stippled areas symbolize the areas of increased conductance – when activated synaptically (upper) or antidromically (lower) – which would allow secondary leakage of current (L), thus reducing the initial E.G current

FIG. 24b. Hypothetical paths of an *intrinsic generator current* eliciting the spontaneous S–W activity. The current is supposed to enter the apical pole of the soma and to emerge across the axon-hillock area and act there as an intrinsic generator (I.G) current for S–W's. The stippled areas of the axon-branches and the leakage current (L) have the same significance as in Fig. 24a in reducing the initial intensity of the I.G current

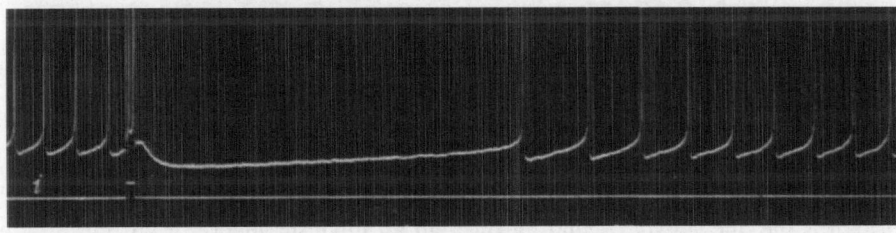

FIG. 25. In a regularly autorhythmic neuron of *Helix pomatia*, a long-lasting inhibition follows a brief discharge elicited by a short (200-msec) depolarizing dc intracellularly injected

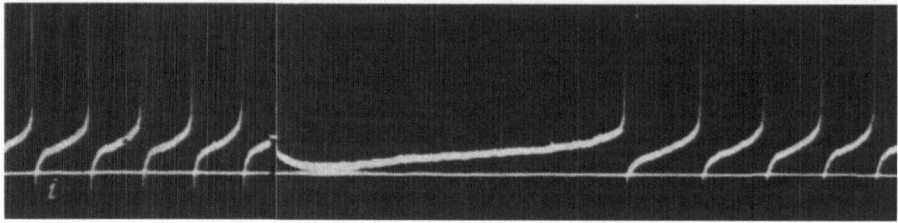

FIG. 26. In a regularly autorhythmic neuron *(Helix)*, a long-lasting inhibition follows a brief extracellular application of depolarizing dc pulse (200 msec) through an extracellular micro-electrode

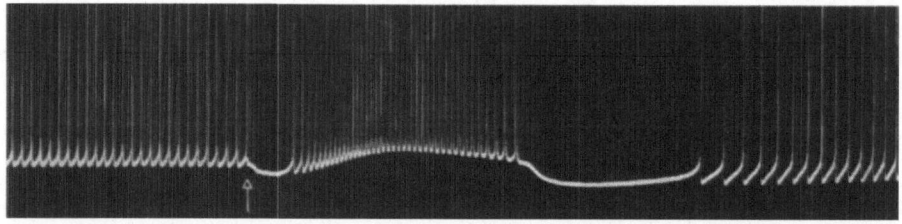

FIG. 27. Regularly autoactive *Helix* neuron. After a brief inhibitory synaptic input (arrow) hyperpolarization followed by a consecutive S–W display (Takeuchi and Chalazonitis, un published)

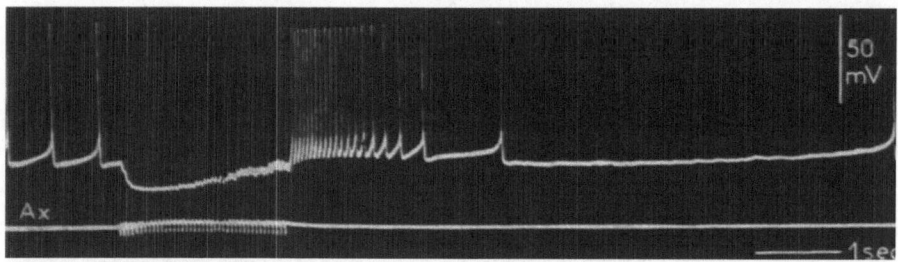

FIG. 28. A series of IPSP's invade a regularly autoactive neuron. In Ax – lower trace – a recording of the afferent spikes from the stimulated axon. Following the cessation of the IPSP's, the cell displays a consecutive waving and a high-frequency burst of spikes (Chalazonitis and Arvanitaki 1961)

through antidromic input. Comparatively, as suggested above, the single-cell system is the one slowly oscillating through an intrinsic generator current action (Fig. 24). Its inhibition would result from some loss of the intrinsic current.

However, even if the two-cell system has to be adopted in the mammalian centers, the oscillatory potentialities of the rhythmic cell cannot be overlooked. For instance, when IPSP's summate on a pyramidal cell, not only do they stop the activity, but they elicit a conspicuous post-inhibitory S–W rebound. On that S–W, the frequency of the spikes is extremely high. The genesis of that S–W denotes highly oscillatory properties of the pyramidal cell, which should indeed respond in the same way at 'off' of any hyperpolarizing extrinsic current.

Recent results pointed out intracellular oscillatory rhythms in pyramidal tract neurons in the cat (Jasper and Stefanis 1965).

In conclusion, whatever the mechanism of rhythmic discharge formation in higher nerve centers (cortical, thalamic, etc.), the electrical patterns obtained, at least suggest oscillatory potentialities of the neurons.

II. POTENTIALLY S–W OSCILLATING NEURONS

Many neurons firing normally at regular frequencies are able to display slow oscillations of their MP during or after a synaptic or a transmembrane current input. Such neurons are qualified as potentially S–W oscillating neurons. The development of prolonged depolarization and after-discharge, after the cessation of a synaptic excitatory input as well as the development of prolonged hyperpolarization were known in *Aplysia* neurons (Arvanitaki and Chalazonitis 1956; Tauc 1958; Fessard and Tauc 1960; Chalazonitis and Arvanitaki 1961a).

In apparently nonoscillating neurons, prolonged hyperpolarization outlasting any type of inhibitory postsynaptic input, is just an early sign allowing the expectation that the neuron has latent oscillatory properties. As a matter of fact, in order for it to be called potentially oscillatory, a cell must be able to display the whole behavior of an oscillatory unit, i.e. to display metasynaptic* long-lasting waving. It is thus worth while to discriminate between long-lasting hyperpolarization and post-inhibitory oscillations. When a given inhibitory action only leads to a long-lasting interval between the increase of the MP and its recovery, without any further change, the effect is merely a LLH. Yet – whatever its presynaptic cause – an LLH seems to be somehow related to the oscillating processes of the postsynaptic membrane. Such a conclusion is drawn from the possibility of eliciting an LLH by injections of a short *depolarizing* pulse of current (Figs. 25 and 26).

In oscillating and potentially oscillating neurons, an LLH may outlast a brief inhibitory synaptic input. It is very often followed by a depolarizing rebound. This is a real post-inhibitory S–W, characterizing the oscillatory potentialities of the neuromembrane (Fig. 27).

In the case of prolonged IPSP occurrence the depolarizing rebound develops immediately at off and is followed by long-lasting oscillations (Fig. 28).

As a general rule, any prolonged 'metasynaptic' reaction of the membrane (i.e., subsequent to a synaptic input) constitutes the first phase of a metasynaptic oscillatory behavior.

A S–W rebound may be provoked by the injection of a short inward-current pulse into the neuron. The membrane is said to display a poststimulative depolarizing rebound imitative of the metasynaptic changes.

* The term postsynaptic implying a spatial significance, the term metasynaptic is here adopted in its temporal significance: subsequent to the time of the occurrence of the synaptic event.

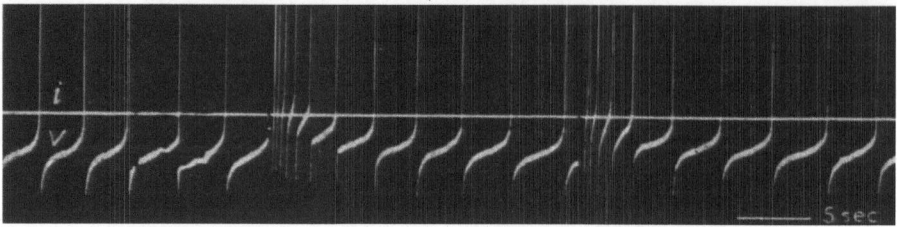

FIG. 29. In a *Helix pomatia* neuron, a depolarizing wave of the MP with a superimposed burst is developed following a short hyperpolarizing pulse (200-msec duration) through an extra-cellular microelectrode. At the second trial, the intensity of the current pulse being lower, the amplitude of the ensuing depolarization wave and the number of the spikes in the burst are minor (Chalazonitis and Takeuchi, unpublished)

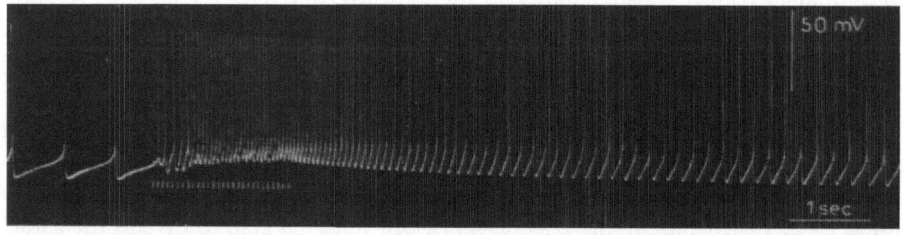

FIG. 30. In a regularly autoactive neuron of *Aplysia*, a long depolarization and high-frequency spikes outlast a synaptic excitatory input (lower vertical bars)

More recent results have shown that repeated short inward current pulses — injected dc — in potentially O–N's give rise to recurring poststimulative changes imitating the metasynaptic ones (Fig. 29).

It was, moreover, shown that in neurons of the S–W type, it is very easy to evoke a S–W extra pattern by injecting a sequence of short outward-current pulses (Fig. 3).

On regularly-firing neurons endowed with oscillatory potentialities, repetitive excitatory synaptic activation may give rise to a prolonged depolarization (Fig. 30). However, on firing neurons of the same type, direct injection of outward-current pulses does not always elicit prolonged depolarization. Sometimes the poststimulative prolonged depolarization is rather limited (see Fig. 31).

It is thus necessary to inquire further into the optimum membrane potential and duration of the impulse for such a response. As is already known, extracellular application of depolarizing dc (of 200-msec duration and 1-μA intensity) can, imitating the metasynaptically evoked prolonged depolarization, elicit a poststimulative S–W.

DISCUSSION

Studies of the S–W neuron normal behavior have thus led to the concept of 'oscillability' as the most general property of the neuronal membrane – although distributed to an extremely variable degree, according to the type of neuron considered and to the instantaneous values of its environmental parameters.

A neuron regularly autoactive at constant frequency, if submitted to a sudden short action – whether this is a synaptic event (either exctitatory or inhibitory) or a current pulse (either outward or inward) – reacts at the cessation of the stimulus by exhibiting damped slow oscillations of its MP, the first term being of a sign opposite to that of the MP displacement effected by the stimulus. Such an 'off' reaction – the so-called metasynaptic or poststimulative response – is merely an expression of the dynamic properties inherent in central neurons.

A sound concept thus results: whatever may be the pre- and postsynaptic mechanisms (neurotransmitters and the like), the metasynaptic effects due to the dynamic characteristics of the membrane can sufficiently explain or predict prolonged post- and metasynaptic activities.

They may be at the origin of the functional aspects discussed below.

1. *Frequency-modulator function*

It has already been emphasized that other neurons besides the S–W neuron have oscillatory potentialities, but are apparently devoid of S–W activity. Normally, near their firing level, these neurons are regularly autoactive with eventually some very slight modulation of their frequency.

If a very brief synaptic excitatory input is afferent until the appearance of a first spike, the cessation of the input does not prevent the further development of the depolarization (single-wave response) surmounted by a spike discharge.

This LLD, or single-wave response, may result in a 'frequency-multiplier' property in neurons endowed with oscillatory potentialities (Chalazonitis and Arvanitaki 1961). Another aspect of the 'frequency-multiplier' effect is the already considered post-inhibitory one-wave depolarizing response (of mammalian neurons and of molluscan neurons) in which the frequency is also highly multiplied (Figs. 27 and 28).

When compared to the self-sustained oscillatory neurons, what is remarkable in the P–O–N's is their 'damped' oscillability; in fact, these neurons do not oscillate continuously, but can display a single wave or several waves of decremental amplitude following each repetitive synaptic input.

2. *Systems of periodic massive discharge generators*

In the so-called after-discharge following a strong local stimulation of the brain (Chang 1959), the participation of single, self-sustained, autoactive neurons was recognized early. Furthermore, it was thought that short- and/or long-reverberating neuronal circuits might participate in the generation of the discharges.

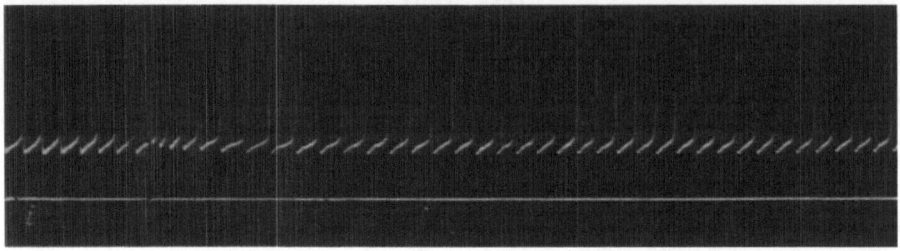

FIG. 31. In a regularly autoactive neuron of *Helix*, a transient outlasting depolarization follows a short depolarizing dc pulse (lower trace)

Concerning massive discharge production of periodic-style — particularly, of the epileptiform or convulsive type — one could here advance the possibility of the conjunction of a pacemaker undamped (or self-sustained) oscillatory cell with slightly-damped frequency-multiplier cells (Fig. 32).

The frequency of the periodic discharge of the pacemaker S–W cell will be multiplied by all other excitatorily driven cells of the chain, which will determine prolonged discharges of higher frequencies. The firing frequency of the S–W cells will also multiply through the post-inhibitory rebounds displayed by the inhibitorily driven cells — i.e., controlled by the inhibitory interneuron. It is evident that with such a system powerful output reciprocal discharges will be displayed periodically.

3. *Functional significance of oscillatory neurons conjugated to current-generating cells (receptor cells)*

It is often accepted that in many cases the transduction of different stimuli in the receptor cells leads to generator current production. This current in turn electrically stimulates the nerve cell conjugated to the receptor cell (Chalazonitis 1964; Goldman 1965; Hagins 1965; Byzov 1965).

It seems evident that the responses of the conjugated nerve cells will depend not only on the direction, shape, and intensity of the generator current crossing through them, but also on their own potentially oscillatory behavior. If, for example, the current is depolarizing, the nerve cell will display an extra burst at 'on' and will be silent at 'off'. If the current is hyperpolarizing and the nerve cell autoactive, the latter will be inhibited at 'on' and will display a burst at 'off'.

The interesting and far-reaching concept of high-plasticity reactions to generator currents — if the connected neurons are of the oscillatory type — is thus put forward.

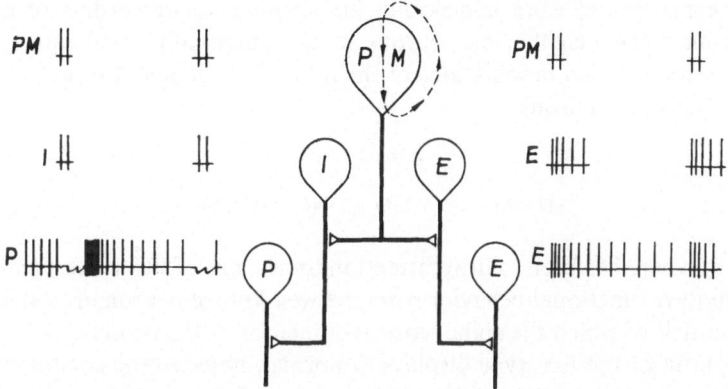

Fig. 32. Schematic representation of a hypothetical chain of five neurons securing a massive discharging system. PM is the pacemaker neuron of the *Br* type periodically displaying slow-waves and bursts under the action of an intrinsic generator current. *I* and *E* are, respectively, an inhibitory interneuron and a neuron driven excitatorily by the PM. *P* and *E* are receptor neurons of the frequency-multiplier type. The *P* neuron responds with a long-lasting depolarizing wave after the inhibitory arrest. The terminal *E* neuron responds with a LLD subsequent to the excitatory input resulting from the upper *E* neuron

4. *Potentially oscillatory neurons as modulators of wave frequency*

A periodic excitatory synaptic input can intermittently drive a silent O–N to its waving level. The shape and frequency of the secondary waves may be predicted by the laws governing the shape and frequency of S–W's as a function of the level of the membrane potential (see Arvanitaki and Chalazonitis, this Symposium).

Suppose that an oscillating W–N excitatorily controls another one which, because of its high membrane potential, is silent (Si). It is evident that the discharge of the silent, controlled neuron, will start later than that of the driver excitatory neuron. If the duration of the Si–N discharge is shorter than the excitatory W–N discharge, the Si–N may be considered to act as a filter for the periodic activity of the W–N.

If the duration of the Si–N discharge is longer than that of the W–N discharge, the Si–N discharge will each time substantiate a prolonged information output, and the input periodicity will thus smooth down and will tend to a continuous firing.

In summary, a potentially oscillatory neuron (and *a fortiori*, a chain of P–O–N) driven by a pacemaker O–N would be capable of modulating the frequency of the input S–W's, their amplitude (controlling the spike frequency), and their duration in a highly predictable manner if the input characteristics and those of the P–O–N oscillability are given.

Recent experimental work concerning the simultaneous recording of periodic activity from many neighboring neurons of the potentially oscillating type demonstrates the very rich modulations of the input S–W activity through chains of these very dynamic neurons.

SUMMARY AND CONCLUSIONS

(1) The aim here has been to draw attention to neurons of two types significantly differing in their functional behavior from the well-known neuron of 'stable membrane potential' of which the spinal cord motoneuron is the prototype.

(2) A neuron of the first type displays in normal, homeostatic conditions (normothermic, normoxic, etc.) periodic MP changes of the S–W type determining bursts of spikes. These changes are considered as autochthonous, for they may be evoked, modified, or supressed at will simply by transmembrane currents.

(3) Slow-waves are thought to be due to recurring intrinsic generator currents outward through some axon-hillock area near the soma.

(4) Neurons of the second type are autoactive at regular, constant frequencies, but under any synaptic input — either excitatory or inhibitory — they display slow, periodic MP changes. Such activity, outlasting the cessation of the synaptic input, is termed metasynaptic.

(5) In normally oscillating neurons, the synaptic excitatory input is mostly of the unitary type. A single EPSP may give rise to the whole S–W pattern. Summation of EPSP's evokes repetitive S W patterns.

(6) Neurons of the S–W type are devoid of classical IPSP's. Inhibition in such neurons is effected through two main processes to which MP transients of two main classes correspond:

(a) The gradually established, LLH either by small amphoteric synaptic potentials or by antidromic activation of axonal branches.

(b) The abrupt 10–20 mV hyperpolarizing potential of hundreds of seconds duration effected by the giant inhibitory potential or by a burst of 'amphoteric' postsynaptic potentials. These peculiar inhibitory potentials are all concomitant with a severe decrease in the resistance of remote axonic membrane areas. Such sudden increases in conductance would provide sinks in these remote areas, permitting the loss of the S–W generator currents and hence a net repolarization of the cell.

(7) The main functional significance attributable to the normally oscillating neuron is its pacemaking ability. Any normal S–W neuron can drive pairs of cells simultaneously and reciprocally: i.e., excitatorily and inhibitorily.

(8) The second type of neuron, called 'potentially oscillatory', may effect prolonged depolarizations or hyperpolarizations, respectively, to excitatory or inhibitory synaptic inputs. If these are 'dense' enough, the metasynaptic (following in time a synaptic action) behavior of the membrane may be slowly oscillating. Generally, the synaptic and/or metasynaptic oscillatory behavior of the MP of such neurons may be achieved by just injecting a transmembrane current — of

appropriate intensity, time course, and direction — equivalent to the synaptic input. It is thus suggested that the oscillatory potentialities of these cells are inherent to some specific molecular structure of their membrane, independently or not of specific presynaptic mechanisms.

(9) The overall functional implications of the P–O–N's have been discussed:

(a) from the point of view of their frequency-multiplier capabilities.

(b) considered as 'massive discharge' generators when driven by normally oscillating pacemakers.

(c) as systems excitable by the generator current, i.e., as receptor cells of electric current; the plasticity of their responsiveness at the onset and cutoff of any directional current input has been examined.

(d) as modulators of S–W frequency in controlling other neurons of the same type.

In conclusion, it can be said that the functional properties of these dynamic neurons are of the highest bearing in any interpretation of periodic activities in the higher nervous centers.

REFERENCES

AMASSIAN, V. E. and DE VITO, R. V. (1954): Unit activity in reticular formation and nearby structures. *J. Neurophysiol.* **17** 575—603.

ANDERSEN, P. and ECCLES, J. C. (1962): Inhibitory phasing of neuronal discharge. *Nature (Lond.)* **161** 214—215.

ANDERSEN, P. and SEARS, T. A. (1964): The role of inhibition in the phasing of spontaneous thalamocortical discharge. *J. Physiol. (Lond.)* **173** 459—480.

ARVANITAKI, A. and CHALAZONITIS, N. (1949): Prototypes d'interactions neuroniques. Données bioélectriques de préparations cellulaires. *Arch. Sci. physiol.* **3** 547—566.

ARVANITAKI, A. and CHALAZONITIS, N (1955a): Variations lentes et périodiques du potentiel de membrane associées à des groupes de pointes (Neurone géant d'*Aplysia*). *C. R. Acad. Sci. (Paris)* **240** 462—464.

ARVANITAKI, A. and CHALAZONITIS, N. (1955b): Potentiels d'activité du soma neuronique géant *(Aplysia)*. *Arch. Sci. physiol.* **9** 115—144.

ARVANITAKI, A. and CHALAZONITIS, N. (1956): Activations du soma géant d'*Aplysia* par voie orthodrome et par voie antidrome (dérivation endocytaire). *Arch. Sci. physiol.* **10** 95—128.

ARVANITAKI, A. and CHALAZONITIS, N. (1958): Configurations modales de l'activité propres à différents neurones d'un même centre. *J. Physiol.* **50** 122—125.

ARVANITAKI, A. and CHALAZONITIS, N. (1961a): Phases d'inhibition prolongées consécutives aux potentiels postsynaptiques d'excitation ou d'inhibition (neurones d'*Aplysia*). *J. Physiol. (Paris)* **53** 253—254.

ARVANITAKI, A. and CHALAZONITIS, N. (1961b): Slow waves and associated spiking in nerve cells of *Aplysia*. *Bull. Inst. océanogr. Monaco* **58** (1224) 1—15.

ARVANITAKI, A. and CHALAZONITIS, N. (1964a): Processus d'excitation d'un neurone autoactif sur ondes lentes. *C. R. Soc. Biol. (Paris)* **158** 1119—1123.

ARVANITAKI, A. and CHALAZONITIS, N. (1964b): Processus inhibiteurs du "courant générateur intrinséque" des neurones autorythmiques sur onde lente. *C. R. Soc. Biol. (Paris)* **158** 1674—1677.

ARVANITAKI, A. and CHALAZONITIS, N. (1965a): Les oscillations de basse fréquence du potentiel de membrane somatique (neurone d'*Aplysia*). *C. R. Soc. Biol. (Paris)* **159** 1179—1184.

ARVANITAKI, A. and CHALAZONITIS, N. (1965b): Différenciations modales des activités électriques, par variation de la pression partielle de l'oxygène sur trois neurones identifiables (*Aplysia depilans*). *C. R. Acad. Sci. (Paris)* **261** 548—551.

ARVANITAKI, A. and CHALAZONITIS, N. (1965c): Intéractions entre onde lente du potentiel de la membrane neuronique et potentiels post-synaptiques d'excitation (neurone "branchial" d'*Aplysia*). *C. R. Soc. Biol. (Paris)* **159** 1783—1787.

ARVANITAKI, A. and CHALAZONITIS, N. (1968): Electrical properties and temporal organization in oscillatory neurons. In: *Neurobiology of Invertebrates*. pp. 169—199.

BUSER, P. and ALBE-FESSARD, D. (1957): Explorations extracellulaires au niveau du cortex sensori-moteur du Chat. *In* "Microphysiologie comparée des éléments excitables". *Coll. internat. CNRS*, Ed. CNRS, Paris, 332—352.

BYZOV, A. L. (1965): Functional properties of different cells in the retina of cold-blooded vertebrates. *In* "Sensory receptors". *Cold Spring Harb. Symp. quant. Biol.* **30** 547—558.

CHALAZONITIS, N. (1962): Inhibition thermique des ondes électriques lentes d'un neurone géant identifiable (Neurone Br d'*Aplysia fasciata*). *C. R. Acad. Sci. (Paris)* **255** 1652—1653.

CHALAZONITIS, N. (1963): Effects of changes in pCO_2 and pO_2 on rhythmic potentials from giant neurons. *Ann. N. Y. Acad. Sci.* **109** 451—479.

CHALAZONITIS, N. (1964): Light energy conversion in neuronal membranes. *Photochem. Photobiol.* **3** 539—549.

CHALAZONITIS, N. and ARVANITAKI, A. (1961): Slow changes during and following repetitive synaptic activation in ganglion nerve cells. *Bull. Inst. océanogr. Monaco* **58** (1225) 1—23.

CHALAZONITIS, N., GOLA, M. and ARVANITAKI, A. (1965): Oscillations lentes du potentiel de membrane neuronique, fonction de la pO_2 intracellulaire (Neurones autoactifs d'*Aplysia depilans*). *C. R. Soc. Biol. (Paris)* **159** 2451—2453.

CHANG, H. T. (1959): The evoked potentials. *In "Handbook of Physiology. Neurophysiology I"*, Ed.: J. Field, American Physiological Society, Washington, 299—313.

ECCLES, R. M., SEARS, T. A. and SHEALY, C. N. (1962): Intracellular recording from respiratory motoneurones of the thoracic spinal cord of the cat. *Nature (Lond.)* **193** 844—846.

FESSARD, A. and TAUC, L. (1960): Variations prolongées du rhythme de neurones autoactifs provoquées par la stimulation synaptique. *J. Physiol. (Paris)* **52** 101.

GOLDMAN, D. E. (1965): *In* "Sensory receptors" *Cold Spring Harb. Symp. quant. Biol.* **30** 59—68.

HAGINS, W. (1965): Electrical signs of information flow in photoreceptors. *In* "Sensory receptors". *Cold Spring Harb. Symp. quant. Biol.* **30** 403—418.

HUGHES, G. M. and TAUC, L. (1965): A unitary biphasic post-synaptic potential (BPSP) in *Aplysia* brain. *J. Physiol. (Lond.)* **179** 27—288.

JASPER, H. and STEFANIS, C. (1965): Intracellular oscillatory rhythms in pyramidal tract neurons in the cat. *Electroenceph. clin. Neurophysiol.* **18** 541—553.

LI, C. L. and JASPER, H. (1953): Microelectrode studies of the electrical activity of the cerebral cortex in the cat. *J. Physiol.* **121** 117—140.

SEARS, T. A. (1964): The slow potentials of thoracic respiratory motoneurones and their relations to breathing. *J. Physiol. (Lond.)* **175** 404—424.

TAKEUCHI, H. and CHALAZONITIS, N. (1967): Hyperpolarisations ou dépolarisations prolongées post-stimulatives de la neuromembrane. *C. R. Soc. Biol. (Paris)* (in press).

TAKEUCHI, H. and CHALAZONITIS, N. (unpublished experiments).

TAUC, L. (1958): Processus post-synaptiques d'excitation et d'inhibition dans le soma neuronique de l'*Aplysia* et de l'Escargot. *Arch. ital. Biol.* **96** 78—110.

TAUC, L. (1966): Physiology of the nervous system. In *"Physiology of Mollusca* II", Ed.: K. M. Wilbur and C.M. Yonge, Academic Press, New York, 387—454.

DISCUSSION

M. Mirolli: When you obtain the giant inhibitory potential by axonal activation, what kind of stimuli do you use?

N. Chalazonitis: Long-lasting hyperpolarizations (LLH) of any kind are established gradually by axonal antidromic activation or abruptly by giant IPSP or by amphoteric postsynaptic potentials. These may all be obtained by single or repetitive stimulation of the pleurobranchial nerve in *Aplysia* and of the 'visceral nerve' in *Helix*. The duration of each stimulation is usually 5 msec at threshold intensity.

Sometimes, the giant IPSP may appear spontaneously, after a gradual LLH elicited by axonal stimulation *(Aplysia)*.

A. O. D. Willows: The cell you describe appears similar to one often studied by Strumwasser in the U.S.A. Do you suppose your *Br* cell is the same individual that Strumwasser calls cell-3, and Tauc calls 'Oberon'?

N. Chalazonitis: What I describe here, are the synaptic properties of the O-N's. However, the activity on S-W's of these neurons was first described by Arvanitaki and myself in 1950 based on extracellular recordings [*Arch. Sci. Physiol.* (1950) 3 547] and in 1955 on intracellular recordings [*C. R. Acad. Sci.* (1955) **240** 462].

If Strumwasser's cell-3 displays the same patterns as the *Br* cell, it could be of the same type. On the other hand, I am not aware of what *Aplysia* cell Tauc named 'Oberon', but I am afraid it sounds like a famous opera by Weber.

G. M. Hughes: I believe that you recorded the biphasic responses following stimulation of a whole nerve trunk. This and the fact that the amplitude of your response is much greater than that of the BPSP recorded by Tauc and myself suggests that they may be unrelated phenomena.

N. Chalazonitis: I agree. The giant IPSP and the amphoteric PSP already described by Arvanitaki and myself are totally different in shape and function from the BPSP's.

J. Salánki: Are the changes in the O_2 level you used in your experiments present in the intact animal? In other words, is this type of reaction to be considered as a physiological phenomenon or an artificial one?

N. Chalazonitis: The large pO_2 transition described in Dr. Arvanitaki's paper is used as a tool in studying the so-called biophysical — or better, molecular — properties of a neuronal membrane. It turned out that oxygen fluxes are equivalent to electrical currents in that the pO_2 controls the membrane resistance as well as the membrane potential. The limits of oxygen transitions here are rather pathological (hypoxia, hyperoxia, etc.).

In strictly physiological conditions, the normal pO_2 fluctuations in the blood of the ganglia are not yet known, and research is under way in this regard. Nevertheless, the high sensitivity of some *Aplysia* neurons, displaying ± 1 mV change in MP for 1-3 mmHg change in pO_2, could promote them to real chemoreceptor range in physiological conditions.

In connection with this, it is worth while to quote Clark's work on mammalian nerve centers, in which rather large pO_2 fluctuations seem to occur even in normal conditions [Clark, L. C., Jr., Misrahy, G. and Fox, R. P. *J. Physiol.* (1958) **13** 85].

B. Glaizner: Do you always find periodic activity such as you have described in this cell? The reason for my asking is that in my own experience with *Helix*, the occurrence of periodic activity is very variable.

N. Chalazonitis: In *Aplysia*, the typical periodic S–W neuron, the so-called *Br* cell, is easily identifiable, being near the origin of the branchial nerve [*J. Physiol. (Paris)* (1958) **50** 120].

In *Helix*, some cells are of the same type. They are visible on the dorsal side of the visceral ganglion toward the caudal part. If oscillatory cells are synaptically controlled, the occurrence of their periodic activity may be extremely variable.

Symposium on Neurobiology of Invertebrates 1967 (227—253)

CERTAIN ASPECTS OF THE PHARMACOLOGY OF
HELIX AND *HIRUDO* NEURONS

R. J. WALKER

Department of Physiology and Biochemistry, University of Southampton
Southampton, England

INTRODUCTION

A study of invertebrate pharmacology can lead to observations which are of value both to vertebrate and to comparative studies. A number of invertebrate preparations are used as biological assays for compounds which are of interest to vertebrate physiologists and pharmacologists — for example, the leech *(Hirudo)* dorsal muscle strip as an assay procedure for acetylcholine (Fuehner 1918; Minz 1932); the quahaug [*Mercenaria (Venus)*] heart as an assay for 5-hydroxytryptamine [5-HT' (serotonin)] (Welsh 1953); the crayfish stretch receptor neuron as an assay for gamma-aminobutyric acid (GABA) and for Factor I (Wiersma, Furshpan and Florey 1953; Florey 1954); and the crayfish hind gut for the assay of GABA and Factor I (Florey 1961).

On the whole, the pharmacologists who have employed these tissues for assay purposes have paid little attention to the natural role, if any, of these compounds in the normal physiology of the tissues used. Their main concern has been in the sensitivity and specificity of the tissue for the compound which they wish to measure. However, interest in the physiology of certain of these tissues has been shown. For example, Loveland (1953) has presented evidence for 5-HT as the mediator of excitation in the *Mercenaria* heart. The physiology of another molluscan heart, that of *Helix*, which has also been used as an assay tissue for 5-HT, has been investigated by S.-Rózsa and Perényi (1966). These authors demonstrated that 5-HT was released from the heart during stimulation of the extracardial nerve.

Invertebrate preparations can also be used in structure-activity studies in an attempt to elucidate the form of the receptor for a particular compound. Probably the most successful of these investigations was performed by Greenberg (1960) using the heart of *Mercenaria*. This author studied the effects of series of agonists structurally related to 5-HT in an attempt to find the particular configuration of 5-HT required to fit the receptor. The danger in such studies is that the final paper may look rather like a series of pages from a chemical catolog with little scientific merit or significance. Invertebrates often offer a simple alternative to the complex vertebrate nervous system when studying the fundamental properties of neurons and of axons. The most widely used preparation in this context is the giant axon of the squid, described by Young in 1935 (Young 1935, 1936, 1938). Certain annelids also possess equally large axons (Nicol 1948). In the heart of the decapod crustaceans there is a ganglion with only nine neurons which was described by Alexandrowicz (1932). This makes an ideal preparation for a study of the interaction of neurons. The stomatogastric ganglion, with about 35 neurons, offers another relatively simple system (Maynard 1966). More complex, but of great value due to the large size of some of their neurons, are the ganglia of the gastropod molluscs. Certain of the large neurons, $150-500\ \mu$ in diameter, in these animals can be identified by their size and position in relation to surrounding neurons, their electrical responses, and their pharmacological responses. Another preparation in which it is possible to recognize neurons due to their relative size is the segmental ganglion of the leech. On the ventral surface of each ganglion there are two relatively large neurons, 60–80 μ in diameter, the so-called giant cells of Retzius.

In this paper, the results from neuropharmacological and fluorescence-microscope studies on the snail *Helix* (Kerkut, Sedden and Walker 1967b; Walker and Hedges 1967a, 1968; Walker, Woodruff, Glaizner, Sedden and Kerkut 1968) and on the leech *Hirudo* (Kerkut, Sedden and Walker 1967a; Kerkut and Walker 1967) will be described.

METHODS

The experimental animals used in this study were the garden snail, *Helix aspersa*, and the medicinal leech, *Hirudo medicinalis*. The snails were obtained locally and the leeches from a dealer. Both animals were kept in a constant-temperature room at 15 °C. The snails were activated before use by being placed in about 2 cm of water in a beaker at room temperature. This procedure hydrates the animals and they begin to crawl actively around the container.

The dissection procedure for the snail is as follows (Walker 1968): The animal is pinned on a wax block and the brain, comprizing the supra- and suboesophageal ganglionic masses, is exposed. The tough outer connective tissue from over the upper surface of the suboesophageal ganglionic mass is removed with the brain still *in situ*. The larger neurons can now clearly be seen with the aid of a × 40 magnification binocular microscope. The neurons lie beneath an inner transparent layer of tissue. In some experiments, this inner layer was carefully slit to expose the neuron soma. Both the supra- and suboesophageal ganglionic masses are removed from the animal and placed on a glass slide. The brain can be held in position by means of rubber bands.

The dissection procedure for the leech is as follows: The animal is pinned on a wax block, dorsal side uppermost. A medial slit is made in the dorsal body wall along the entire length of the animal. The dorsal body wall is then pinned back to expose the ventral blood sinus. The ventral nerve cord lies within this ventral sinus. The whole of the cord and ventral blood sinus is then removed and placed in leech Ringer. The cord is then divided into three parts, each containing approximately seven ganglia. A group of seven ganglia are then placed on a glass slide, ventral side uppermost. A slit is made in the wall of the blood sinus to expose the ganglia and cord. The neurons can then be seen beneath the transparent capsule and within the giant glial cells. If required, the capsule can carefully be slit to expose the soma of the neurons.

The snail Ringer used had the following formula: NaCl, 80 mM; KCl, 4 mM; $CaCl_2$, 7 mM; $MgCl_2$, 5 mM; Tris HCl buffer, 5 mM; the pH was 8.

The leech Ringer used had the following formula: NaCl, 115 mM; KCl, 4 mM; $CaCl_2$, 2 mM; maleic acid, 10 mM; glucose, 10 mM; Tris HCl buffer, 10 mM; NaOH ~ 10 mM to give a pH of 7.4.

The recording electrodes were pulled from Pyrex glass tubing using a vertical electrode puller and were filled with a molar solution of potassium acetate. The electrode tips ranged in diameter from 0.1 to 0.5 μ, and had a resistance of between 20 and 60 MΩ. The spontaneous activity of the neurons was recorded by means of a Medistor negative-capacity electrometer amplifier and displayed on a Tetronix 502A oscilloscope. The activity was then filmed or recorded on an Ediswan pen oscillograph.

The volume of the bath was 20 ml. The value quoted for the agonists is the amount added to the bath expressed in μg, while the value quoted for the antagonists is the final concentration present in the bath expressed as μg/ml. The sections for examination in the fluorescence microscope were prepared as follows: The nervous tissue was quickly dissected out, coated in talc, and frozen in liquid nitrogen (Moline and Glenner 1964). The pieces of frozen tissue were treated according to the method of Falck and Owman (1965). In earlier experiments, they were held in small depressions in a metal block 35 mm from a cold finger which was maintained at -70 °C by a mixture of acetone and solid carbon dioxide. The temperature of the tissues was kept at -40 °C by a cooling mixture of solid carbon dioxide and ethyl oxalate. This temperature was stable as long as solid ethyl oxalate remained in the system. The tissues were dried for 2–5 days. In later experiments, the tissues were dried overnight in a Speedivac-Pearce tissue dryer. The tissues were then exposed to paraformaldehyde vapor (relative humidity 47%) for $1-3$ h at 80 °C. This was followed by vacuum embedding in paraffin wax (melting point 58 °C) and then sectioning at 10 μ. Controls were treated similarly but they were not exposed to the paraformaldehyde vapor. The sections were examined in a Zeiss fluorescence microscope, where the filter inserted in the exciting light beam was a Schott BG 12 (4-mm thick) and the barrier filters were Zeiss 50 and 44. The specificity of the fluorescence was checked by subjecting some of the slides to the sodium borohydride reduction test (Corrodi, Hillarp and Jonsson 1964).

RESULTS

Studies on the snail, Helix aspersa

1. The action of cholinergic antagonists

Acetylcholine can have one of two actions on *Helix* neurons (Tauc and Gerschenfeld 1960; Kerkut and Walker 1961, 1962). It will depolarize and excite some neurons and hyperpolarize and inhibit the activity of other neurons. The former cells are termed *D* cells and the latter cells are called *H* cells. Both the excitatory and

R. J. WALKER

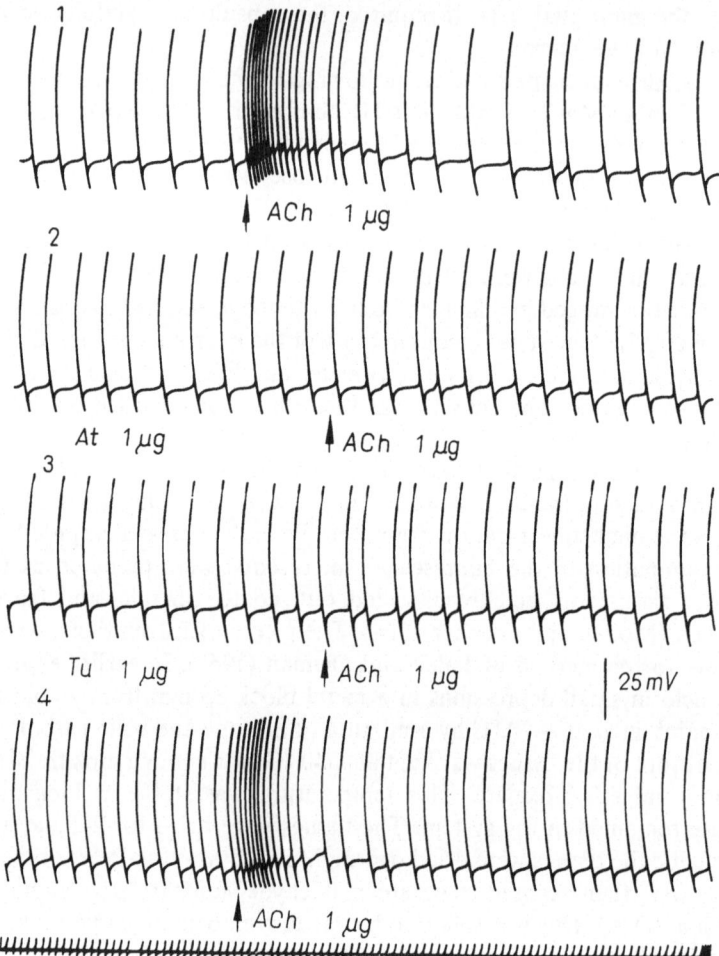

FIG. 1. The effect of pretreatment with 1 µg/ml atropine (At) or 1 µg/ml tubo-curarine (Tu) on the acetylcholine (ACh) response of a *Helix D* cell. The time trace in this and subsequent figures is in intervals of 1 sec (from Walker and Hedges 1967a)

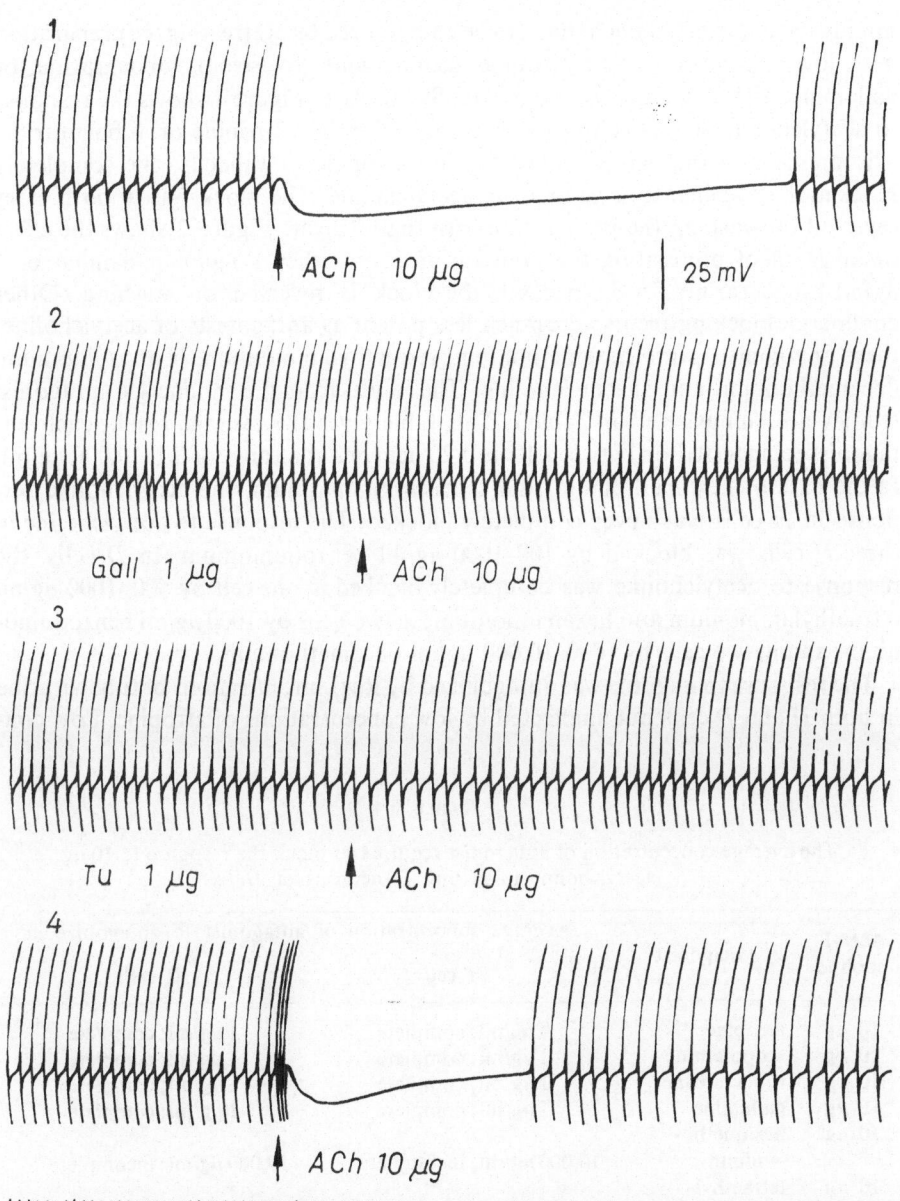

FIG. 2. The effect of pretreatment with 1 μg/ml gallamine (Gall) or 1 μg/ml tubocurarine (Tu) on the acetylcholine (ACh) response of a *Helix H* cell (from Walker and Hedges 1967a)

inhibitory action of acetylcholine can be antagonized by pretreating the preparation with low concentrations of atropine, scopolamine (hyoscine), tubocurarine, or gallamine. All four antagonists were equally effective. Figure 1 shows the response to acetylcholine of a D cell in the presence of either atropine or tubocurarine. The presence in the bath of either 1 μg/ml atropine or tubocurarine completely blocked the response to 1 μg or 10 μg acetylcholine. The block was in both cases reversed on washing the preparation with snail Ringer. Figure 2 shows the effect on an H cell of pretreating the preparation with either 1 μg/ml gallamine or 1 μg/ml tubocurarine. In both cases, the block is reversed on washing. Other cholinergic blocking agents were much less potent as antagonists of acetylcholine; these antagonists were hexamethonium, tetraethylammonium, benzoquinonium, decamethonium, and succinylcholine. The results from the cholinergic antagonist studies are summarized in Table 1. This table gives the average result, each antagonist being tested on between 10 and 20 neurons at a range of 0.1–10,000 μg/ml. In the case of hexamethonium and tehraetyltammonium, the acetylcholine response in H cells was never completely blocked. The response to acetylcholine in three H cells was blocked by 100–1000 μg/ml benzoquinonium. In D cells, the response to acetylcholine was completely blocked in one cell by 500–1000 μg/ml tetraethylammonium and hexamethonium; in two cells by 1000 μg/ml benzoquinonium; and in two cells by 1000–10,000 μg/ml decamethonium.

The response to nicotine was antagonized by low concentrations of tubocurarine in both H and D cells but unaffected by low concentrations of atropine. Low con-

TABLE 1

The average concentration of antagonist required to block the response to 10 μg acetylcholine in H and D neurons of *Helix**

Acetyl-choline	Antagonist	Average concentration of antagonist to antagonize in	
		H cells	D cells
10 μg	atropine	1 μg/ml; complete	1 μg/ml; complete
10 μg	scopolamine	1 μg/ml; complete	1 μg/ml; complete
10 μg	tubocurarine	1 μg/ml; complete	1 μg/ml; complete
10 μg	gallamine	1 μg/ml; complete	1 μg/ml; complete
10 μg	hexamethonium	10,000 μg/ml; incomplete	10,000 μg/ml; incomplete
10 μg	tetraethylammonium	10,000 μg/ml; incomplete	10,000 μg/ml; incomplete
10 μg	decamethonium	1 000 μg/ml; complete	10,000 μg/ml; incomplete
10 μg	benzoquinonium	10,000 μg/ml; incomplete	10,000 μg/ml; incomplete
10 μg	succinylcholine	1 000 μg/ml; complete	10,000 μg/ml; incomplete

* Where the response to acetylcholine is only reduced and not blocked, the antagonism is termed incomplete.

TABLE 2

The mean threshold dose required to excite or inhibit *Helix* neurons*

Compound	Number of cells tested	Mean dose in μg to		Number of cells tested	Compound
		Excite	Inhibit		
Carbachol	7	2.1 (1–5)	2.5 (0.5–10)	11	Nicotine
Nicotine	11	2.5 (0.5–10)	3.6 (1–10)	12	Butyryl-choline
Butyrylcholine	10	3.2 (1–10)	3.9 (1–10)	9	Carbachol
Propionylcho-line	10	3.6 (1–10)	4.6 (1–10)	250+	Acetylcho-line
Acetylcholine	250+	4.7 (1–10)	7.0 (1–50)	11	Propionyl-choline
Muscarine	12	8.4 (0.5–50)	10.9 (1–50)	9	Muscarine
Muscarone	10	23.8 (0.5–100)	11.1 (1–50)	12	Succinyl-choline
Mecholine	17	89.4 (1–1000)	20.2 (1–100)	10	Muscarone
Dimethylphenyl-piperazinium	7	117.1 (10–500)	27.8 (1–50)	4	Dimethyl-amino-ethanol
Pilocarpine	8	160.0 (1–1000)	37.5 (5–100)	6	Benzoyl-choline
Tetramethyl-ammonium	6	195.0 (10–500)	46.4 (1–100)	8	Tetrame-thylammo-nium
Benzoylcholine	6	222.6 (10–1000)	47.5 (10–100)	8	Dimethyl-phenylpi-perazinium
Choline	7	226.5 (10–1000)	52.2 (1–1000)	5	Pilocarpine
Arecoline	6	301.7 (10–1000)	58.8 (10–100)	8	Choline
Dimethylamino-ethanol	4	312.5 (50–1000)	74.0 (1–500)	13	Mecholine
Oxotremorine	7	550.0 (100–1000)	124.3 (10–500)	7	Oxotremo-rine
Succinylcholine	9	666.6 (500–1000)	194.3 (10–500)	7	Arecoline

* The cholinergic agonists are arranged in decreasing order of potency for both *H* and *D* cells. The figures in brackets refer to the range for the threshold response.

centrations of atropine completely antagonized the response in *H* and *D* cells to mecholine. The mecholine response was partially antagonized by tubocurarine.

Sixteen agonist compounds of acetylcholine were tested for their effect on *H* and *D* cells. The results are shown in Table 2. From this table it can be seen that a number of compounds were similar in potency to acetylcholine. These compounds were carbachol, nicotine, butyrylcholine, propionylcholine, and muscarine. Muscarone was slightly less potent than acetylcholine. The remaining agonists were

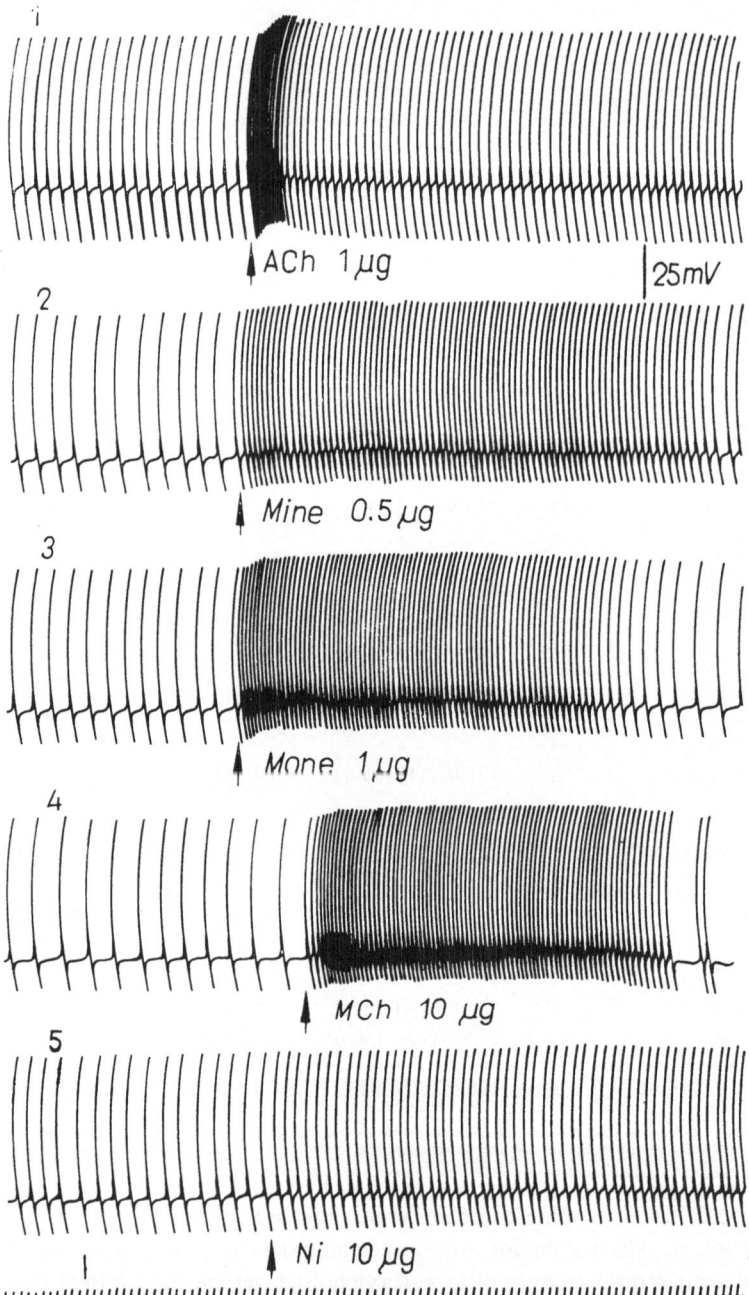

FIG. 3. The muscarinic cell; the effect of 1 μg acetylcholine (ACh), 0.5 μg muscarine (Mine), 1 μg muscarone (Mone), 10 μg mecholine (MCh), or 10 μg nicotine (Ni) on the activity of a *Helix D* cell (from Walker and Hedges 1968)

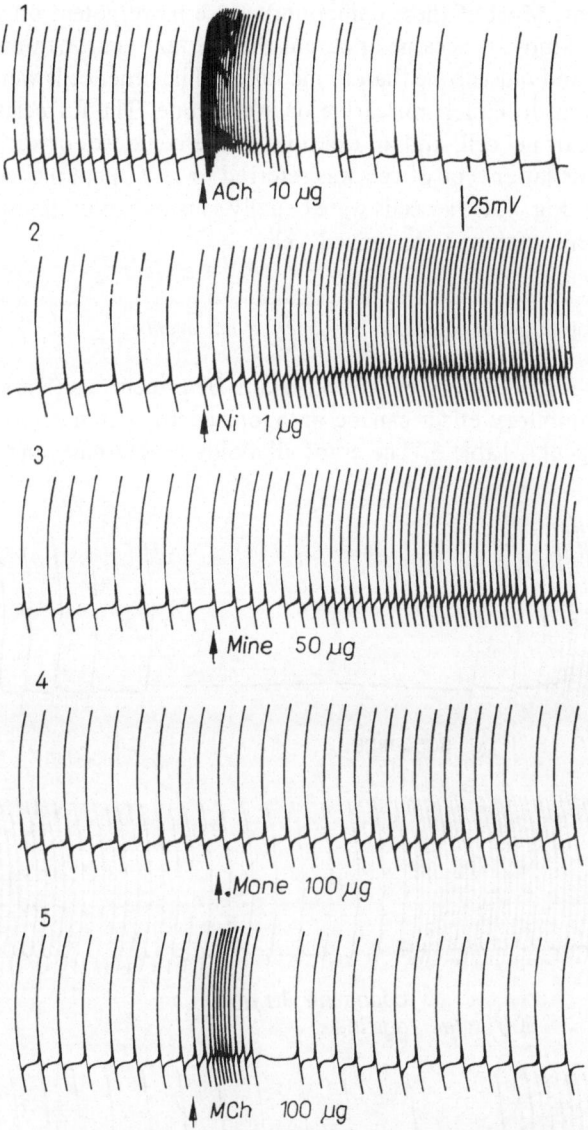

FIG. 4. The nicotinic cell; the effect of 10 μg acetylcholine (ACh), 1 μg nicotine (Ni), 50 μg muscarine (Mine), 100 μg muscarone (Mone), or 100 μg mecholine (MCh) on the activity of a *Helix D* cell (from Walker and Hedges 1968)

considerably less potent. Most of these compounds were more potent on *H* than on *D* cells. During the study, it became apparent that certain neurons were more sensitive to muscarine and muscarone than to nicotine, while other cells were more sensitive to nicotine than to either muscarine or muscarone. The former type of cell was called a 'muscarinic' cell, and an example of the responses of such a cell are shown in Fig. 3. The latter type of cell was referred to as a 'nicotinic' cell, and an example is shown in Fig. 4. Other cells were equally sensitive to both muscarine and nicotine. These were referred to as 'mixed' cells.

2. *The action of dopamine antagonists and agonists*

Dopamine has a strong inhibitory effect on certain *Helix* neurons (Kerkut and Walker 1961). This inhibitory effect can be antagonized by a number of alpha adrenergic blocking agents, Table 3. The ergot alkaloids ergotamine and ergot-

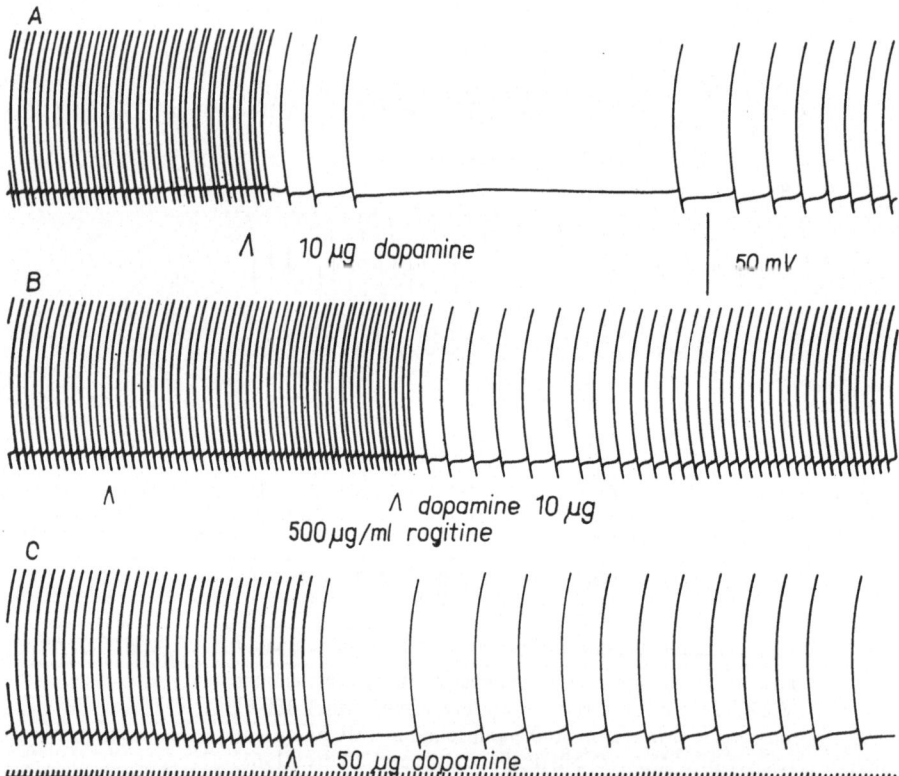

FIG. 5. (A) The effect of 10 μg dopamine on the activity of a *Helix* neuron; (B) the effect of pretreatment with 500 μg/ml Rogitine on the response of the cell to 10 μg dopamine; (C) after 30 min washing the preparation with Ringer, the effect of 50 μg dopamine on the activity of the cell. The inhibition of the dopamine effect by this very high concentration of Rogitine could not be completely reversed

TABLE 3

The threshold range of antagonist required to block the response
to 1 μg dopamine on *Helix* neurons

Dopamine	Antagonist threshold	
1 μg	Ergometrine	0.001 > 0.1 μg/ml
1 μg	Ergotamine	1 > 10 μg/ml
1 μg	Ergotoxine	1 > 10 μg/ml
1 μg	Dibenyline	1 > 10 μg/ml
1 μg	Rogitine	1 > 50 μg/ml
1 μg	Yohimbine	100 > 1000 μg/ml

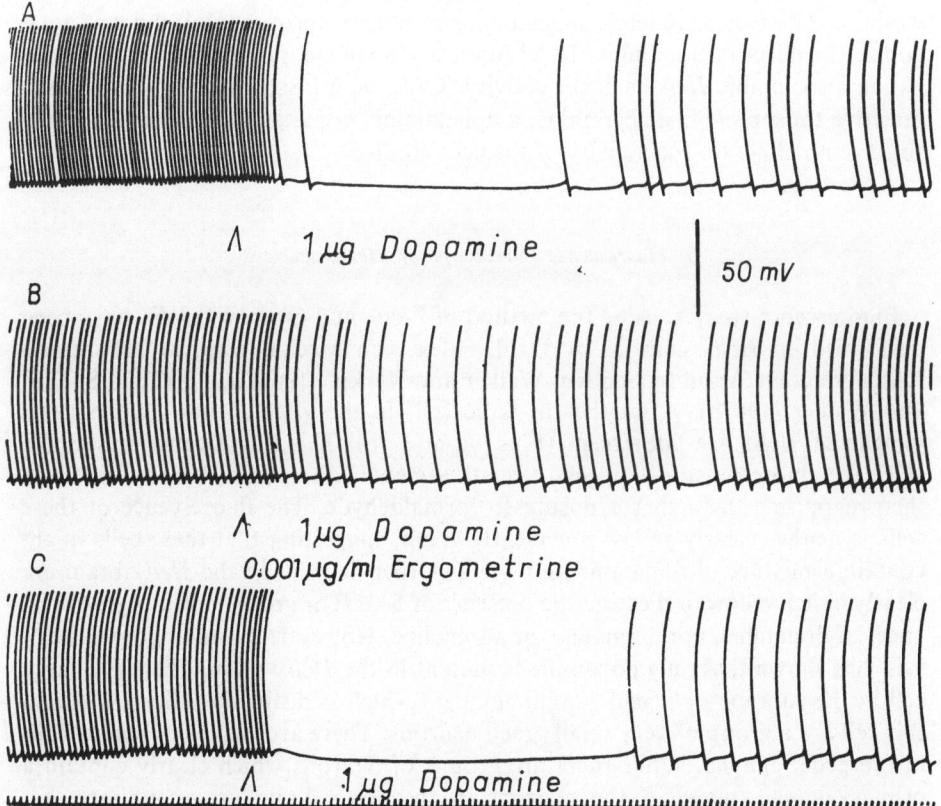

FIG. 6. (A) The effect of 1 μg dopamine on the activity of a *Helix* neuron; (B) the effect of
pretreatment with 0.001 μg/ml ergometrine on the response of the cell to 1 μg dopamine;
(C) after washing with Ringer the response to 1 μg dopamine returns completely (from Walker
et al. 1968)

oxine antagonize the effect of 1 µg dopamine when present at a concentration of between 1 and 10 µg/ml. Dibenyline (phenoxybenzamine) is also equally effective as an antagonist. Rogitine (phentolamine) and Priscol (tolazoline) are less effective, and yohimbine is a poor antagonist.

Figure 5 shows the blocking action of Rogitine. In this preparation, the effect of 10 µg dopamine was greatly reduced by the presence in the bath of 500 µg/ml Rogitine. This block could not be fully reversed, as can be seen in Fig. 5C, in which after 30 min 50 µg dopamine was still less effective than 10 µg dopamine added before the addition of Rogitine. The antagonism of Rogitine was fully reversible when it was used at lower concentrations. By far the most potent antagonist of the inhibitory dopamine response was the ergot alkaloid ergometrine, which is not an adrenergic blocking agent. The presence in the bath of 0.001 µg/ml ergometrine reversibly antagonized the response to 1 µg dopamine (Fig. 6). The blocking action of higher concentrations of ergometrine was only partially reversed on prolonged washing. The beta adrenergic blocking agent inderal (propranolol) did not antagonize the effect of dopamine. In addition to dopamine, noradrenaline and adrenaline also inhibit *Helix* neuron activity. Cells were insensitive or only slightly sensitive to isoprenaline. Tyramine, amphetamine, dopa, and mescaline appeared to have no effect on the activity of the cells studied.

3. *Fluorescence microscope of Helix neurons*

Fluorescence studies, using the method of Falck and Owman (1965), have been performed on *Helix* neurons by Dahl, Falck, von Mecklenberg, Myhrberg and Rosengren (1966) and by Sedden, Walker and Kerkut (1967) and Kerkut, Sedden and Walker (1967b). A number of large cells in the visceral and right parietal ganglia of *Helix* are fluorescent (Figs. 7a and 7b). This fluorescence is specific, since the fluorescence disappears after treatment with sodium borohydride and then reappears on further exposure to formaldehyde. The fluorescence of these cells is neither clearly yellow nor clearly green, suggesting that these cells might contain a mixture of dopamine and 5–HT. Other neurons in the *Helix* brain are clearly either yellow, indicating the presence of 5–HT, or green, indicating the presence of dopamine, noradrenaline, or adrenaline. However, chromatographic analysis has shown that only dopamine is present in the *Helix* brain. There is a large cell in the supraoesophageal ganglionic mass which is distinctly yellow. Close to this cell are a group of very small green neurons. There are groups of yellow cells in the pedal ganglia. While there are groups of neurons which clearly contain a primary amine, many of the neurons in the *Helix* brain are nonfluorescent (Fig. 7a).

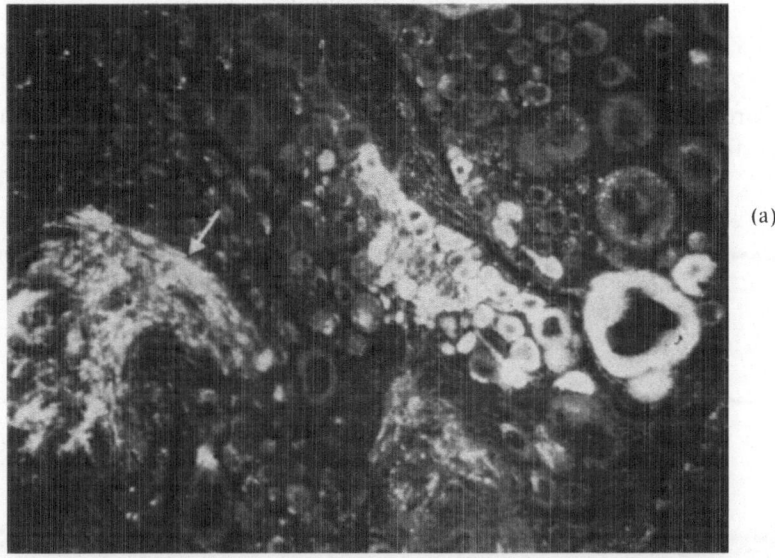

(a)

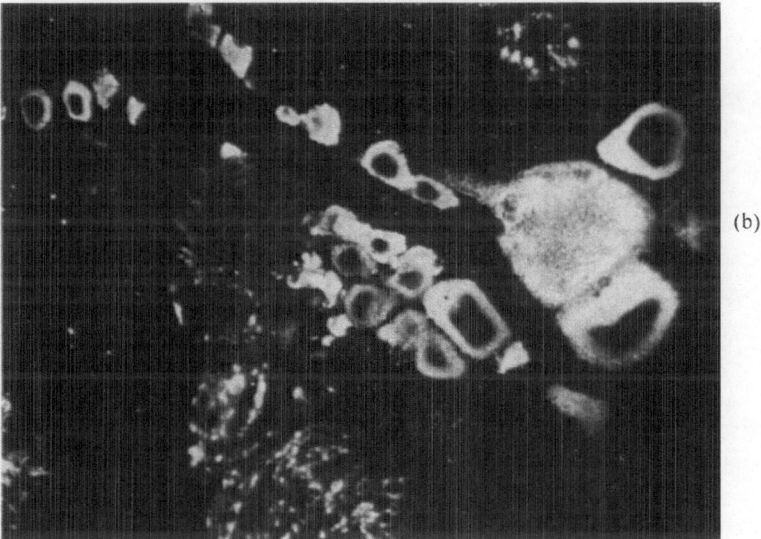

(b)

FIG. 7. (a) Fluorescence micrograph showing the right parietal ganglion, the visceral ganglion, and the left parietal ganglion of *Helix*. The neuropile joining the visceral and the left parietal ganglia can be seen. The scale is 100 μ. (b) Fluorescence micrograph of the visceral ganglion (lower) and right parietal ganglion (upper) showing fluorescent cells and the weaker but still distinctly fluorescent axons. Varicosities can be seen in the neuropile. The scale is 100 μ

Studies on the leech, Hirudo medicinalis

1. The action of acetylcholine and its antagonists

All recordings were made from the two large cells of Retzius located on the ventral surface of the segmental ganglia of *Hirudo*. These cells have resting potentials which are generally between 30 and 50 mV, the inside being negative. They

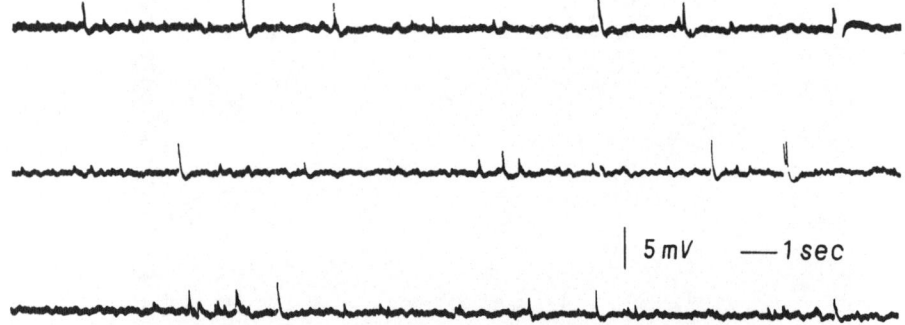

| 5 mV   —1 sec

FIG. 8. Recording from a leech Retzius cell to show small excitatory postsynaptic potentials

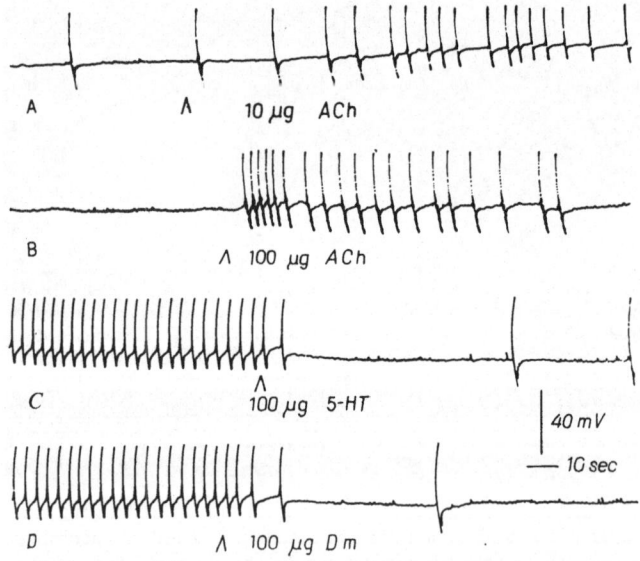

FIG. 9. A—The effect of 10 μg acetylcholine on the spontaneous activity of a Retzius cell of *Hirudo*: B—the effect of 100 μg acetylcholine on the spontaneous activity; C—inhibition by 100 μg 5–HT of activity induced by acetylcholine; D—inhibition by 100 μg dopamine of activity induced by acetylcholine (from Kerkut and Walker 1967)

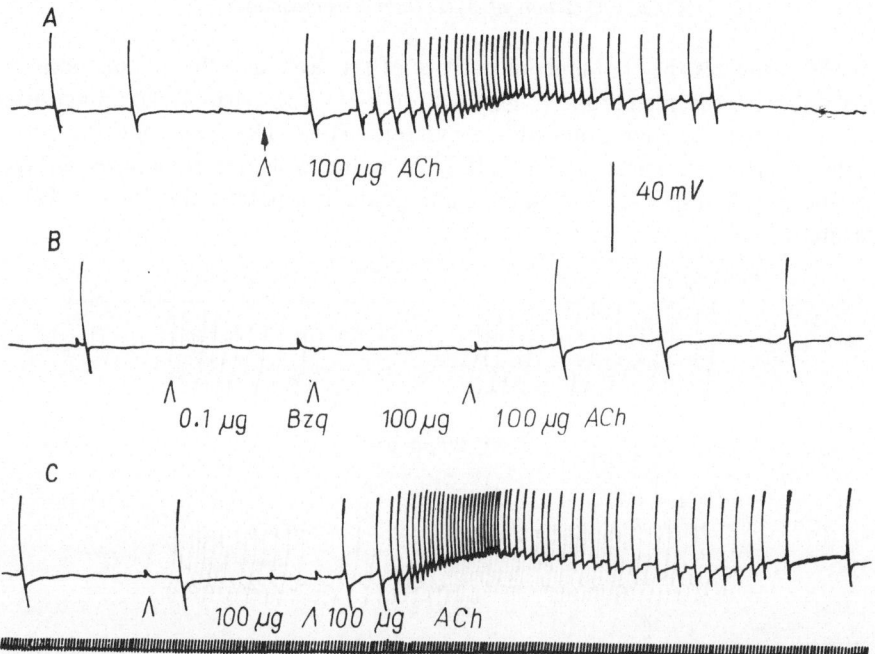

FIG. 10. A — The effect of 100 μg acetylcholine on the activity of a Retzius cell of *Hirudo;* B — the effect of the presence in the bath of 0.1 μg/ml benzoquinonium on the response to two applications of 100 μg acetylcholine; C — the recovery of the response to 100 μg acetylcholine following washing with leech Ringer (from Kerkut and Walker 1967)

usually exhibit spontaneous action potentials of between 20 and 40 mV in amplitude. The action potentials have a large after-potential of up to 20 mV. In addition to these action potentials, the cells also have many small depolarizations which are probably excitatory postsynaptic potentials (Fig. 8). The synaptic connections of these cells do not occur on the soma but are located on the axon hillock in the central neuropile (Coggeshall and Fawcett 1964).

Acetylcholine, when applied to the bath or when applied iontophoretically by means of an electrode containing acetylcholine, depolarized and excited the Retzius cells (Figs. 9A and 9B). This activity can be depressed by 5–HT or dopamine (Figs. 9C and 9D). The excitatory effect induced by acetylcholine can be antagonized by a number of cholinergic antagonists. Figure 10B shows the block of the acetylcholine response when benzoquinonium is present in the bath at a concentration of 0.1 μg/ml. This concentration is able to block the response to 100 μg acetylcholine. This block can be reversed on washing (Fig. 10C). Decamethonium is also a potent antagonist of acetylcholine. Atropine, hexamethonium, and tubocurarine are less potent as antagonists at this site.

2. *The action of 5–HT and its antagonists*

5–HT depresses the spontaneous activity of the Retzius cell and on occasion hyperpolarizes the membrane. This effect of 5–HT can partially be antagonized by pretreating the preparation with methysergide (Fig. 11B). This blocking action can be reversed on washing (Fig. 11C). Atropine will also sometimes antagonize the 5–HT response. But neither antagonist is a potent blocker of 5–HT at this site.

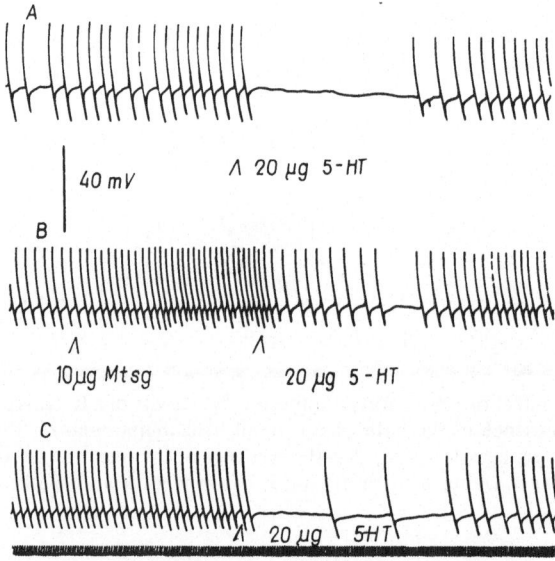

FIG. 11. A—The effect of 20 μg 5–HT on the spontaneous activity of a *Hirudo* Retzius cell; B—the effect of the presence in the bath of 10 μg/ml methysergide (Mtsg) on the response to 20 μg 5–HT; C—the recovery of the 5–HT response following washing with leech Ringer (from Kerkut and Walker 1967)

3. *Fluorescence microscopy of Hirudo neurons*

The ganglia of *Hirudo* after treatment with paraformaldehyde vapor for one hour contain six very strongly fluorescent yellow cells. Paper chromatographic analysis shows the presence of 5–HT but not 5-hydroxytryptophan in the ganglia. This indicates that the substance present in the ganglia is 5–HT rather than 5-hydroxytryptophan. Two of the fluorescent cells are located on the ventral surface of the ganglion and are 60–80 μ in diameter. These two cells have been identified as the two Retzius cells. These cells are shown in Fig. 12.

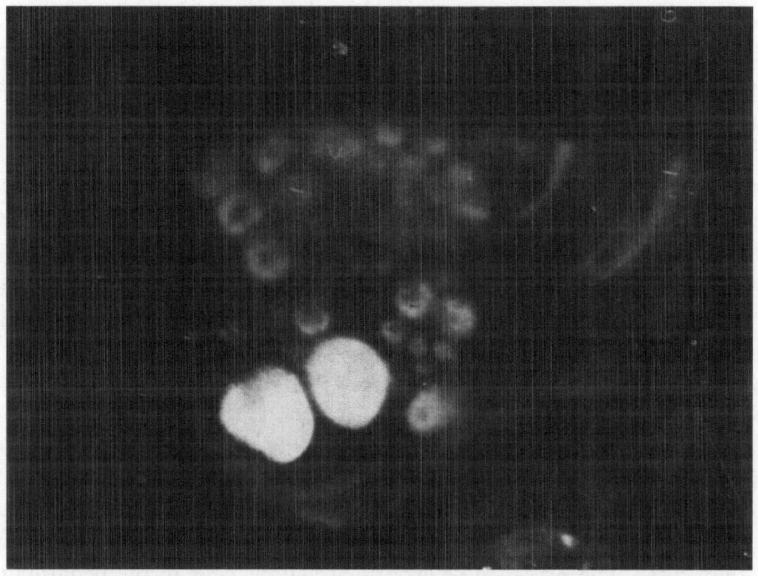

FIG. 12. Fluorescence micrograph of the segmental ganglion of *Hirudo* showing the two strongly fluorescent yellow cells of Retzius. The other cells in the micrograph are non-fluorescent. The scale is $100\,\mu$. (from Kerkut, Sedden and Walker 1967a)

DISCUSSION

The present study has attempted to obtain new information regarding the functioning of the nervous system by making use of electrophysiological, pharmacological, and histochemical techniques. The results will be discussed in three sections.

1. *Evidence for the nature of transmitter chemicals in the invertebrate central nervous system*

The best evidence for chemical transmission in the invertebrate central nervous system is the work of Kerkut and Thomas (1964) for acetylcholine as the inhibitory transmitter onto certain neurons in *Helix*. These neurons exhibit spontaneous inhibitory postsynaptic potentials. It is possible to demonstrate that the reversal potential for acetylcholine and for these inhibitory potentials changes in a similar manner when the ionic concentration is changed and that both inhibitions are associated with an increased permeability to chloride ions. However, not all inhibitory postsynaptic potentials can be associated with acetylcholine, and certain neurons which are excited by acetylcholine possess inhibitory postsynaptic potentials (Gerschenfeld 1964). A number of compounds have been shown to inhibit the activity of *Helix* neurons; these include 5–HT (Glaizner 1967) and dopamine

(Kerkut and Walker 1961). Both compounds have been shown to be present in *Helix* nervous tissue (Kerkut and Cottrell 1963; Kerkut, Sedden and Walker 1966). The cellular localization of both compounds using fluorescence microscopy has been performed by Dahl, Falck, von Mecklenberg, Myhrberg and Rosengren (1966) and by Kerkut, Sedden, and Walker (1967b). It is tempting to speculate on the function of these monoamines in molluscan neurons. It has been suggested by Myhrberg (1967) and Clark (1966) that 5-HT in invertebrates may be associated with sensory function. It is known that 5-HT is the excitatory transmitter on the snail heart (S.-Rózsa and Perényi 1966) and that it has an effect on the snail gut at low concentrations (Gryglewski and Supniewski 1963). Dopamine, in contrast, has less effect on snail peripheral organs. On the other hand, cells containing dopamine or noradrenaline have been found in the skin of invertebrates, and it has been suggested that these may be sensory receptors (Clark 1966; Myhrberg 1967; Rude 1966). Sensory cell axons would then terminate onto neurons in the central nervous system. From the present study, it is clear that a number of neurons in the snail brain are sensitive to low concentrations of dopamine. It would be necessary to demonstrate the presence of green varicosities in the region of the axon hillocks of these neurons before postulating dopamine as a chemical transmitter agent at these sites. The presence of groups of yellow cells in the pedal ganglia is suggestive of a motor role for 5-HT, since axons from cells in the pedal ganglia pass to the pedal musculature. 5-HT increases the rate of relaxation of *Helix* muscle (Kerkut and Leake 1966). 5-HT may not exclusively be associated with motor function, since certain of the central neurons of *Helix* and allied species respond to 5-HT (Kerkut and Walker 1961; Gerschenfeld and Stefani 1966) and it is a possible candidate as a transmitter within the snail central nervous system.

According to Retzius (1891), the giant neurons in the segmental ganglia of *Hirudo* have three axons; two pass down the ipsilateral segmental nerves and the third down the main connective to the next ganglion. This has been confirmed by Hagiwara and Morita (1962). These authors stimulated the segmental nerves and drove action potentials antidromically in the corresponding Retzius cell. This suggests that axons from the Retzius cell pass down to the body-wall musculature and possibly release 5-HT onto the muscles. It has been shown by Poloni (1955) and by Schain (1961) that 5-HT causes the leech muscle to relax and reduces the response to acetylcholine. Since only six neurons in each segmental ganglion would appear to contain 5-HT (Kerkut, Sedden, and Walker 1967a) this is further evidence that if 5-HT naturally inhibits the contraction of the body-wall musculature, then it is likely to come from the Retzius cells. 5-HT has been found to depress small excitatory postsynaptic potentials recorded from leech muscle (Walker, Woodruff, and Kerkut 1968). In addition to containing 5-HT and possibly releasing it, the Retzius cells themselves respond to 5-HT.

If an inhibitory synaptic input could be found onto the Retzius cells and this be shown to have similar characteristics to the 5-HT response — for example, the same reversal potential or the same ionic requirements — then this would be evidence for a physiological role for 5-HT acting as an inhibitory transmitter in the

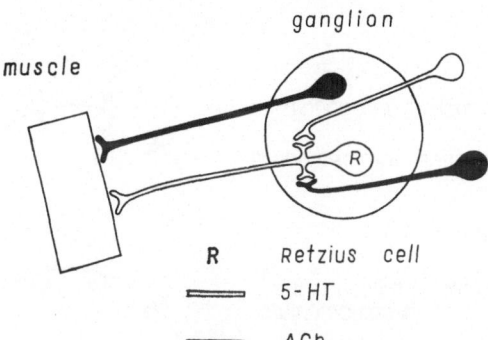

FIG. 13. A hypothetical scheme for nerve-nerve and nerve-muscle transmission in the leech; the Retzius cell having a cholinergic and a 5-HT input; the dorsal muscle receiving an excitatory cholinergic input and an inhibitory 5-HT input

leech central nervous system. This would mean that a single cell would itself release 5-HT onto the dorsal body-wall musculature and would have a 5-HT-containing axon terminating onto it. As there are only six neurons (two of which are the Retzius cells themselves) in each ganglion that contain 5-HT, then there is the possibility that the inhibitory synapse arises from a Retzius cell in the adjacent ganglion. A hypothetical scheme for synaptic connections onto the Retzius cell and onto the dorsal body-wall muscle fibers is shown in Fig. 13.

2. *Evidence for chemical heterogeneity of nerve cells*

There is now considerable evidence that the neurons present in the snail ganglia are not a homogeneous population of neurons and must therefore be considered either as individual neurons or as groups of neurons with similar characteristics but differing from neighboring groups of nerve cells. This means that it is important to work on either known cells or on cells in a certain restricted area of the brain. Identification of single cells depends on their size and position in relation to surrounding neurons, their electrical characteristics, their pharmacological responses, and on their contents. It is clear from the present studies on the snail that its neurons respond differently to cholinergic agonists. This has led us to suggest that there are both nicotinic and muscarinic neurons. These terms are taken from vertebrate cholinergic studies on autonomic ganglia, nerve-striated muscle junctions, and nerve-smooth muscle junctions, and are summarized in Fig. 14. The effect of acetylcholine is mimicked by nicotine at nerve-nerve junctions in autonomic ganglia and at nerve-striated muscle junctions. The effect of acetylcholine

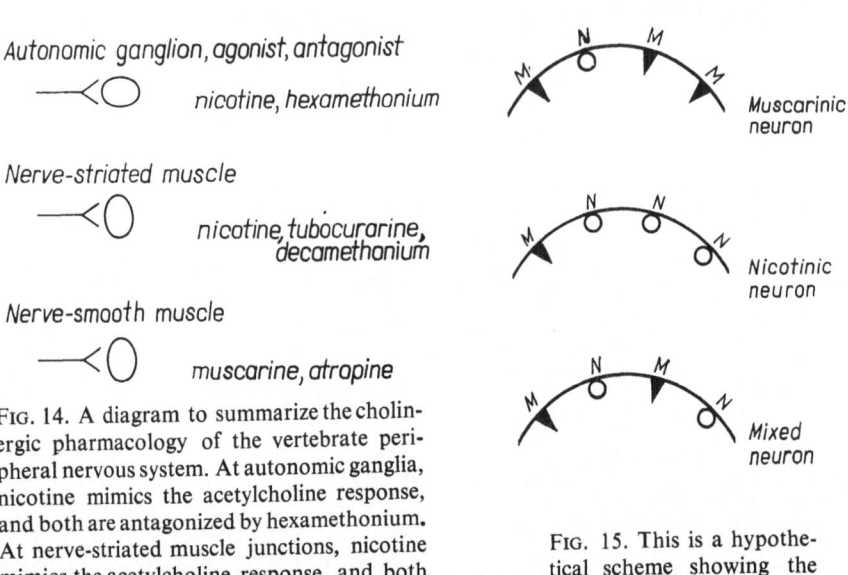

Autonomic ganglion, agonist, antagonist

⎯⟨○   nicotine, hexamethonium

Nerve-striated muscle

⎯⟨○   nicotine, tubocurarine,
                    decamethonium

Nerve-smooth muscle

⎯⟨○   muscarine, atropine

                                        Muscarinic
                                        neuron

                                        Nicotinic
                                        neuron

                                        Mixed
                                        neuron

FIG. 14. A diagram to summarize the cholinergic pharmacology of the vertebrate peripheral nervous system. At autonomic ganglia, nicotine mimics the acetylcholine response, and both are antagonized by hexamethonium. At nerve-striated muscle junctions, nicotine mimics the acetylcholine response, and both are antagonized by tubocurarine and decamethonium. At nerve-smooth muscle junctions, muscarine mimics the acetylcholine response, and both are antagonized by atropine

FIG. 15. This is a hypothetical scheme showing the distribution of nicotinic and muscarinic receptor sites on the membrane of 'nicotinic' cells, 'muscarinic' cells, and 'mixed' cells in *Helix*

is mimicked by muscarine at nerve-smooth muscle junctions. The action of both acetylcholine and nicotine is antagonized by hexamethonium and tetraethylammonium at autonomic ganglia and by tubocurarine, gallamine, benzoquinonium, decamethonium, and succinylcholine at nerve-striated muscle junctions. The action of both acetylcholine and muscarine are antagonized by atropine and scopolamine at nerve-smooth muscle junctions. In the snail, there are neurons which are more sensitive to muscarine than to nicotine, (the muscarinic cells), while other neurons are more sensitive to nicotine than to muscarine (the nicotinic cells). Other neurons appear to be equally sensitive to both nicotine and to muscarine (the 'mixed' cells). One might suggest that there are more nicotinic receptor sites on nicotinic neurons and more muscarinic receptor sites on muscarinic neurons (Fig. 15). No cells have been found that have only nicotinic or only muscarinic receptor sites.

Further evidence for neuron heterogeneity in the snail comes from the recent studies of Chiarandini, Stefani, and Gerschenfeld (1967). These authors present evidence for two types of excitation by acetylcholine of neurons from the land snail *Cryptomphallus aspersa*. Certain neurons are excited by acetylcholine by an increase in permeability to sodium ions, while others are excited by an increase in permeability to chloride ions. In *Helix*, the action potentials of certain neurons disappear in chloride-free Ringer, suggesting their dependence on chloride. It would be of interest to measure the internal chloride concentration of these two types of D cell of *Cryptomphallus*, since it has been shown by Kerkut and Meech (1966) that the chloride concentration in H cells is low, while the concentration in D cells is high. The acetylcholine response in these D cells was mediated via an increase in permeability to sodium ions and not to chloride ions. In addition to differences in the concentration of chloride, there is also evidence for differences in the concentration of potassium in a single cell at different times of the year (Kerkut and Meech 1967). This suggests that there may be a seasonal variation in the levels of ions in snail neurons.

Fluorescence microscopy studies on *Helix* neurons indicate the presence of a large number of neurons which contain a monoamine, probably restricted to dopamine and 5–HT, although there may also be some noradrenaline present in the brain (Cottrell 1967). The fluorescent neurons appear to be present in groups. For example, there are a group of large and small neurons on the right of the visceral ganglion and to the left of the right parietal ganglion. These are cells whose contents is difficult to identify in terms of either 5–HT and/or dopamine. These neurons would appear to be capable of taking up both dopa and 5–hydroxytryptophan, since, when the snail is pretreated with dopa, the neurons appear green, and when the snail is pretreated with 5–hydroxytryptophan, the neurons appear yellow. Groups of neurons in the pedal ganglia of *Helix* are clearly yellow, and this yellow color is intensified when the animals are exposed to 5–hydroxytryptophan pretreatment. Groups of very small cells surround a large yellow neuron in the cerebral ganglion. Thus, there is evidence for possibly three types of monoamine-containing neuron: cells containing dopamine, cells containing 5–HT, and cells containing both dopamine and 5–HT.

3. *The nature of the receptor membrane as determined by agonist and antagonist studies*

There would appear to be little difference between the form of the cholinergic receptor on H and on D cells. In both H and D cells, the most effective antagonists are atropine, tubocurarine, and gallamine. Decamenthonium and succinylcholine would appear to be more effective as antagonists on H than on D cells, but even in H cells they are very weak, a concentration of 1000 μg/ml being required to block the response to 10 μg acetylcholine. This contrasts with the complete block of the response to 10 μg acetylcholine when 1 μg/ml atropine, tubocurarine, or gallamine is present in the bath. In *Helix*, hexamethonium is equally ineffective on both H and D cells. This finding contrasts with the observation on *Aplysia* by Tauc and Gerschenfeld (1962). These authors found that 100 μg/ml hexamethonium antagonized the excitatory response to acetylcholine while having no effect on the inhibitory effect on H cells. The weak antagonism exhibited by succinylcholine, decamethonium, and hexamethonium to acetylcholine observed in *Helix* agrees with the findings of Vulfius and Zeimal (1966) for *Planorbis*. These authors found that acetylcholine was blocked by low concentrations of compounds with either 14 atoms between the terminal nitrogens such as suberyldicholine or 16 atoms between the terminal nitrogens such as KB–72. It would be interesting to test these compounds on the acetylcholine response of *Helix* neurons. In both gallamine and tubocurarine, the distance between the two terminal nitrogen atoms is around 10 atoms. However, the distance between the quaternary nitrogens in decamethonium is also 10 atoms. Atropine has only one quaternary nitrogen atom. In terms of antagonist studies, the cholinergic receptor in the leech would appear to be different. Here, the most potent antagonists are benzoquinonium, with a distance between the two terminal nitrogen atoms of 12 atoms, and decamethonium, in which the distance is 10 atoms. It may be of significance that both benzoquinonium and decamethonium are depolarizing and blocking agents while atropine, tubocurarine, and gallamine are competitive antagonists. These observations are summarized in Table 4.

TABLE 4

A comparison between the cholinergic antagonists
for *Helix* and *Hirudo* neurons*

| Antagonist | Distance between terminal N atoms | Potency | |
		Helix	*Hirudo*
Atropine	One nitrogen	++++	+++
Tetraethyl-ammonium	One nitrogen	+	Not tested
Hexamethonium	Six atoms	+	++
Tubocurarine	Nine-ten atoms	++++	++
Gallamine	Ten atoms	++++	Not tested
Decamethonium	Ten atoms	++	++++
Succinylcholine	Eleven atoms	++	Not tested
Benzoquinonium	Twelve atoms	+	++++

* The intensity of the antagonism is indicated as follows: + very weak; + + weak; + + + medium; + + + + strong. The number of atoms between the terminal quaternary nitrogen atoms is indicated.

The agonist studies also stress the similarities between H and D cholinergic receptors. Thus, the six most potent agonists in D cells are the same most potent agonists in H cells. All these agonists have a quaternary nitrogen group and, with the exception of nicotine, a –C–C–O– chain attached to the nitrogen. Increasing the number of carbons attached to the acetyl group of acetylcholine, as in the case of propionylcholine or butyrylcholine, does not appear to influence potency. Substitution on the beta carbon, as in the case of mecholine (acetyl-β-methylcholine), clearly reduces the potency in both H and D cells compared with acetylcholine or muscarine. There are some differences between H and D cells. For example, succinylcholine is a very weak agonist on D cells, being about 100 times less potent than acetylcholine, but on H cells it is only two times weaker than acetylcholine. Succinylcholine has a quaternary nitrogen group with a $-C-C-O-C-C-$ side chain attached to it, as in acetylcholine. This structure may be more important for the H than for the D response.

Dimethylaminoethanol was also more active on H than on D cells, but this compound inhibited the activity of both types of cell. Dimethylaminoethanol does not have a quaternary nitrogen group and clearly warrants further investigation. Many of the cholinergic agonists are more potent on H cells than on D cells (Table 2). For example, oxotremorine, which was among the weakest of the cholinergic agonists tested, was four times more potent on H than on D cells. The relative lack of potency of oxotremorine of *Helix* neurons compared with the vertebrate studies of Cho, Haslett, and Jenden (1962) is unexpected. Oxotremorine does not possess a quaternary nitrogen. The quaternary nitrogen group is not on its own sufficient for potency, since tetramethylammonium is relatively inactive as an agonist on *Helix* neurons. A surprising observation was the sensitivity of certain neurons in *Helix* to choline. Some cells were inhibited by as little as 1 μg choline, while others were excited by 10 μg. Other cells required 1000 μg or more to produce an effect on cell activity. Khromov-Borisov and Michelson (1966) have suggested a scheme for the possible active sites of a cholinoceptive receptor together with the important end groups of acetylcholine for linking between acetylcholine and the receptor (Fig. 16).

The finding that the most potent antagonist of the dopamine response in *Helix* neurons is neither an alpha nor a beta adrenergic blocking agent suggests the possibility that the dopamine receptor may be different from the vertebrate adrenergic receptor. However, studies on the dopamine response in the caudate nucleus of the cat by McLennan and York (1967) have shown that dibenyline, an alpha adrenergic antagonist, can block the dopamine response. Dibenyline is also capable of blocking the dopamine response in *Helix*. It would be interesting to test the effect of ergometrine on the mammalian dopamine response.

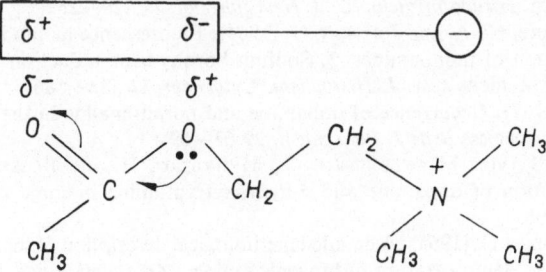

FIG. 16. Diagram of a possible cholinoceptive receptor showing the active sites for interaction with the acetylcholine molecule

R. J. WALKER

SUMMARY

1. Intracellular recordings have been made from neurons in the suboesophageal ganglionic mass of the snail *Helix aspersa* and from neurons in the segmental ganglia of the leech *Hirudo medicinalis*. Fluorescence-microscopy studies have also been performed on both tissues.

2. The most effective antagonists of the response to acetylcholine on *Helix* neurons are atropine, scopolamine, tubocurarine, and gallamine. The most effective cholinergic antagonists on *Hirudo* neurons are decamethonium and benzoquinonium.

3. Antagonist and agonist studies suggest that the cholinoceptive receptors of *Helix H* and *D* cells are similar. However, the proportion of nicotinic and muscarinic receptor sites may vary from cell to cell.

4. The most potent antagonist of the dopamine response on *Helix* neurons is ergometrine. Alpha adrenergic antagonists are less potent as blockers of the dopamine response.

5. Certain neurons in *Helix* contain 5–HT, while others contain dopamine. There is evidence that some neurons may contain both 5–HT and dopamine.

6. Six cells in the leech segmental ganglion contain 5–HT. Two of these cells are the Retzius cells.

7. The Retzius cells of *Hirudo* are depolarized and excited by acetylcholine and hyperpolarized and inhibited by 5–HT. It is suggested that 5–HT may be an inhibitory transmitter onto the leech body-wall musculature.

REFERENCES

ALEXANDROWICZ, J. S. (1932): The innervation of the heart of the Crustacea. 1. Decapoda. *Quart. J. micr. Sci.* **75** 181—249.

CHIARANDINI, D. J., STEFANI, E. and GERSCHENFELD, H. M. (1967): Ionic mechanisms of cholinergic excitation in Molluscan neurons. *Science* **156** 1597—1599.

CHO, A. K., HASLETT, W. L. and JENDEN, D. J. (1962): The peripheral actions of oxotremorine, a metabolite of tremorine. *J. Pharmacol. exp. Ther.* **138** 249—257.

CLARK, M. E. (1966): Histochemical localization of monoamines in the nervous system of the polychaete *Nephtys*. *Proc. roy. Soc. B.* **165** 308—325.

COGGESHALL, R. E. and FAWCETT, D. W. (1964): The fine structure of the central nervous system of the leech, *Hirudo medicinalis*. *J. Neurophysiol.* **27** 229—289.

CORRODI, H., HILLARP, N. A. and JONSSON, G. (1964): Fluorescence methods for the histochemical demonstration of monoamines. 3. Sodium borohydride reduction of the fluorescent compounds as a specificity test. *J. Histochem. Cytochem.* **12** 582—586.

COTTRELL, G. A. (1967): Occurrence of dopamine and noradrenaline in the nervous tissue of some invertebrate species. *Brit. J. Pharmacol.* **29** 63—69.

DAHL, E., FALCK, B., VON MECKLENBERG, C., MYHRBERG, H. and ROSENGREN, E. (1966): Neuronal localization of dopamine and 5-hydroxytryptamine in some mollusca. *Z. Zellforsch.* **71** 489—498.

FALCK, B. and OWMAN, C. (1965): A detailed methological description of the fluorescent method for the cellular demonstration of biogenic amines. *Acta Univ. Lund.* Section II, No. 7. 1—23.

FLOREY, E. (1954): An inhibitory and an excitatory factor of mammalian brain and their action on a simple sensory neurone. *Arch. int. Physiol.* **62** 33—53.

FLOREY, E. (1961): A new test preparation for bio-assay of Factor I and gamma-amino-butyric acid. *J. Physiol.* **156** 1—7.

FUEHNER, H. (1918): Untersuchungen über den Synergismus von Giften Chemische Erregbar-keits-steigerung glatter Muskeln. *Arch. exp. Path. Pharmak.* **82** 57—80.

GERSCHENFELD, H. M. (1964): A non-cholinergic synaptic inhibition in the central nervous system of a Mollusc. *Nature (Lond.)* **203** 415—416.

GERSCHENFELD, H. M. and STEFANI, E. (1966): An electrophysiological study of 5-hydroxy-tryptamine receptors of neurones in the molluscan nervous system. *J. Physiol.* **185** 684—700.

GLAIZNER, B. (1967): Pharmacological mapping of cells in the sub-oesophageal ganglia of *Helix aspersa. Neurobiology of Invertebrates.* pp. 267—283.

GREENBERG, M. J. (1960): Structure-activity relationship of tryptamine analogues on the heart of *Venus mercenaria. Brit. J. Pharmacol.* **15** 375—388.

GRYGLEWSKI, R. and SUPNIEWSKI, J. (1963). Influence of 5-hydroxytryptamine and other biolo-gically active substances on the movements of the isolated stomach of *Helix pomatia. Bull. Acad. pol. Sci.* **II** 53—56.

HAGIWARA, S. and MORITA, H. (1962). Electrotonic transmission between two nerve cells in leech ganglion. *J. Neurophysiol.* **25** 721—731.

KERKUT, G. A. and COTTRELL, G. A. (1963): Acetylcholine and 5-hydroxytryptamine in the snail brain. *Comp. Biochem. Physiol.* **8** 53—63.

KERKUT, G. A. and LEAKE, L. D. (1966): The effect of drugs on the snail pharyngeal retractor muscle. *Comp. Biochem. Physiol.* **17** 623—633.

KERKUT, G. A. and MEECH, R. W. (1966): The internal chloride concentration of H and D cells in the snail brain. *Comp. Biochem. Physiol.* **19** 819—832.

KERKUT, G. A. and MEECH, R. W. (1967): The effect of ions on the membrane potentials of snail neurones. *Comp. Biochem. Physiol.* **20** 411—429.

KERKUT, G. A., SEDDEN, C. B. and WALKER, R. J. (1966): The effect of DOPA, α-methylDOPA, and reserpine on the dopamine content of the brain of the snail, *Helix aspersa. Comp. Bio-chem. Physiol.* **18** 921—930.

KERKUT, G. A., SEDDEN, C. B. and WALKER, R. J. (1967a): Cellular localization of mono-amines by fluorescence microscopy in *Hirudo medicinalis* and *Lumbricus terrestris. Comp. Bio-chem. Physiol.* **21** 687—690.

KERKUT, G. A., SEDDEN, C. B. and WALKER, R. J. (1967b): Uptake of DOPA and 5-hydroxy-trytophan by monoamine-forming neurones in the brain of *Helix aspersa. Comp. Biochem. Physiol.* **22 23** 159—162

KERKUT, G. A. and THOMAS, R. C. (1964): The effect of anion injection and changes in the external potassium and chloride concentration on the reversal potentials of the IPSP and acetylcholine. *Comp. Biochem. Physiol.* **11** 199—213.

KERKUT, G. A. and WALKER, R. J. (1961): The effect of drugs on the neurones of the snail *Helix aspersa. Comp. Biochem. Physiol.* **3** 143—160.

KERKUT, G. A. and WALKER, R. J. (1962): The specific chemical sensitivity of *Helix* nerve cells. *Comp. Biochem. Physiol.* **7** 277—288.

KERKUT, G. A. and WALKER, R. J. (1967): The action of acetylcholine, dopamine and 5-hy-droxytryptamine on the spontaneous activity of the cells of Retzius of the leech, *Hirudo medi-cinalis. Brit. J. Pharmacol.* **30** 644—654.

KHROMOV-BORISOV, N. V. and MICHELSON, M. J. (1966): The mutual disposition of cholino-receptors of locomotor muscles, and the changes in their disposition in the course of evolu-tion. *Pharmacol. Rev.* **18** 1051—1090.

LOVELAND, R. E. (1963): 5-Hydroxytryptamine, the probable mediator of excitation in the heart of *Mercenaria (Venus)* mercenaria. *Comp. Biochem. Physiol.* **9** 95—104.

MAYNARD, D. M. (1966): Integration in crustacean ganglia. *Symp. Soc. exp. Biol.* **20** 111—149.

McLENNAN, H. and YORK, D. H. (1967): The action of dopamine on neurones of the caudate nucleus. *J. Physiol.* **189** 393—402.

MINZ, B. (1932): Pharmakologische Untersuchungen am Blutegelpräparat, zugleich eine Methode zum biologischen Nachweis von Acetylcholin bei Anwesenheit anderer pharmakologisch wirksamer körpereigener Stoffe. *Arch. exp. Path. Pharmak.* **168** 292—304.

MOLINE, S. W. and GLENNER, G. G. (1964): Ultrarapid tissue freezing in liquid nitrogen. *J. Histochem. Cytochem.* **12** 777—783.

MYHRBERG, H. E. (1967): Monoaminergic mechanisms in the nervous system of *Lumbricus terrestris* (L.). *Z. Zellforsch.* **81** 311—343.

NICOL, J. A. C. (1948): Giant axons of annelids. *Quart. Rev. Biol.* 23. 291—323.

POLONI, A. (1955): Il muscolo dorsale di sanguisuga quale test bilogico per l'evidenziamento dell'attivita serotoninica nei liquidi organici. *Cervello* **31** 472—476.

RETZIUS, G. (1891): Zur Kenntnis des centralen Nervensystems der Würmer. *Biol. Unters. (N. F.)* **2** 1—28.

RUDE, S. (1966): Monoamine containing neurons in the nerve cord and body wall of *Lumbricus terrestris. J. comp. Neurol.* **128** 397—412.

S.-RÓZSA, K. and PERÉNYI, L. (1966): Chemical identification of the excitatory substance released in *Helix* heart during stimulation of the extracardial nerve. *Comp. Biochem. Physiol.* **19** 105—113.

SCHAIN, R. J. (1961): Effects of 5-hydroxytryptamine on the dorsal muscle of the leech *(Hirudo medicinalis). Brit. J. Pharmacol.* **16** 257—261.

SEDDEN, C. B., WALKER, R. J. and KERKUT, G. A. (1967): The localization of dopamine and 5-hydroxytryptamine in neurones of *Helix aspersa.* Symposium on studies on the structure physiology and ecology of molluscs. Ed. by FRETTER, V. *Symp. Zool. Soc. London* (in press),

TAUC, L. and GERSCHENFELD, H. M. (1960): Effect excitateur ou inhibiteur du chlorure d'acetylcholine sur le neurone d'Escargot. *J. Physiol. Path. gén.* **52** 236.

TAUC, L. and GERSCHENFELD, H. M. (1962). A cholinergic mechanism of inhibitory synaptic transmission in a molluscan nervous system. *J. Neurophysiol.* **25** 236—262.

VULFIUS, F. A. and ZEIMAL, E. V. (1967): The effects of acetylcholine and cholinomimetics on giant neurons of the mollusc *Limnea stagnalis.* In: "Problems of evolutionary neurophysiology and neurochemistry" pp. 17—25. Supplement to the *J. evol. biochem. physiol.* (in Russian).

WALKER, R. J. (1968): In: *Laboratory Experiments in Physiology and Biochemistry.* Ed. KERKUT, G. A., Academic Press, London.

WALKER, R. J. and HEDGES, A. (1968a): The effect of cholinergic antagonists on the response to acetylcholine, acetyl-ß-methylcholine and nicotine of neurones of *Helix aspersa. Comp. Biochem. Physiol.* **23** 977—989.

WALKER, R. J. and HEDGES, A. (1968b): The effect of cholinergic agonists on the spontaneous activity of neurones of *Helix aspersa. Comp. Biochem. Physiol.* **24** 355—876

WALKER, R. J., WOODRUFF, G. N., GALIZNER, B., SEDDEN, C. B. and KERKUT, G. A. (1967): The pharmacology of the dopamine receptor of specific neurones in the snail, *Helix aspersa. Comp. Biochem. Physiol.* **24** 455—469

WALKER, R. J., WOODRUFF, G. N. and KERKUT, G. A. (1968): The effect of acetylcholine and 5-hydroxytryptamine on electrophysiological recordings from muscle fibres of the leech, *Hirudo medicinalis. Comp. Biochem. Physiol.* **24** 987—990

WELSH, J. H. (1953): Excitation of the heart of *Venus mercenaria. Arch. exp. Path. Pharmak.* **219** 23—29.

WIERSMA, C. A. G., FURSHPAN, F. and FLOREY, E. (1953): Physiological and pharmacological observations on muscle receptor organs of the crayfish, *Cambarus clarkii* Girard. *J. exp. Biol.* **30** 136—150.

YOUNG, J. Z. (1935): Structure of nerve fibres in *Sepia. J. Physiol.* **83** 27—28 P.

YOUNG, J. Z. (1936): The giant nerve fibres and epistellar body of Cephalopods. *Quart. J. micr. Sci.* **78** 367—386.

YOUNG, J. Z. (1938): The functioning of the giant nerve fibres of the squid. *J. exp. Biol.* **15** 170—185.

DISCUSSION

G. A. Cottrell: Have you tried other methods for investigating receptors, for example, using autoradiography?

R. J. Walker: This sort of approach has been used by Waser for locating cholinoceptive receptors on the mouse diaphragm muscle and it would be worth trying to develop a similar technique for the receptors on *Helix* neurons.

E. V. Zeimal: What do you measure when you determine the response to a drug?

R. J. Walker: With the results described we were only interested in the threshold amount of the compound for a response. However, the best quantitative method for measuring the response to a compound is the change in the time interval between successive action potentials. This can be used for either excitatory or inhibitory responses.

E. V. Zeimal: How potent is decamethonium as a cholinergic antagonist on *Helix* neurons?

R. J. Walker: Decamethonium is a poor antagonist of acetylcholine of *Helix* neurons.

I. Zs.-Nagy: Have you evidence that 5–HT is present in the axons as well, and not only in the perikaryon?

R. J. Walker: We have examined sections of heart and foot muscle and have found some indication of a yellow fluorescence but we have not checked its specificity. On treatment with 5-hydroxytryptophan there was no obvious increase in the amount of yellow fluorescence.

S. I. Plotnikova: Is there any evidence for a catecholamine in the leech nervous system?

R. J. Walker: We have not seen any green neurons in the leech ganglion, but there are green cells in the peripheral tissue.

V. D. Gerasimov: Which muscles do the Retzius cells innervate?

R. J. Walker: I do not know to which body-wall muscles the axons from the Retzius cells go.

Symposium on Neurobiology of Invertebrates 1967 (255—265)

THE ACTION OF CHOLINOMIMETICS AND CHOLINOLYTICS ON THE GASTROPOD NEURONS

E. V. Zeimal* and E. A. Vulfius†

*Sechenov Institute of Evolutionary Physiology and Biochemistry, Leningrad
†Institute of Biological Physics (Moscow region, Putchino on Oka)
of the Academy of Sciences of the USSR
Moscow, USSR

Cholinergic transmission in the ganglia of gastropod molluscs was recently shown by Tauc and Gerschenfeld (1962), Kerkut and Cottrell (1963), and Kerkut and Thomas (1963). However, very little is known about the properties of cholino-receptors of mollusc neurons. For studying these cholinoreceptors, cholinergic drugs of different chemical structure have been used as a tool (Vulfius and Zeimal 1967a, 1967b). Cholinergic drugs with well-known pharmacological properties which have been widely studied in experiments on vertebrates have been chosen. In this way, we hope to compare the properties of mollusc cholinoreceptors with those of vertebrates and get some information about the occurrence and arrangement of active groups on cholinoreceptive membranes.

The experiments were performed on the freshwater gastropod molluscs *Planorbis corneus* and *Lymnaea stagnalis*. The isolated perioesophageal ganglionic ring was placed in a bath (about 1 ml) and perfused with saline solution (15 ml/min). The visceral and the large parietal ganglia were opened and a large neuron (100μ–$200\,\mu$) was impaled by a microelectrode filled with 2.5 M KCl (diameter of the tip about 0.5 μ; R = 15–60 MΩ). The electrode was connected through a cathode follower to a dc amplifier and then to a cathode oscilloscope. The compounds studied were added to the perfusion fluid; in some experiments, acetylcholine (ACh) was electrophoretically microapplied.

The majority of neurons studied fire spontaneously. Different types of responses to ACh are shown in Fig. 1. The addition of ACh to the perfusion fluid or its electrophoretical microapplication usually caused depolarization and an increase the rate of firing. In some cases, in cells without spontaneous firing, ACh induced only the depolarization. Few cells responded to ACh with a hyperpolarization and a decrease in spike frequency. Some cells are not affected by ACh.

The phenomenon of desensitization was observed with each single application of ACh (Fig. 1). The concentration of ACh remained constant during each test due to permanent perfusion. Despite this, the depolarization and increase in the rate of firing caused by ACh gradually disappeared. This desensitization is apparently

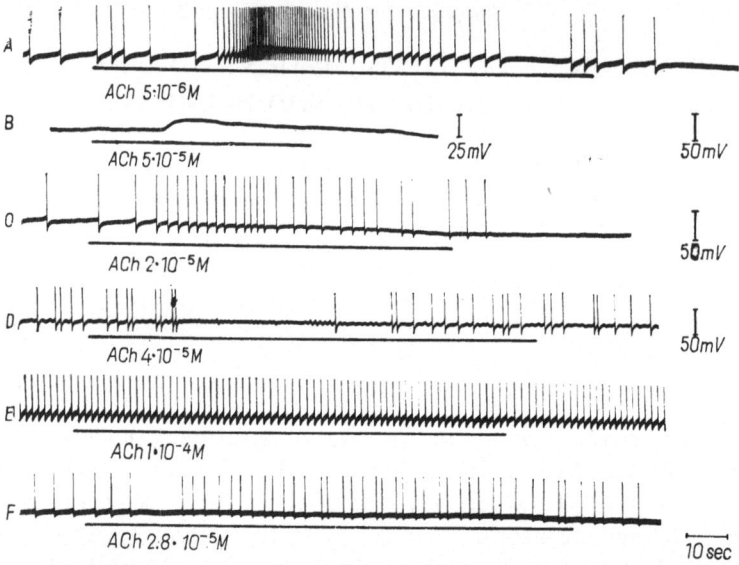

FIG. 1. Different types of responses of mollusc neurons to acetylcholine. Application of ACh to the perfusion fluid is shown (here and in other figures) by the horizontal line. *A, B, D, E, F* — neurons of *Lymnaea stagnalis, C* — neuron of *Planorbis corneus*

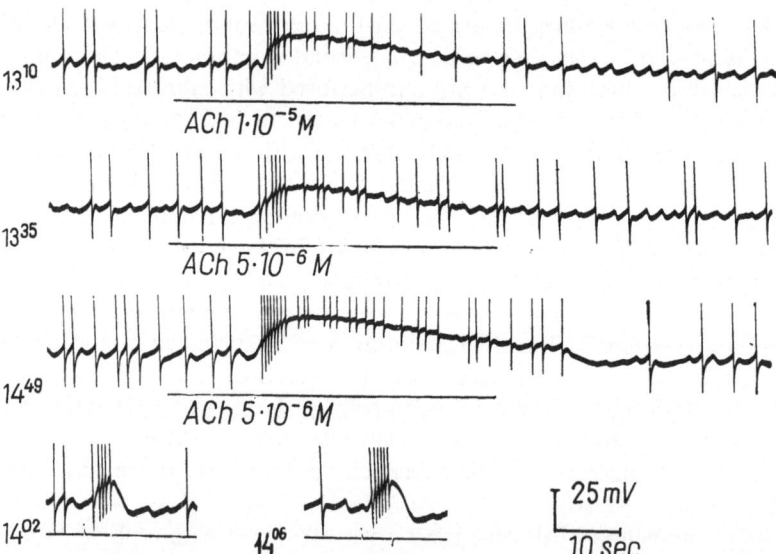

FIG. 2. The increase of sensitivity to ACh. The neuron of *Lymnaea stagnalis*. Large amplification was used: only the lower part of spike can be seen. At 13¹⁰, 13³⁵, and 14⁴⁹, ACh was applied to the perfusion fluid; at 14⁰² and 14⁰⁶, it was microapplied electrophoretically

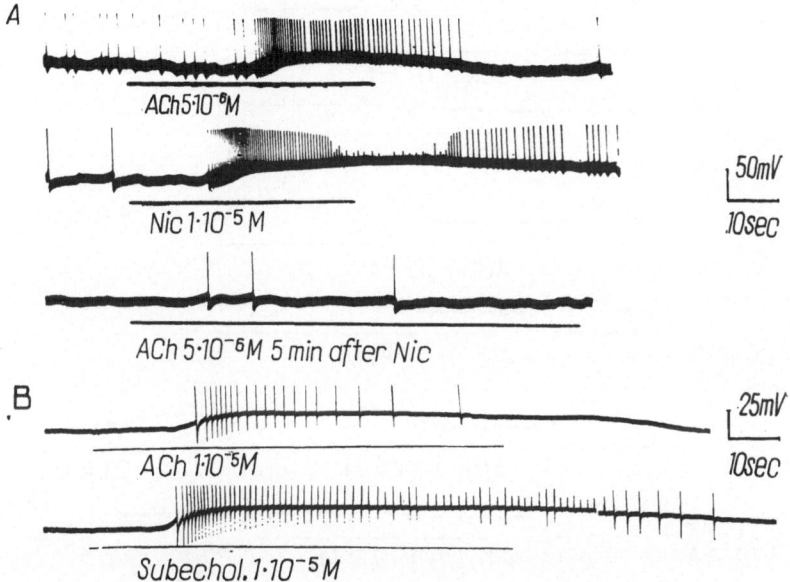

FIG. 3. The action of ACh, nicotine, and suberyldicholine A — the neuron of *Lymnaea stagnalis*, B — the neuron of *Planorbis corneus*

similar to that described by Katz and Thesleff (1957) for skeletal muscle endplate and by Tauc and Bruner (1963) for *Aplysia* neurons.

With repetitive applications of ACh, another interesting phenomenon, an increase of sensitivity to the ACh, was observed. We tested the sensitivity to ACh many times in the course of the experiment, and found that the response to the ACh increased as result of the repetitive testing. As is shown in Fig. 2, ACh of 10^{-5} M (at 13^{10}) produced the depolarization and increase of firing; 25 min later, nearly the same effect was produced by a concentration twice as weak (at 13^{35}, ACh of $5 \cdot 10^{-6}$ M); after several more tests the effect of ACh (of $5 \cdot 10^{-6}$ M at 14^{49}) became still greater. One can see that the phenomenon of desensitization remained but the initial sensitivity increased. The effect of ACh of $5 \cdot 10^{-6}$ M at 14^{49} was even greater than that of ACh of 10^{-5} M at the beginning (13^{10}). The increase in sensitivity to ACh when applied electrophoretically was also observed: the depolarization and the rate of firing were greater at 14^{06} than at 14^{02} (the same electrophoretical current was used for ACh ejection). The nature of the increase of sensitivity needs further investigation.

Among the cholinomimetics studied in the mollusc neurons, the most active were those which are known to excite the nicotinic cholinoreceptors of vertebrates. A comparison of the action ACh and of nicotine is shown in Fig. 3A. The addition of ACh caused the depolarization and increase in spike frequency. Nicotine produced a similar effect. The concentration of nicotine used was twice as great as

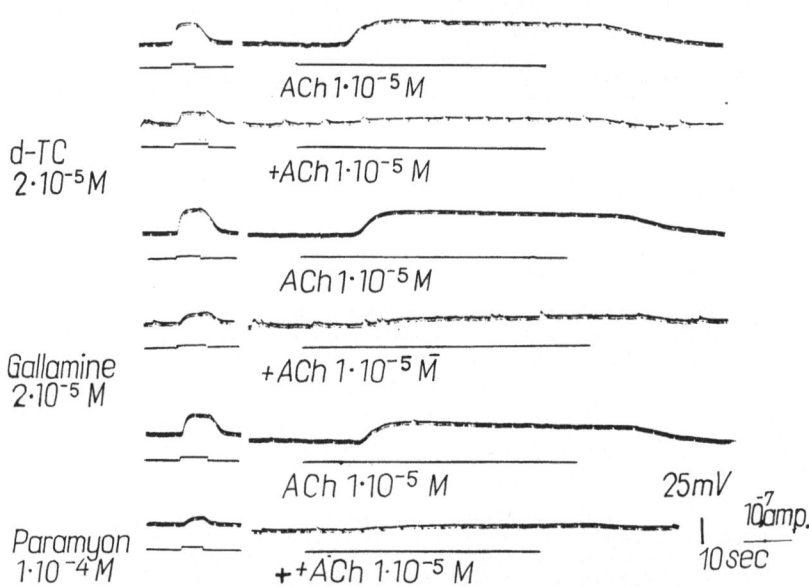

FIG. 4. The antiacetylcholine action of curare-like drugs. The neuron of *Lymnaea stagnalis*. To the left — electrophoretical microapplication of ACh; to the right — addition to the perfusion fluid. Reading from top to bottom: control testing of acetylcholine; the action of ACh after 5-min perfusion with d-tubocurarine (d-TC) $2 \cdot 10^{-5}$ M; control testing of ACh; the action of ACh after 5-min perfusion with gallamine $2 \cdot 10^{-5}$ M; control testing of ACh; the action of ACh after 5-min perfusion with paramyon 10^{-4} M

that of ACh, and so is the effect. Thus, nicotine is approximately as effective as ACh. As in the experiments on vertebrates, the cholinomimetic action of nicotine on the mollusc neurons is followed by the second, blocking phase. Five minutes after the application of nicotine, the same concentration of ACh did not produce any effect. In Fig. 3B, the action of the potent nicotinomimetic dicholinic ester of suberic acid is shown. Equimolar concentrations of ACh and suberyldicholine have been used. The depolarization and the rate of firing were greater in the case of suberyldicholine, and its action developed more quickly.

Table 1 summarizes the action of cholinomimetics of different groups. Those known to excite nicotinic cholinoreceptors were potent on the mollusc neuron. Muscarinomimetics were weak or not effective at all. This suggests nicotinic properties of the cholinoreceptors of gastropod neurons.

This suggestion was further supported when the action of cholinolytics was studied. The most effective were curare-like drugs. The antiacetylcholine action of d-tubocurarine, gallamine, and paramyon on the cell without spontaneous firing is shown in Fig. 4. One can see that in the presence of d-tubocurarine $2 \cdot 10^{-5}$ M, gallamine $2 \cdot 10^{-5}$ M or paramyon $1 \cdot 10^{-4}$ M, the action of ACh is significantly less than that in the control.

TABLE 1

Cholinomimetics

Drug	Action in vertebrates	Potency on the mollusc neuron
Acetylcholine	Excite nicotinic and muscarinic cholinoreceptors	+++++
Nicotine		+++++
Dicholine ester of suberic acid		++++++
Dicholinic ester of sebacinic acid	Excite nicotinic cholinoreceptors	++++++
Butyrylcholine		+++
Arecoline methiodide		++
Arecoline hydrochloride	Excite muscarinic cholinoreceptors	—
Methacholine		++

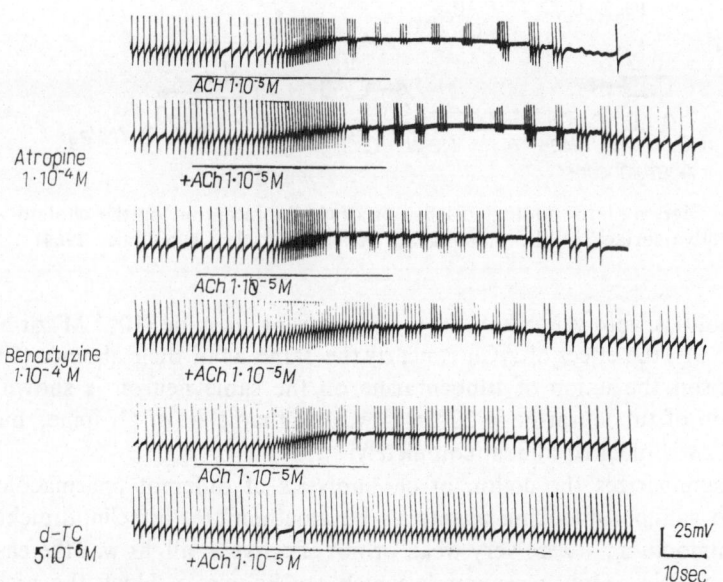

FIG. 5. The antiacetylcholine action of atropine, benactyzine, and d-tubocurarine. The neuron of *Planorbis*. Reading from top to bottom: control testing of ACh 10^{-5} M; the action of ACh 10^{-5} M after 5-min perfusion with atropine 10^{-5} M; the control testing of ACh 10^{-5} M; the action of ACh 10^{-5} M after 5-min perfusion with benactyzine 10^{-4} M; control testing of ACh 10^{-5} M; the action of ACh 10^{-5} M after 5-min perfusion with d-tubocurarine (d-TC) $5 \cdot 10^{-6}$ M

TABLE 2

Cholinolytics

Drug	Action in vertebrates	Potency on the mollusc neuron
Nicotine (second phase)	Blocks nicotinic cholino-receptors	+ + + +
d-Tubocurarine	Block nicotinic	+ + + +
Gallamine	cholinoreceptors of	+ + + +
Paramyon	skeletal muscles	+ + +
Hexamethonium	Block nicotinic	+
Dimelin	cholinoreceptors of	+
Dimekamin	autonomic ganglia	−
Tetraethylammonium		−
Atropine	Block muscarinic	+
Benactyzine	cholinoreceptors	+

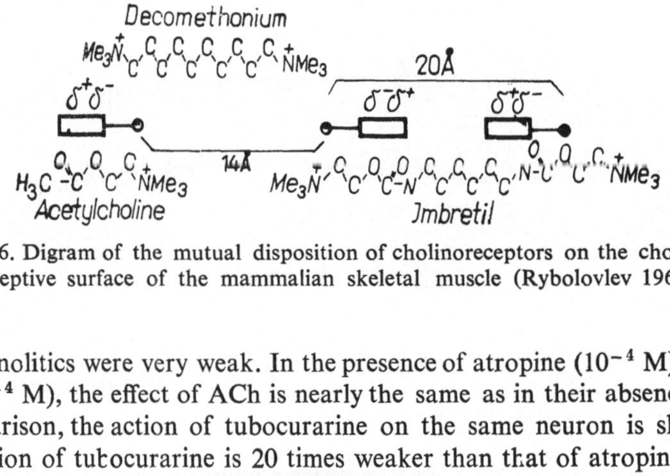

FIG. 6. Digram of the mutual disposition of cholinoreceptors on the cholino-receptive surface of the mammalian skeletal muscle (Rybolovlev 1964)

Muscarinolitics were very weak. In the presence of atropine (10^{-4} M) or benactyzine (10^{-4} M), the effect of ACh is nearly the same as in their absence (Fig. 5). For comparison, the action of tubocurarine on the same neuron is shown. The concentration of tubocurarine is 20 times weaker than that of atropine, but the action of ACh is blocked almost completely.

Table 2 summarizes the action of cholinolytics of different pharmacological groups. The ganglion-blocking agents − hexamethonium, dimelin, dimekamin, tetraethylammonium − were very weak or not effective at all, as was the case for the muscarinolytics. Only compounds which are known to block the nicotinic cholinoreceptors of skeletal muscles were effective on the cholinoreceptors of *Planorbis* and *Lymnaea* neurons. Thus cholinoreceptors of neurons of these molluscs have some common features with cholinoreceptors of the skeletal muscles of vertebrates.

The mammalian skeletal muscles are known to be very sensitive to bisquaternary compounds which have an internitrogen chain containing 10 atoms (about 14 Å)

– such as decamethonium and suxamethonium (Paton and Zaimis 1949). Recently high activity was also shown for bisquaternary compounds which have an internitrogen chain containing 14–16 atoms (about 20 Å) (Brücke 1956; Barlow and Zoller 1964). The suggestion was made by Rybolovlev (Rybolovlev 1964; Magazanik *et al.* 1965) and by Khromov–Borisov and Michelson (1966) that there exists a definite arrangement of single cholinoreceptors on the postsynaptic membrane of mammalian skeletal muscles (Fig. 6). In some cases the distance between the anionic points of neighboring receptors is about 14 Å (C–10 structure), in the other cases – about 20 Å (C–16 structure).

It is of interest to know if some definite arrangement of receptors on the neuron membrane of molluscs exists? To elucidate this question the action of some bisquaternary compounds with internitrogen distance 14 Å and 20–22 Å has been studied (Fig. 7). Decamethonium and suxamethonium (containing ten atoms between nitrogens) caused only a poor depolarization and a slight increase in spike frequency even in the concentration as high as 10^{-4} M. ACh produced the same effect in the concentration 20 times weaker. It is interesting that the monoquaternary compound, TMA, is approximately as active as bisquaternary com-

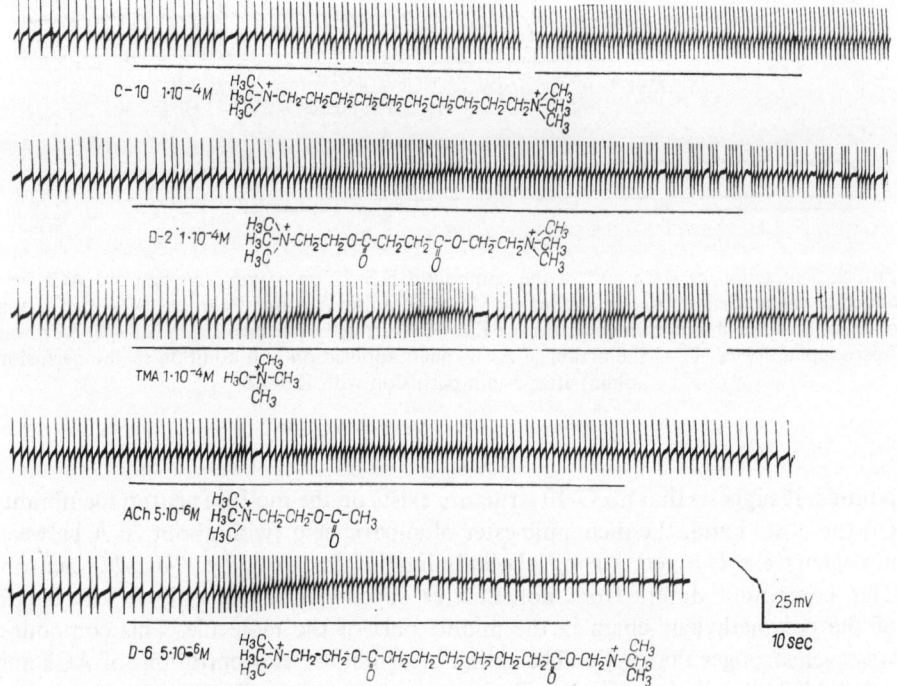

FIG. 7. The action of decamethonium (C–10), suxamethonium (D–2), tetramethylammonium (TMA), suberyldicholine (D–6), and acetylcholine (ACh) on the same *Planorbis* neuron. The interruptions in the lines and the intervals in the recordings indicate: in the case of C–10, 90 sec; in the case of TMA, 30 sec. Here and in Fig. 9, large amplification was used; only the lower part of the spikes can be seen

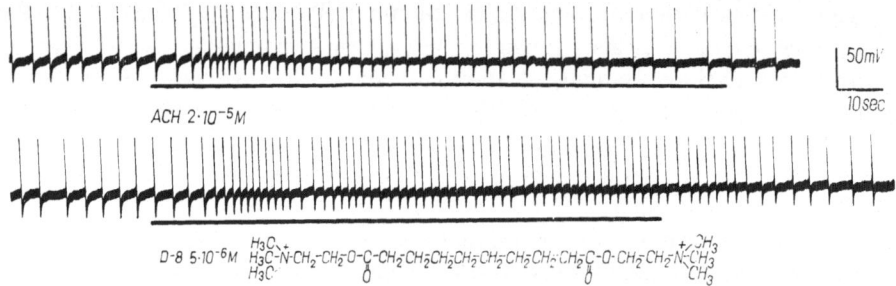

FIG. 8. The action of sebacinyldicholine (D-8) and acetylcholine on a *Planorbis* neuron

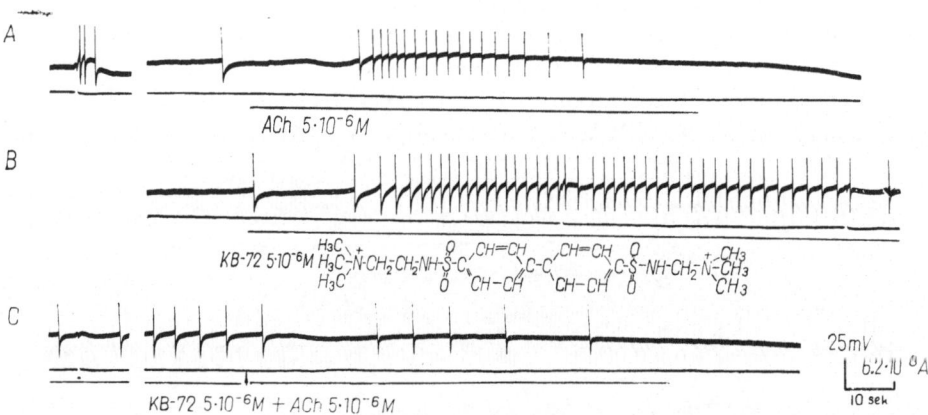

FIG. 9. The action of ACh and of the compound KB–72 on a *Planorbis* neuron. ACh was added to the perfusion fluid or applied by 600-msec pulses (see middle trace). A — the action of ACh, B — the cholinomimetic action of KB–72. During the perfusion with KB–72, ACh was microapplied twice, C — the action of ACh (microapplication and addition to the perfusion fluid) after 5-min perfusion with KB–72

pounds. It suggests that no C–10 structure exists on the mollusc neuron membrane. On the other hand, the dicholinic ester of suberic acid (with about 20 Å between nitrogens) exerts a very strong cholinomimetic effect on the *Planorbis* neuron. This compound differs from noneffective succinyldicholine only in the length of the polymethylene chain in the middle part of the molecule. This compound was even stronger than ACh. The action of equimolar concentrations of ACh and suberyldicholine is shown in Fig. 7. One can see that the effect of suberyldicholine is greater. The action of another compound with 16 atoms between nitrogens, dicholinic ester of sebacinic acid, is shown in Fig. 8. Approximately equieffective concentrations have been chosen. The concentration of ACh is four times greater; this means that sebacinyldicholine is about four times as strong as ACh.

The action of one more compound with an internitrogen distance about 20 Å, a depolarizing myorelaxant KB–72, has been also studied on the *Planorbis* neuron (Fig.9). With the same concentration, KB–72 produced nearly the same effect as ACh However, the response developed later and lasted longer than that of ACh. After 5 min perfusion with KB–72, neither electrophoretical microapplication of ACh nor addition to the perfusion fluid produced any effect. So KB–72 also exerted a strong blocking action on *Planorbis* neurons cholinoreceptors.

Thus, bisquaternary compounds containing 16 atoms between nitrogens were very strong on the *Planorbis* neurons while those with 10 atoms nearly ineffective. These data suggest the existence of C–16 structure on the neuronal membrane of *Planorbis:* a mutual disposition of receptors with the distance about 20 Å between anionic points of the neighboring receptors. A mutual disposition at the distance of 14 Å (C–10 structure) apparently does not exist on the *Planorbis* neurons membrane.

On the neurons of *Lymnaea stagnalis* different results were obtained: suberyldicholine was five times less active than ACh. So in this species no clear signs of the existence of C–16 structure were obtained as yet. Maybe only a small part of the cholinoreceptors is arranged in C–16 structure.

SUMMARY

The action of ACh, some cholinomimetics and cholinolytics on giant neurons of gastropod molluscs *(Planorbis corneus* and *Lymnaea stagnalis)* has been studied using intracellular microelectrodes. The phenomenon of increasing of sensitivity to ACh due to repetitive testing of it was observed.

Among cholinomimetics studied, the most active were those known to excite the nicotinic cholinoreceptors of vertebrates. Muscarinomimetics were weak or not active at all.

Among cholinolytics studied, the most active were those known to block cholinoreceptors of the skeletal muscles of vertebrates. Muscarinolytics and compounds known to block the cholinoreceptors of autonomic ganglia of vertebrates proved to be ineffective. This suggests some common features of mollusc cholinoreceptors and those of vertebrates. Just as is the case for mammalian skeletal muscles, the *Planorbis* neurons were found to be very sensitive to bisquaternary compounds with internitrogen chains of 14–16 atoms (20–22 Å). However, in contrast to the cholinoreceptors of skeletal muscles, those of gastropod neurons proved to be insensitive to bisquaternary compounds with 9–10 atoms (14 Å) between nitrogens.

This suggests the mutual disposition of cholinoreceptors on the membrane of *Planorbis* neurons with a distance of 20–22 Å between anionic sites. An arrangement of cholinoreceptors with anionic sites at the distance 14 Å seems not to exist.

REFERENCES

BARLOW, R. B. and ZOLLER, A. (1964): Some effects of long chain polymethylene bis-onium salts on junctional transmission in the peripheral nervous system. *Brit. J. Pharmacol.* **23** 131.

BRÜCKE, F. (1956): Dicholinesters of α, ω-dicarboxylic acids and related substances. *Pharmacol. Rev.* **8** 265—335.

KATZ, B. and THESLEFF, S. (1957): A study of the 'desensitization' produced by acetylcholine at the motor endplate. *J. Physiol.* **138** 63.

KERKUT, G. A. and COTTRELL, G. A. (1963): Acetylcholine and 5-hydroxytryptamine in the snail brain. *Comp. Biochem. Physiol.* **8** 53.

KERKUT, G. A. and THOMAS, R. C. (1963): Acetylcholine and the spontaneous inhibitory postsynaptic potentials in the snail neurone. *Comp. Biochem. Physiol.* **8** 39.

KHROMOV-BORISOV, N. V. and MICHELSON, M. J. (1966): On the mutual disposition of cholinoreceptors of locomotor muscles and on the changes in this disposition in the course of evolution. *Pharmacol. Rev.* **18** 1051.

MAGAZANIK, L. G., FRUENTOV, N. R., ROSHKOVA, E. K., RYBOLOVLEV, R. S. and MICHELSON, M. J. (1963): On the evolution of cholinoreceptive sites of locomotor muscles. In: *Pharmacology of Cholinergic and Adrenergic Transmission*, Ed.: G.B. Koelle et al., 113—127. *Proc. II. Internat. Pharmacol. Meeting*, Pergamon Press, 1965.

PATON, W. D. M. and ZAIMIS, E. (1949): The pharmacological actions of polymethylene bistrimethylammonium salts. *J. Pharmacol.* **4** 381.

RYBOVLEV, R. S. (1964): The mechanism of stimulating and blocking action of dicholinic esters of dicarboxylic acids on nicotinosensitive cholinoreceptors (in Russian). *Doctor's Thesis*, Pavlov's First Medical Institute, Leningrad.

TAUC, L. and BRUNER, J. (1963): 'Desensitization' of cholinergic receptors by acetylcholine in molluscan central neurones. *Nature* **199** 33.

TAUC, L. and GERSCHENFELD, H. (1962): A cholinergic mechanism of inhibitory synaptic transmission in a molluscan nervous system. *J. Neurophysiol.* **25** 236.

VULFIUS, E. A. and ZEIMAL, E. V. (1967a): The effects of acetylcholine and cholinomimetics on giant neurons of the mollusc *Limnaea stagnalis*. In: "Problems of evolutionary neurophysiology and neurochemistry", pp. 17—25. *Supplement* to the *J. evol. biochem. physiol.* (in Russian).

VULFIUS, E. A. and ZEIMAL, E. V. (1967b): Characteristic of cholinoreceptors of the giant nerve cells of Gastropod molluscs. In: *Second Symposium on the Problems of General Physiology.* "Synaptic Processes", Kiev (in press) (in Russian).

DISCUSSION

O. Fehér: You use drugs like decamethonium for judging the distance between the anionic points of cholinoreceptors. But the real distance between nitrogens may change due to the flexibility of the molecule.

E. V. Zeimal: I believe that the more probable conformation of the molecule of decamethonium is a straight one, with a distance of about 14 Å between the nitrogens, because of mutual repulsion of positive charges. As for compounds with a distance of about 20 Å, we have among them, for example, compound KB–72, in which the distance between nitrogens is more fixed due to the rigidity of the middle part of the molecules.

R. J. Walker: You believe that no C–10 structure exists on the cholinoreceptive membrane of molluscs. But tubocurarine and gallamine were effective in the blocking of cholinoreceptors in your experiments.

How does tubocurarine fit into the cholinoreceptive receptor you propose for *Lymnaea* and *Planorbis?*

E. V. Zeimal: You are quite right that a contradiction exists in results obtained with decamethonium and succinyldicholine on the one hand and with tubocurarine and gallamine on the other. The distance between quaternary nitrogens is approximately the same (about 14 Å) in both cases. At the same time, decamethonium and succinyldicholine produced a weak effect, but tubocurarine and gallamine were potent. We believe that results with depolarizing compounds — decamethonium and succinyldicholine — are more valuable for judging the structure of the cholinoreceptive surface. Indeed, a more precise molecular fit is apparently required for mimicking acetylcholine action than for preventing it. For example, tubocurarine effectively blocks cholinoreceptors of autonomic ganglia of vertebrates, but no one speaks of a C–10 structure there. Probably only one quaternary nitrogen is involved in the cholinolytic action of tubocurarine on the mollusc neurons. The potency of tubocurarine can be due in this case not to the two quaternary nitrogens at the suitable distance, but to the fact that the molecule of tubocurarine is bulky and contains flat rings, and therefore can prevent the access of acetylcholine to the receptor.

Symposium on Neurobiology of Invertebrates 1967 (267— 284)

PHARMACOLOGICAL MAPPING OF CELLS IN THE SUBOESOPHAGEAL GANGLIA OF *HELIX ASPERSA*

B. GLAIZNER

Department of Zoology, University of Southampton
Southampton, England

INTRODUCTION

For some time past, certain pharmacological reagents which modify neuronal activity in the vertebrates, have been found to be reactive on invertebrate nervous systems (Eccles 1964; Kerkut 1967).

Tauc and Gerschenfeld (1960) found that acetylcholine either inhibited or excited ganglion cells in *Helix pomatia,* and in *Aplysia depilans* (1962) a similar situation exists. They referred to excited cells as *D* cells and inhibited cells as *H* cells, a notation which will be used throughout this paper. The cells of *H. aspersa* were shown to be similar in their response to ACh by Kerkut and Walker (1961).

Of other substances which are regarded as neurotransmitters in vertebrate brains, indole-alkylamines and catecholamines are finding favor as possible neurotransmitters in invertebrates.

Welsh and Moorhead (1960) have found that 5–HT is widely distributed among invertebrates. The nervous systems of annelids and molluscs containing particularly large quantities of this substance.

Kerkut and Walker (1961) investigated the action of ACh, dopamine, noradrenaline, and 5–HT on the cells of the ganglia of *H. aspersa*. They suggested that a more detailed study of identifiable neurons would add to our understanding of the chemical heterogeneity of the snail brain and throw more light on role of pharmacological reagents as neurotransmitter substances.

In this paper, taken from work published elsewhere, (Glaizner 1967), an attempt is made to show that their hypothesis was valid.

METHODS

The brains from unanaesthetized snails were removed from the animals and the overlying connective tissue dissected away. The preparation was then pinned to a wax block and placed in a suitable tissue bath of 5 ml capacity. The cells being investigated were impaled with glass microelectrodes. These were drawn on a Palmer electrode puller. The electrodes were filled with either 3 M KCl or 1 M

KAc, depending on the type of experiment being performed. Electrode resistances were within the range of 2–10 MΩ.

Signals from the recording electrodes were fed to a M. E. 1400 cathode follower system, which was coupled to a modified Koketsu and Nishi bridge circuit. The output from this system was fed to a Tektronix 502 oscilloscope, modified to give a vertical output. An A. E. I. pen recorder was coupled to the oscilloscope vertical output for the purpose of making permanent records.

Ordinarily, the preparation was bathed in a Ringer of the following composition: NaCl 64.34 mM; KCl 5.76 mM; $CaCl_2$ 7 mM; $MgCl_2$ 16.5 mM; Tris HCl buffer 5 mM; pH 7.9–8.4.

Drug solutions perfused through the bath were made up from concentrates and added to 100 ml of Ringer at the appropriate dilution. When a compound was applied iontophoretically, it was from a drug electrode with a tip diameter of 1–2 μ. The ACh concentration in the drug electrodes was 10 mg/ml and dopamine, noradrenaline, and 5–HT were at concentrations of 100 mg/ml. All the drugs were dissolved in an HCl solution to give a pH of between 3 and 5.

After drug perfusion, the preparation was washed in at least 2/00 ml of Ringer. A minimum time of 5 min elapsed between the application of different compounds, and the interval was usually 15 min.

RESULTS

Acetylcholine

Sixty-two cells in the suboesophageal ganglia were tested with ACh (including eight by Walker and Hedges, 1967a, 1967b). The results are shown in Fig. 1.

The predominant response was H (Fig. 2), 34 cells showing a hyperpolarization; 24 cells were depolarized. A typical D effect is shown in Fig. 3. ACh was applied to these cells only by iontophoresis.

Significantly, only four out of the 62 cells examined failed to respond in any way to ACh.

In the right parietal ganglion, 17 out of the 29 cells investigated were H cells, while 10 were D. In the abdominal ganglion, 14 H cells were found, against 12 D cells out of 28 cells studied, a less marked difference. No marked grouping of the H and D cells, into areas, was found.

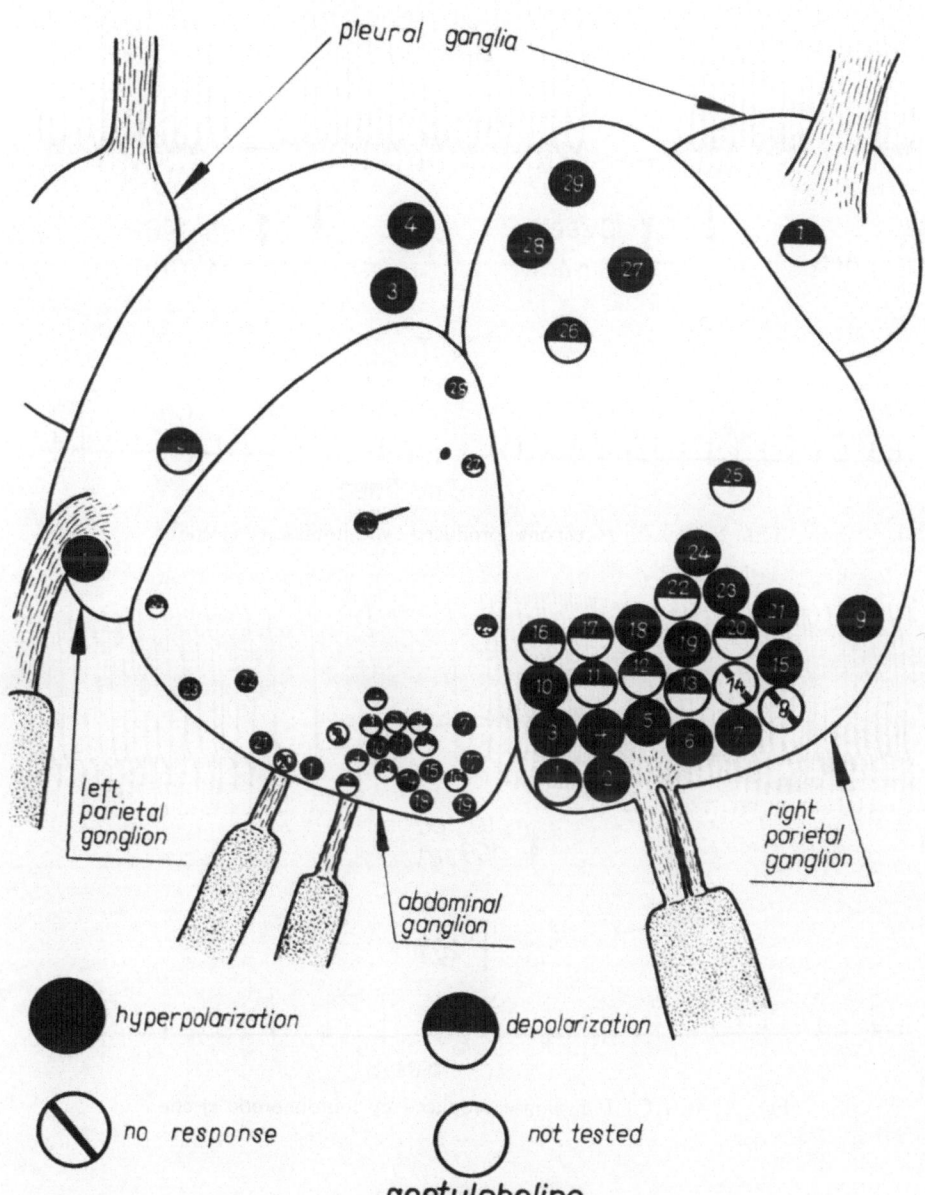

FIG. 1. Map of cells tested with ACh

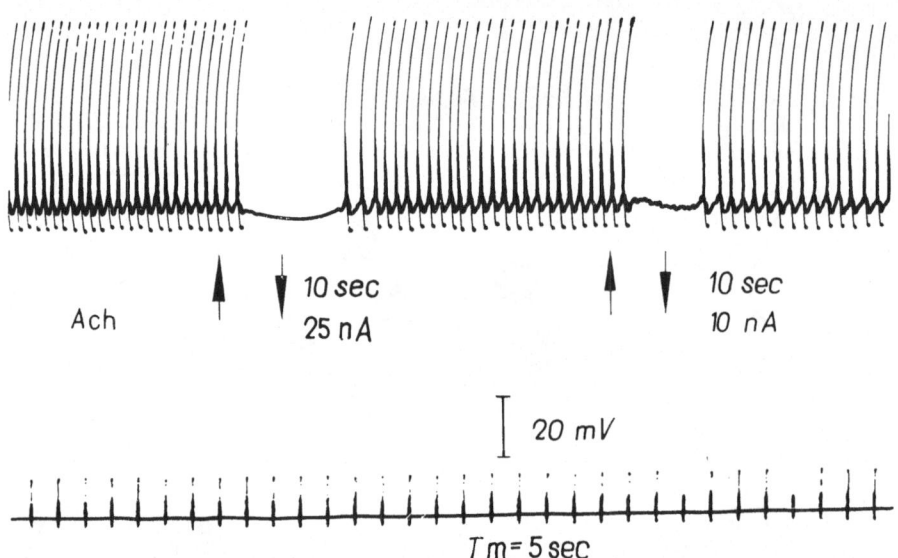

Fig. 2. An ACh *H* response produced by iontophoretic ejection

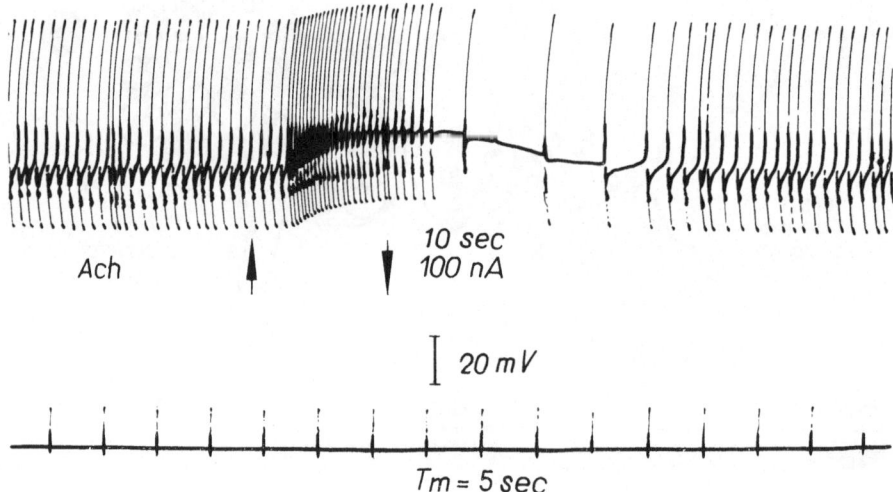

Fig. 3. An ACh *D* response produced by iontophoretic ejection

Catecholamines

Dopamine acts almost exclusively as an inhibitory reagent in *H. aspersa*. Figure 4 shows the distribution of cells which responded, and others which failed to respond, to dopamine. Of the 44 cells tested, 22 were sensitive to this substance. Twenty-one were hyperpolarized by dopamine and one depolarized.

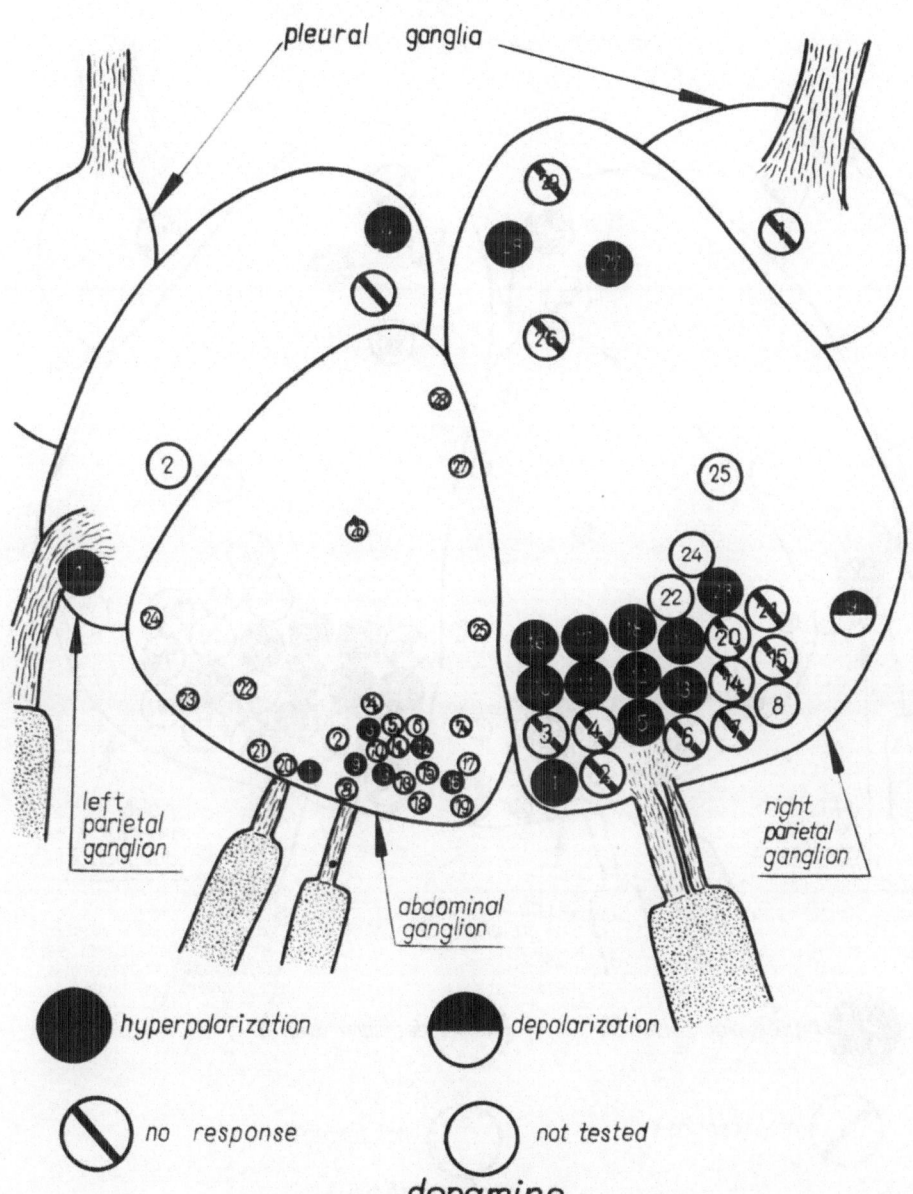

FIG. 4. Map of cells tested with dopamine

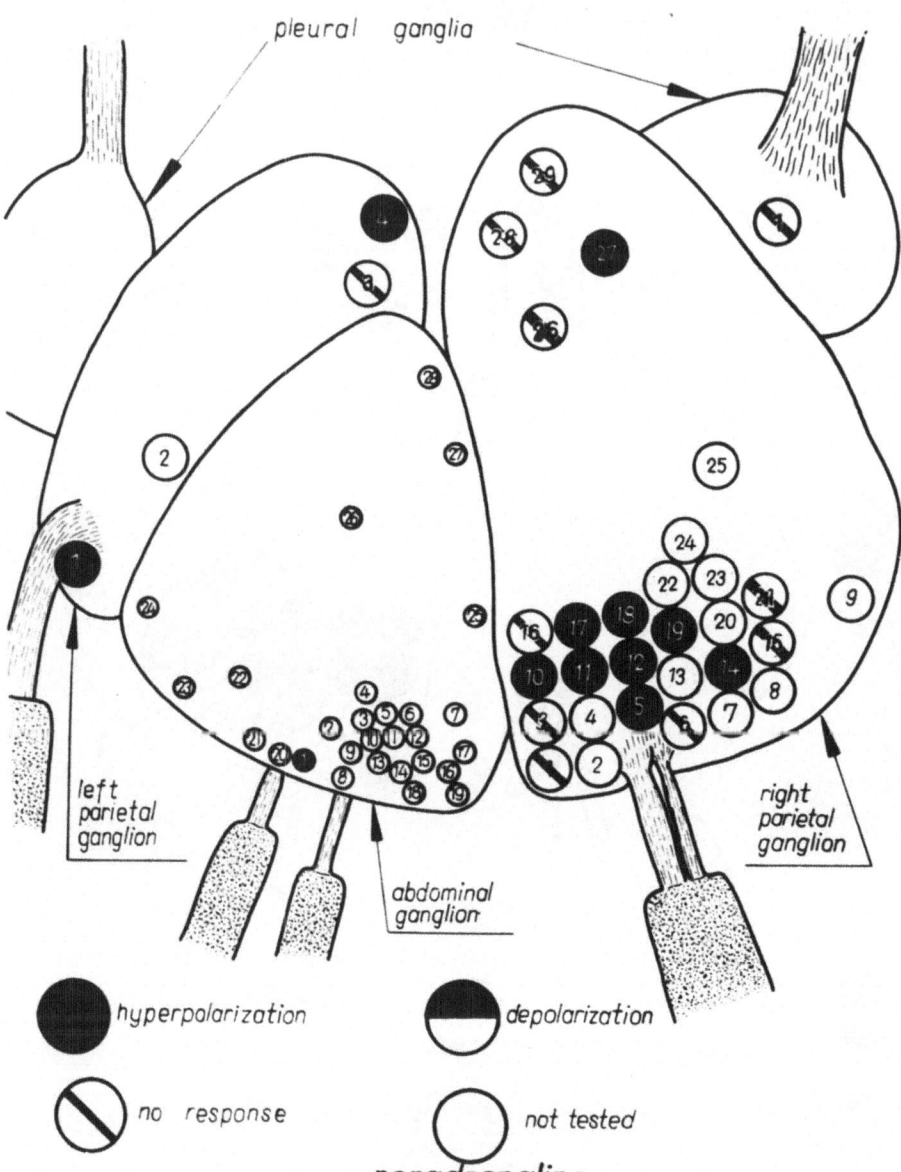

FIG. 5. Map of cells tested with noradrenaline

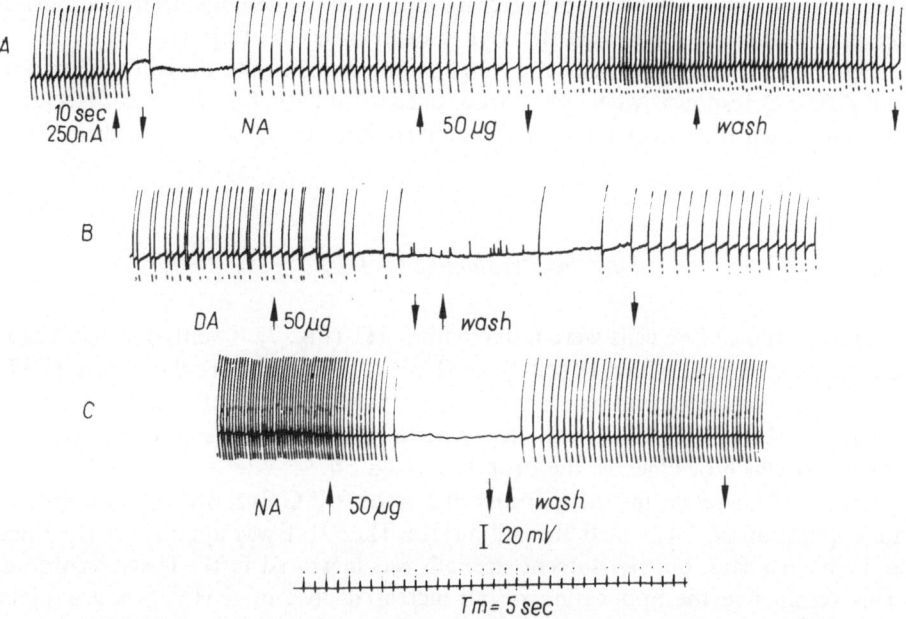

FIG. 6. Spurious *D* effect produced by gross application of noradrenaline, together with true cell response to noradrenaline and dopamine

Noradrenaline was slightly less effective as a pharmacological reagent (see Fig. 5). Twenty-three cells were investigated. Twelve responded with a hyperpolarization, 11 gave no detectable result.

Of the cells tested with dopamine and noradrenaline, more reacted to dopamine than to the latter compound. Not all the cells which reacted to dopamine gave a response to noradrenaline, and in one notable example, cell 14 in the R. P. G., responded to noradrenaline and showed no reaction to dopamine.

Occasionally a *D* response was observed on the application of weak solutions of dopamine, circa 10^{-7}, and both weak and strong solutions of noradrenaline (10^{-5}). This effect was found to be spurious, and was not seen if the catecholamines were applied iontophoretically. In the case of cells which were receptive to the catecholamine, hyperpolarization took place.

Figure 6 shows recordings taken from cell 5 in the R. P. G., Fig. 6A shows how this cell was hyperpolarized by the application of noradrenaline for 10 sec with a current of 250 nA, 125 sec later, 10^{-5} noradrenaline was added to the bath. Excitation occurred, an effect previously noticed on several occasions. The same concentration of dopamine applied to the cell in the same preparation gave a hyperpolarization, with the marked appearance of E. P. S. P. (Fig. 6B). Figure 6C is another recording from cell 5, but this time from a different preparation. It shows the true inhibitory effect of noradrenaline when it was added to the bath. Presumably the interneuron(s) causing the excitation were either insensitive on this occasion, or, more likely, the interneurons were accidentally destroyed during the preparation, thus allowing this cell to express its true response.

Cells 3 and 6 in the right parietal ganglion are insensitive to the catecholamines, and give a *D* response when they are applied to the bath. But give no detectable response when the catecholamines are applied from previously tested drug electrodes.

5-hydroxytryptamine (5-HT)

The reactions of 46 cells were tested with 5–HT (Fig. 7). Twenty-four cells produced a response, of which 18 cells were *D*. Six cells were hyperpolarized by 5–HT. Figure 8 shows a typical *D* response when 5–HT is applied iontophoretically. Figure 9 taken from cell 1 in the left parietal ganglion shows an excitation of a silent cell which declines as the drug is washed off.

Figure 10 is a recording taken from cell 2 in the R.P.G. It shows the iontophoretic application of 5–HT near the cell surface. The 5–HT was applied for the same time in each case, but the current strength was increased in the lower recording. This resulted in the application of an increased dose of 5–HT, giving a slight increase in hyperpolarization and a lengthened period of inhibition. An electrical artifact is ruled out as an explanation for this result, since cationic ejection currents can be excitatory if the electrode is applied too close to the cell surface. A pH effect is also ruled out, because an excess of hydrogen ions is also excitatory.

DISCUSSION

The chemical heterogeneity of the snail brain is well shown by the results of this investigation. Figures 11, 12 and 13 summarize the data obtained using ACh, dopamine, noradrenaline, and 5–HT. They show that contiguous cells often differ from one another pharmacologically. Thus, although it may be difficult to distinguish cells anatomically, a certain diagnosis can be made if the pharmacological data is known. Figure 14 is a photograph of part of the snail brain (the numbers placed on the cells refer to the numbers given in Figs. 1, 4, 5 and 8).

Not only are the cells different in their pharmacological responses, but in the ionic effects produced by the drugs. Thus, Kerkut and Thomas (1963) found that the ACh *H* effect was due to chloride and potassium. Kerkut and Meech (1966) showed that the ACh *D* response was due to the influx of sodium ions. More recently, Chiarandini, Stefani, and Gerschenfeld (1967) have shown that in a species related to *H. aspersa*, *Cryptomphallus aspersa*, certain cells were *D* to ACh because of an efflux of chloride ions. Glaizner (1967) has found that the catecholamine response is due to the efflux of potassium.

Gerschenfeld and Stefani (1966) noted that 5–HT was effective only on CILDA

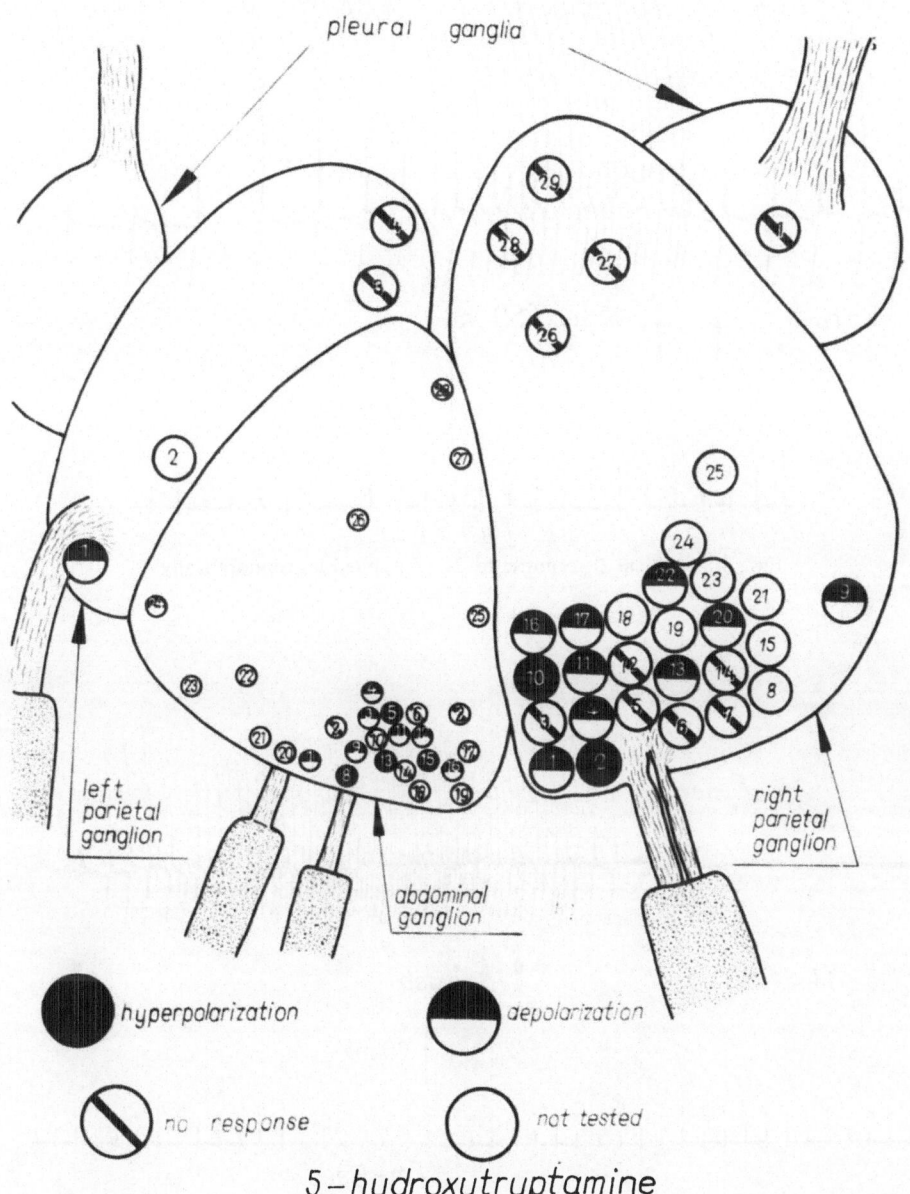

5-hydroxytryptamine

FIG. 7. Map of cells tested with 5-HT

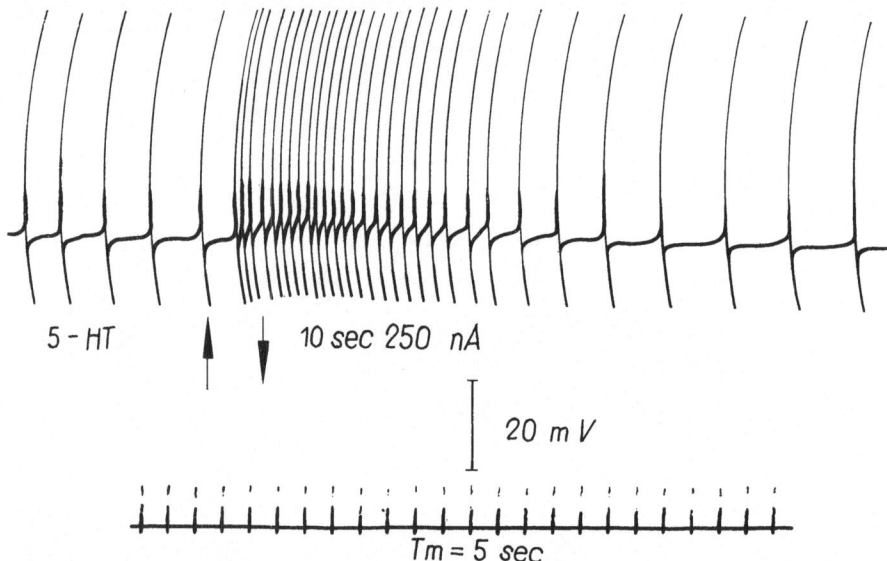

FIG. 8. A typical *D* response to 5–HT applied iontophoretically

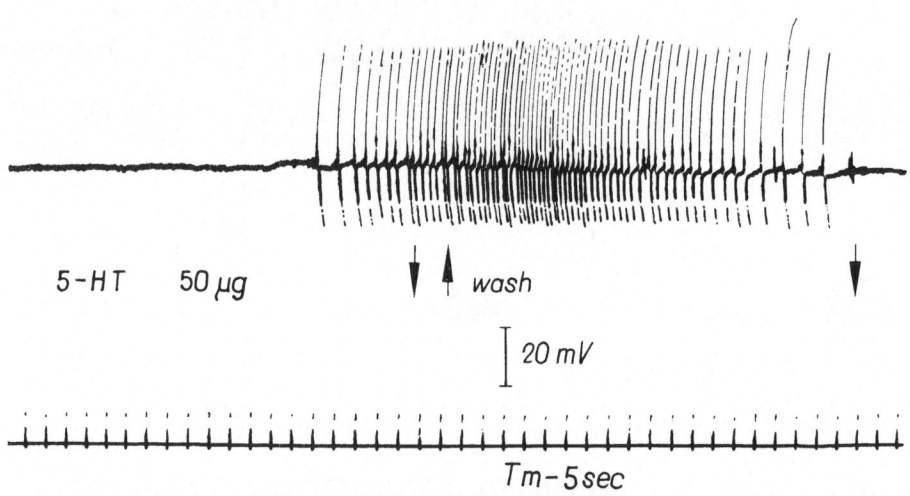

FIG. 9. 5–HT added to the bath and exciting a silent cell. Cell 1, left parietal ganglion

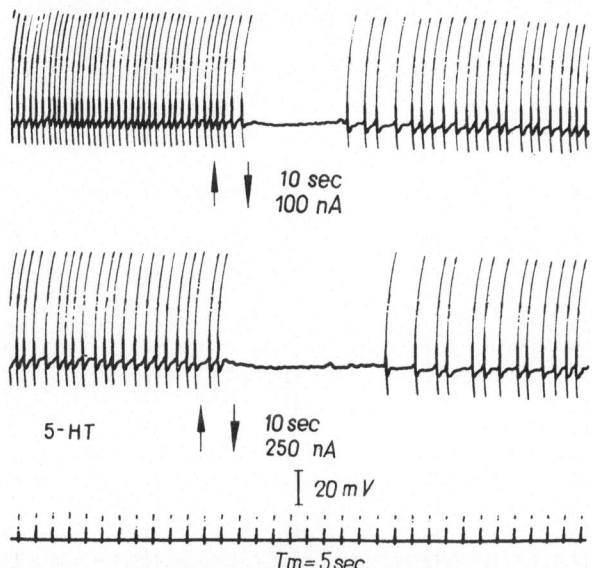

FIG. 10. 5–HT *H* response from cell 2, right parietal ganglion.
Iontophoretically applied

cells. The term CILDA was coined by Tauc (1959), and was referred to by Kerkut and Meech (1966) when they applied it to their *A* cell (believed to correspond with cell 11 in the R.P.G.). In this investigation, 5–HT was effective both on so-called CILDA cells and many others. Also, 5–HT was effective on both ACh *H* and ACh *D* cells, and not merely on the *D* cells. The rhythmic activity is only sometimes seen and may come and go in the same cell preparation; its occurrence is very variable and therefore not a useful diagnostic feature.

Hitherto, 5–HT has been regarded as exclusively excitatory with respect to molluscan neurons. The finding of six cells which are inhibited by this reagent is of interest, and 5–HT therefore joins ACh and dopamine in having both *H* and *D* effects.

A number of criteria must be satisfied before a transmitter-like chemical can be regarded as a true neurotransmitter. Paton (1958), Curtis (1961), and McLennan (1963) have discussed the evidence which must be produced to justify such a claim. Kerkut and Cottrell (1963) found both ACh and 5–HT in snail brain extracts. Kerkut, Sedden, and Walker (1966) and Sedden, Walker, and Kerkut (in press) have confirmed the presence of both 5–HT and dopamine in nerve cells in the snail brain. Welsh and Moorhead (1960) have found considerable quantities of 5–HT in molluscan tissues. The criterion that the presumptive neurotransmitter should be present in the nervous tissue is thus well established. S.-Rózsa and Perényi (1966) have been able to show that 5–HT is released from the ends of the cardiac nerves in *Helix* heart. Thus, a second criterion is established for 5–HT.

The widespread reactivity of the reagents tested is one further indication that

B. GLAIZNER

Right parietal ganglion

Cell No	ACh	DA	NA	5-HT
1	D	H	O	D
2	H	O	—	H
3	H	O	O	O
4	H	O	—	D
5	H	H	H	O
6	H	O	O	O
7	H	O	—	O
8	O	—	—	—
9	H	D	—	D
10	H	H	H	H
11	D	H	H	D
12	D	H	H	O
13	D	H	—	D
14	O	O	H	O
15	H	O	O	—
16	D	H	O	D
17	D	H	H	D
18	H	H	H	—
19	H	H	H	—
20	D	O	—	D
21	H	O	O	—
22	D	—	—	D
23	H	H	—	—
24	H	—	—	—
25	D	—	—	—
26	D	O	O	O
27	H	H	H	O
28	H	H	O	O
29	H	O	O	O

Abreviations: ACh = Acetylcholine; DA = Dopamine (3-Hydroxytyramine); NA = Noradrenaline; 5–HT = 5–Hydroxytryptamine; H = Hyperpolarization; D = Depolarization; O = No Response

FIG. 11. Summary of data for right parietal ganglion

they are neurotransmitters. ACh is a particularly strong candidate in this respect. It seems unlikely that such a high proportion of the nerve cells should possess cholinergic receptors and for ACh not to be a true transmitter substance.

Kerkut, Sedden and Walker (1966) found dopamine in the snail brain, but could not find either adrenaline or noradrenaline. This result was confirmed in their 1967 paper. Both dopamine and noradrenaline react with a large proportion of the cells tested. The case for considering dopamine as a transmitter is better made out, since it has also been shown to be present in the brain. The fact that noradrenaline has not been found does tell against it. However, its apparent absence may be due to the fact that present methods for its detection are insufficiently sensitive. Re-

Abdominal ganglion

Cell No	ACh	Da	Na	5-HT	
1	H	H	H	D	
2	O	—	—	O	
3	D	H	—	D	
4	D	O	—	D	
5	D	O	—	H	
6	D	O	—	O	
7	H	O	—	O	
8	D	O	—	H	
9	D	H	—	D	
10	H	O	—	O	
11	H	O	—	D	
12	D	H	—	D	
13	D	H	—	H	
14	H	O	—	O	
15	H	O	—	H	
16	D	H	—	D	
17	H	—	—	O	
18	H	O	—	O	
19	H	—	—	O	
20	O	—	—	—	Cell No*
21	H	—	—	—	12
22	H	—	—	—	11
23	H	—	—	—	13
24	D	—	—	O	14
25	D	—	—	—	8
26	H	—	—	—	10
27	D	—	—	—	9
28	H	—	—	D	

* ACh responses Vide Walker and Hedges 1967a and b

FIG. 12. Summary of data for abdominal ganglion

Right pleural ganglion

Cell No	ACh	D	Na	5-HT
1	D	O	O	O

Left parietal ganglion

	ACh	D	Na	5-HT
1	H	H	H	D
2	D	—	—	—*
3	H	O	O	O
4	H	H	H	O

* Vide Walker and Hedges 1967a and b, Cell 16

FIG. 13. Summary of data for right pleural and left parietal ganglia

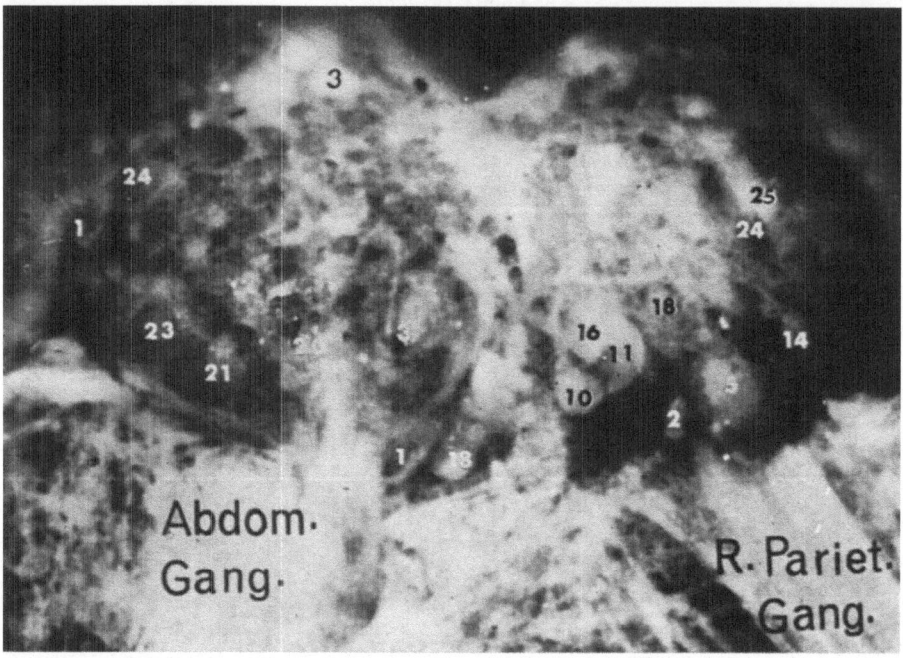

FIG. 14. Photograph of snail brain. Posterior dorsal view

cently, Cottrell (1967) found noradrenaline in the nervous tissue of *Spisula*, a marine pelecypod. More cells in the snail brain are sensitive to dopamine than are to noradrenaline. Some cells, while sensitive to the former, are insensitive to the latter. It would therefore appear that dopamine plays the more dominant role in *H. aspersa*, and the apparent absence of noradrenaline could be explained by its very low concentration in tissue extracts.

It could be argued that the noradrenaline response is fortuitous, owing its effectiveness to its close structural and chemical similarity to dopamine. However, such an argument cannot explain the fact that cell 14 in the right parietal ganglion responds to noradrenaline while failing to give a response to dopamine. Throughout the course of this work, the impression has been gained that dopamine is more potent on cells which react to both catecholamines. The potency is at least an order of magnitude greater. Cell 14 must have receptors which are capable of discriminating between the two substances. The converse is also true. In the R.P.G., cells 1, 15, 16 and 28 are hyperpolarized by dopamine, but noradrenaline has no effect at the maximum concentration applied ($5 \cdot 10^{-5}$). The fact that there are discrete noradrenaline receptors is confirmed by experiments with blocking agents. Walker, Woodruff, Glaizner, Sedden and Kerkut (1968) have shown that propanolol does not block dopamine. Glaizner (1967) found that while dopamine was not blocked by this reagent, on the same cell noradrenaline was effectively blocked by propanolol. This result clearly shows that discrete receptor sites exist for both dopamine and noradrenaline on the membrane of the same cell. It is proposed that the presence of a distinct receptor site for noradrenaline is also evidence for its being regarded as a possible neurotransmitter.

Kerkut and Walker (1961) found that noradrenaline exerted both inhibitory and excitatory effects. The only excitatory reactions found by the present author were later found to be interneuronal. An excitatory role is not ruled out, but it appears probable that what Kerkut and Walker observed were the interneuronal effects.

One or more interneurons takes part in the excitatory action of noradrenaline and the more dilute applications of dopamine. Since three cells are known to produce this response, the findings of the interneuron(s) involved would allow closely controlled experiments to be performed

SUMMARY

(1) ACh has been tested on 62 cells, 34 of which were hyperpolarized and 24 depolarized. Only four cells failed to respond. ACh is effective on more than 95% of the cells tested.

(2) Dopamine was tested on 44 cells; 21 cells were hyperpolarized and only one depolarized. Twenty-two cells gave no response.

(3) Noradrenaline was applied to 23 cells and 12 gave a response. The effect produced was entirely inhibitory. Spurious excitatory effects on three cells due to interneuronal action were found, however.

(4) Three cells which responded to dopamine did not respond to noradrenaline. One cell responded only to noradrenaline.

(5) 5-HT was investigated in 46 cells. Twenty-four cells were receptive; 18 gave a depolarizing response and 6 were hyperpolarized. A hyperpolarizing response is unusual, as hitherto this substance was regarded as an excitatory reagent.

(6) It is proposed that separate receptors exist for noradrenaline and for dopamine, even where both react with the same cell. The evidence for this proposal is discussed.

Acknowledgments: The author wishes to acknowledge a special debt of gratitude to Professor L. Brent of the Department of Zoology, University of Southampton, for his encouragement and generous assistance throughout the course of this work.

He also wishes to thank Professor G. A. Kerkut and Dr. R. J. Walker of the Department of Physiology and Biochemistry, University of Southampton, for their encouragement and assistance.

Thanks are also due to the Medical Research Council for grants provided.

REFERENCES

CHIARANDINI, D. J., STEFANI, E. and GERSCHENFELD, H. M. (1967): Ionic mechanisms of cholinergic excitation in molluscan neurones. *Science.* **156** 1597—1599.

COTTRELL, G. A. (1967): Occurrence of dopamine and noradrenaline in nervous tissues of some invertebrates. *Brit. J. Pharmacol.* **29** 63—69.

CURTIS, D. R. (1961): The identification of mammalian inhibitory transmitters. In: *Nervous inhibition,* Ed.: F. Florey, Pergamon Press, Oxford.

CURTIS, D. R. and WATKINS, J. C. (1960): The excitation and inhibition of spinal neurones by structurally related amino acids. *J. Neurochem.* **6** 117—141.

ECCLES, J. C. (1964): *The Physiology of Synapses.* Springer-Verlag, Berlin.

GERSCHENFELD, H. M. and STEFANI, E. (1966): An electrophysiological study of 5-Hydroxytryptamine receptors of neurones in the molluscan nervous system. *J. Physiol.* **185** 684—700.

GLAIZNER, B. (1967). The pharmacological responses of identified nerve cells of *Helix aspersa.* (in press).

KANDEL, E. R. and TAUC, L. (1965): Heterosynaptic facilitation in neurones of the abdominal ganglion of *Aplysia depilans. J. Physiol.* **181** 1—27.

KANDEL, and TAUC, L. (1965): Mechanisms of heterosynaptic facilitation in the giant cell of the abdominal ganglion of *A. depilans. J. Physiol.* **181** 28—47.

KERKUT, G. A. (1967): Biochemical aspects of invertebrate nerve cells. In: *Invertebrate Nervous Systems.* Ed.: , C. A. G. Wiersma, Chicago, University of Chicago Press, 5—37.

KERKUT, G. A. and MEECH, R. W. (1966): The internal chlorida concentration of H and D cells in the snail brain. *Comp. Biochem. Physiol.* **19** 819—832.

KERKUT, G. A. and THOMAS, R. C. (1963): Ach and spontaneous inhibitory post-synaptic potentials in the snail brain. *Comp. Biochem. Physiol.* **8** 38—45.

KERKUT, G. A. and COTTRELL, G. A. (1963): Acetylcholine and 5-Hydroxytryptamine in the snail brain. *Comp. Biochem. Physiol.* **8** 53—63.

KERKUT, G. A., SEDDEN, C. B. and WALKER, R. J. (1966): The effect of DOPA, a-methyl-dopa, and reserpine on the dopamine content in the brain of the snail *Helix aspersa. Comp. Biochem. Physiol.* **18** 921—930.

KERKUT, G. A. and WALKER, R. J. (1961): The effect of drugs on the neurones of the snail *Helix aspersa. Comp. Biochem. Physiol.* **3** 143—160.

KERKUT, G. A. and WALKER, R. J. (1962): The specific chemical sensitivity of *Helix* nerve cells. *Comp. Biochem. Physiol.* **7** 277—288.

KOKETSU, K. and NISHI, S. (1957): Action potentials of single intra-fusal muscle fibres of frogs. *J. Physiol.* **137** 193—209.

MCLENNAN, H. (1963): *Synaptic Transmission.* W. B. Saunders and Co., Philadelphia.

PATON, W. D. M. (1958): Central and synaptic transmission in the nervous system. (Pharmacol. aspects). *Ann. Rev. Physiol.* **20** 431—470.

S.-RÓZSA, K. and PERÉNYI, L. (1966): Chemical identification of the excitatory substance released in *Helix* heart during stimulation of the extracardial nerve. *Comp. Biochem. Physiol.* **19** 105—113.

SEDDEN, C. B., WALKER, R. J. and KERKUT, G. A. (1967): Localisation of dopamine and 5-HT in nerves of *Helix aspersa.* Symposium on the studies of the structure, physiology and ecology of molluscs. Ed.: V. Fretter, *Sympos. of Zoo. Soc. of London.* (in press).

TAUC, L. (1965): Presynaptic inhibition in the abdominal ganglion of *Aplysia. J. Physiol.* **181** 282—307.

TAUC, L. and GERSCHENFELD, H. M. (1960): Cholinergic transmission mechanisms for both excitation and inhibition in molluscan central synapses: *Nature. (Lond.)* **92** 366—367.

TAUC, L. and GERSCHENFELD, H. M. (1962): A cholinergic mechanism of inhibitory synaptic transmission in a molluscan nervous system. *J. Neurophysiol.* **25** 236—262.

WALKER, R. J. and HEDGES, A. (1967): The effect of cholinergic antagonists on the response of acetylcholine, acetyl-ß-methylcholine and nicotine of neurones of *Helix aspersa. Comp. Biochem. Physiol.* **23** 977—990.

WALKER, R. J. and HEDGES, A. (1968): The effect of cholinergic agonists on the spontaneous activity of neurones of *Helix aspersa. Comp. Biochem. Physiol.* **24** 355—376.

WALKER, R. J., WOODRUFF, G. N., GLAIZNER, B., SEDDEN, C. B. and KERKUT, G. A. (1968): The pharmacology of *Helix* dopamine receptor of specific neurones in the snail *Helix aspersa. Comp. Biochem. Physiol,* **24** 455—469.

WELSH, J. H. and MOORHEAD, M. (1960): The quantitative distribution of 5-HT in the invertebrates, especially their nervous systems. *J. Neurochem.* **6** 146—169.

DISCUSSION

G. A. Cottrell: Is it possible that the effects of catecholamines, on at least some of the neurons, are mediated by the endogenous release of ACh? Have you tried the effect of ACh blockers on the responses to catecholamines?

B. Glaizner: I do not know if ACh is released by the catecholamines; they do, however, produce an *H* effect on both ACh *D* and ACh *H* cells. I have done no work with ACh blockers.

R. J. Walker: I have applied dopamine in the presence of atropine and the dopamine effect is the same with atropine and without it.

Symposium on Neurobiology of Invertebrates 1967 (285—292)

ELECTRICAL PROPERTIES AND CONNECTIONS
OF CNS GIANT NERVE CELLS OF *HIRUDO MEDICINALIS*

V. D. GERASIMOV

A. A. Bogomolets Institute of Physiology,
Academy of Sciences of the USSR
Kiev, USSR

In each ganglion of the leech abdominal nerve chain there are two giant neurons close by. They are the so-called 'colossal' cells of Retzius. These neurons are globe-shaped with a diameter of up to 80 μ. The size of these cells is not the same for different ganglia. Nevertheless, their distinctive electrical properties, their constant position, and their ability to be colored selectively by methylene blue allow them to be distinguished from other surrounding neurons (Fig. 1). Since the giant cells are superficially located in the ganglion, they are as extremely convenient as the giant cells of molluscs for electrophysiological investigations.

It is known that between the two giant cells of a given ganglion there exists a bidirectional electrotonic type of transmission (Hagiwara and Morita 1962; Eckert 1963; Gerasimov and Akoev 1967).

In the present work, the electrical properties and connections between the giant neurons of one ganglion and between those of different ganglia were studied.

METHOD

Leeches kept under laboratory conditions were used for the investigations. The experiments were carried out on the single isolated ganglion or isolated abdominal nerve chain deprived of the cerebral and caudal ganglionic masses. It is impossible to investigate the interrelation between CNS neurons of an intact animal because of the continuous muscle contractions of the body wall. Therefore, instead of the whole animal, the nerve-muscle-skin preparation consisting of the isolated abdominal nerve chain with the anterior and posterior suckers was used. After successful preparation, the anterior and posterior sections of the body maintain the coordinated movements resembling the locomotory activity of an intact leech. The anterior sucker responds to the slight mechanical stimulation of the posterior sucker and vice versa.

In order to have a chance of observing the muscle contractions of the body wall simultaneously with the neuron electrical activity, the skin-muscle strip (with three ganglia) was left in the middle part of the isolated nerve chain. The right and left

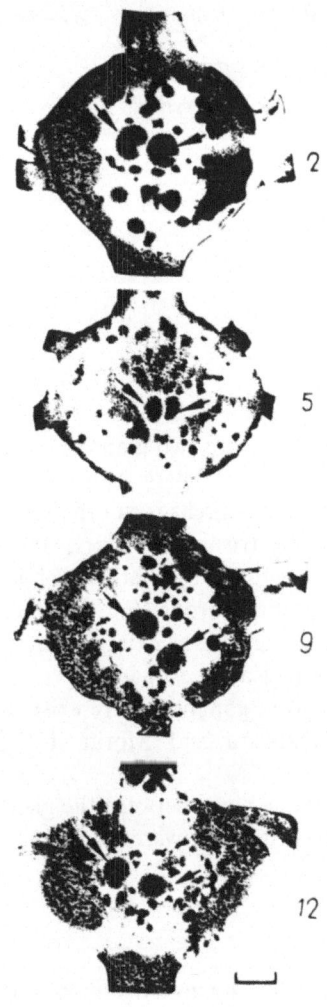

FIG. 1. Photograph of some isolated ganglia of the leech abdominal nerve chain. Arrows indicate the giant neurons in each ganglion. The figures at the side are the ordinal numbers of the ganglion beginning from the cerebral ganglionic mass. Scale — $100\,\mu$

lateral nerve branches of both extreme ganglia were cut across so that the remaining body strip received the innervation only from one middle ganglion. The recording and stimulating electrodes were inserted into the giant cells of this ganglion.

The whole nerve chain and especially the ganglia under investigation were carefully fixed on the bottom of the plexiglas chamber with the corresponding saline solution. The intracellular potentials were recorded simultaneously from two giant cells. For artificial depolarization or hyperpolarization, a third microelectrode was inserted in one of the cells. The direct current of the necessary strength, duration, and direction was passed through it.

The nerve fibers were stimulated directly by a single rectangular pulse or indirectly by the adequate stimulation of the skin mechano- or photoreceptors.

For testing the chemical sensitivity of the cells, acetylcholine chloride, adrenaline, or noradrenaline were added to the saline solution surrounding them. All the other experimental conditions were the same as those described elsewhere (Gerasimov and Akoev 1967a, 1967b).

ELECTRICAL PROPERTIES AND CONNECTIONS
OF THE GIANT CELLS IN THE SAME GANGLION

The amplitude of the giant neuron action potentials (AP's) in the normal saline solution is usually small. The AP's range from 15 to 30 mV. Since the resting potential (RP) of the cell ranges from 30 to 50 mV, no obvious overshoot occurs. The AP's of the surrounding smaller neurons have a large overshoot.

Between two giant neurons of a given ganglion there exists a bidirectional electrical type of transmission. The electrical response of one cell is transmitted electrotonically to another cell. In Fig. 2 (1), the anelectrotonic and catelectrotonic potentials of two giant neurons are demonstrated. The direct polarization is applied only to one of them (cell I, upper trace). These records show that the potential change of the cell II is a result of electrotonic spread from the directly polarized cell I. When the AP of one cell evokes suprathreshold electrotonic depolarization in another cell, the latter cell generates its own AP. Because of the abrupt rising phase of the AP's, the attenuation factor for transmitting AP's are bigger then for spreading the electrotonic potentials.

In normal conditions, the frequencies of spontaneous electrical discharges in both giant cells are similar. The AP's are generated synchronously, the AP of one cell corresponding to the AP of the other [Fig. 2 (2)].

Because of the electrical type of connection between the two cells, the slow-wave of depolarization (prepotential) rises in both cells simultaneously. Nevertheless, sometimes the AP's in both cells may not be generated at the same time, since the

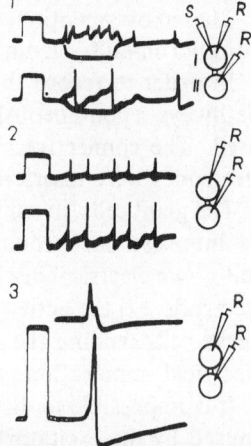

FIG. 2. Electrotonic coupling of the pair of giant neurons in a ganglion. 1 — Electrotonic transmission of the potentials of cell I to cell II. Current: $3 \cdot 10^{-9}$ A. 2, 3 — Spontaneous action potentials from two cells recorded simultaneously. R — recording electrode; S — stimulating electrode. Voltage, time calibration: 20 mV, 50 msec (1, 3) 20 mV, 500 msec (2)

times required for the depolarization to reach the threshold level can be different for each cell. Therefore, the AP of one cell can appear during the descending phase of the other cell AP and give an electrotonic response as in Fig. 2 (3).

Sometimes the amplitude of an AP-rising phase of one cell can be added to the electrotonic response produced by the AP of the other cell. Then, an increase of AP amplitude occurs. An especially marked addition occurs during electrotonic transmission of the barium prolonged action potentials (PAP's). The barium PAP's of the leech giant cells reach a large amplitude that exceeds the zero level of the RP.

CONNECTIONS BETWEEN GIANT CELLS OF DIFFERENT GANGLIA

The intracellular stimulation applied to the giant neurons of one ganglion does not produce synaptic potentials and AP's in the analogous neurons of the adjacent ganglia. Also, no electrotonic transmission is observed between these cells [Fig. 4B (b, c)]. However, a brief electrical stimulus applied to a cerebral or caudal connective, evokes excitatory postsynaptic potentials (EPSP's) and action potentials in each of the giant neuron pairs (Fig. 3). A complete block of the excitation of any pair of neuron by a hyperpolarizing current applied through the intracellular micro-electrode does not prevent the excitatory wave from propagating to other giant neurons [Fig. 3A(b), B(d)].

Thus, the wave of excitation is conducted along the abdominal nerve chain through the descending as well as ascending pathways.

The mechanical or light stimulations applied to the skin of anterior or posterior suckers produce the same EPSP and AP discharges in each giant neuron. The excitation is conducted along the nerve chain with the same velocity as follows the electrical stimulation of the connectives. The direction of the propagating excitatory wave depends on the place of the stimulation. When any of the giant neuron pairs is artificially hyperpolarized (to prevent their firing) the propagation of an excitatory wave to other giant cells is not blocked.

The above-mentioned descending and ascending nerve pathways apparently conduct impulses from skin mechano- and photoreceptors.

In order to record the excitatory wave arising in the descending and ascending pathways 'spontaneously', the nerve-muscle-skin preparation of the animal was used. The connective tissue capsule of two ganglia was opened and the micro-electrodes were inserted into the giant cells of these ganglia.

The giant cells, in contrast to some other cells of the leech CNS, are characterized by intensive background electrical activity. This activity can depend on: (1) the cell's own electrical discharges, caused in particular by its puncture with the micro-electrode, (2) the activity of the neighboring giant cell, (3) the activity of the small nerve cells sending the excitatory axon terminals to the giant neuron, and (4) the electrical impulsation running along the descending and ascending pathways.

It is impossible to distinguish from such background electrical activity the AP's caused by the excitatory wave propagating along the nerve chain. The excitatory

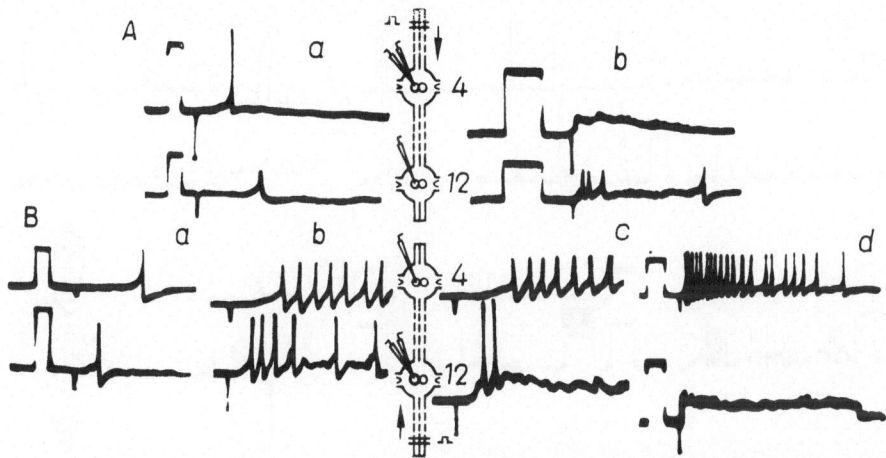

FIG. 3. Electrical responses of the giant cells in different ganglia during propagation of the excitatory wave along the abdominal nerve chain. A — electrical stimulation of the descending nerve pathways. a, b — electrical activity recorded simultaneously from the cells in the fourth and twelfth ganglia. Distance between these ganglia was 2.5 cm. The cells in the fourth ganglion were artificially hyperpolarized to avoid the spike firing (b); B — electrical stimulation of the ascending nerve pathways; a, b, c, d — electrical cell activity recorded simultaneously from the same ganglia of another animal; b, c, d — the amplitude of the stimulating current shock was increased. Calibration (b, c) is the same as in (a); c, d — gradual artificial hyperpolarization of the cells in the twelfth ganglion to avoid spike generation. The scheme of recording from the ganglion cells is shown in the middle part of the picture. Calibration: 20 mV, 50 msec; A(b) B(d) 500 msec

wave propagating along the nerve chain is lost in the 'noise' of the giant cell background activity. Therefore, the background electrical activity in our experiments was partially depressed either by passing direct hyperpolarizing current or by adding small concentrations of barium chloride to the surrounding saline solution. The barium ions increase the membrane cell resistance, increase the threshold of the cell and produce PAP's. The spontaneous PAP's are generated by the giant neurons with a slow rhythm. On such a background, the PAP's evoked by the impulses running periodically along the whole nerve chain are easily observed. These PAP's are always initiated first in the cells of one ganglion and then, only after a constant interval, in the cells of another ganglion (Fig. 4A). The time interval corresponds to the conduction velocity of the excitation along the nerve chain.

Similar results are obtained in giant cells partially hyperpolarized by the direct current. In these conditions, bursts of spike discharges are developed instead of PAP's [Fig. 4B(a)].

Thus, the giant neurons do not take part in the propagation of the excitation along the abdominal nerve chain. Evidently, they transmit it to the periphery of each body segment. It is quite possible that excitation is conducted along the nerve chain by special nerve pathways. These pathways conduct the impulses with slow velocity (0.3–0.5 m/sec). They are probably polysynaptic pathways. Perhaps the

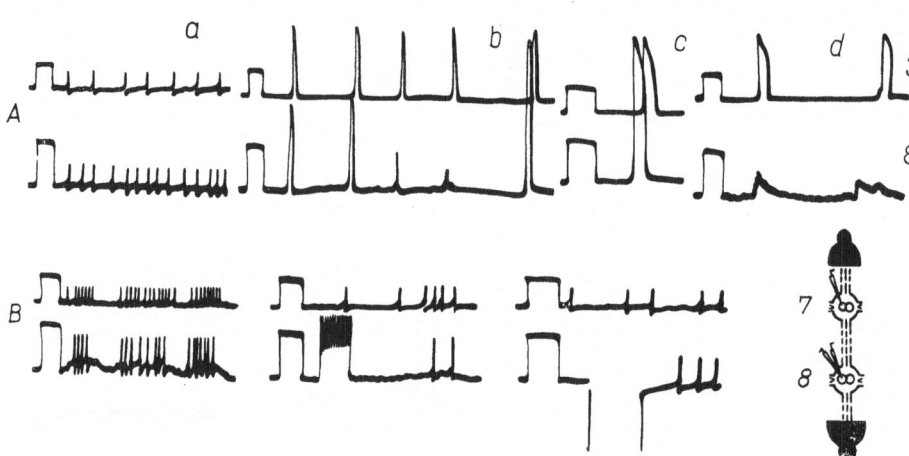

FIG. 4. Spontaneous electrical activity of the giant cells from different ganglia. A — spontaneous discharges of spike potentials recorded simultaneously from the cells of the third and eighth ganglia. Distance between these ganglia was 2 cm; b, c, d — after 20 mM $BaCl_2$ was added; d — the cells in the eighth ganglion were artificially hyperpolarized to avoid the PAP generations. Only EPSP's are seen in the oscillogram; B — the electrical responses recorded simultaneously from the neurons of the seventh and eighth ganglia; a — the spontaneous AP burst recorded after a partial depression of the cell background activity by the current hyperpolarization; b, c — the depolarized and hyperpolarized cell response induced by a rectangular current pulse applied to the cells of the eighth ganglion through the intracellular electrode; d — schematic diagram of the leech nerve-muscle-skin preparation. Figures — the ordinal number of the ganglion. Calibration: 20 mV, 500 msec

interneurons of these tracks send axon collaterals to the giant cells. Because of the electrical connection between the two giant cells in each ganglion, the excitation is transmitted simultaneously to both sides of the body. This is probably essential to the locomotory activity of the animal. Electrical stimulation of connectives produce giant neuron AP discharges synchronously with longitudinal muscle contractions of the body wall. In spite of this, we did not observe any visible muscle contractions in response to direct giant neuron stimulation. Thus, the functional role of these neurons remains unclear for the present.

CHEMICAL SENSITIVITY OF THE GIANT CELLS

The giant cells are sensitive to acetylcholine (ACh) in concentrations of 10^{-5} g/ml. ACh depolarizes these cells and produces impulse discharges. The giant neurons adapt to these concentrations very quickly. After transient depolarization, the cell recovers both the RP and the frequency of spontaneous AP's (Fig. 5a). The increase of the acetylcholine concentration by one order depresses the cell electrical activity for 4–5 min and significantly decreases the cell membrane resistance (Fig. 5B). After that action, the normal functional properties can be recovered only by prolonged washing with normal saline solution. A 10^{-6} g/ml solution of acetylcholine does not produce any visible effect.

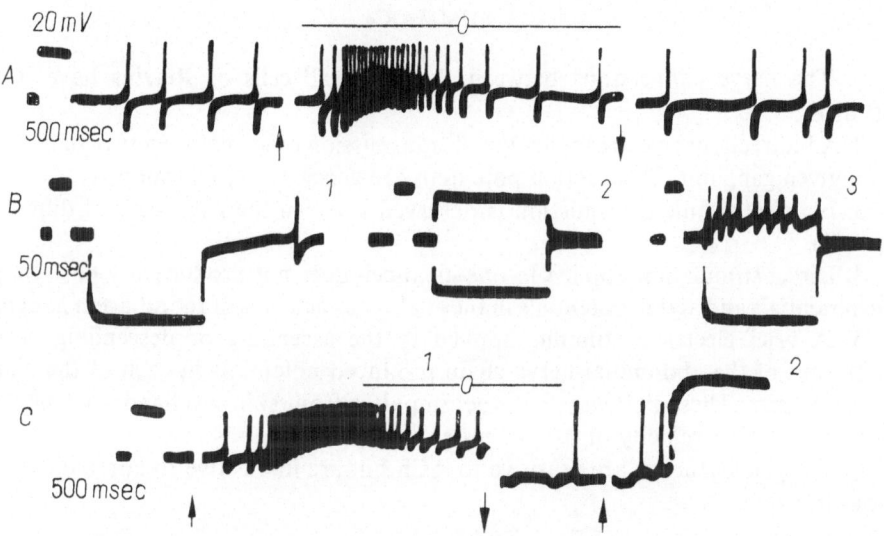

FIG. 5. Chemical sensitivity of the leech giant cells. A—action of a 10^{-5} g/ml solution of ACh on the cell electrical activity (between the arrows); B — action of a solution of $4 \cdot 10^{-1}$ g/ml ACh on the cell membrane resistance. Current: $8 \cdot 10^{-9}$ A; 1 — in normal solution; 2 — in ACh solution; 3 — the ACh solution was changed back to normal solutions; C — action of a solution of $4 \cdot 10^{-5}$ g/ml ACh on the spontaneous cell activity in a Ca-free solution; 1 — at the beginning of perfusion with Ca-free solution; 2 — after perfusion for 30 min with Ca-free solution

As was found in our previous work (Gerasimov and Akoev 1967c) the leech giant cells generate spontaneous PAP's in a Ca-free solution. These PAP's can be evoked artificially by adding threshold ACh concentrations to the Ca-free solution (Fig. 5C).

The giant neurons in each ganglion are enveloped by a large glial cell (glial cell of the packet). This neuron position can be the reason for their insensitivity to the small ACh concentrations. It is possible that before the ACh reaches the neuron it is partially destroyed. We can make this suggestion because the leech glial cells are sensitive to the ACh as well (our own observation). A solution of 10^{-4} g/ml ACh depolarizes them up to 25–30 mV. The ACh effect on the glial cell RP is reversible, so that after washing it returns to the initial level (50–60 mV).

The giant cells are absolutely insensitive to adrenaline and noradrenaline.

It is possible that an ACh-like substance is a transmitter for these cells. This substance can be liberated by synaptic terminals of the ascending and descending nerve pathways of the abdominal chain.

It should be noted that 10^{-4} g/ml solution of atropine sulphate blocks the orthodromic responses of the giant neurons, whereas 10^{-3} g/ml solution of d-tubocurarine does not produce any visible effect.

SUMMARY

1. The nerve connections between the 'colossal' cells of Retzius have been studied.

2. A bidirectional electrotonic type of transmission exists between two giant cells of a given ganglion. Their action potentials are generated synchronously.

3. No electrotonic transmission is observed between the giant cells of different ganglia.

4. Direct stimulation applied to one giant cell does not produce any postsynaptic potentials and action potentials in the analogous neurons of the adjacent ganglia.

5. A brief electrical stimulus applied to the ascending or descending nerve pathways of the abdominal nerve chain produced potentials in each of the giant neuron pairs. These pathways conduct impulses from skin mechano- and photo-receptors with a velocity of 0.3–0.5 m/sec.

6. The giant neurons are sensitive to ACh but are insensitive to adrenaline and noradrenaline.

REFERENCES

ECKERT, R. (1963): Electrical interaction of paired ganglion cells in the leech. *J. Gen. Physiol.* **46** 573.

GERASIMOV, V. D., AKOEV, G. N. (1967a): Features of the giant neuron electrical activity of the leech *(Hirudo medicinalis)* in various saline solutions. *J. evol. biochem. physiol.* **3** 234–240.

GERASIMOV, V. D. and AKOEV, G. N. (1967b): Electrical potentials of different neurons of the leech in Ca-free solution. *Dokl. Akad. Nauk USSR* **172** 494–497.

GERASIMOV, V. D. and AKOEV, G. N. (1967c): Effect of various ions on the resting and action potentials of the leech *Hirudo medicinalis. Nature* **214** 1351–1352.

HAGIWARA, S. and MORITA, H. (1962): Electrotonic transmission between two nerve cells in leech ganglion. *J. Neurophysiol.* **25** 721.

Symposium on Neurobiology of Invertebrates 1967 (293—301)

ON THE MECHANISM OF ANISOTROPIC EXCITABILITY FOR THE ADDUCTOR RESPONSE OF GLOCHIDIA

E. Lábos

Biological Research Institute of the Hungarian Academy of Sciences
Tihany, Hungary

Different anisotropic properties of excitable tissues are known; the mechanisms behind these properties can be intercellular or subcellular. Such phenomena include, for example, rectification in membranes (Cole and Curtis 1941; Hutter and Noble 1960; Hecht *et al.* 1964; Nakajima et al. 1961), synaptic transmission itself, and the exclusive effectiveness of a longitudinal stimulatory current in cable-like excitable tissues (Rushton 1930).

The present paper reports on the complex anisotropic behavior of mussel-larval adductor-muscle excitability (Lábos 1964). After demonstrating the experimental facts, some assumptions for explaining the observed spatial organization will be discussed.

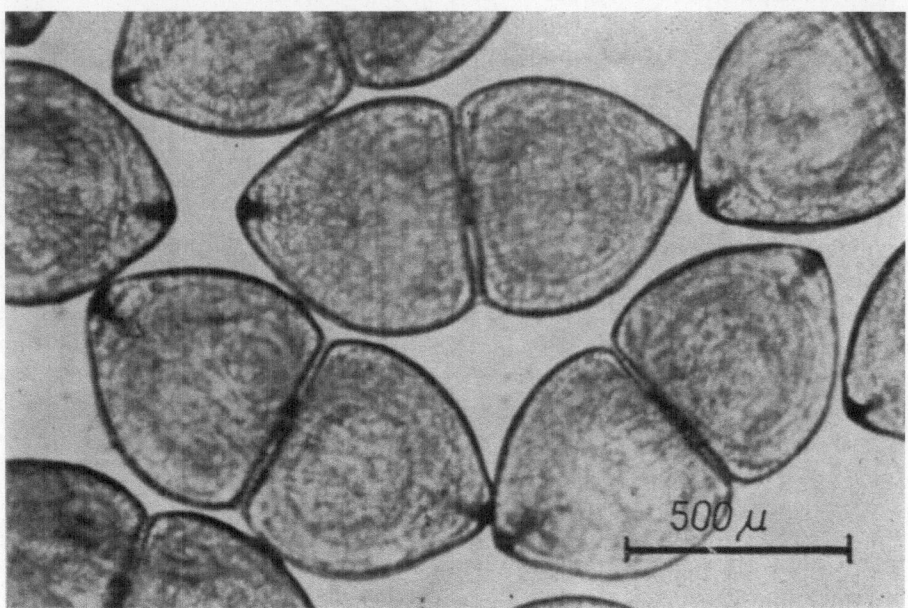

FIG. 1. The glochidium, the postembryonal larva of *Anodonta*

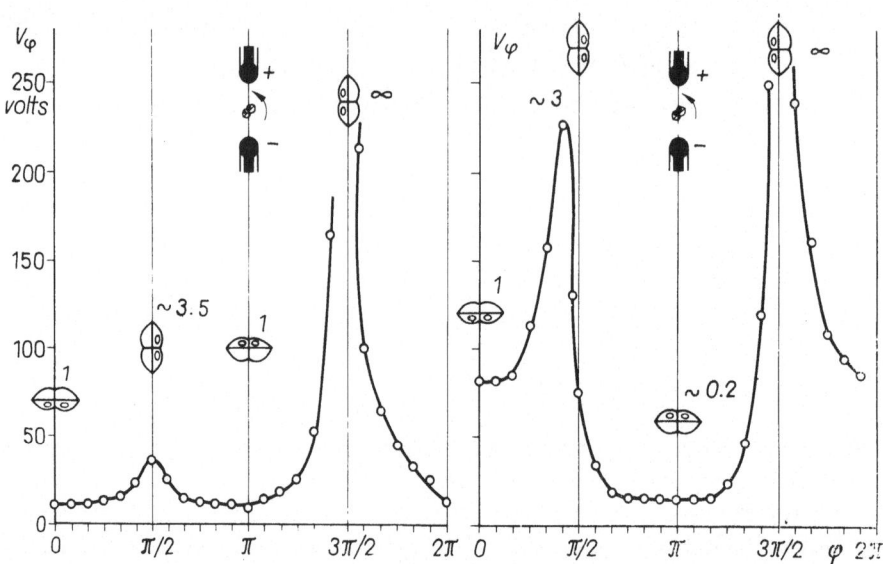

FIG. 2. The threshold voltage (Vφ) is plotted against the angle φ between the muscle longitudinal axis and the direction of current field (see Fig. 4). The orientation of the animal is
drawn in the four main directions, as are the relative values of threshold. Symmetric (V$_o$ = V$_\pi$)
and asymmetric (V$_o$ ≠ V$_\pi$) animals are shown

Glochidia, the postembryonal larval form of the freshwater mussel, were used
in the experiments (Fig. 1). Under microscopic control, a precise orientation of
glochidia in relation to the stimulating electrodes was achieved with the help of
micromanipulators. The stimulation by square pulses was carried out through
physiological bathing fluid (Lapicque 1931; Rushton 1927, 1930), and, with
careful orientation and by precise control experiments, some possible artifacts
due to the effect of field distortions or due to rectifying layers on the electrodes
were succesfully excluded.

Figure 2 demonstrates the two kinds of glochidia. The threshold voltages
(V$_\varphi$) for phasic muscle response are plotted against the angle (φ) between the
muscle's longitudinal axis and the direction of excitatory current field. It can be
seen that the transverse thresholds are always different for stimulation from
opposite directions. If the lateral cavity of the larvae is made positive, the threshold is smaller; when this side looks at the negative pole, the threshold is extremely high or could not be measured. The thresholds in longitudinal fields are
often equal, but sometimes the thresholds at 0° and 180° are unequal. The latter
type of glochidia will be referred to as the asymmetric type.

The ratio of thresholds in the four main directions (0, $\pi/2$, π, $3\pi/2$) is about
1 : 4 : 1 : ∞ for symmetric and 1 : 16 : 4 : ∞ for the asymmetric animal.

In the case of asymmetric animals, the measure of the longitudinal polarization,
that is, the ratio of the two longitudinally measured thresholds α, depends on

the impulse duration. It has a definite optimum at about 1–3 msec. With long pulses, the asymmetry is decreased or abolished.

Figure 3 shows that the asymmetry factor varies in different solutions. Both the asymmetry of the animals and the symmetry itself are sensitive to the ions in the medium. It can be made symmetric or asymmetric by LiCl, $MgCl_2$, $CaCl_2$, or quaternary amines. In Fig. 3, the α asymmetry factor is plotted against the time spent in the medium.

The nature of anisotropically organized excitability can be well studied by an electrode system consisting of two pairs of electrodes and by orienting them at right angles to each other (Fig. 4).

In Fig. 5, the threshold values are plotted against the strength of a subthreshold prestimulus. When the prepulse has a direction at right angles to the test pulse, an inhibition is generally observed. The transverse prepolarization increases the longitudinal threshold by a quantity proportional to the intensity of the prepulse.

On the contrary, the parallel prepulse does not inhibit, but facilitates the test stimulation. However, the linearity could be observed only if the delay is very small. In this latter case, the threshold decrease is equal to the prepulse. However, after longer delays, the amount of facilitation decreases.

In Fig. 6, the decrease of the threshold (that is, the facilitation caused by longitudinal stimulation) and the inhibition, (that is, the increase evoked by a right-

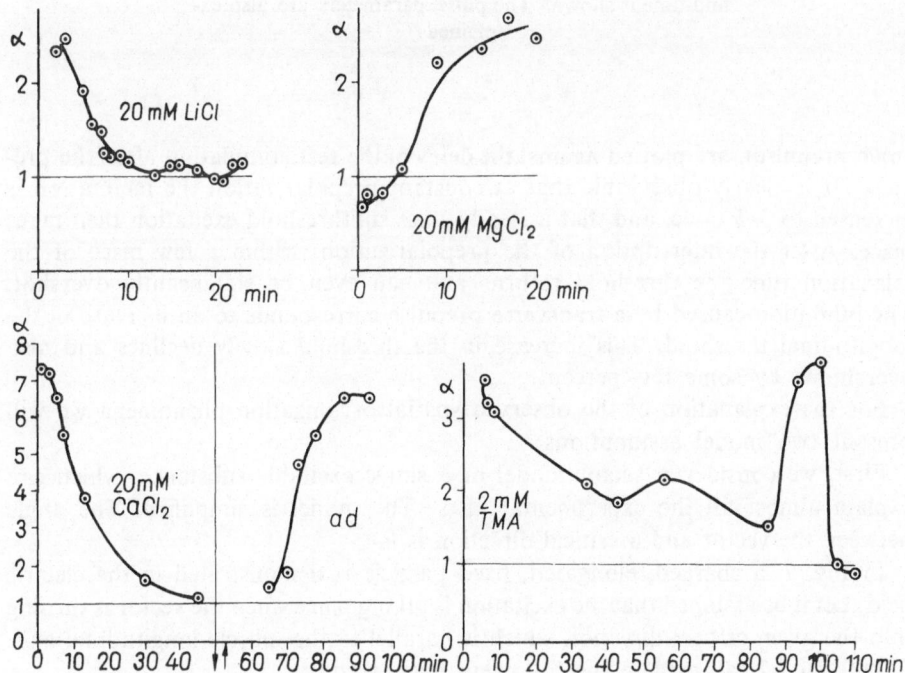

FIG. 3. The α asymmetry factor plotted against the time spent in different bathing fluids.
ad—aqua. dest.

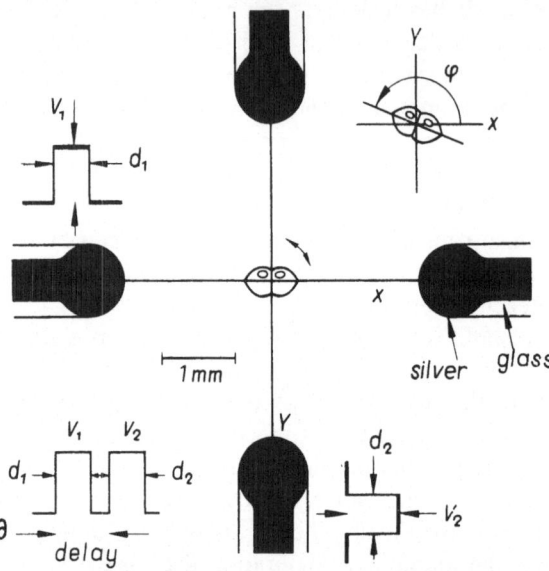

FIG. 4. The experimental situation used to investigate the effect of prefield. Four AgAgCl electrodes were used. The order of the square-pulse stimulation space and time is shown. The pulse parameters are also explained

angle prepulse), are plotted against the delay of the test stimulation after the prepulse. It is clearly observable that at constant prepolarization the facilitation is increased to 3–5 msec, and that a steady-state subthreshold excitation then takes place. After the interruption of the prepolarization, within a few msec of the relaxation time the threshold returns and can even be significantly overshot. The inhibition caused by a transverse prepulse corresponds to an increase in the longitudinal threshold. This increase in the threshold slowly declines and also overshoots by some few percent.

For the explanation of the observed spatial organization phenomena we will present two model assumptions:

First, we consider a vector model of a single excitable substance, which can explain almost all the experimental facts. The model is simplified. The angle between the vector and a critical direction is β.

In Fig. 7, a charged, elongated, fixed particle is demonstrated in the electric field. Let it be assumed that the excitation is taking place when the vector is turning into the given critical direction which is parallel to the muscle longitudinal axis.

This model can explain the following phenomena:

1. Only the parallel current is effective because it evokes the steady-state parallel and critical orientation.

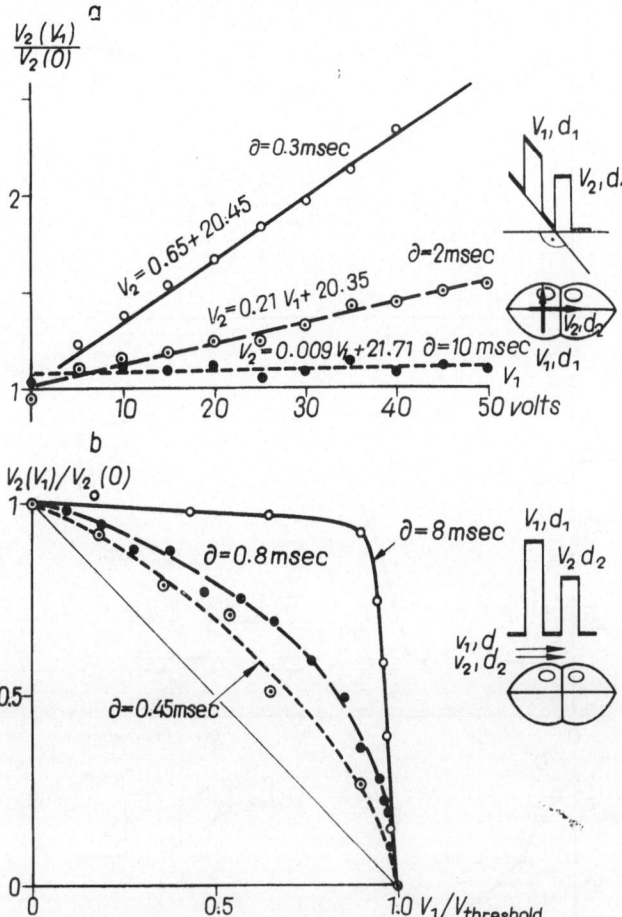

FIG. 5. The increased or decreased threshold voltage plotted against the strength (V_1) of prestimulus. In (a) the prepulse was at right angles to the test pulse, and in (b) it was parallel to the second pulse. The three curves are derived from experiments at different delays

2. The transverse current is not effective when it evokes right-angled orientation of the vector to the critical direction.

3. However, the transverse current is effective when evoking a critical but non-steady-state orientation. The transient character of this orientation is thought to be responsible for the 3–4-fold increase in the transverse threshold at $\phi = \pi/2$.

4. The equation of motion for such a vector postulates a damped oscillatory behavior of the threshold. There are solutions of elliptic integrals.

5. The asymmetric position of the supposed vectors allows a longitudinal asymmetry or a symmetry, depending on whether the forces are against the

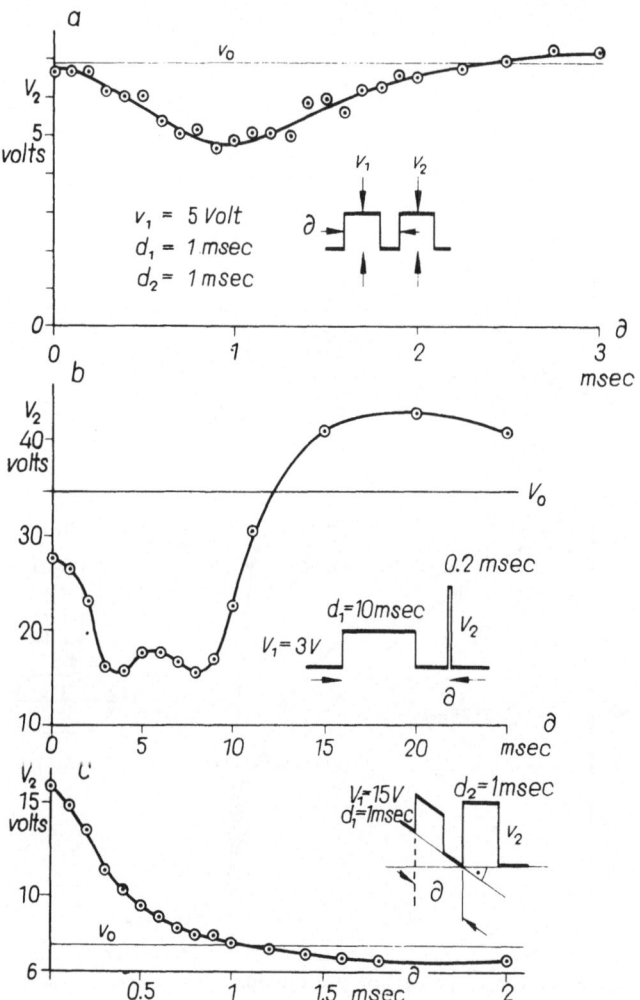

FIG. 6. The threshold voltage (V_2) for contraction is plotted against the delay (δ) after a prestimulus (V_1, d_1) of a given direction. The prestimulus was parallel to the test pulse at (a) and (b) and was perpendicular at (c). The V_o indicates the control threshold without prefield

stimulatory field. If the forces acting against the stimulus are of higher values, the motion becomes direction-dependent. A high effective countereffect could be achieved by salts.

6. Applying a subthreshold prepulse which directs the vector against the test pulse results in an inhibition, but turning the vector in the same direction as the test pulse results in a facilitation. The model postulates a linear function depending

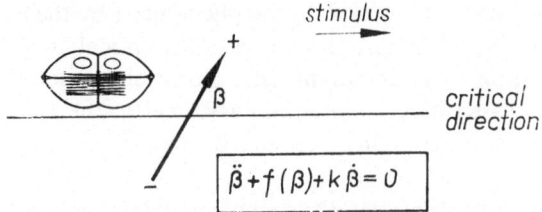

FIG. 7. Vector model for the explanation of anisotropic phenomena. The β angle is the angle between the positions of the vector in rest and under
excitation. The nonlinear equation of motion is also
given

on the strength of the prefield when the delay is zero. When the delay is not zero,
the facilitation function is notlinear.

The physicochemical reality of such a model could be an orientation in the excitable membrane dipoles. However, when considering measurements of membrane
dipoles, the observed relaxation times, and the known formula of Debye (1929),
the effective viscosity of the medium in which the orientation could take place
is found to be very high. Therefore, the location and state could only be the structural gel viscosity of the excitable cell cortex (Heilbrunn 1958). Naturally, the
supposed charge orientation is only a simplified equivalent of the complex con-
figurational changes in the membrane (Goldman 1964).

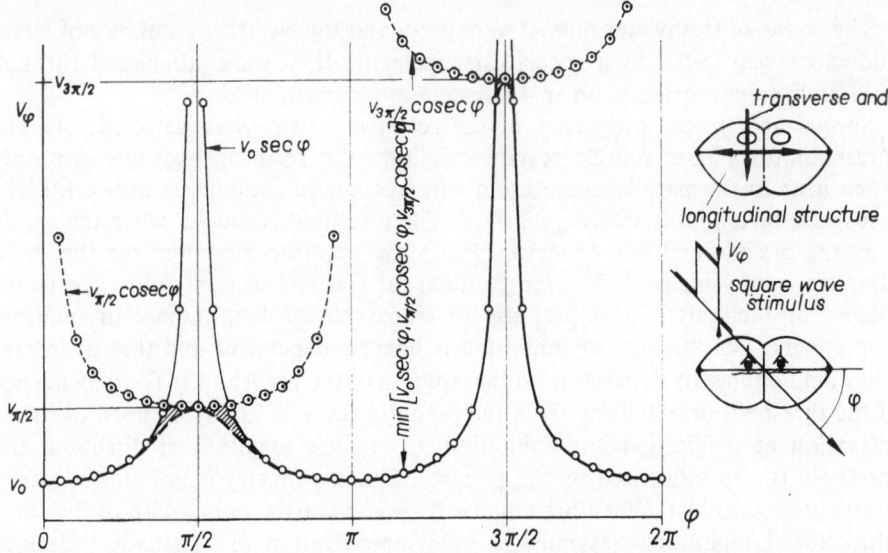

FIG. 8. Two excitable structures. The hypothetical φ-angle dependence of threshold for
adductor contraction. The first (longitudinal) structure is described by $V_0 \sec \varphi$, but the
second (transverse) excitable substance is shifted by 90° and is described by an asymmetric V_0
cosec φ function. The resultant function (min-function) selects the smaller threshold at a given
orientation angle. A possible neural structure is shown in the lower right corner

The above model attempts to explain the phenomena by the assumption of one excitable structure. Nevertheless, there is another model, which takes into account the innervation of adductor muscle. However, some of the phenomena, for example, the salt-sensitive longitudinal asymmetry, the transverse inhibition, and the transverse nonexcitability, cannot be explained simply by supposing innervation.

In Fig. 8, the curves of two separate excitable structures were taken into account. Both structures are cable-like and are excitable only with current parallel to the longitudinal axis. Such structures show the reciprocal-cosinus relation (Du Bois Reymond, cited by Rushton 1930) when the threshold is plotted against the ϕ angle between the critical axis and the stimulatory current. Let us assume two such excitable structures, one of which is asymmetric and parallel to the muscle longitudinal axis and is at a right angle to the first. The experimental threshold-angle characteristics could then be explained in the following way: The muscle response is always evoked through that structure which is more excitable at a given orientation. However, only some of the experimental facts can be interpreted by this assumption of innervation by one of two mutually perpendicular structures. For the interpretation of the unexplained phenomena, we cannot neglect to take into account the charged-particle orientation.

SUMMARY

The larvae of freshwater mussel were used, and the electric excitability of larval adductors was tested in a special arrangement; they were stimulated through fluid at a precise orientation in the excitatory current field.

Special anisotropic properties of the reactive system were detected: (1) The threshold for phasic muscle response follows the reciprocal-cosinus law only when near to the parallel orientation with respect to the muscle fibers. (2) The transverse threshold is always polarized. The threshold is smaller when the anode is at the mesodermal line of larvae. From the opposite direction, the threshold obtained is very high. (3) The longitudinal (or transverse) excitability can be inhibited or facilitated by a prepulse of transverse or longitudinal orientation, respectively. The measure of inhibition is linearly dependent and that of the facilitation nonlinearly dependent on the strength of the prefield. (4) The dependence of the threshold on the delay from the prepulse shows a highly damped overshot relaxation of facilitation and inhibition. (5) In the longitudinal direction, the threshold is also influenced by the polarity. This asymmetry is not an obligatory property of glochidia. When present, it is dependent on the pulse width of the stimulus. (6) Longitudinal asymmetry or symmetrization of originally polarized threshold can be evoked by $CaCl_2$, $LiCl$, $MgCl_2$, TEA and TMA.

The origin of these phenomena is discussed with special reference to the rectification property of the membrane or protoplasma, the orientation of charged particles and the innervation of larval adductor.

REFERENCES

COLE, K. S. and CURTIS, H. J. (1941): Membrane potential of the squid giant axon during current flow. *J. gen. Physiol.* **24** 551—563.

DEBYE, P. (1929): *Polar Molecules.* Chemical Catalog Co., New York. Cited by Kittel (1961), p. 191.

GOLDMAN, D. E. (1964): A molecular structural basis for the excitation. Properties of axons. *Biophysic. J.* **4** 167—188.

HECHT, H. H., HUTTER, O. F. and LYWOOD, D. W. (1964): Voltage-current relation of short Purkinje fibres in sodium-deficient solution. *J. Physiol.* **170** 5—6.

HEILBRUNN, L. V. (1958): The viscosity of protoplasma. Protoplasmatologia. *Handbuch der Protoplasmaforschung.* vol. II C. 1. Springer, Wien.

HUTTER, O. F. and NOBLE, D. (1960): Rectifying properties of heart muscle. *Nature* **188** 495.

KITTEL, C. (1961): *Introduction to Solid State Physics.* John Wiley and Sons. Inc., New York, 2nd ed.

LAPICQUE, L. (1931): On the electric stimulation of muscle through Ringer's solution. *J. Physiol.* **73** 219—246.

LÁBOS, E. (1964): Studies on the electric excitability of the adductor muscle of glochidia. *Annal. Biol. Tihany* **31** 27—37.

NAKAJIMA, S., IWASAKI, S. and OBATA, K. (1961): Delayed and anomalous rectification in skeletal muscle membrane. *Proc. Jap. Acad.* **37** 505—508.

RUSHTON, W. A. H. (1927): The effect upon the threshold for nervous excitation of the length of nerve exposed and the angle between current and nerve. *J. Physiol.* **63** 357 (cited by Rushton, 1930).

RUSHTON, W. A. H. (1930): Excitable substances in the nerve-muscle complex. *J. Physiol.* **70** 317—337.

NEUROHUMORS AND ENZYMES

Symposium on Neurobiology of Invertebrates 1967 (305—314)

EVOLUTION OF CHOLINERGIC TRANSMISSION

D. A. SAKHAROV and T. M. TURPAEV

Institute of Developmental Biology, Academy of Sciences of the USSR
Moscow, USSR

Although there are several substances which are considered to be neural transmitters, it is unlikely that the physiological diversity of synapses is based on the chemical diversity of transmitters. One cannot a priori ascribe to a synapse any physiological features even if we know by which substance transmission is mediated. Thus, a transmission mediated by ACh* may be excitatory or inhibitory, with slow or fast adaptation to orthodromic input, long or short duration of transmission process, etc.

Yet, specific physiological properties of different junctions are sure to correlate with their morphological and chemical organization. To find out the correlations is an intriguing problem, and comparative physiology aids in this endeavor.

As long ago as 1941 a parallelism between ChE activity and electric events was detected in fish electric organs by Nachmansohn and Meyerhof (1941), whose attention was also attracted to the extremely high activity of this enzyme in the brain of cephalopods. There is no need to recall here all investigations related to the problem. The recent finding of Kerkut and Meech (1966) should be mentioned, that molluscan nerve cells excited under the action of ACh differ neurochemically from those inhibited by the same transmitter.

This report is concerned with another attempt to use comparative data to advance our knowledge of the correlation between the neurochemical and functional properties of cholinergic junctions. A comparative method allows us to elucidate a tendency of evolution by which the transmission achieves its perfection. It should be noted that an evolutionary view is a rather traditional one for Soviet physiologists to take, having been established by the late Professors Kh. S. Koshtoyants and L. A. Orbeli.

Indeed, in any line of animal evolution, the higher representatives have more elaborated and effective synaptic transmission than lower ones. The junction itself seems to undergo evolution — chemical, structural, or both.

The molluscs are a good example. The complicated problem of the actual

---

* Abbreviations used in this work: ACh = acetylcholine, AChE = acetylcholinesterase, ChE = cholinesterase, PSP = postsynaptic potential, R = cholinergic receptor, ATP = adenosine triphosphate.

phylogeny of the molluscs can be avoided. In a limited neurobiological respect, 'the primates of the ocean', squids and octopuses, undoubtedly have the highest success achieved by the type. Their superiority is manifested not only by a larger brain with its elaborate architectonics, but by more elaborate elementary nervous mechanisms as well. The obvious behavioral difference between swift cephalopods and sluggish snails and mussels is translatable into the language of electrophysiology. The postsynaptic potentials (PSP) of a squid synapse are known to be very short, lasting no more than a few msec. On the contrary, PSP of central synapses of snails and slugs are notable for their long duration, up to tenths of a second, the falling phase being more prolonged than the ascending one; the long duration of PSP is, at least partly, an actual manifestation of prolonged action of the transmitter; again, the central junctions of gastropods, unlike those of cephalopods, fail to maintain transmittory activity for a long time, being blocked by repetitive impulses, and thus affect the whole behavior of the animal (see Tauc 1966).

The next step is then to translate these phenomena from electrophysiological into neurochemical language.

Probably the most important link in the chain of chemical events constituting a single act of transmission is the interaction between the transmitter and its receptor. This interaction may be expressed by the equation of reversible equilibrium:

$$ACh + R \rightleftharpoons ACh - R$$

It is not ACh, but its combination with the receptor, $ACh - R$, which is the member of the equation affecting the ionic permeability of the postsynaptic membrane. Quantitatively, the combination of ACh with the receptor depends on (1) the quantity of ACh molecules in the vicinity of the receptive membrane, and (2) the readiness of the receptor protein to interact with the transmitter. Thus, the equilibrium may be affected in two ways: (i) by influencing free ACh in the synaptic cleft, and (ii) by influencing the sensitivity of the receptor. Interneuronic junctions of molluscs do not lose the chance of using both these opportunities.

The tissue of the central nervous system of gastropods differs strikingly from that of cephalopods in near to complete lacking histochemically revealable ChE in its synaptic zone, the neuropile. Our data, obtained with the 'direct thiocholine' procedure of Karnovski and Roots, confirm and amplify the results of Zs.-Nagy and Salánki (1965). The localization of the enzyme is somewhat variable in ganglia of different gastropod molluscs. In a land snail, *Caucasotachea atrolabiata*, rather strong activity of the enzyme is revealed in nerve trunks leaving the ganglia; inside the ganglionic ring, only slight positivity is located in interganglionic tracts. A very prolonged incubation makes it possible to notice some precipitate in restricted regions of the neuropile of *Helix lucorum*. Conversely, in the freshwater snail *Lymnaea stagnalis*, the cortical zone of ganglia is occasionally positive, the loci of enzyme activity being clustered around perikarya of nerve cells. It should be emphasized that no evidence for the presence of strong or at least moderate activity of ChE located in synaptic neuropile was detected by light histochemical methods in any of the gastropod species studied by Hungarian authors and by us. Neither was it found in Pelecypoda (Salánki and Zs.-Nagy 1965).

The presence of an inactivating enzyme localized in junctions is often regarded as a necessary criterion for the identification of a neural transmitter. At the same time, ACh is considered to be a transmitter in central synapses of gastropods, with a good many facts being in favor of this conclusion. It is not a real contradiction, but most likely an indication of an unusual chemical mechanism of cholinergic transmission.

It seems quite reasonable to consider the mechanism of cholinergic transmission in central synapses of gastropods to be identical with one discovered in this laboratory when studying the neural regulation of the molluscan heart. The detailed evidence obtained in experiments with heart preparation are presented elsewhere in this volume (Nistratova 1968), and only the updated final results will be given here (Fig. 1).

The basic preparation is an isolated ventricle of the heart of the freshwater mussel *Anodonta*. The inhibitory action of ACh is terminated, though the transmitter is not destroyed by tissue ChE, which has negligible activity. This is the basic phenomenon. This termination of ACh effect turned out to be caused by a substance influencing the cholinergic receptor.

The action of an antagonistic substance on the receptor was shown to have a formally competitive character (Nistratova 1968). The study of such substances which prevent receptors from corresponding transmitters was started in this laboratory after Turpaev found a substance antagonistic to ACh in the blood of frog (see Turpaev 1965). The next finding was that of Sakharov and Nistratova (1963) experimenting on mussel heart; it was shown that the appearance of an antagonistic substance was evoked by the transmitter action.

Thus, there are two alternative paths of transmitter reaction in cholinergic

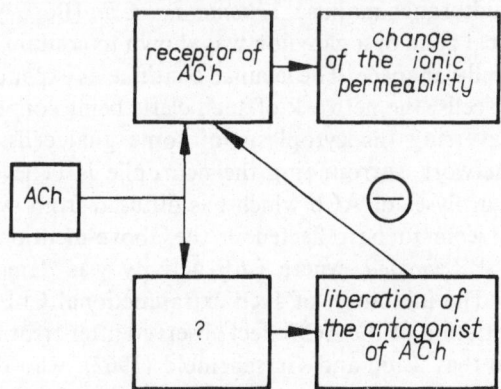

FIG. 1. The negative feedback scheme of regulation of a cholinergic receptor by a substance liberated into the synaptic cleft under the action of the transmitter. (Modified from Turpaev 1965)

junctions: one in accordance with the classic scheme, involving junctional ChE when chemical inactivation of the transmitter occurs, and another involving the chemical inactivation of the receptor by an antagonistic substance. Knowledge of this fact helps us to reconstruct the sequence of evolutionary elaboration of cholinergic transmission in molluscs.

The diffusion of ACh to extrajunctional spaces seems to be the simplest way removing the transmitter from the junctional area. In general, diffusion as a factor terminating the action of any extracellular controlling substance, was probably the initial point in the development of mechanisms of chemical regulation. Some data of embryophysiology should be mentioned in this respect − namely, the fact that the action of 5-HT, which induces motor activity in the development of some gastropod embryos, is stopped because of the diffusion of the substance rather than because of its chemical inactivation (Buznitsov and Manukhin 1961). Mathematical calculations that show that diffusion can account for the decay of transmitter action are referred to by Werman (1966), who supposes this mechanism is widespread in central nervous systems.

The presence of active ChE in the body fluid of many animals, including the hemolymph of Gastropoda and Pelecypoda, seems to be in accordance with this point of view. Molluscs other than Cephalopoda have no capillaries in their ganglia, the perineurium being washed on its outer surface by the hemolymph. Thick perineuriums which contain vessels are occasionally found, but here again the vessels do not penetrate the nervous tissue. It seems that a long path must be covered by the transmitter before its meeting with the inactivating enzyme.

However, other factors simplify the conditions. The perineurium of gastropods was shown to consist mainly of an extracellular ground substance impregnated with relatively rare cells. As such, the sheath offers the only passage for nutritive and excretive metabolites of nervous tissue, and it is assumed to be permeable to macromolecules. The material taken for electron-microscopic study was ganglionic tissue of a nudibranch mollusc, *Tritonia diomedia* (Borovyagin and Sakharov 1968). The cortical zone of a ganglion was shown to contain extensive lacunar expansions of extracellular space. The lacunae continue as expanded clefts between neighboring satellite cells, the network of such clefts being connected in turn with special channels traversing the cytoplasm of some glial cells. This elaborated lymph-containing network surrounding the neuropile is believed to be the site of the enzymatic hydrolyses of ACh which has diffused from synapses.

These conditions seem to be reflected in the above-mentioned histochemical picture in ganglia of *Lymnaea*, where ChE activity was demonstrated around neuronal perikarya. The inhibition of such extrajunctional ChE may well be the cause of some potentiation of the ACh effect observed after treatment with eserine. It is very suggestive that Tauc and Gerschenfeld (1962), who described this potentiation, failed to obtain any influence of eserine on the duration of PSP in transmission mediated by ACh in *Aplysia*. The prolongation of PSP was expected by these authors, who thought that cholinergic interneuronic junctions in gastropods were provided with specific ChE.

In addition, the hemolymph ChE of gastropods and pelecypods is close to the AChE of vertebrates with respect to substrate and inhibitor specificity (Grigorieva, Rosengart and Turpaev 1968), thus indirectly indicating support of the proposed relation of the enzyme to the junctional ACh.

In real cholinergic synapses of gastropods, diffusion of ACh is unlikely to act alone, and the liberation of an antagonistic substance seems to improve the transmission. Such an added factor should shorten the lifetime of the combination of transmitter with receptor, thus making the transmission more mobile, though at the cost of its stability. As the accumulation of antagonistic substance in the junctional area proceeds, the sensitivity of receptor molecules decreases. In the end, the transmission can be completely blocked.

The desensitization of cholinergic receptors by ACh was demonstrated in gastropod central neurons by Tauc and Bruner (1963). These authors suggested the effect to be essentially similar to that occurring on the subsynaptic membrane during repetitive synaptic action, where the efficacy of transmission progressively diminished. It is interesting to recall that analogous progressive diminution of inhibition of molluscan heart was observed by Prosser (1940), who supposed that ACh was depleted from endings of cardiac nerve. This interpretation was wrong. It was later demonstrated in this laboratory that repeatedly injected ACh gradually lost its inhibitory action as well, the loss having been connected with the liberation of the above-mentioned antagonistic substance (Sakharov and Nistratova 1963). These experiments made possible the presentation in 1963 of a view regarding the desensitization as a peculiar type of termination of transmitter action on heart cells.

Thus, phenomena demonstrated with the molluscan heart preparation have their equivalents at the level of single nerve cells of gastropods. This has led to the conclusion that active chemical regulation of the sensitivity of receptors is an essential feature of cholinergic transmission in the central nervous system of gastropods (Nistratova and Sakharov 1964; Sakharov 1965). The physiological role of desensitization was also proposed by Gerschenfeld and Stefani (1965) for receptors of 5–HT in molluscan nerve cells, although these authors were not concerned with the chemical mechanism of the desensitization. It would be interesting to discover if the latter transmission involves an antagonistic substance.

Some recent findings also support our interpretation of the cholinergic transmission in gastropods. Wulfius and Zeimal (1967) demonstrated that sensitivity of *Lymnaea* ganglionic cells to ACh rises during the perfusion of an isolated ganglion with Ringer solution; a sharp increase of sensitivity can be evoked by a short treatment with ACh in high concentration. Similar effects are known to be connected with an antagonistic substance in heart preparation (Turpaev 1965; and unpublished data). Both kinds of treatment are believed to deplete the stored antagonist from the tissue, thus diminishing the spontaneous liberation that influences the sensitivity.

Direct indications of the chemical resemblance of a cholinergic antagonist liberated in heart of *Anodonta* and a substance of analogous function which is supposed to operate in interneuronal junctions of gastropods have been obtained. It should be mentioned that earlier, Nistratova and Malinovskaya identified the antagonist of ACh in mussel heart as being ATP (see Nistratova 1968). That is why 2,4-dinitrophenol, which decreases the stores of tissue ATP, strongly increases the inhibitory action of ACh on mussel heart and counteracts the desensitization (Sakharov and Nistratova 1963). Proceeding from these data, the action of ACh on nerve cells of *Helix* before and after treatment with dinitrophenol was compared. The sensitivity of *D* cells to ACh was shown to increase markedly (Sakharov and Korobtsov, unpublished).

A mechanism of transmission involving the liberation of an antagonist of the transmitter is not yet capable of providing significant improvement in synaptic efficacy as compared with mere diffusion.* Moreover, a new complication arises: how is the antagonist to be inactivated? This problem does not seem to have been solved by gastropods. The recovery of cellular receptors from a desensitization effect requires a long time, from a few minutes to half an hour and more (Tauc and Bruner 1963). It is not known what happens to the antagonist in the course of reestablishing sensitivity in receptors.

Much more effective is an alternative mechanism involving the binding of ChE in the vicinity of the junction. This chemical addition to primitive cholinergic transmission occurs in cephalopods. It may be that cephalopods were forced to adopt this improvement after the closing of blood circulation, because these new conditions made the contact of junctional ACh with the ChE of the blood more difficult. In any event, the location of the inactivating enzyme near to the site of action of the transmitter does provide fast and repetitive transmission.

Thus, the blood ChE becomes ineffective in this respect, and, accordingly, its activity is lower in cephalopods than in gastropods and pelecypods (Turpaev, Nistratova and Sakharov 1967). On the other hand, the blood ChE of cephalopods is less specific than the hemolymph enzymes of gastropods and pelecypods (Grigorieva, Rosengart and Turpaev 1968).

As to the enzyme of the nervous tissue of cephalopods, its localization is very suggestive. Histochemical investigation of a regularly organized nervous center, the cortex of optic lobes, has shown that ChE was located mainly at synapses, with only a certain portion of synapses being cholinergic. The synaptic plexiform zone of the cortex includes eight layers which appear to be similar in octopi, cuttlefish, and squid. In these three animals, the positive band seems to be located in the same layer or neighboring layers (Drukker and Schadé 1964; Sakharov in press). One may conclude that homologous synapses in cephalopods possess a similar transmitter, such as is known to occur in vertebrates.

---

* Recently, Prof. G. Kerkut succeeded in revealing junctional esterase in *Helix* brain by means of electron-microscope histochemistry (personal communication). Thus, the speculation presented here needs some correction. Nevertheless, it could be agreed that ChE activity in the gastropod neuropile is at least extremely low as compared to that of junctions known to be cholinergic.

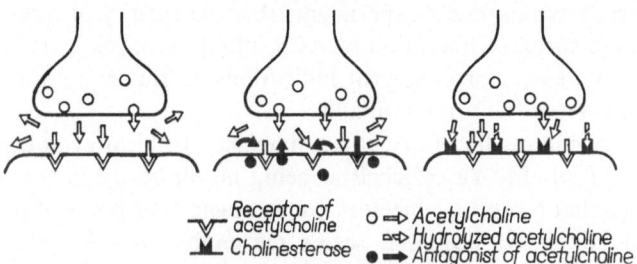

FIG. 2. Schematic drawing of alternative mechanisms for termination of the synaptic action of ACh

The location of the ChE near to the junction was demonstrated by Bryant and Brzin (1966) in their microchemical investigation of the squid giant synapse. Even at the level of light-microscopic histochemistry it is seen that in the stellate ganglion the most positive region is the neuropile, the area of interneuronal contacts.

Some conclusions can now be drawn (Fig. 2). Nervous systems of molluscs demonstrate that a cholinergic junction provided with ChE at or adjacent to the site of action of the transmitter is far from being a general rule. Rather, such transmission seems to be the highest success achieved by molluscan cholinergic junctions in the course of their evolution.

Another attempt to improve the efficacy of transmission involves operation by means of an antagonist of the transmitter. The latter attempt does not permit marked increase of the speed of transmission. Nevertheless, it is also a useful achievement because it is a good neurochemical basis for controlling the adaptivity.

This evolutionary acquisition from lower animals seems to be used in those synapses of higher animals which are to be adapted to orthodromic input. The recent finding of the anti-ACh action of a substance liberated during the cholinergic transmission in a vertebrate junction (Putintseva and Turpaev 1966) supports this view.

Chemical factors controlling the sensitivity of receptors can be of positive as well as negative action, thus providing both facilitation and adaptation. They may control not only PSP, but also generator potentials in corresponding receptor cells of sense organs. The results of experiments in progress support this generalization.

We deliberately limited the problem in question. Undoubtedly, there are other manifestations of the chemical evolution of cholinergic transmission. Thus, the fine localization of synaptic ChE needs to be discussed in this respect. It is not difficult to imagine that the dynamics of the accumulation and disappearance of ACh in the vicinity of the receptive membrane should not be the same, the ChE being located on pre- and postsynaptic membranes, respectively, for these cases.

Again, it was shown in model experiments that the efficacy of transmission was dependent on the speed of liberation of ACh into the synaptic cleft (Turpaev and Nikitin 1962). At least morphological indications of the evolution of secretory apparatus of nerve endings are available.

There is a voluminous literature concerning the chemical evolution of cholinergic receptors, Prof. M. Ya. Michelson being an authority in this field. If one accepts the view that receptor molecules are arranged over pores of the membrane (Turpaev and Nistratova 1965), it seems reasonable to think that alternative effects of ACh on ionic permeability in different membranes depend on the pattern of receptor molecules.

Physiological properties of cholinergic synapses seem to be controlled by relatively independent features of their chemical organization. Different combinations of these changeable components may result in a wide diversity of junctions, giving variable physiological manifestations of transmission even though a similar transmitter is used.

SUMMARY

Nervous systems of molluscs demonstrate that cholinergic junctions provided with ChE at or adjacent to the site of action of the transmitter are far from being a general rule. A point of view is discussed according to which the lability of cholinergic transmission is dependent on the way that ACh is removed from the junctional area.

The initial mode seems to be the diffusion of ACh from the synaptic cleft and its subsequent hydrolysis by ChE of the lymph. Indeed, in the neuropile of gastropods and pelecypods, ChE is present in small amounts, if at all. At the same time hemolymph ChE activity is relatively high, and the enzyme resembles AChE. The long falling phase of the PSP, which is the manifestation of a prolonged action of the transmitter, also suggests that the latter is not destroyed by a specific enzyme at the site of transmission.

The removal of ACh from receptors on the postsynaptic membrane in gastropods is probably accomplished not only by diffusion but also by an additional mechanism − competition between the transmitter and its antagonist, which is released into the synaptic cleft under the influence of the transmitter itself. Such a mechanism of desensitization of the cholinergic receptor was discovered in this laboratory and studied in experiments on a molluscan heart. Habituation of neurons in gastropods to repeated orthodromic stimuli and desensitization to repeatedly applied ACh are supposed to be conditioned by the same competitive mechanism.

Only in cephalopods does ChE seem to be bound in the vicinity of the junction. The blood ChE in cephalopods has lower activity and is less specific as compared to the hemolymph enzyme of gastropods, due to the existence of this more advanced method of transmitter removal.

Similar suggestions are believed to be applicable to cholinergic transmission in vertebrates, where junctions are also variable in respect to transmission lability and adaptivity. Operation by means of an antagonist of the transmitter seems to be used in those synapses of higher animals which are to be adapted to orthodromic input.

REFERENCES

BOROVYAGIN, V. L. and SAKHAROV, D. A. (1968): *Ultrastructure of Giant Nerve Cells of Tritonia* (An atlas). Nauka, Moscow (in Russian).

BRYANT, S. H. and BRZIN, M. (1966): Cholinesterase activity of isolated giant synapses. *J. cell. Physiol.* **68** 107—108.

BUZNICOV, G. A. and MANUKHIN, B. N. (1961): Serotonin-like substances in the embryogenesis of some gastropods. *Zh. obsch. Biol.* **22** 226—232 (in Russian).

DRUKKER, J. and SCHADÉ, J. R. (1964): Neurobiological studies on Cephalopods. III. Histochemistry of 24 enzymes in the optic system. *Netherl. J. Sea Res.* **2** 155—182.

GERSCHENFELD, H. M. and STEFANI, E. (1965): 5-hydroxytryptamine receptors and synaptic transmission in molluscan neurons. *Nature (Lond.)* **205** 1216—1218.

GRIGORIEVA, G. M., ROSENGART, E. V. and TURPAEV, T. M. (1968): The specificity of cholinesterases from molluscan hearts and hemolymphs. In *"Physiology and Biochemistry of Invertebrates"*. Nauka, Leningrad (in Russian).

KERKUT, G. A. and MEECH, R. W. (1966): The internal chloride concentration of H and D cells in the snail brain. *Comp. Biochem. Physiol.* **19** 819—832.

NACHMANSOHN, D. and MEYERHOF, B. (1941): Relation between electrical changes during nerve activity and concentration of cholinesterase. *J. Neurophysiol.* **4** 348—361.

NISTRATOVA, S. N. (1968): New steps in studying the mechanism of cholinergic inhibition. In *"Neurobiology of Invertebrates."* pp. 315—326.

NISTRATOVA, S. N. and SAKHAROV, D. A. (1964): The cholinesteraseless cholinergic transmission in molluscs. *10th Congress of the Physiol. Soc. of the USSR* (abstracts). Nauka Moscow—Leningrad, **2** part 2, 131—132 (in Russian).

PROSSER, C. L. (1940): Acetylcholine and nervous inhibition in the heart of *Venus mercenaria*. *Biol. Bull. Woods Hole* **79** 92—102.

PUTINTSEVA, T. G. and TURPAEV, T. M. (1966): Participation of the stimulating agent released from the frog heart under the effect of acetylcholine in controlling the activity of choline receptors. *Fiziol. Zh. (Mosk.)* **52** 1093—1099 (in Russian).

SAKHAROV, D. A. (1965): The functional organization of giant neurons of molluscs. *Uspekhi sovr. Biol.* **60** 386—403 (in Russian).

SAKHAROV, D. A. (in press): Contributions to comparative architectonics of the optic center in cephalopods. In *"Biology of Far East Seas"*. Siberian Dept. of the Acad. Sci. of the USSR (in Russian).

SAKHAROV, D. A. and NISTRATOVA, S. N. (1963): Peculiarities of cholinergic response in the heart of *Anodonta*. *Fiziol. Zh. (Mosk.)* **49** 1475—1481 (in Russian).

SALÁNKI, J. and ZS.-NAGY, I. (1965): Histochemical investigation of cholinesterase in different molluscs with reference to functional conditions. *Nature (Lond.)* **206** 842—843.

TAUC, L. (1966): Physiology of the nervous system. In *"Physiology of Mollusca"*. Acad. Press, N. Y. -London. **2** 387—454.

TAUC, L. and BRUNER, J. (1963): "Desensitization" of cholinergic receptors by acetylcholine in molluscan central neurones. *Nature (Lond.)* **198** 33—34.

TAUC, L. and GERSCHENFELD, H. M. (1962): A cholinergic mechanism of inhibitory synaptic transmission in a molluscan nervous system. *J. Neurophysiol.* **25** 236—262.

TURPAEV, T. M. (1965): Participation of macroergs in control of the activity of cholinoreceptive substance. *Zh. evol. Biochim. Fiziol.* **1** 500—506 (in Russian).

TURPAEV, T. M. and NIKITIN, O. A. (1962): On the velocity of neuroeffector transmission (Experiments with a "biochemical model of the synapse"). *Fiziol. Zh. (Mosk.)* **48** 936—941 (in Russian).

TURPAEV, T. M. and NISTRATOVA, S. N. (1965): Biochemical properties of cholinoreceptive substance of cellular membrane and the mechanism of acetylcholine action. In *"Protoplasmic membranes and their functional role"*. Naukova Dumka, Kiev, 197—210 (in Russian).

TURPAEV, T. M., NISTRATOVA, S. N. and SAKHAROV, D. A. (1967): Evolution of the cholinergic regulation of the cardiac activity in molluscs. *Zh. obsch. Biol.* **28** 618—626 (in Russian).

VULFIUS, E. A. and ZEIMAL, E. V (1967): The effects of acetylcholine and cholinomimetics on the giant neurons of the mollusc *Limnaea stagnalis*. In *"Evolutionary Neurophysiology and Neurochemistry"*. Nauka, Leningrad. 17—25 (in Russian).

WERMAN, R. (1966): A review-criteria for the identification of a central nervous system transmitter. *Comp. Biochem. Physiol.* **18** 745—766.

DISCUSSION

B. Csillik: I am not sure that the method of Karnovsky does give useful results for analyzing the fine localization of ChE in the junction.

D. A. Sakharov: We used the method of Karnovsky and Roots for revealing ChE by light-microscope histochemistry, rather than Karnovsky's procedure for electron microscopy. Our results obtained with molluscan nervous system are in accordance with those of other authors who used other procedures.

J. Salánki: Is the ChE activity you found on gastropods a specific one or not? If not, what do you think about its role? Anyway, it is well known that the ChE found in invertebrates does not always correspond to all the criteria of AChE. Actually, Lee and Florey could not show substrate inhibition of the cholinesterase isolated from Cephalopod brain. Our investigations on gastropods are also very indefinite in this respect.

D. A. Sakharov: I am sure that esterases from the molluscan nervuos system cannot be classified in terms of vertebrate neurochemistry. The situation in a vertebrate brain is rather simple: AChE in nerve cells, and 'nonspecific' ChE in the glia. One would therefore obtain different localization of activity in histochemical preparations using acetylthiocholine and butyrylthiocholine as substrates. The picture in the molluscan nervous system is quite different. Thus, in the squid brain, both substrates give the activity at similar locations. Recently, the enzyme from the squid optic lobes was carefully investigated by a group of neurochemists from Leningrad and Moscow headed by Turpaev. They found that the enzyme possessed peculiar features distinct from those of the AChE of vertebrates. The esterases from hemolymph of different molluscs are neither AChE nor nonspecific ChE in the exact sense, though in some respects they may be more or less similar to these vertebrate enzymes.

Symposium on Neurobiology of Invertebrates 1967 (315—326)

NEW STEPS IN STUDYING THE MECHANISM
OF CHOLINERGIC INHIBITION

S. N. Nistratova

Institute of Developmental Biology, Academy of Sciences of the USSR
Moscow, USSR

During the four decades since the discovery by O. Loewy of the mediatory function of acetylcholine, a distinct idea has been elaborated concerning the main links of the cholinergic mediator process. According to this idea, the end of the cholinergic reaction is determined by the enzyme cholinesterase which hydrolyzes free acetylcholine and promotes the liberation of the receptor from the mediator. The data leading to these conclusions were obtained mainly on vertebrate animals. When studying mediation in molluscs, however, information was accumulated long ago which did not keep within this scheme.

Many authors mention the fact that eserine and other inhibitors of cholinesterase either had no effect on the heart of molluscs or only weakly increased the acetylcholine effect (Ungar 1937; Prosser 1940; Ghiretti 1948; Welsh and Taub 1948, 1953; Gaddum and Paasonen 1955; Krijgsman and Krijgsman 1959, etc.). This fact may indicate a low content of cholinesterase or its absence. The direct determination of the activity of this enzyme in myocardium, especially in Lamellibranchia, confirmed this suggestion (see Smith and Glick 1939; Turpaev, Nistratova and Sakharov 1967). At the same time, in the molluscan heart, acetylcholine is apparently a universal inhibiting mediator (see Krijgsman and Divaris 1955; Crescitelli and Geissman 1962; Greenberg 1965; Chong and Phillis 1965; Hill and Welsh 1966).

The combination of these two facts, namely the liberation of acetylcholine under nervous stimulation and, at the same time, the absence (complete or partial) of cholinesterase, gives rise to the question of the mechanism of the release from inhibition. An attempt to answer this question is undertaken in the present study.

The experiments were carried out on the heart of the freshwater mussel *Anodonta sp.* Specific experiments have shown that acetylcholine plays the role of mediator of nervous impulses (Ten Cate 1955; Nistratova and Yuzhanskaya 1966). At the same time, it was also shown that the ventricle of *Anodonta* either did not contain cholinesterase at all (Nistratova and Yuzhanskaya 1966; Turpaev, Nistratova and Putintseva 1967), or contained it in amounts too insignificant to explain the rapid release of the heart from inhibition (Salánki, Varanka and Hiripi 1967).

The stimulation of the visceral ganglion by electric stimuli (6 V magnitude,

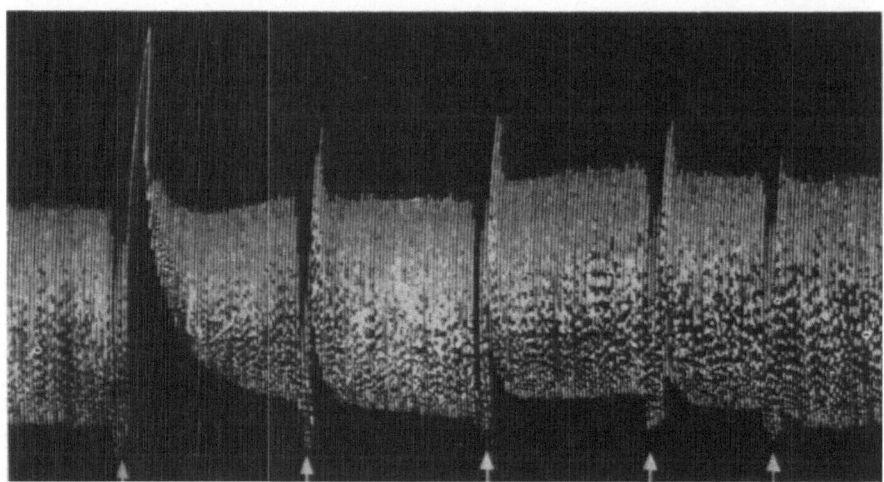

FIG. 1. The effect of acetylcholine on the isolated ventricle of *Anodonta*. Arrows indicate the addition of acetylcholine at a concentration of 10^{-6} g/ml

2 msec duration, and 10/sec frequency) or the addition of acetylcholine at concentrations of $10^{-8}-10^{-7}$ g/ml intothe heartled to the increase of the heart muscle tonus and reduction in the amplitude of the beat. An increase in stimulus strength or in concentration of acetylcholine resulted in the complete abolition of ventricular contractions. However, at all concentrations of acetylcholine, the heart overcomes the inhibition and soon the strength of contractions not only attains its initial level, but even exceeds it by somewhat. Under repeated action of acetylcholine, the inhibition reaction becomes weaker, until finally the sensitivity to acetylcholine disappears entirely: 'desensitization' occurs (Fig. 1).

What mechanism determines the end of the cholinergic process and such a rapid desensitization in *Anodonta*?

In 1959, Putintseva and Turpaev showed in the frog heart that under the action of a high concentration of acetylcholine (10^{-5} g/ml), a substance was found in the perfusion fluid which exerted an excitatory influence uponthe atropine-treated heart-recipient of the frog (Putintseva and Turpaev 1959). Later, some properties of this substance were determined and its nature was somewhat delineated. It appeared to be a derivative of uridine diphosphates (Putintseva 1966). However, it remained unclear whether this substance was only a side product of the reaction between acetylcholine and the acetylcholine receptor or played some functional role. In 1963, in experiments carried out together with Sakharov (Sakharov and Nistratova 1963), we succesfully showed that a substance of macroergic nature was found in the perfusion fluid from the heart of *Anodonta* in the presence of acetylcholine. We have suggested that this substance is ATP. The following facts corroborated this suggestion:

1. The stimulation of the visceral ganglion or the application of acetylcholine

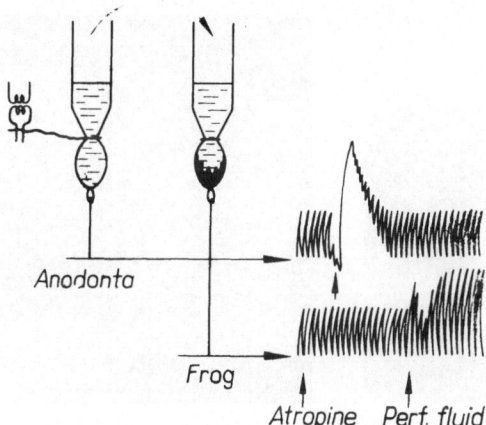

Fig. 2. The experiment for testing the perfusion fluid from the heart of *Anodonta* on the
isolated atropine-treated frog heart

was accompanied after inhibition by an increase of contractions. ATP exerts the same effect upon the heart of *Anodonta*.

2. If the perfusion fluid from the heart of *Anodonta*, collected after the release from inhibition, was added into the atropine-treated frog heart, a two-phase increase of the beat amplitude was observed, which is also characteristic of the ATP effect. Figure 3 represents the action of the perfusion fluid from the heart of *Anodonta* on the isolated frog heart (the ionic composition of the perfusion fluid from the heart of *Anodonta* being previously brought to a state equivalent to Ringer solution). After the stimulation of the visceral ganglion, this perfusion fluid induced the inhibition of heart contractions due to the presence of acetylcholine. After atropinization, the inhibiting action was, however, changed to an excitatory one which resembled that of ATP. It is interesting that in the perfusion fluid from the heart of *Anodonta* which had already overcome the inhibition, the content of acetylcholine remained almost the same as before.

3. Finally, direct evidence of the presence of ATP in the perfusion fluid was obtained by means of the method of Strehler and McElroy (1957). This method, based on activation of the system luciferinluciferase by ATP, is highly specific, and such substances as ADP, AMP, UTP, CTP, GTP, creatinephosphate, and inorganic polyphosphates do not cause luminescence. It was shown by means of this method that under the action of acetylcholine at concentrations of 10^{-7}–10^{-5} g/ml, ATP was liberated into the heart cavity at concentrations of 10^{-8}–10^{-7} g/ml. The higher the concentration of acetylcholine, the sooner was the macroergic substance liberated and the greater was the quantity. The accumulation of ATP in the perfusion fluid takes place gradually: its smallest quantity is found at the stage of maximal tonic reaction and its greatest at the period of full restoration of heart contractions and under desensitization (Nistratova and Malinovskaya 1967).

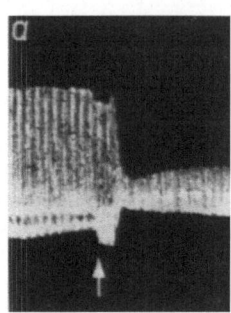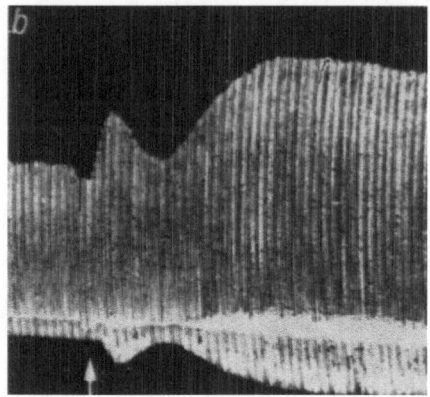

FIG. 3. The effect of the perfusion fluid from the heart of *Anodonta*
a upon the isolated frog heart. a —effect of the perfusion fluid after
ganglion stimulation (6-V magnitude, 2-msec duration, 10/sec frequency,
60 sec); b — the same after the treatment of the frog heart with atropine at a concentration of 10^{-5} g/ml. Arrows indicate the addition of
the perfusion fluid

Evidently, then, the liberation of ATP into the heart cavity is not a consequence
of the reaction between acetylcholine and the acetylcholine receptor. The fact of
the matter is that when the receptor is blocked by such cholinolytic substances as
mytolon (Win 2747), tetraethylammonium iodide (TEA), or arpenal, an excitatory
substance is still liberated under the influence of acetylcholine (Nistratova and
Pécsi 1966). As is shown in Fig. 4, after the treatment of the heart of *Anodonta*
with arpenal at the concentration of 10^{-5} g/ml, the inhibition phase in the action
of acetylcholine disappears. The arpenal itself increases the amplitude of the beat,
and on this background acetylcholine increases the amplitude even more. A large
amount of ATP was found in the perfusion fluid collected at this stage.

Many authors who also observed the excitatory effect under the action of acetylcholine or stimulation of the visceral ganglion after treatment of the heart with
mytolon attributed this effect to serotonin (5–HT) (Luduena and Brown 1952;
Welsh 1953; Loveland 1963; Chong and Phillis 1965; Pécsi 1966; Phillis 1966). In
our case, the excitatory effect cannot be explained in such a way, because BOL–148
at the concentration of 10^{-5} g/ml, which completely blocked the effect of 5–HT
at the concentration of 10^{-5} g/ml, does not diminish the excitatory effect. Besides,
the perfusion fluid obtained from the heart of *Anodonta* after the treatment of
this heart with mytolon and its subsequent treatment with acetylcholine still
exerts a distinct excitatory action upon the frog heart insensitive to 5–HT.

This allowed us to suggest that it is ATP that determined the end of the cholinergic process in *Anodonta*. A suggestion was put forward that acetylcholine and ATP
competed for the receptor: either the inhibition or the stimulation reaction would
dominate, depending on the ratio of these substances.

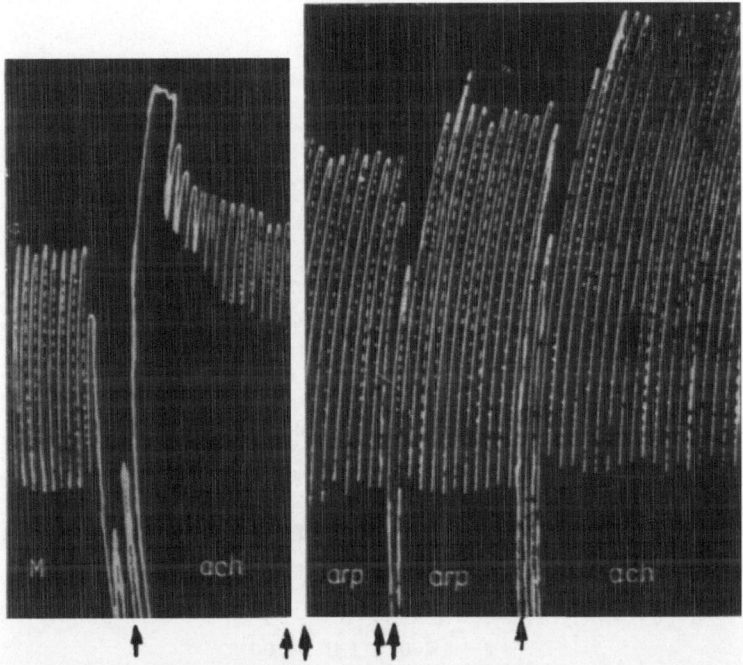

FIG. 4. Reaction of the isolated mussel heart to acetylcholine at a concentration of 10^{-6} g/ml (a) before and (b) after the treatment of the heart with arpenal at a concentration of 10^{-5} g/ml. ach — acetylcholine; arp—arpenal

In order to solve this problem, special experiments were undertaken to study the kinetics of the reaction between acetylcholine and the acetylcholine receptor in the presence of ATP (Nistratova 1965; Nistratova and Turpaev 1965). At first, the dependence of the acetylcholine effect on its concentration was determined on the isolated ventricle of *Anodonta*. The heart was then perfused by a solution of ATP at a concentration of 10^{-5} g/ml and the efficiency of different concentrations of acetylcholine prepared on a solution of ATP (production of 'Reanal') was again determined. Figure 5 represents the curves relating the size of the acetylcholine effect to the concentration of acetylcholine under a normal reaction and in the presence of ATP in the system of double reverse coordination. The position of the curves beginning from one point on the ordinate axis demonstrates competitive relations between ATP and acetylcholine under their simultaneous action upon the receptor.

The experiments with 2,4-dinitrophenol (DNP), which is known to inhibit ATP synthesis and destroy the already present store of the macroergic substance, served as further evidence of the direct ATP participation in the release of the heart from inhibition and in desensitization. After the treatment of the ventricle of *Anodonta*

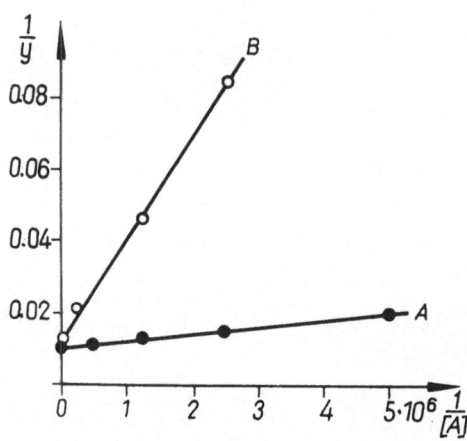

Fig. 5. The dependence of the acetylcholine effect on (A) its concentration in the normal heart and (B) in the presence of ATP. $\dfrac{1}{y}$ — inverse value of the efficiency of the acetylcholine tonotropic effect (in per cent); $\dfrac{1}{[A]}$ — inverse value of the square of the acetylcholine concentration (in g/ml). The graph is calculated according to the equation

$$\frac{1}{y} = \frac{K}{100} \cdot \frac{1}{[A]} + \frac{I}{100}$$

with the solution of DNP at concentrations of 10^{-6}–10^{-5} g/ml, the sensitivity to acetylcholine and duration of the inhibition phase increase, and cessation of inhibition and the phenomenon of 'desensitization' are eliminated (Sakharov and Nistratova 1963).

Finally, the experiments repeatedly testing the action of acetylcholine may also show the competitive antagonism between acetylcholine and ATP. As already mentioned, the accumulation of ATP in the perfusion fluid takes place gradually and the largest amount of ATP is found under the complete release from inhibition and at the period of 'desensitization'. Thus, if there exists a competition between acetylcholine and ATP, the effect of the repeated action of acetylcholine will be weaker the later the mediator is added. It is shown in Fig. 6 that the addition at the stage of inhibition of acetylcholine at a concentration of $5 \cdot 10^{-6}$ g/ml has the same systolic effect (Fig. 6a). At the beginning of the release of the heart from inhibition, a subsequent addition of acetylcholine induces a weaker reaction (Fig. 6b). Finally, at the last stage, the repeated addition of acetylcholine appears to be almost ineffective: by this time the heart has already overcome the inhibition and the ventricular beats become normal (Fig. 6c).

On the basis of all these data, the cholinergic process in the heart of *Anodonta* may be represented in the following way: Under stimulation of the visceral ganglion, the liberation of acetylcholine occurs in the cardiac nerve endings. Acetylcholine on the one hand interacts with the receptor and induces increase of the tone and inhibition of heart contractions, and on the other hand leads to the liberation of ATP.

In the perfusion fluid of the normally working heart, an insignificant amount of ATP (at concentrations of 10^{-9} g/ml) can also be detected. Acetylcholine apparently only strengthens the process of ATP liberation, influencing some system connected with ATP synthesis or hydrolysis. In particular, we have been able in the histo- and biochemical experiments to show that acetylcholine suppresses by about 30% the activity of the superficially situated ATP-ase of the myocardium of *Anodonta* (Zs.-Nagy and Nistratova, in press) and promotes accumulation of ATP in perfusion fluid. One more possibility that should not be excluded was mentioned by Frey (cited in Kruta 1936; also see Welsh and Taub 1948; Ditadi 1964; Chong and Phillis 1965; etc.) — the possibility of coexistence of two types of acetylcholine receptors. It is possible that under interaction with one of them a positive tonotropic and negative inotropic reaction takes place, while the interaction with the other one is accompanied by an increase of the beat amplitude at the expense of ATP release. If such is the case, one can easily imagine that in the heart of *Anodonta*, acetylcholine plays a role both of inhibitor and of excitatory mediator: either inhibition or the excitatory effect is demonstrated, depending on the strength and frequency of incoming impulses. It seems to us more probable that in *Anodonta* a weak stimulation of the visceral ganglion at subthreshold concentrations of acetylcholine on the order of 10^{-9}–10^{-8} g/ml does not induce inhibition, but instead, stimulation of the heart activity. The same excitatory effect is demonstrated as well under intense stimulation after treatment of the heart with cholinolytic substances (mytolon, arpenal, or TEA). This effect is not eliminated on the heart of *Anodonta* by BOL–148 or methylsergide. Besides, we have not succeeded in finding 5-HT in the perfusion fluid after electric stimulation.

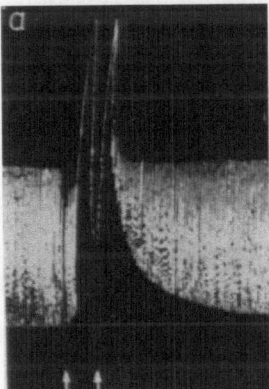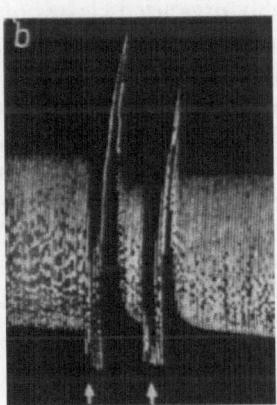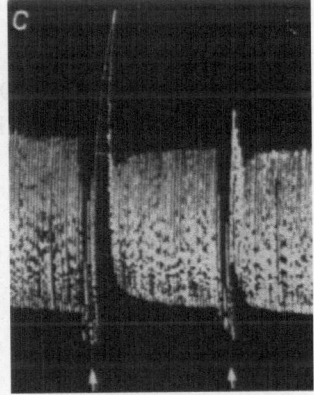

FIG. 6. The effect of the repeated addition of acetylcholine at a concentration of $5 \cdot 10^{-6}$ g/ml at different stages of inhibition. a — at the systolic stop; b — at the beginning of restoration of the ventricle beats; c — at the stage when the heart has almost completely overcome the inhibition. Arrows indicate the addition of acetylcholine

But, at all events, it is certain that ATP is released under visceral ganglion stimulation or acetylcholine addition. Initially, when acetylcholine concentration exceeds that of ATP, a positive tonotropic and negative inotropic effect is observed. Under the influence of acetylcholine, however, ATP is accumulated, and acts as a competitive antagonist of acetylcholine. As in the presence of the macroergic substance, the affinity of the receptor to acetylcholine decreases, so that ATP at a sufficiently high concentration knocks acetylcholine out of its bond to the receptor, inducing, as a result, the cessation of cholinergic inhibition. At the same time, the acetylcholine receptor appears to be blocked by the macroergic substance, so that the repeated addition of acetylcholine cannot induce inhibition: the so-called 'desensitization' occurs.

It remains unclear whether ATP reacts directly with the active center of the acetylcholine receptor or changes the sensitivity to the mediator in an allosteric way. It is possible that sulfhydril (SH-) groups of the receptor represent the common site of the action, as both substances (acetylcholine and ATP) are capable of reducing the reactivity of SH-groups by means of either formation of a 'closed' model (Katz 1963), or direct reaction of SH-groups with the adenylic part of ATP (Barany 1956; Tonomura and Yoshimura 1962) or with 'cation head' of acetylcholine (Turpaev 1962).

In the heart of the freshwater mussel *Anodonta*, the possibility has therefore been shown of a new mechanism for the cessation of heart inhibition which acts not at the expense of cholinesterase, but of a substance which acts as a competitive antagonist of the mediator. A similar mechanism of control of the receptor protein activity takes place in other molluscs as well. In particular, we have observed the participation of an excitatory substance of the macroergic type in the release of the heart from inhibition in *Mytilus grayanus, Helix lucorum, Rapana bezoar, Neptunea eulimata, Ommatostrephes sloanei-pacificus*, and *Octopus sp.* At the same time, in representatives of more highly organized classes of Mollusca, the cessation of inhibition at the expense of competitive antagonism between ATP and acetylcholine is completed by the destruction of the mediator by cholinesterase (Turpaev, Nistratova and Putintseva 1967; Turpaev, Nistratova and Sakharov 1967).

Putintseva and Turpaev (1966) have also demonstrated the participation of a macroergic substance of the uridinediphosphate type in the control of the acetylcholine receptor activity in the heart of vertebrate animals (frog). One can think, however, that in vertebrates, where the mediator is destroyed by cholinesterase at an enormous rate, the cessation of inhibition by means of macroergic substances is of value mainly for high concentrations of acetylcholine which induce 'desensitization'.

SUMMARY

Until recently, the destruction of acetylcholine by cholinesterase was considered to be the only cause of cessation of the cholinergic process. On the ventricle of the freshwater mussel *Anodonta*, it was shown that the release of the heart from inhibition could occur because of ATP liberation into the perfusion fluid under the stimulation of the visceral ganglion or action by acetylcholine. This macroergic substance is not accumulated as a result of the reaction between acetylcholine and acetylcholine receptor because in the receptor being blocked by mytolon, TEA, or arpenal, ATP liberation goes on in the presence of acetylcholine. Acetylcholine appears to simultaneously influence both the receptor and the systems responsible for ATP liberation.

The study of the kinetics of the reaction between acetylcholine (A) and acetylcholine receptor (R) in the heart of *Anodonta* has shown that the major part of the receptor reacts according to: $2A + R \rightleftarrows A_2R$. ATP is a competitive antagonist of acetylcholine. It remains unclear whether ATP reacts directly with the active center of acetylcholine receptor or changes the sensitivity to the mediator in a completely allosteric way.

The data obtained allow a representation of the cholinergic process in the heart of *Anodonta* as follows: Under visceral ganglion stimulation, the liberation of acetylcholine in the cardiac nerve endings occurs. Acetylcholine on one hand interacts with the receptor and inhibits heart contractions, and on the other hand liberates ATP from the nerve endings or from the muscle tissue. ATP accumulation in the perfusion fluid is also promoted by the inhibition of ATP-ase activity in the presence of acetylcholine. As the affinity of the receptor to acetylcholine decreases in the presence of ATP, sufficiently high concentrations of ATP knock acetylcholine out of its bond with the receptor, inducing, as a result, the cessation of cholinergic inhibition.

At the same time, acetylcholine appears to be bound by means of the macroerg, so that the repeated addition of acetylcholine cannot induce inhibition: 'desensitization' to acetylcholine occurs.

A similar mechanism of the control of the acetylcholine receptor activity is observed in other molluscs as well as in vertebrates, in which it exists side by side with the destruction of acetylcholine by cholinesterase.

REFERENCES

BARANY, M. (1956): Studies on the actin-actin bonding. *Biochim. Biophys. Acta* **19** 560.

CHONG, C. C. and PHILLIS, J. W. (1965): Pharmacological studies on the heart of *Tapes watlingi*: a mollusc of the family Venerida. *Brit. J. Pharmacol.* **25** 481—496.

CRESCITELLI, F. and GEISMANN, T. A. (1962): Invertebrate pharmacology. Selected topics. *Ann. Rev. Pharmacol.* **2** 143—192.

DITADI, A. S. F. (1964): Effects of acetylcholine on the heart of a fresh-water mussel. *Rev. bras. Biol.* **24** 297—314.

GADDUM, J. H. and PAASONEN, M. K. (1955): The use of some molluscan hearts for the estimation of 5-hydroxytryptamine. *Brit. J. Pharmacol.* **10** 474—483.

GHIRETTI, F. (1948): Studio comparativo dell'azione della colina e dell'acetilcolina sur cuore isolato dei molluschi gastropodi e cefalopodi. *Arch. Sci. biol. (Napoli)* **32** 239—251.

GREENBERG, M. J. (1965): A compendium of responses of bicalve hearts to acetylcholine. *Comp. Biochem. Physiol.* **14** 513—539.

HILL, R. B. and WELSH, J. H. (1966): Heart, circulation and blood cells. I: *Physiology of Mollusca.* Ed. by Wilbur and Yonge, Acad. Press, New York–London, **2** 125—174.

KATZ, A. M. (1963): The influence of cations on the reactivity of the sulfhydril groups of actin. *Biochim. Biophys. Acta* **71** 397—407.

KRIJGSMAN, B. J. and DIVARIS, G. A. (1955): Contractile and pacemaker mechanisms of the heart of molluscs. *Biol. Rev.* **30** 1—39.

KRIJGSMAN, B. J. and KRIJGSMAN, N. E. (1959): Investigations into the heart function of *Ciona intestinalis.* The action of acetylcholine and eserine. *Arch int. Physiol. Biochem.* **67** 567—578.

KRUTA, V. (1936): Effects de l'excitation des nerfs visceraux sur l'activité cardiaque chez cephalopodes. *C. R. Soc. Biol. (Paris)* **122** 582—585.

LOVELAND, R. E. (1963): 5-hydroxytryptamine, the probable mediator of excitation in the heart of *Mercenaria (Venus) mercenaria. Comp. Biochem. Physiol.* **9** 95—104.

LUDUENA, F. P. and BROWN, T. G. (1952): Mytolon and related compounds as antagonists of acetylcholine on the heart of *Venus mercenaria. J. Pharmacol. exp. Ther.* **105** 232—239.

NISTRATOVA, S. N. (1965): Possible mechanism of release of the *Anodonta* heart from the acetylcholine effect. *Physiol. J. USSR* **51** 1012—1016 (in Russian).

NISTRATOVA, S. N. and TURPAEV, T. M. (1965): Mechanism of inactivation of cholinereceptors on exposure of the heart of *Anodonta* to the effect of acetylcholine. *J. evol. biochim. physiol.* **1** 158—165 (in Russian).

NISTRATOVA, S. N. and PÉCSI, T. (1966): Analysis of the action of some cholinergic compounds on the heart of freshwater mussel *(Anodonta sp.). Annal. Biol. Tihany* **33** 111—123.

NISTRATOVA, S. N. and YUZHANSKAYA, M. G. (1966): Acetylcholine as transmitter of nervous influences on the heart of *Anodonta. J. evol. biochim. physiol.* **2** 214—220 (in Russian).

NISTRATOVA, S. N. and MALINOVSKAYA, K. J. (1968): ATP as a participant of the cholinergic reaction in the heart of *Anodonta. J. evol. biochim. physiol.* in press (in Russian).

PÉCSI, T. (1968): Contributions to the innervation of the heart on a fresh-water mussel, *Anodonta cygnea* L. *Acta biol. Acad. Sci. Hung.,* (in press).

PHILLIS, J. W. (1966): Innervation and control of a molluscan *(Tapes)* heart. *Comp. Biochem. Physiol.* **17** 719—739.

PROSSER, C. L. (1940): Acetylcholine and nervous inhibition in the heart of *Venus mercenaria. Biol. Bull.* **78** 92—102.

PUTINTSEVA, T. G. (1966): On the biochemical nature of the "x-factor" released by the frog heart under the effect of acetylcholine. *Physiol. J. USSR* **52** 734—740 (in Russian).

PUTINTSEVA, T. G. and TURPAEV, T. M. (1959): On a cardiac stimulant secreted by the frog's heart ventricle under the influence of acetylcholine. *Dokl. Akad. Nauk SSSR* **129** 1442—1444 (in Russian).

PUTINTSEVA, T. G. and TURPAEV, T. M. (1966): Participation of the stimulating agent released from the frog heart under the effect of acetylcholine in controlling the activity of choline receptors. *Physiol. J. USSR* **52** 1093—1099 (in Russian).

SAKHAROV, D. A. and NISTRATOVA, S. N. (1963): Pecularities of cholinergic response in the heart of *Anodonta. Physiol. J. USSR* **49** 1475—1480 (in Russian).

SALÁNKI, J., VARANKA, I. and HIRIPI, L. (1967): Comparative study on the cholinesterase activity of different tissues of fresh-water mussel *(Anodonta cygnea* L.). *Annal. Biol. Tihany* **34** 99—116.

SMITH, C. C. and GLICK, D. (1939): Some observations on cholinesterase in invertebrates. *Biol. Bull.* **77** 321—322.

STREHLER, B. L. and MCELROY, W. D. (1957): Assay of adenosine triphosphate. In: *Methods in Enzymology* **3** 871—873.

TEN CATE (1955): Contribution à la question de l'innervation cholinergique du coeur de l'*Anodonta cygnea. Publ. staz. zool. (Napoli)* **27** 199—203.

TONOMURA, Y. and YOSHIMURA, J. (1962): Binding of p-chloremercuribenzoate to actin. *J. Biochim. (Tokyo)* **51** 259—266.

TURPAEV, T. M. (1962): *The Mediator Function of Acetylcholine and the Nature of the Acetylcholine Receptor.* Moscow, Akad. Nauk SSSR (in Russian).

TURPAEV, T. M., NISTRATOVA, S. N. and PUTINTSEVA, T. G. (1967): Peculiarities of the release of the heart of *Anodonta sp.* and *Helix pomatia* from cholinergic inhibition. *J. evol. biochim. physiol.* **3** 40—46 (in Russian).

DISCUSSION

G. A. Cottrell: What is the concentration of ATP that is required to excite the *Anodonta* heart? Also, would you please tell me what the relative amounts are of ATP required to antagonize the inhibitory effects of different concentrations of acetylcholine?

S. N. Nistratova: The excitatory effect upon the heart of *Anodonta* is exerted by ATP at concentrations of $5 \cdot 10^{-7} - 1 \cdot 10^{-5}$ g/ml. At the same time, under visceral ganglion stimulation ATP is found in the perfusion fluid at concentrations of $10^{-8} - 10^{-7}$ g/ml. The stronger the stimulation (or the higher the concentration of acetylcholine), the greater is the amount of ATP liberated. A study of the kinetics of the reaction between acetylcholine and the acetylcholine receptor in the presence of ATP has shown that the affinity of the acetylcholine receptor to acetylcholine is higher by one order than it is to ATP. It is this fact, apparently, which determines the character of the heart reaction to the visceral ganglion stimulation; initially acetylcholine interacts with the receptor and induces the systolic stop. Under its influence, however, the liberation of ATP occurs, the latter acting as competetive antagonist of acetylcholine and promoting the cessation of inhibition.

O. Fehér: Have you done an electrophysiological control to these experiments?

S. N. Nistratova: No, we have not.

O. Fehér: What is the source of ATP in the heart of *Anodonta?*

S. N. Nistratova: At present, I cannot answer this question exactly. However, one can believe, that analogy to the case for vertebrate animals, this role is played either by the nerve endings or by muscles. In 1962, Abood, Koketsu and Myamoto found in experiments on the single muscle fiber that the content of ATP in the surrounding fluid during depolarization increased by more than 1000% and the amount of macroergic substance liberated did not depend on the degree of the muscle fiber contraction. On the other hand, the stimulation of the isolated nerve can also be accompanied by the liberation of ATP, as shown by Holton (1959), Koketsu and Myamoto (1961), Kuperman, Volpert, and Okamoto (1964), and others.

O. Fehér: How can the competition between ATP and acetylcholine for the receptor be imagined?

S. N. Nistratova: Earlier, it was shown by T. M. Turpaev that in the frog heart, under the acetylcholine effect the decrease to the reactive capacity of SH-groups occurred, which groups enter the composition of the active center of the acetylcholine receptor. On the other hand, the ATP has capacity for reducing the reactivity of SH-groups in proteins, and in myosin in particular. It is SH-groups which can be the common site of the action of ATP and acetylcholine. It is difficult to say, however, whether ATP interacts directly with the active center of the acetylcholine receptor or alters its affinity to acetylcholine in an allosteric way.

J. Salánki: Your speculation is very interesting. However, there are several problems. First: as was found in our biochemical investigations, the ventricle of *Anodonta* heart possesses cholinesterase activity to about the same degree as the ganglia. The auricle splits acetylcholine twice as much. Second: if you are right, and acetylcholine is not eliminated in the heart, what happens to it outside the heart, where the cholinesterase activity is also not higher?

Does the sensitivity to acetylcholine show seasonal changes, since we found well-expressed seasonal variations in the cholinesterase activity?

S. N. Nistratova: Neither in histo- nor biochemical experiments did we succeed in finding cholinesterase in the ventricle of *Anodonta*. The difference in the results obtained may be explained either by the species difference, or by the fact that experiments were carried out in different seasons. In any case, however, the rapid cessation of inhibition in the heart of *Anodonta* cannot be explained by the destruction of acetylcholine by cholinesterase because of the following facts: (1) At the complete cessation of inhibition in the heart or at the moment of 'desensitization', the amount of acetylcholine in the perfusion fluid remains almost the same as under its addition. (2) Carbocholine and other cholinomimetics which are not destroyed by cholinesterase induce the same desensitization as acetylcholine and at the same rate. (3) Inhibitors of cholinesterase (eserine, GD-42, etc.) do not alter the character of the cessation of inhibition in the heart of *Anodonta*.

What is the further fate of acetylcholine? After the recommencement of the heart beat, acetylcholine is transferred together with hemolymph from the heart into surrounding tissues, where it is destroyed by the cholinesterase of the latter. If a portion of the mediator remains undestroyed, it is then destroyed by the cholinesterase of the auricle. So, in any case, acetylcholine which was transferred into the lymphatic system cannot exert its action upon the ventricle twice.

Symposium on Neurobiology of Invertebrates 1967 (327—334)

CYCLIC 3',5'-AMP AS A SECOND MESSENGER
OF EXCITATORY INFLUENCES ON THE HEART OF *HELIX POMATIA*

K. S.-Rózsa

Biological Research Institute of the Hungarian Academy of Sciences
Tihany, Hungary

Lately, it has become more and more evident that in the regulation of the molluscan heart one has to deal with more than a single excitatory factor; as in the process of excitatory effects, a chain-like reaction of several different factors is involved. The data that reported the existence of several cardio-accelerators in the nervous system of gastropods (Kerkut and Laverack 1960; Jaeger 1966; Cottrell 1966; Welsh and Frontali 1966; S.-Rózsa and Zs.-Nagy 1967) have already indicated that for the realization of excitatory effects on molluscan heart more than a single excitatory transmitter is responsible.

According to our earlier data (S.-Rózsa and Perényi 1966) on *Helix* heart, during the excitation of the extracardiac nerves, another substance is liberated besides 5–HT which can be regarded as a transmitter. Later on we also demonstrated that in *Helix* heart all applied bioactive amines (5–HT, dopamine, adrenaline, noradrenaline, tryptamine, glutamine, γ-aminobutyric acid (GABA) and histamine) evoke excitatory effects with the exception of acetylcholine (ACh) (S.-Rózsa and Pécsi 1967). The differences among the agents is seen, not in the type of the effect nor in the threshold concentration, but in the maximal values of excitation they can evoke. In this respect, 5–HT invariably proved to be the most effective of the amines.

The uniformity of response-types obtained by different amines drew attention to the fact that the process taking place at identical molecular levels can be responsible in all cases for the excitatory effects. Taking this hypothesis as a starting point, we began to investigate the role of adenyl cyclase and cyclic 3',5'-AMP in the realization of the excitatory effects. Recently their connection with the catecholamines and other bioactive amines was suggested by Sutherland and Robinson (1966) and by Hangaard and Hess (1966).

The experiments were carried out on isolated *Helix* heart. The method of investigation was described earlier (S.-Rózsa and Pécsi 1967). The following substances were used: 5-hydroxytryptamine (5–HT), Fluka; dopamine, Fluka; adrenaline, Rhone-Paulenc; noradrenaline, Calbiochem; tryptamine, Fluka; tyramine, Fluka, glutamine, Fluka; GABA, Calbiochem; histamine, Calbiochem; cyclic 3',5'-AMP, Sigma; theophylline, B.D.H. The effect of the last two agents was investigated

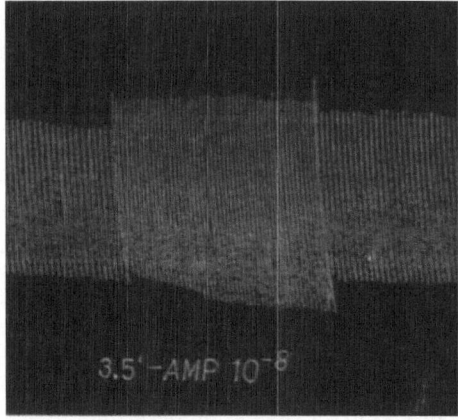 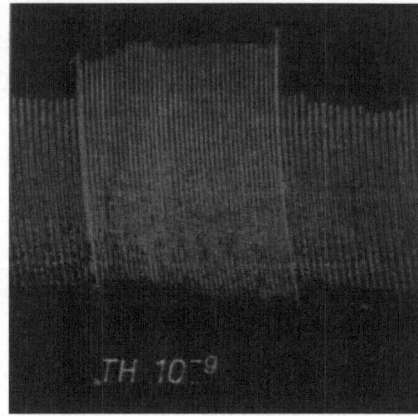

FIG. 1. Threshold concentrations of cyclic 3',5'-AMP (10^{-8} M) and theophylline (TH) (10^{-9} M) on the isolated *Helix* heart

after dichlorisoproterenol (DCI) and alderlin treatment in a study of the interaction with excitatory factors. All applied agents were dissolved in Meng's solution (Meng 1958).

It was found that cyclic 3',5'-AMP produces an excitatory effect in isolated *Helix* heart. This effect takes place at 10^{-8} M concentration and does not become inhibiting even at a high concentration (10^{-2} M). The excitatory effect appears quickly after application and diminishes slightly after washing (Fig. 1a and 1b). By applying theophylline, an excitatory effect was also obtained at 10^{-9}–10^{-2} M concentrations (Fig. 1). The excitatory effect of theophylline surpasses that of the cyclic 3',5'-AMP. We can explain the difference by the permeability conditions.

While investigating the interaction of excitatory effects and that of theophylline and cyclic 3',5'-AMP we found that they potentiate the effect of several substances to different degrees. The effects of this potentiation are calculated in per cent of the amplitude preceding application. In the case of applying theophylline, the rate of potentiation is higher than of cyclic 3',5'-AMP (Table 1).

From Table 1 it can be seen that both agents potentiate the excitatory effect of 5–HT to the highest degree. By using theophylline, this potentiation is around 80–100%, while in the case of cyclic 3',5'-AMP, it is 20–50%. The excitatory effect of tryptamine is not increased considerably by these two agents (5–10%) and they do not modify the effect of tyramine, glutamine, GABA, and histamine, although their original excitatory effect was lower than that of 5–HT and catecholamines. The interactions noted here are characteristic for concentrations around the threshold.

Following pre-incubation with high concentrations of 3',5'-AMP (10^{-3}–10^{-2} M), 5–HT becomes ineffective, but after pre-incubation with theophylline it often produces an inhibition which has never been observed in normal heart (Fig. 2).

TABLE 1

Effects of different excitatory agents on the interaction
of theophylline and cyclic 3',5'-AMP in the *Helix* heart

Material	Maximal effect, %	Increase of maximal effect in % after 10 min theophylline treatment $(10^{-4}$ M)	Increase of maximal effect in % after 10 min cyclic 3',5'-AMP treatment $(10^{-4}$ M)
5-HT	90	80–100	50–80
Dopamine	40	30–50	20–40
Adrenaline	40	30–50	20–40
Noradrenaline	40	30–50	20–40
Tryptamine	30	5–10	5–10
Tyramine	30	—	—
Glutamine	10	—	—
GABA	10	—	—
Histamine	8	—	—

Although from time to time the heart is freed from inhibition, 5-HT fails to evoke excitation, but the heart contraction never reaches even the original amplitude of the excitatory factor. This type of response, evoked by pre-incubation with theophylline in high concentrations, is characteristic among the excitatory factors only for 5-HT.

The β-blocking DCI in 10^{-4} M concentration diminishes or inhibits the excitatory effect of both cyclic 3',5'-AMP $(10^{-5}$ M) and theophylline, but for the latter, only at very low concentrations $(10^{-8}$ M) near to the threshold values. In higher concentrations, the excitatory effect of theophylline cannot be inhibited by other β-blocking agents, such as alderlin.

Our data indicate interesting interactions among the excitatory agents known as transmitters (5-HT, catecholamines) and the cyclic nucleotides.

It is known that the adrenergic receptors are formed around the phosphorylating enzymes, which utilize ATP as substrate. Taking into account that 5-HT and catecholamines are stored in the corresponding structures being bound to ATP (Roberts 1966), it is suggested that the realization of the effect of excitatory transmitters at a molecular level takes place as at Table 2.

As this scheme shows, the substrate of the enzyme, ATP, and also the activator of the enzyme, the bioactive amine (in our case, 5-HT or catecholamines) are simultaneously liberated as a consequence of nerve excitation. Of course, in the process there are other substrate reserves present, but, supposedly, the reactivity of ATP just liberated from binding is higher, and thus it starts the reaction.

From the scheme it can clearly be seen that the excitatory effect is achieved in all cases through the cyclic 3',5'-AMP.

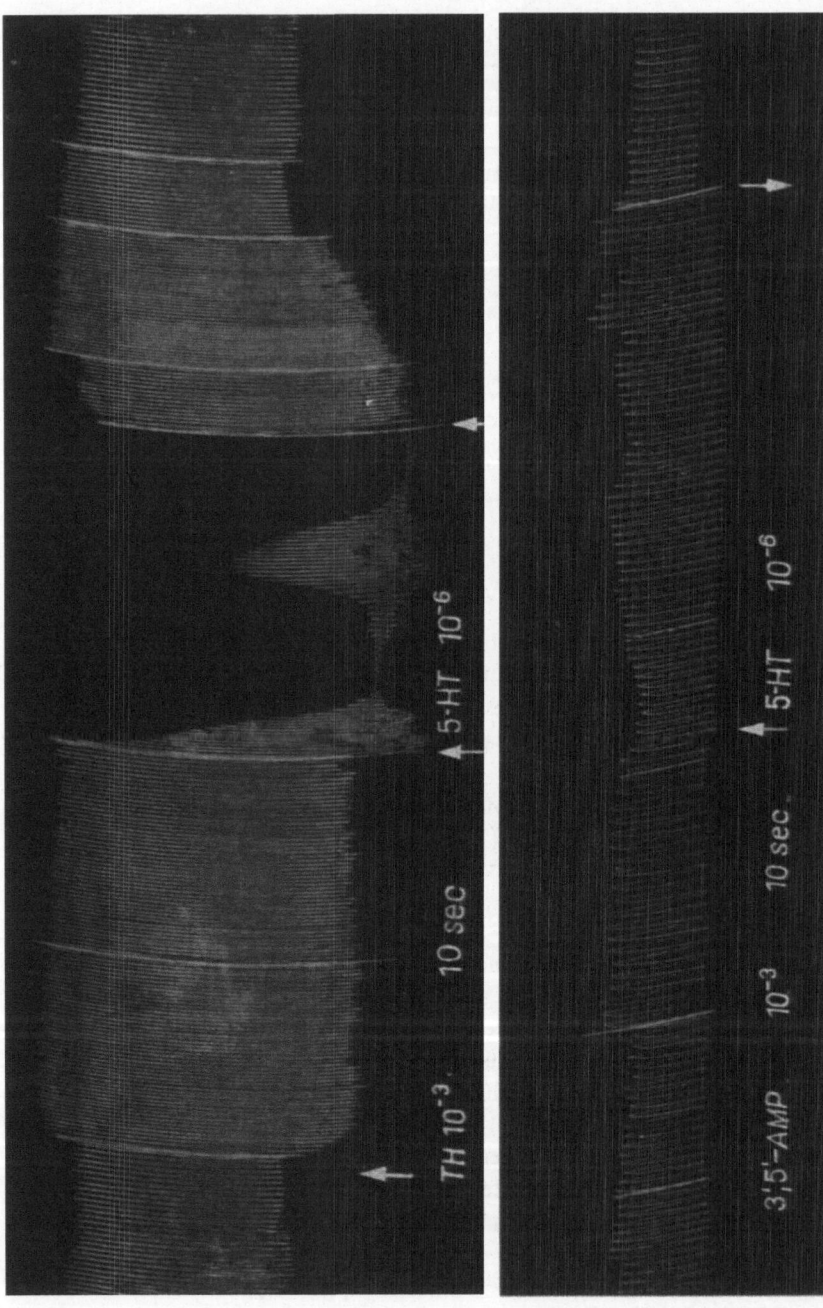

FIG. 2. Influence of the pre-incubation with theophylline $(10^{-3}$ M$)$ and with cyclic 3',5'-AMP $(10^{-3}$ M$)$ in high concentrations on the effect of 5–HT $(10^{-6}$ M$)$

Consequently, adenyl cyclase is the first phosphorylating enzyme which may interfere in snail heart with the excitatory regulation of the heart contraction. The facilitation of adenyl cyclase produced by 5–HT and catecholamines results in the β-type response, which is also demonstrated by the fact that the β-blocking agents

TABLE 2

The realization of the effect of excitatory transmitters

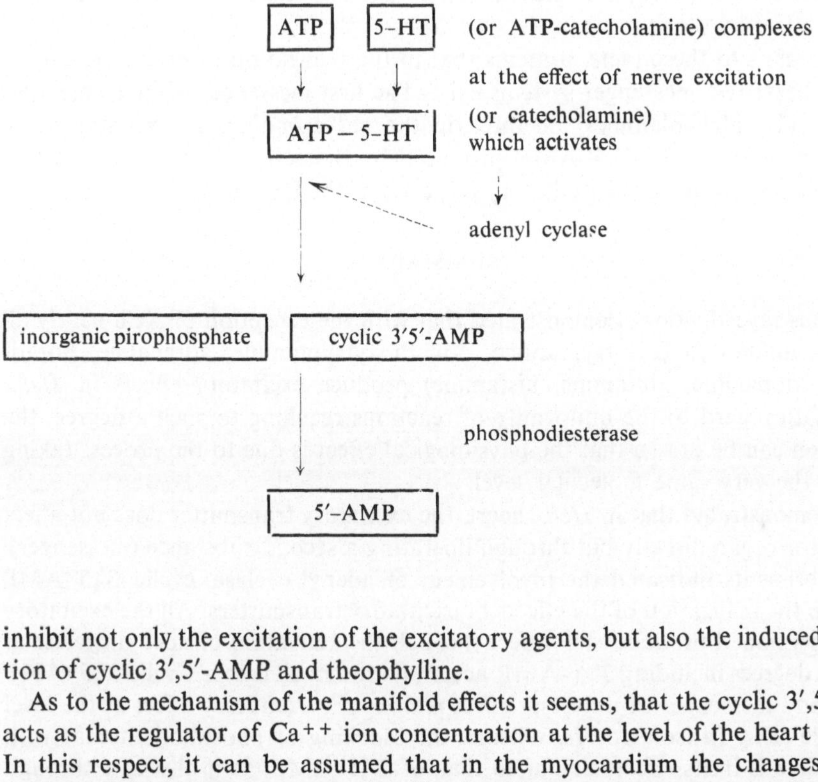

| ATP | | 5–HT | (or ATP-catecholamine) complexes |

↓ ↓ at the effect of nerve excitation

| ATP − 5–HT | (or catecholamine) which activates

adenyl cyclase

| inorganic pirophosphate | | cyclic 3′5′-AMP |

phosphodiesterase

| 5′–AMP |

inhibit not only the excitation of the excitatory agents, but also the induced excitation of cyclic 3′,5′-AMP and theophylline.

As to the mechanism of the manifold effects it seems, that the cyclic 3′,5′-AMP acts as the regulator of Ca^{++} ion concentration at the level of the heart muscle. In this respect, it can be assumed that in the myocardium the changes of the cyclic 3′,5′-AMP level are responsible for the different degrees of contractions when excitatory nervous and humoral effects are present. It can also be assumed that it is the role of cyclic nucleotides to sensitize the actomiozin system to Ca^{++} ions (Stam and Honig 1965).

The process, similar to other enzymatic processes, possesses a negative feedback too. This is demonstrated by our data, according to which 5–HT results in inhibition after treatment with high concentrations of cyclic 3′,5′-AMP and theophylline. This seems to indicate that the product, i.e., 3′,5′-AMP, over certain concentrations inhibits the activity of adenyl cyclase and that this inhibition can not be totally compensated by the introduced excitatory agent.

The configuration of the adenyl cyclase may be different in single tissues, and so this enzyme may be affected by different compounds, thus insuring the specificity of effects. At the same time it is also obvious that changes of different degrees can be caused by different compounds in the enzyme configuration. This can serve as an explanation for the different degrees of potentiation after incubation with cyclic 3′,5′-AMP.

On the basis of our data it can be supposed that the adenyl cyclase and cyclic 3′,5′-AMP play a role in the realization of excitatory effects of 5–HT and catecholamines (dopamine, adrenaline, noradrenaline). The other, less specific excitatory agents (tryptamine, tyramine, glutamine, GABA and histamine) realize their excitatory effect by other means.

With regard to these facts, it seems that in the realization of excitatory effects on snail heart two messenger systems exist. The first messenger is the transmitter itself (5–HT, catecholamines) liberated in the nerve endings at excitation. The second one is the cyclic 3′,5′-AMP produced by the activating effect of the first messenger and insuring the excitatory response at the level of muscle cells.

SUMMARY

Previous investigations demonstrated that with the exception of ACh nearly all bioactive amines (5–HT, tryptamine, 5-methoxytryptamine, adrenaline, noradrenaline, dopamine, glutamine, histamine) produce excitatory effects in *Helix* heart. With regard to the uniformity of reactions reaching to such a degree, the conclusion can be drawn that the physiological effect is due to the process taking place at the very same molecular level.

It is demonstrated that in *Helix* heart, the excitatory transmitter does not affect the effector organ directly but through liberating a second substance (messenger). The experiments indicated the involvement of adenyl cyclase, cyclic 3′,5′-AMP system in the realization of the effects of excitatory transmitters. All the excitatory agents mentioned produce their effect through the activation of adenyl cyclase at different degrees including 3′,5′-AMP accumulation in all cases. Cyclic-3′,5′-AMP as a second messenger of universal effect realizes the excitatory effect at the level of muscle cells. In isolated *Helix* heart, the blocking of phosphodiesterase with theophylline is in its effect, similar to that of 3′,5′-AMP, as it also acts through the accumulation of this nucleotide.

On the basis of experimental data, the excitatory process viewed as a system of several messengers seems to be acceptable in case of a snail heart. In this case, the first messenger is the transmitter itself liberated in the nerve endings at stimulation. Through adenyl cyclase, it produces the accumulation of 3′,5′-AMP, thus insuring the excitatory response of muscle cells.

REFERENCES

COTTRELL, G. A. (1966): Separation and properties of subcellular particles associated with 5-hydroxytryptamine, with acetylcholine and with an unidentified cardioexcitatory substance from *Mercenaria* nervous tissue. *Comp. Biochem. Physiol.* **17** 891–907

HANGAARD, N. and HESS, M. E. (1966): The influence of catecholamines on heart function and phosphorylase activity. Second. Symp. Catecholamines. *Pharmacol. Rev.* **18** 145—161.

JAEGER, C. P. (1966): Neuroendocrine regulation of cardiac activity in the snail *Strophocheilus oblongus*. *Comp. Biochem. Physiol.* **17** 409—415.

KERKUT, G. A. and LAVERACK, M. S. (1960): A cardioaccelerator present in tissue extracts on the snail *Helix aspersa*. *Comp. Biochem. Physiol.* **1** 62—71.

MENG, K. (1958): 5-hydroxytryptamine und Acetylcholine als Wirkungsantagonisten beim *Helix*-Herzen. *Naturwissenschaften*, **19** 470.

ROBERTS, G. C. K. (1966): The formation of complexes between 5-hydroxytryptamine, adenosine triphosphate and bivalent cations *in vitro*. *Biochem. J.* **100** 30P.

S.-RÓZSA, K. and PERÉNYI, L. (1966): Chemical identification on the excitatory substance released in *Helix* heart during stimulation of the extracardial nerve. *Comp. Biochem. Physiol.* **19** 105—113.

S.-RÓZSA, K. and ZS.-NAGY, I. (1967): Physiological and histochemical evidence for neuroendocrine regulation of heart activity in the snail *Lymnaea stagnalis* L. *Comp. Biochem. Physiol.* **23** 373—382.

S.-RÓZSA, K. and PÉCSI, T. (1967): Comparative studies on the effect produced by biologically active agents on the isolated hearts of *Helix pomatia* L. and *Anodonta cygnea* L. *Annal. Biol. Tihany* **34** 59—72.

STAM, A. C. JR. and HONIG, C. P. (1965): A biochemical mechanism by which adrenergic mediators modify cardiac contraction. *Amer. J. Physiol.* **209** 8—16.

SUTHERLAND, E. W. and ROBINSON, G. A. (1966): The role of cyclic-3',5'-AMP in responses to catecholamines and other hormones. Sec. Symp. Catecholamines, *Pharmacol. Rev.* **18** 145—161.

WELSH, J. H. and FRONTALI, N. (1966): Cardioregulator substances of the heart of *Mercenaria (Venus) mercenaria. III. Internat. Pharmacol. Congr.–S. Paolo–Brasil*, July 24—3 640.

DISCUSSION

G. A. Cottrell: Is it possible that the inhibitory effect of ACh on the *Helix* heart is also mediated via the adenyl cyclase system, as has been proposed for certain mammalian tissues?

K. S.-Rózsa: We did not investigate the realization of the inhibitory effects, we only studied the role of the adenyl cyclase system in connection with the stimulatory agents on *Helix* heart.

R. J. Walker: The heart muscle contains 5–HT. Is it possible that 5–HT released from the nerve ending following nerve stimulation releases 5–HT from the heart muscle which is present in the muscle bound with phosphate?

K. S.-Rózsa: Earlier, we found that in *Helix* heart, 5–HT is localized in the muscle cells themselves. by Falk's method, we were not able to detect any 5–HT in the neurons or nerve endings [S.-Rózsa, Zs.-Nagy, *Comp. Biochem. Physiol.*(1967) **23** 373]; the yellow fluorescence characteristic for 5–HT is always localized in the heart muscle. We also demonstrated that the nervous elements of the heart contain catecholamines which are liberated during the stimulation of the extracardial nerve. For this reason, we are sure that 5–HT is liberated only from heart muscle during the stimulation of the intestinal nerve and takes part in the realization of excitatory effects only at the level of myocardial elements.

S. N. Nistratova: In experiments on *Fasciola hepatica*, it was shown that 5–HT promotes the rapid transformation of ATP into cyclic AMP. Is it possible that under nervous stimulation ATP is the transmitter which transforms into the AMP detected by you after the liberation of 5–HT from tissue?

K. S.-Rózsa: We think that ATP plays a role in the realization of excitatory effects, but only as a substrate in the formation of cyclic 3′,5′-AMP. In this respect, it is interesting that the activator (5–HT) and the substrate (ATP) of the enzyme are liberated in the heart at the same time. But this process is only one step in the realization of the excitatory effects, which have several links in the chain. We found interactions between the stimulatory agents and the adenyl cyclase system. It is pointed out that cyclic nucleotides play a specific role in the stimulatory processes and not ATP itself. The 5–HT liberated under nervous stimulation increases the activity of the adenyl cyclase and the accumulation of cyclic 3′,5′-AMP, which in turn takes part in the realization of excitatory influences by transporting Ca^{++} ions into the contractile muscle fibers of the heart. Thus, this process serves as molecular basis for the increased work of the heart under the effects of excitatory nerve or stimulatory agents.

Symposium on Neurobiology of Invertebrates 1967 (335—339)

INFLUENCE OF 5-HT AND ACh ON THE HEART OF MUSSELS AT DIFFERENT TEMPERATURES

H. KUZIEMSKI

Department of Physiology, Medical Academy
Gdansk, Poland

The effect of neurohormones on the isolated organs of poikilothermal animals has thus far received only scant attention. It has often been possible to come across various authors who reported that their experiments were performed at room temperature or that the temperature changed by several degrees. The heart of mussels as well as the heart of other poikilothermal animals can contract vigorously over a wide range of temperature; however, this does not mean that it reacts to neurohormones similarly at lower and higher temperatures. This is especially true for the threshold concentrations of substances or when testing the expected physiological role of neurohormones of the examined heart.

When I was examining the effect of serotonin (5-HT) on isolated heart of *Unio pictorum* (Kuziemski 1962) and *Anodonta cygnaea* (Kuziemski, unpublished), I found that they showed a very high sensitivity to serotonin at 15 °C and a very low sensitivity at a higher temperature (25 °C). The lowering of the temperature from 25 °C to 20 °C and then to 15 °C caused an increase in the sensitivity to serotonin. The increase of the bath fluid temperature from 15 °C to 20 °C and then to 25 °C lowered the sensitivity to serotonin. The threshold concentration for serotonin was 10^{-16}–10^{-13} M at 15 °C and $5 \cdot 10^{-9}$–10^{-8} M at 25 °C.

These changes in the concentration of serotonin at higher and lower temperatures are similar to the seasonal changes—namely, in winter, the sensitivity of *Unio pictorum* heart to serotonin is at its greatest, and corresponds to the threshold concentrations 10^{-16}–10^{-13} M, whereas in the summer months, the sensitivity is at its lowest, and is 10^{-10}–$5 \cdot 10^{-10}$ M at 15 °C.

Greenberg (1960) also found that the high temperature of bath fluids lowers the heart reactivity to serotonin in *Mercenaria mercenaria (Venus)*. At 30–35 °C, serotonin concentration of 10^{-7} M reacts very insignificantly. In addition, Twarog and Page (1953) found that in the heart preparation of *Mercenaria mercenaria* the sensitivity increases with the lowering of temperature.

From the sensitivity changes of mussel heart it may be supposed that the increase of heart sensitivity to serotonin with the lowering of the organ's temperature is a compensatory reaction for the small amplitude of contraction and for the slow rhythm of the heart muscle at physiological conditions.

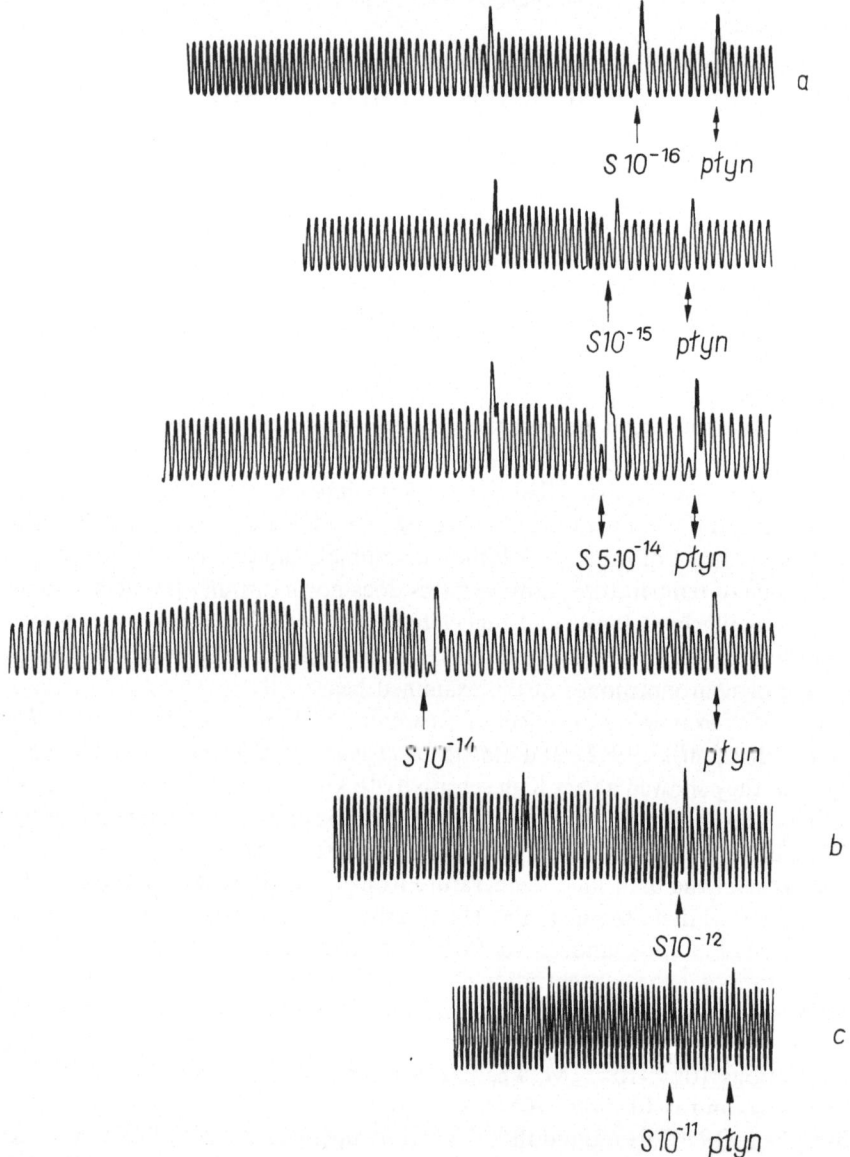

FIG. 1. Dependence of the size of the threshold concentration and increase of the contraction amplitude of *Unio pictorum* heart on serotonin at different temperatures. Photoelectric recording method. Speed of paper tape, 10 mm/min. Temperature: a — 15 °C, b — 20 °C, c — 25 °C. płyn — control exchange of bath fluid

The effect of acetylcholine (ACh) on the mussel heart action is a more complex phenomenon. Wait (1943) carried out experiments on the sensitivity of *Mercenaria mercenaria* heart to acetylcholine in relation to temperature changes. He found that in order to obtain a 50% reduction in amplitude of contractions at a temperature of 30 °C the concentrations of ACh needed were one hundred times greater than for a similar reduction at 10 °C. Luduena and Brown (1952) have also reported that the lowering of temperature increases the sensitivity of the heart of *Venus mercenaria* to ACh. If the lowering of the temperature was simultaneous with the increase of mussel heart sensitivity to serotonin as well as to ACh, then the compensatory effect of serotonin at low temperatures would be much less. However, it was observed by Welsh and Taub (1948) that ACh has a two-phase action on the isolated heart of *Mercenaria mercenaria*. The threshold concentration of the order of 10^{-11}–10^{-10} M increases the amplitude of the heart contractions, whereas greater concentrations, of the order of 10^{-9}–10^{-8} M, have a negative inotropic effect. The analogous phenomenon was observed by Hill (1958) in the heart of the snail *Busycon canaliculatum*.

This unexplained two-phase action of ACh would become clear if we assume that lowering of the mussel temperature also produces a decrease of ACh synthesis. Accordingly, at low temperature and under physiological conditions the ACh would have a stimulatory and synergic action with serotonin.

It is probable that there is a close correlation between the changes of heart sensitivity to neurohormones at different temperatures and the change of sponta-

TABLE 1

Dependence of the size of the contraction
amplitude of *Unio pictorum* heart
on serotonin at different temperatures

Temperature	Mole concentration of 5–HT	Average of the amplitude of heart contractions, %
	10^{-16}	10
	10^{-15}	16
15 °C	$5 \cdot 10^{-14}$	17
	10^{-14}	12
	10^{-13}	4
20 °C	10^{-12}	9
	10^{-11}	13
	10^{-11}	0
	10^{-9}	0
25 °C	10^{-8}	8
	10^{-7}	31

neous activity of neurons. Kerkut and Taylor (1958) have shown in isolated neurons from several species of poikilothermal animals that the lowering of the temperature increased the spontaneous activity and that an increase in the temperature diminished the spontaneous activity of isolated neurons for some time.

It has been known for some time that the poikilothermal animals show active movements over a wide range of temperature (Bullock 1955), although in most cases the body temperature of these animals is the same as the temperature of the environment (Gunn 1942). It appears that this fact can be partly explained by the changes with changing temperature in the excitability of the organs of the poikilothermal animals toward the inhibiting and stimulating neurohormones.

In conclusion, we may form two hypotheses:

1. With lowered temperature in the eurythermal poikilothermal animal, there is an increase of the organ excitability toward the neurohormones which stimulate as well as inhibit the animal's movements.

2. Accordingly, at low temperature, the substances which are known to have an inhibiting action will have a synergetic action with the stimulating substances.

REFERENCES

BULLOCK, H. T. (1955): Compensation for temperature in the metabolism and activity of poikilotherms. *Biol. Rev.* **30** 311—342.

GREENBERG, J. M. (1960): The responses of the *Venus* heart to catecholamines and high concentrations of 5-hydroxytryptamine. *Brit. J. Pharmacol.* **15** 365—374.

GUNN, L. D. (1942): Body temperature in poikilothermal animals. *Biol. Rev.* **17** 293—314.

IIILL, B. R. (1958): The effect of certain neurohumors and of other drugs on the ventricle and radula protractor of *Busycon canaliculatum* and on the ventricle of *Strombus gigas*. *Biol. Bull.* **115** 471—482.

KERKUT, A. G. and TAYLOR, K. J. B. (1958): The effect of temperature changes on the activity of poikilotherms. *Behaviour* **13** 259—279.

KUZIEMSKI, H. (1962): Izolowane serce skojki *(Unio pictorum* L.*)* i jego zastosowanie do oznaczen serotoniny. *Acta Biol. et Med. Soc. Gedan.* **6** 429—454.

LUDUENA, F. P. and BROWN, JR., G. T. (1952): Mytolon and related compounds as antagonists of acetylcholine on the heart of *Venus mercenaria*. *J. Pharmacol. exp. Ther.* **105** 232—293.

TWAROG, M. B. and PAGE, H. J. (1953): Serotonin content of some mammalian tissues and urine and a method for its determination. *Amer. J. Physiol.* **175** 157—161.

WAIT, B. R. (1943): The action of acetylcholine on the isolated heart of *Venus mercenaria*. *Biol. Bull.* **85** 79—85.

WELSH, H. J. and TAUB, R. (1948): The action of choline and related compounds on the heart of *Venus mercenaria*. *Biol. Bull.* **95** 346—353.

DISCUSSION

J. Salánki: Do you think that 4–10% increases in the amplitude are mathematically significant? It seems to me that it is difficult enough to tell whether or not an effect occurs at all at concentrations as low as 10^{-14} or 10^{-16}.

H. Kuziemski: I did not carry out statistical analyses on the changes of mussel heart sensitivity of serotonin in relation to the temperature. I am inferring these changes from the investigations with serotonin on the hearts with temperature changes of 25 to 20 to 15 °C, and from the other preparations when changing from a lower to higher temperatures (15–20–25 °C). I found that the greater the temperature difference, the greater the sensitivity difference of the hearts to serotonin and to temperature changes. These differences had a certain direction and, depending upon the preparation, had values from the 4th power to the 8th power when the limits of the temperature range were 15 and 25 °C (or 25 and 15 °C).

Determination of the subtlety of the reaction is dependent on the method used to record the heart contractions. The most accurate method of recording isotonic contractions is the photoelectric method which I used. In this method, there is a great amplification (to about 90 times) and a great fidelity of registration. It was possible to record more subtle reactions than was in the case with the kymograph. In some cases, I could record the reaction of *Unio* heart to the concentrations of serotonin at 10^{-14}–10^{-16} M at a temperature of 15 °C in winter time.

O. Fehér: Did you make an attempt to calculate the dissociation constants of your substances to the receptors at different temperatures? I think it would be a valuable approach to the problem and would make it possible to express the changes you reported here in terms of physical chemistry.

H. Kuziemski: No, but such an approach might prove very interesting.

Symposium on Neurobiology of Invertebrates 1967 (341—352)

THE MOVEMENTS OF CHEMICALS IN SNAIL NEURONS

G. A. KERKUT

Department of Physiology and Biochemistry, University of Southampton
Southampton, England

There are two main methods that have been used to study the movements of substances across cells; these are (1) the study of the electrical characteristics of the cell under various specified conditions and (2) the study of the movements of radioactive substances. The application of these two methods to the nervous system of the snail *Helix aspersa* will be described in the present paper.

IONIC MOVEMENTS IN *HELIX* NEURONS DURING
THE ACTION POTENTIAL

A standard method of determining the effect of ions on the action potential is to place the nerve cell in solutions of different ionic composition and see the effect that this has on the rise time and the height of the action potential. This technique has been carried out by various workers on gastropod neurons (Oomura *et al.* 1962; Gerasimov, Kostyuk and Aiskii 1964; Meves 1966; Kerkut and Gardner 1967; Junge 1967). In all cases, the action potential continued in Na-free solutions, but if the Ca ions were also removed the action potential size was often affected and the action potentials later stopped. It was suggested that Ca could be carrying the current during the action potential or that both Na and Ca carried the current. This view was supported when Meves (1966) and Junge (1967) showed that the potentials were very insensitive to tetrodotoxin, an inhibitor of the inward movements of Na ions during the squid action potential.

We had been worried by the role that Ca played in the snail neuron, since we had not been able to get as clear a relationship between the increased height of the action potential or the rise time with an increase in the external Na or Ca concentration as should have followed from the Nernst equation (Kerkut and Gardner 1967). There also appeared to be considerable variability between neurons according to the extent to which they were affected by Na or Ca removal. It was therefore determined to use the voltage clamp technique to study the extent to which the inward current was dependent upon the external Na and Ca concentration. The experiments described in the present section were carried out together with Dr. S. G. Chamberlain (Kerkut and Chamberlain 1967).

VOLTAGE CLAMP STUDY ON IONIC MOVEMENTS
IN *HELIX ASPERSA* NEURONS

A block schematic diagram of the circuit that we used to establish voltage clamp conditions and control the membrane potential of the neurons is shown in Fig. 1. Two glass microelectrodes were inserted into a selected neuron. The action potentials were recorded by high input impedance amplifiers A1 and A5, both of which had provision for input capacitance neutralization. The output of the potential-recording electrode was fed through A1 to an operational control amplifier A2. This supplied current of the correct polarity to the current electrode so as to maintain the neuron membrane potential at any desired level. The error between the membrane potential and the command signal is sensed by this amplifier and in turn causes it to achieve and maintain the control and command potentials $E_c + E_1$. E_c is the command pulse of about 100 msec duration and an off time of 1.5 sec. E_1 is the dc potential necessary to keep the holding potential at the membrane potential. The amplifier A2 has an open loop gain of more than $3 \cdot 10^{-5}$. The open loop gain for the system with the snail neurone was more than 3000. The rise time of the amplifier A1 was 20 μsec and the speed at which the membrane potential was changed was usually limited by the membrane itself. The stability of the system was controlled by correcting networks C_1–R_1 and P_2–C_2. The total clamping current was measured by the operational amplifier A3 while the membrane potential was measured through amplifier A4.

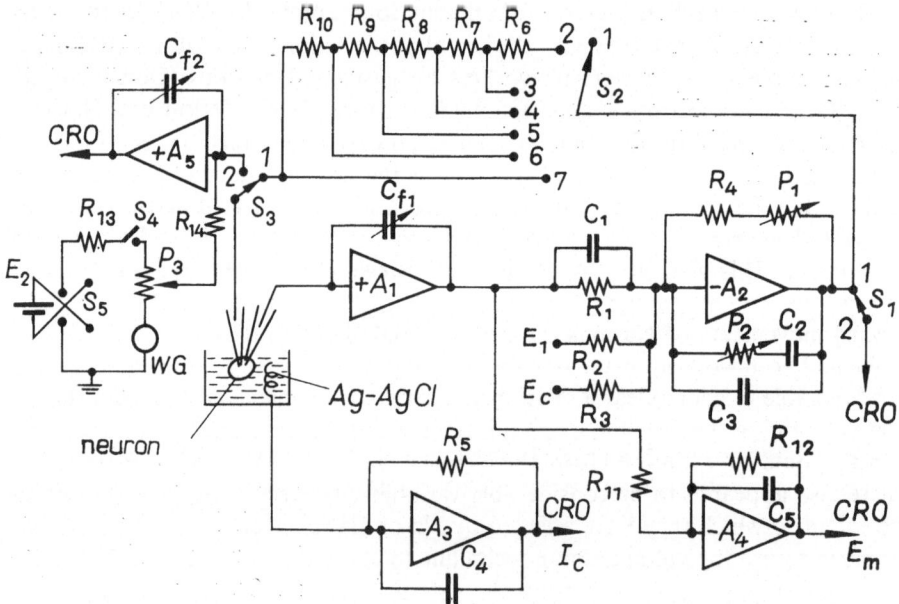

FIG. 1. Schematic diagram of the system used to establish voltage clamp conditions on a *Helix* neuron

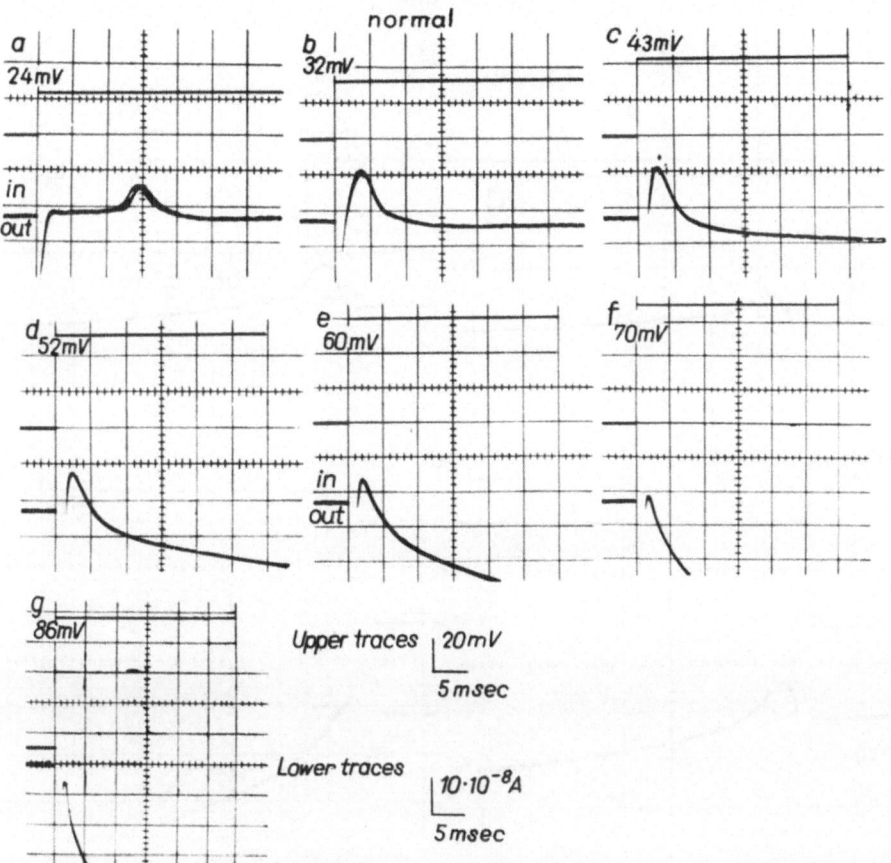

FIG. 2. Voltage clamp of snail neuron in normal Ringer. The value of the command potentia
is shown in the upper line of each figure. The current is shown in the lower line, inward current
being shown as an upward movement, outward current as a downward movement. In a, the
command voltage (24 mV) was sufficient to set up a potential in the axon but not the cell body.
Note the inward current in b and c and that it is reduced in d and e. There is no inward
current at f (86 mV)

Figure 2 is a trace taken of an experiment where the neuron was bathed in normal
snail Ringer solution (Na: 80 mM; Cl: 113 mM; Ca: 7 mM; K: 4 mM; Mg: 5 mM;
Tris 5 mM; pH: 7.8–8.0). For each command voltage there are two currents that
should be noted: (1) The maximum transient current, which is usually inward and
lasts for about 5 msec in the snail neuron, with a maximum value at about 2 msec
after the command pulse onset, and (2) the steady-state current recorded 30 msec
after the onset of the command pulse. Note that as the command voltage increasen
the transient current is initially inward (a–e), but by 70 mV command there si
little inward current and at 86 mV there is no inward current (g).

Following the classical approach of Hodgkin, Huxley and Katz (1952), one cas,

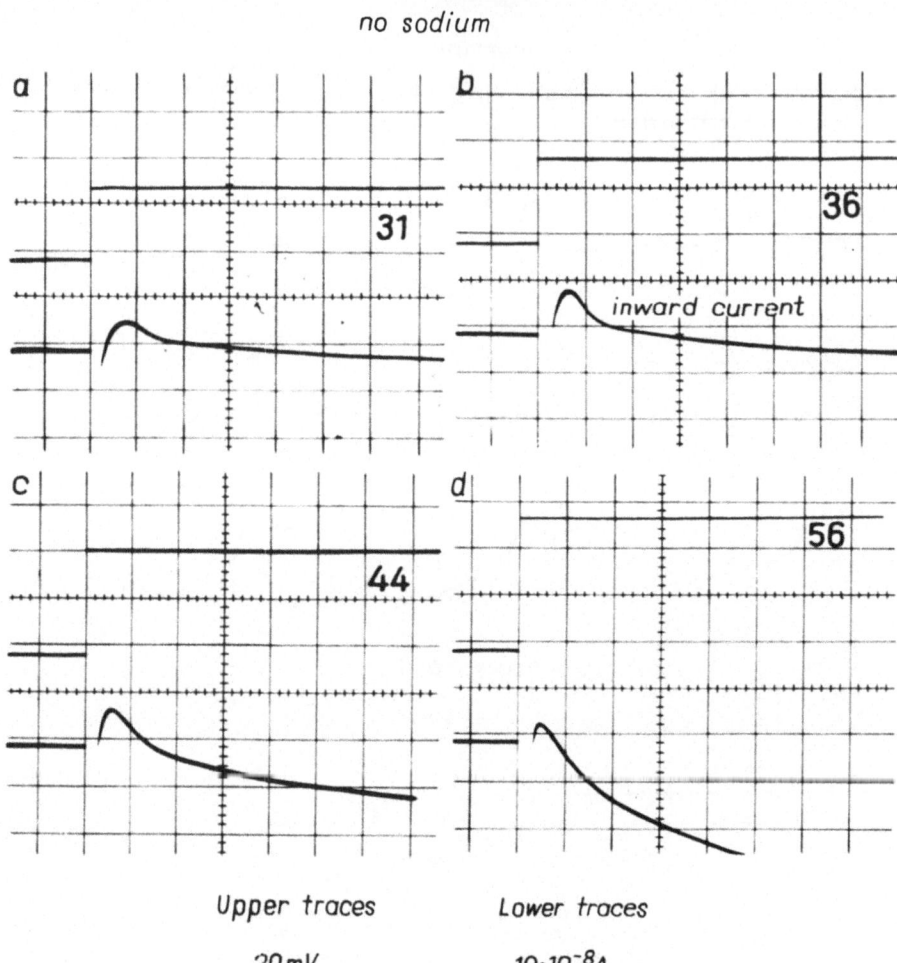

FIG. 3. Voltage clamp of snail neuron in a Na-free Ringer. Note that there is still an inward current present

alter the external solution around the neuron and see what effect this has on the transient and steady-state currents. In our experiments, the bath containing the preparation had a volume of 9 ml and tests with colored solutions showed that 20 ml would change the solution. In practice we used a minimum of 100 ml of solution to change the bath solution and in experiments where we wished to make sure of the removal of Na there was a continuous perfusion of Na-free solution throughout the experiment.

The neuron was first tested in normal Ringer and then tested in Ringer in which the Na had been replaced by Tris HCl (Fig. 3). Note that there is still an inward cur-

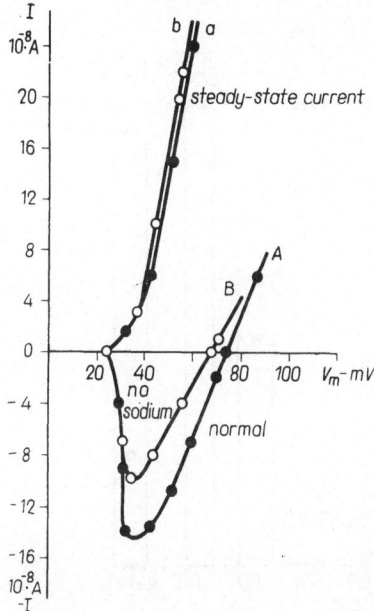

FIG. 4. Current–voltage curve of neuron in normal Ringer and Na-free Ringer. The inward current in Na-free Ringer is less than that in normal Ringer. There is little difference in the steady-state outward currents (b and a)

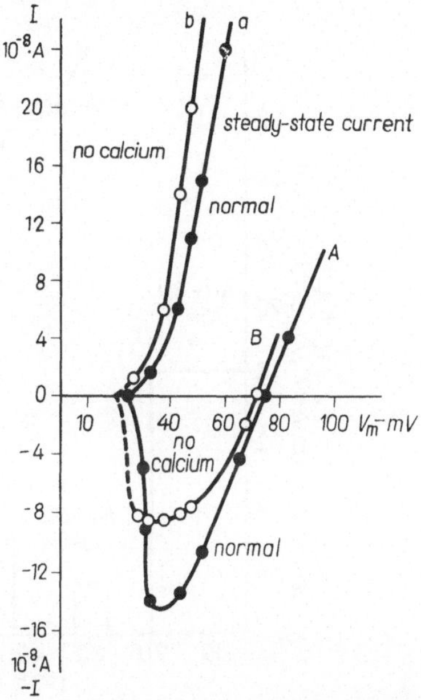

FIG. 5. Current–voltage curve of neuron in normal Ringer and Ca-free Ringer. The inward current in Ca-free Ringer is less than that in normal Ringer. There is also a difference in the steady-state currents

rent even though this is less than when Na is present (compare Fig. 2b with Fig. 3b). The outward steady-state currents had not been affected. The relationship is shown more clearly in Fig. 4, which is a graphical representation of the current (ionic movements through the cell) and the command voltage, the command voltage being the experimentally controlled variable. Two values are shown, the maximum transient current and the steady-state current (a, b). In normal Ringer there is a marked transient current when the membrane potential has become depolarized by about 20–24 mV. This current is an inward one and reaches its maximum when the membrane potential is at 40–47 mV when there is an inward current of $15 \cdot 10^{-8}$ A. At 75 mV depolarization there is a zero inward current. The steady-state current (a) increases from the initial depolarization of 20–24 mV. If the surrounding Ringer is replaced by Na-free Ringer (Tris Ringer or Sucrose Ringer) there is a reduced transient current (B) which has a maximum value of $10 \cdot 10^{-8}$ A. The outward steady-state current (b) has not been appreciably affected by the removal of Na.

The transient current could be restored to its original value by increasing Na (80 mM) but could not be increased by increasing Ca up to five times its normal level (Na: 0 mM; Ca: 35 mM).

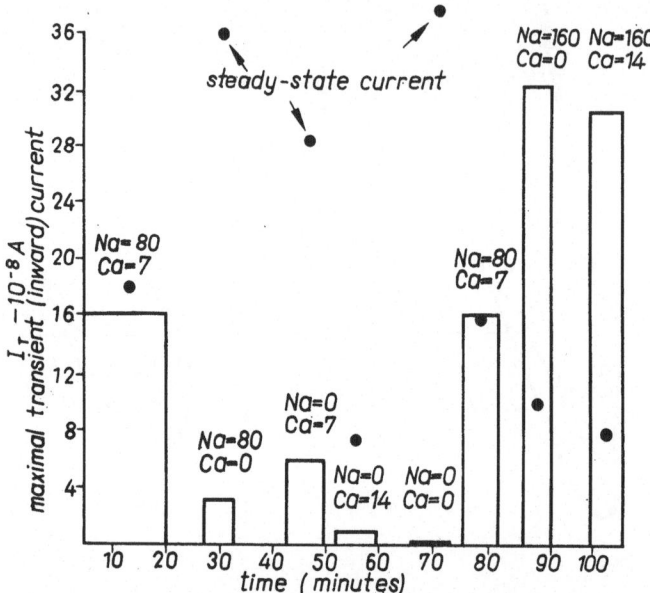

FIG. 6. Effect of calcium and sodium concentrations on the inward and outward currents. The results are taken from a continuous experiment. The inward current is reduced by reducing Ca, but can be restored by increasing Na (Ca = 0, Na = 160). Increasing Ca will not increase the inward current (Na = 0, Ca = 14)

Figure 5 is a graph showing a neuron placed first in normal Ringer and then in Ca-free Ringer (Na: 80 mM; Ca: 0 mM). In the Ca-free solution, the maximum transient current was $8 \cdot 10^{-8}$ A. Note also that the transient current started at a lower depolarization value, i.e., the threshold of the nerve had been lowered and the peak transient came at 32 mV instead of 40 mV depolarization. This movement of the I–V curve along the voltage axis is the stabilizing effect of calcium on the membrane (Frankenhauser and Hodgkin 1957). Increasing the Na content of the Ca-free Ringer would increase the inward current.

The effect of the removal of Ca and/or Na and of the augmentation of the two is shown in Fig. 6. Note that when Ca is removed the inward current can be *increased* by increasing the Na, but if Na is removed the inward current is *decreased* by increasing Ca. This would indicate that most probably the main effect of Ca is to alter the membrane stability (as is also shown by the changes in the steady state current shown in Fig. 6 as 0).

When the neuron is placed in normal Ringer containing tetrodotoxin (10^{-9}–10^{-5} g/ml), there is little or no reduction in the transient inward current. The conclusion is that for the *Helix* neuron that we have studied, tetrodotoxin does not reduce the inward movements of Na ions through the membrane, and hence the nerves

FIG. 7. Effect of injecting EDTA into neuron. The inward current is maintained if Na is present in the Ringer but not if the Na is removed. With sodium Ringer present: A, a. With no sodium in Ringer, calcium present: B, b

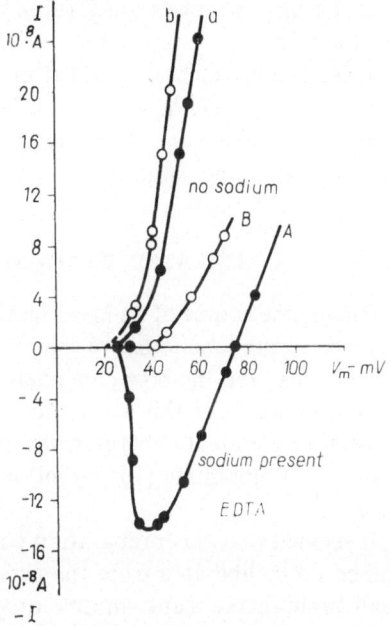

are similar to those of the puffer fish itself, whose neurons also have an inward sodium current unaffected by tetrodotoxin.

Figure 7 shows the effect of injecting EDTA (0.5 mm) into a snail neuron. In normal Ringer, there was a full inward transient current, but if the solution was replaced by Na-free Ringer, there was no inward current even though the solution contained 7 mM Ca.

If the neuron is placed in a Na-free solution it still shows action potentials and an inward current. If the neuron is stimulated many times, the inward current disappears, though it reappears immediately if the Na is replaced (indicating that the neuron had not been damaged and that the inward current is very sodium-sensitive).

In summary of this section, we consider that the inward current is carried by the sodium ions and that the role of Ca is most probably one of controlling the membrane permeability. The difficulty with the preparation lies in the fact that it is difficult to remove all the external Na even in neurons bathed in Na-free Ringer for long period of time; there would appear to be a small store of Na which could be either trapped by the glia or held in the membrane foldings or even in the membrane structure itself. The reasons we think the Na ions to be the important ions are summarized below:

1. Removal of either Na or Ca reduces the inward current.

2. The inward current in Na-free solution cannot be restored by increasing Ca from 7–14 mM.

3. The inward current in Ca-free solution can be restored by increasing Na from 80–160 mM.

4. Tetrodotoxin has no effect on the inward Na current.

5. Injecting EDTA, which should increase the Ca concentration gradient, prevents the development of an inward current in Na-free solution, though allows it in a Na-containing solution.

ISOLATED BRAIN-NERVE-MUSCLE PREPARATION

During the course of an investigation into the probable nerve-muscle transmitter at invertebrate junctions, we found that it was possible to isolate the intact brain, nerve trunk, and the retractor pharangeal muscle from the snail and obtain glutamate coming from the nerve-muscle region when we stimulated the brain; the amount of glutamate liberated was proportional to the number of stimuli, and the addition of glutamate (10^{-8} g/ml) would make the muscle contract (Kerkut *et al.* 1965).

It seemed that this preparation could be used to test whether radioactive material could be liberated from the nerve-muscle junction, and we set up the isolated snail brain–nerve trunk–muscle preparation so that the brain was in one pool of Ringer and the muscle was in another pool of Ringer, the two being connected by the nerve trunk. It was then possible to put labeled material into the brain pool and see if they could later appear in the muscle pool (Fig. 8). Following our experiments in which we had found cold glutamate coming from the nerve-muscle junction, we put C–14 labeled glutamate on the brain to see if we could detect labeled glutamate appearing in the muscle perfusate.

If the brain is incubated in 1 μc of C–14 glutamate for 3 h and then stimulated, labeled material appears in the muscle perfusate. The amount of labeled material that appears is roughly proportional to the number of stimuli given to the brain. This material was shown by thin-layer chromatography to have the same Rf as glutamate in five different solvents, and so we think that it is glutamate. If we repeat the experiment but incubate the brain in either C–14 glucose or C–14 alanine, then on stimulation some radioactive material appears in the muscle perfusate, but this material is glutamate and not glucose or alanine. If we incubate

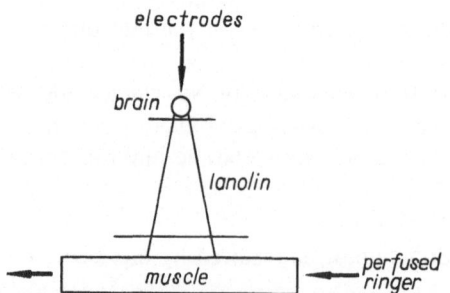

FIG. 8. Diagram of the isolated snail brain, nerve trunks, and muscle preparation. The brain is incubated in labeled C-14 glutamate and then stimulated. It is separated from the muscle by a lanolin barrier. The labeled material later appears in the muscle perfusate and hased travel through the nerve trunks

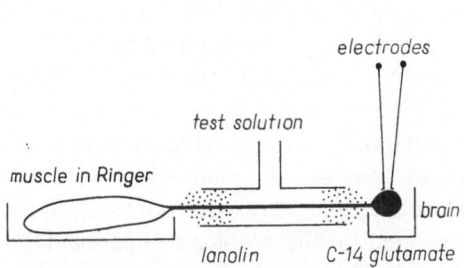

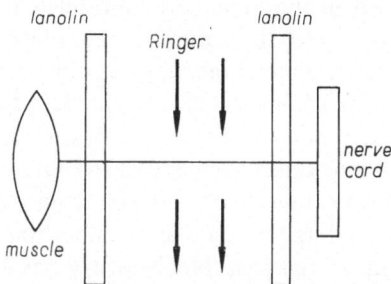

FIG. 9. Diagram of apparatus for the testing of the action of drugs and temperature on the rate of transport of labeled material from the brain to the muscle

FIG. 10. Frog gastrocnemius-nerve trunk-nerve cord preparation. With the long sciatic nerve it is possible to have a layer of Ringer washing between two lanolin barriers and so preventing any direct leakage of labeled material from the muscle compartment to the nerve-cord compartment. If labeled material is put in the muscle compartment, then, after 24 h, the nerve cord shows radioactivity

the brain in C-14 xylose, then we cannot detect any labeled material appearing in the muscle perfusate.

These experiments would support the case that glutamate could be the nerve-muscle transmitter at the snail pharangeal muscle junction.

Rate of transmission. If the brain is incubated in labeled glutamate and immediately stimulated, it takes about 20 min before labeled material appears in the muscle perfusate. The precise time varies according to the preparation, but in general it takes about 20 min for the labeled material to travel the 1 cm of nerve trunk separating the brain and the muscle.

The transport rate is affected by cooling the nerve trunk or by addition of xylocaine or nembutal (10^{-5} g/ml) to the setup as shown in Fig. 9.

If the muscle is incubated in labeled glutamate, then, after 18–24 h, labeled material appears in the brain. This material is not glutamate but is a more complex material that can be separated in starch gel or polyacrylamide electrophoresis.

The short length of nerve trunk between the snail brain and muscle makes the problem of leakage very critical, especially when there is the long incubation time of 24 h. In most experiments, we are quite satisfied that leakage had not taken place. In order to check this more certainly, we set up a different preparation with a longer stretch of nerve trunk: the frog gastrocnemius sciatic nerve–nerve cord preparation. To make even more sure we had a stream of Ringer running between the lanolin barrier (as shown in Fig. 10). After 24 h incubation, there was a transfer of labeled material from the muscle compartment to the nerve cord compartment.

This type of experimental setup, in which large regions of the central nervous system, the isolated snail brain in the one case and a large region of the frog nerve cord in the other, are connected by nerve trunks to the periphery allows us to investigate the nature of the interflow of material between different parts of the animal. One of the major problems in modern biology is the method by which the metazoan body is able to maintain functional connection between the various cells. There must be some system (or even more likely, several systems) by which the various cells of the body are kept in functional order. One possible method would be the transfer of material from one cell to another, this material in some way indicating that there is a connection and also altering the reactions of the recipient cell. In the past, biochemistry has concentrated on the problems of cell metabolism, i.e., the fundamental chemistry of the cell. The stage is now developing where one should be able to concentrate instead on organ biochemistry, i.e., to consider the question, what is it about the nervous system that makes it biochemically distinct from muscle tissue? It should also be possible using functional systems such as the isolated brain–nerve trunk–muscle systems described here to study the interactions between the biochemical systems of nerve and muscle. It should also allow investigation into the differences between known muscles and nerves, and perhaps some insight can thus be gained into the nature of nerve-muscle specificity.

The ready availability of radioactively labeled substances has put a most powerful tool into the hands of the experimental biologist. If this methodology is combined with biochemical, pharmacological, electrophysiological, and anatomical studies, one can look forward to the successful capture of many of the most secure bastions of modern biology, including that of the nature of the integrative-organization of the metazoan body.

This work has been supported by grants from the Science Research Council, the European Office of Research of the United States Army, and the Nuffield Foundation.

SUMMARY

1. Two different techniques are described for studying the movements of substances across cells: (a) a voltage clamp for the study of ionic movements into neurons during the action potential in snail neurons and (b) a snail brain-nerve trunk-muscle preparation where radioactive material can be followed moving from brain to muscle and from muscle to brain.

2. The voltage clamp technique allows the measurement of current flow during the action potential. There is still about 60% of the inward current flow during the action potential in neurons in Na-free solution. This is thought to be due to the presence of sodium ions still trapped in the system.

3. The main inward current is not thought to be carried by Ca ions because (a) the current cannot be increased by increasing the external Ca ion concentration above the normal level, (b) injection of EDTA into the neuron makes it more sensitive to Na_0 and (c) pulsing the neuron stops the inward current in Na-free solution; this returns at once if Na is restored.

4. The sodium current is not affected by tetrodotoxin.

5. Calcium is thought to play an important role in controlling membrane permeability.

6. A preparation can be set up so that the snail CNS lies in one compartment and the muscle lies in another, the two compartments being connected by the nerve trunk. C-14 labeled glutamate placed in the brain compartment is carried to the muscle compartment and liberated on brain stimulation (rate of transport 1 cm/20 min).

7. Labeled materials can be carried from the muscle to the CNS. The rate of transport is about 1 cm/24 h.

REFERENCES

FRANKENHAUSER, B. and HODGKIN, A. L. (1957): The action of calcium on the electrical properties of squid axons. *J. Physiol.* **137** 218–244.

GERASIMOV, V. D., KOSTYUK, P. G. and AISKII, V. A. (1964): Excitability of giant nerve cells of various pulmonate molluscs in sodium-free solution. *Bull. exp. Biol. Med.* **58** 3–6.

HODGKIN, A. L., HUXLEY, A. F. and KATZ, B. (1952): Measurement of current-voltage relations in membrane of the giant axon of the squid. *J. Physiol.* **116** 424–448.

JUNGE, D. (1967): Multi-ionic action potentials in Molluscan giant neurones. *Nature* **215** 546–548.

KERKUT, G. A. and CHAMBERLAIN, S. G. (1967): Voltage clamp study of the inward current in the neurones of *Helix aspersa* during the action potential. *Nature* **216** 89.

KERKUT, G. A. and GARDNER, D. R. (1967): The role of calcium ions in the action potentials of *Helix aspersa* neurones. *Comp. Biochem. Physiol.* **20** 147–162.

KERKUT, G. A., SHAPIRA, A. and WALKER, R. J. (1967): The transport of ^{14}C labelled material from CNS $\rightleftarrows$ muscle along a nerve trunk. *Comp. Biochem. Physiol.* (in press).

KERKUT, G. A., LEAKE, L. D., SHAPIRA, A., COWAN, S. and WALKER, R. J. (1965): The presence of glutamate in nerve-muscle perfusates of *Helix*, *Carcinus*, and *Periplaneta*. *Comp. Biochem. Physiol.* **15** 485–502.

MEVES, H. V. (1966): Das Aktionpotential der Riesennervenzellen der Weinbergschnecke *Helix pomatia. Pflügers Arch. ges. Physiol.* **289**.

OOMURA, Y., MAENO, T., OSZAKI, S. and NAKASHIMA, Y. (1962): Importance of calcium on the excitable membrane. *Setai No Kaguku* **13** 83–90.

DISCUSSION

I. Madarász: I should like to ask if you have any radioautographic data on hand referring to the distribution of the radioactive material in the inside and in the immediate surroundings of the axons?

G. A. Kerkut: Yes. The pictures indicate that the material is mainly in the axons.

J. Salánki: If I understood well, the muscles were in rest during incubation. What happens if they work, even isometrically?

G. A. Kerkut: We have not done this experiment, but hope to do it soon.

P. G. Kostyuk: An important problem arising from your and our experiments is to find what type of *Helix* neurons shows the capability to maintain spike activity in Na-free solutions. There is a big difference among neurons in this capability, and I wonder if you have any data about the possibility of a neuron of the same functional type varying in sensitivity to the absence of sodium ions?

I would also like to indicate that the level of membrane potential at which the inward current during action potential generation decays to zero is close to the equilibrium potential for sodium ions calculated from our measurements of the activity of Na ions, but not of the concentration. This probably indicates that measurements with selective glass microelectrodes give real values of the energy or amount of ions participating in the electrochemical processes at the cell membrane.

G. A. Kerkut: As yet we have no indication concerning sodium sensitivity and the function of the cell.

G. M. Hughes: You suggested that the persistence of action potentials in Na-free solutions might be due to the Na ions remaining in the intracellular spaces which are quite extensive. Have you estimated the conduction time which would be expected in your transmission experiments assuming that the labeled molecules diffused along these and other spaces within the nerve sheath?

G. A. Kerkut: Yes. The fastest rate we get is too fast for diffusion.

D. A. Sakharov: Is there the same glutamate transport in nonmotor nerves of the snail?

G. A. Kerkut: We do not know this, as the nerves are all mixed in the snail.

M. Mirolli: Have you identified the nature of the radioactive material found in the spinal chord?

G. A. Kerkut: Not yet.

Symposium on Neurobiology of Invertebrates 1967 (353—364)

AMINES IN MOLLUSCAN NERVOUS TISSUE AND THEIR SUBCELLULAR LOCALIZATION

G. A. COTTRELL

Wellcome Laboratories of Pharmacology,
Gatty Marine Laboratory, St. Andrews University
Fife, Scotland

INTRODUCTION

The ganglia of bivalve molluscs contain comparatively large amounts of 5-hydro-xytryptamine (5–HT) (Welsh and Moorhead 1960), dopamine (DA) (Sweeney 1963; Cottrell 1967) and, in certain species at least, high levels of acetylcholine (ACh) (Cottrell 1966). Consequently, bivalve ganglia would appear to provide convenient material for investigations on the subcellular localization of these substances in nervous tissue.

In the clam, *Mercenaria mercenaria*, it has already been shown that 70% or more of 5–HT and ACh, as well as Substance X (an unidentified cardioexcitor), is bound to subcellular particles (Welsh 1958; Cottrell 1966). Particles associated with 5–HT and ACh have certain properties in common, although it is clear that the particles binding 5–HT have different sedimentation characteristics than those associated with ACh. The limited number of experiments so far performed on the binding of DA in *Mercenaria* indicate that about the same percentage of this substance is particle-bound as are 5–HT and ACh. Similarly, about two-thirds of the total DA as well as 5–HT are sedimented when homogenates of *Spisula solida* ganglia, in isotonic glucose, are centrifuged at high speed (unpublished observations).

Nerve fibers within neuropile regions of ganglia of *Mercenaria* (Loveland 1963) and *Spisula solida* (unpublished observations) contain different types of small vesicles and granules which may be important in binding these different active substances. At least four different types of small particles have been discerned using conventional methods of fixation and staining for electron microscopy. These are as follows: (1) small clear vesicles of sizes ranging from 300–500 Å in diameter, (2) small vesicles with slightly electron-dense cores (about 500–800 Å overall diameter), (3) large vesicles with strongly electron-dense cores (about 900–2000 Å overall diameter), and (4) large vesicles with comparatively uniform granulation of varying density (1000–3000 Å). Whether the different types of particles listed here represent select populations of functionally different inclusions is not known. The possibility that any one of the types listed is a precursor or more advanced form of another must be borne in mind. From the results of fractionation and electron-microscope studies it has been suggested that the large, uniformly granulated vesicles contain Substance X (Cottrell and Maser 1967).

Wood (1965, 1966) has recently described a method for the histochemical localization of amines with the electron microscope. This technique, with slight modifications, has been used in the present study in an attempt to determine the nature of particles binding 5–HT and DA in the cerebral ganglia of the clam *Spisula solida*.

METHODS

Electron microscopy

Freshly dissected tissue was fixed for 3 h at room temperature in a mixture containing 3 ml of 25% glutaraldehyde solution, 21 ml of 0.2 M cacodylate buffer pH 7.2, and 7 g of sucrose (to adjust the osmotic pressure to approximately that of seawater). After rinsing the tissue for at least 12 h in several changes of cacodylate buffer containing 6.4 g of sucrose (80 ml), it was immersed for 24 h in a solution of the following composition: 1 g $K_2Cr_2O_7$, 0.8 g $Na_2SO_4 \cdot 10H_2O$, and 2 g of sucrose in 40 ml of 0.4 M acetate buffer pH 4.1. The tissue was dehydrated and embedded in Araldite. Sections were cut either on a Porter-Blum microtome or an LKB Ultrotome and examined with a Siemens Elmiskop I, or an AEI 6B electron microscope.

Assays of DA and 5–HT

Levels of DA were estimated by the following methods: (1) Visually by comparing the intensities of tissue spots with controls on developed paper chromatograms which had been exposed to paraformaldehyde (Bell and Somerville 1966), or (2) spectrophotofluorometrically by the iodine oxidation method of Shore and Olin (1958) and also by a similar method described by Carlsson and Waldeck (1958) after butanol extraction (Shore and Olin 1958).

5–HT concentrations were estimated spectrophotofluorometrically according to the method of Quay (1962). An Aminco–Bowman spectrophotofluorometer was used for the fluorescence analyses.

Fluorescence microscopy

Ganglia were rapidly dissected out, frozen in liquid-nitrogen-cooled propane and then dried for two days in a Speedivac–Pearse Tissue Dryer Model 1 at 40 °C and 10^{-5} torr. Next, the ganglia were exposed to paraformaldehyde vapor (relative humidity about 70%) for 1 h at 80 °C before vacuum embedding in paraffin wax (Corrodi and Jonsson 1966). Sections were examined on a Leitz fluorescence microscope. The standard Aristophot–Ortholux microscope assembly was fitted with an HBO 200 mercury lamp, a dark-field condenser, a BG 12 excitation filter, and a 530 mμ barrier filter.

RESULTS

There is a lack of general cellular detail in tissues fixed with glutaraldehyde alone and incubated with potassium dichromate. The only structures which appear electron-dense are those representing accumulations of amines (Wood 1965, 1966).

Figure 1 is an electron micrograph of part of the neuropile region of a *Spisula* cerebral ganglion, and shows a collection of electron-dense granules, the majority of which appear to be confined within a nerve fiber membrane. The granules range from about 250 Å–600 Å in diameter. Some of these consist of an inner area of high electron density surrounded by a slightly less dense peripheral area.

A small area from the neuropile of another ganglion is shown in Fig. 2. Here, several granules are localized along the inside of a small nerve fiber. In some instances, there are indications that the granules may be membrane-bounded. As with some of the granules in Fig. 1, a few exhibit a strongly electron-dense core surrounded by a less dense zone. The overall diameter of the granules which show these characteristics in this micrograph varies between about 450 Å and 650 Å. Some of the stained particles are considerably less dense than others. This may be explained in terms of the position of the plane of section in relation to the positions of each granule in the block of tissue; i.e., by the proportion of the total volume of

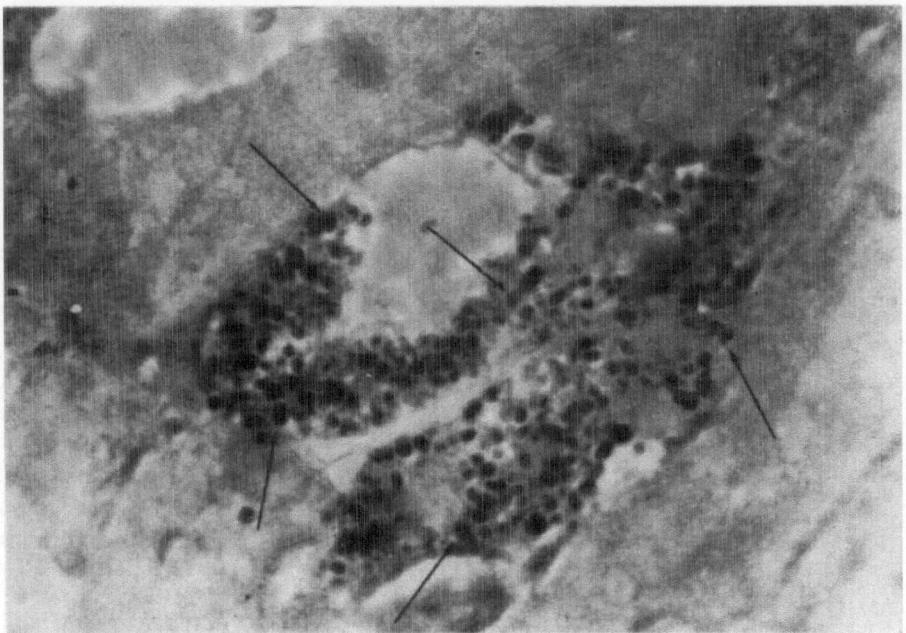

FIG. 1. A collection of electron-dense granules representing sites of amine localization in the neuropile of a cerebral ganglion of *Spisula*. Arrows point to granules which consist of an inner area of high density surrounded by an area which is less dense. × 60,000

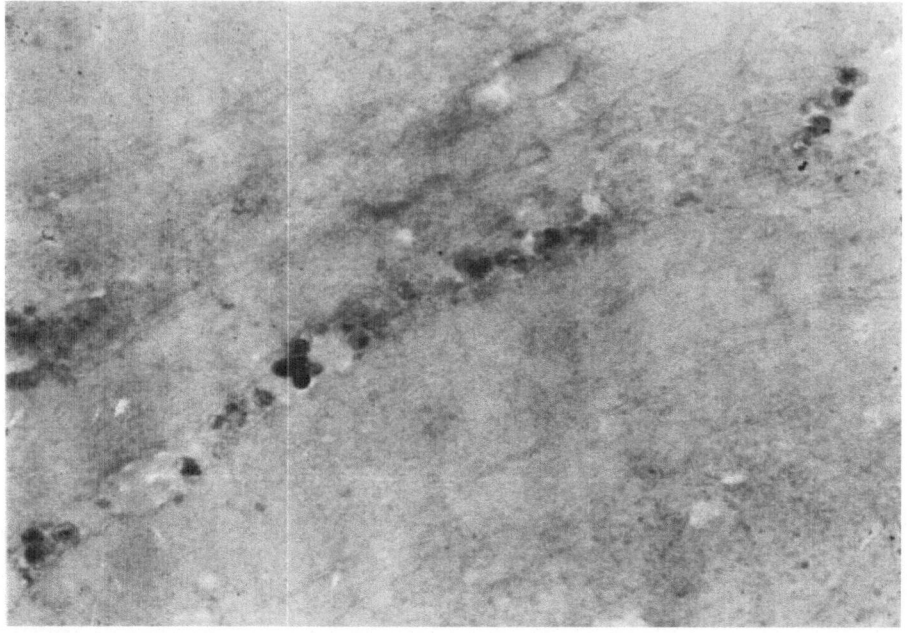

FIG. 2. Restriction of granules along the length of a small nerve fiber. × 47,000

each granule included in the section. At the same time, the variation in the density of the particles may be attributable to the amount of substance they contain.

Figure 3 shows some granules at higher magnification. In this micrograph, membranes can be seen encircling some of the particles. The overall diameter of most of the granules falls within the range 500–700 Å, but one large dense granule (850 Å) is seen towards the top of the figure. Other large granules up to a maximum size of about 1000 Å in diameter have been seen in other micrographs of cerebral and pedal ganglia.

Experiments with reserpine were made in an attempt to determine whether DA or 5–HT or possibly both amines are bound within the small granular vesicles. In order to ensure uniform administration of reserpine to a sufficient number of clams for chemical and electron-microscopic analyses, soluble reserpine phosphate (Ciba) was added to the seawater bathing the animals. The maximum depletion of DA that we have been able to achieve with this method of administration has been 50% of the total of that in control animals, whereas 5–HT was depleted by a maximum of 75% of the total in controls. Table 1 summarizes the results of three separate experiments in which animals were treated with different amounts of reserpine for different periods of time and at different temperatures. Mirolli and Welsh (1963) have shown that depletion of 5–HT in molluscs is dependent on temperature as well as the dose-level of the drug.

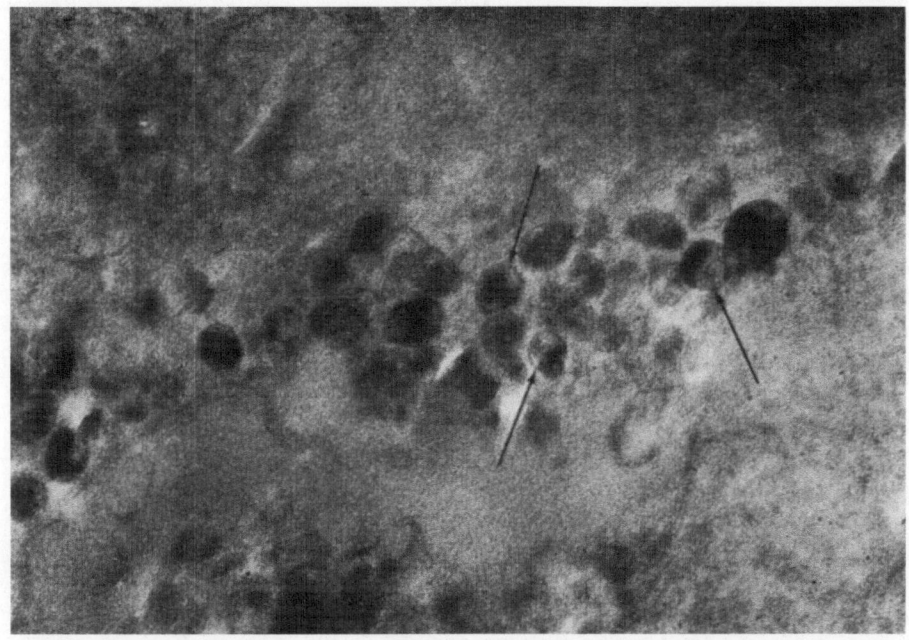

FIG. 3. Part of the neuropile of a *Spisula* ganglion showing some amine-containing granules. Arrows point to granules which appear to be membrane bounded. × 110,000

TABLE 1

Percentage depletion of DA and 5-HT from ganglia
of *Spisula* treated with different amounts
at different temperatures and for different periods of time

Expt. No	Conditions	Percentage depletion	
		DA	5-HT
(1)	2 μg/ml SW at 15 °C for 5 days	40–50*	70–80
(2)	4 μg/ml SW at 16 °C for 6 days	49	70
(3)	8 μg/ml SW at 17 °C for 7 days	49	75

* Paper chomatographic analysis

Neither by increasing the dose of reserpine nor by increasing the temperature and duration of exposure to reserpine were we able to deplete more than half of the total DA. In the third experiment, half of the animals exposed to reserpine died, whereas all of the control animals survived. Thus, it was concluded that it is not possible to lower the level of DA in the ganglia of animals treated in this fashion to a greater extent without killing all the animals.

These results seemed unsatisfactory from the point of view of attempting to observe possible differences in the number of appearances of the granulated vesicles

TABLE 2

Averaged levels of DA and 5–HT in the different
ganglia (μg/g wet wt)

	Visceral	Cerebral	Pedal
DA	80	80	195
5–HT	33	36.5	38

in relation to levels of DA and 5–HT. However, estimates of the concentrations of DA in the different ganglia (cerebral, visceral, and pedal) showed that the pedal ganglia contained more than twice as much of the amine as the other ganglia (Table 2), whereas the concentration of 5–HT was about the same in each of the different ganglia. Thus, assuming reserpine depletes the same percentage of DA from each of the different ganglia, it would be expected that cerebral or visceral ganglia from a reserpinized animal would only contain about 40 μg DA/g compared with 195 μg DA/g in the pedal ganglion of a control animal, i.e., about one-fifth the concentration found in a normal pedal ganglion. By making a similar assumption, the concentration of 5–HT in each of the ganglia from a reserpinized animal would be expected to contain approximately one-quarter that in each ganglia from a control animal.

A comparison was made of the occurrence of the 400–1000 Å granules in both cerebral and pedal ganglia from reserpinized animals with the pedal ganglia of control clams. Although it is very difficult to make a complete quantitative estimation of such small granules in an entire ganglion under the electron microscope, analyses of several sections cut at each of eight different areas of the neuropile of reserpinized cerebral ganglia, and similar analyses of reserpinized and control pedal ganglia, strongly suggest that control pedal ganglia contain many times more granules of 400–1000 Å diameter per unit volume than occur in reserpinized cerebral ganglia. A large proportion of those granulated vesicles that were observed in reserpinized cerebral ganglia had the appearance of being at least partly depleted of their contents (i.e., with a less dense, or uneven and sparse, granular form) (Fig.4). On the other hand, a difference in the number of granules of this size was not discerned in the reserpinized pedal ganglia compared with control pedal ganglia, although in the former case more granules were seen which gave the appearance of being depleted. These results indicate that the granules sequester DA rather than 5–HT.

Further support for this view has come from histochemical studies at the light-microscope level. *Spisula* ganglia sublimated with paraformaldehyde for detecting 5–HT and primary catecholamines according to the method devised by Falck (1962) show intense green fluorescence in the neuropile area and yellow fluorescence in most of the cell bodies at the periphery of the ganglia. The green fluorescence completely disappeared, along with much of the yellow fluorescence, when sections were immersed in sodium borohydride solution, and continuous illumination

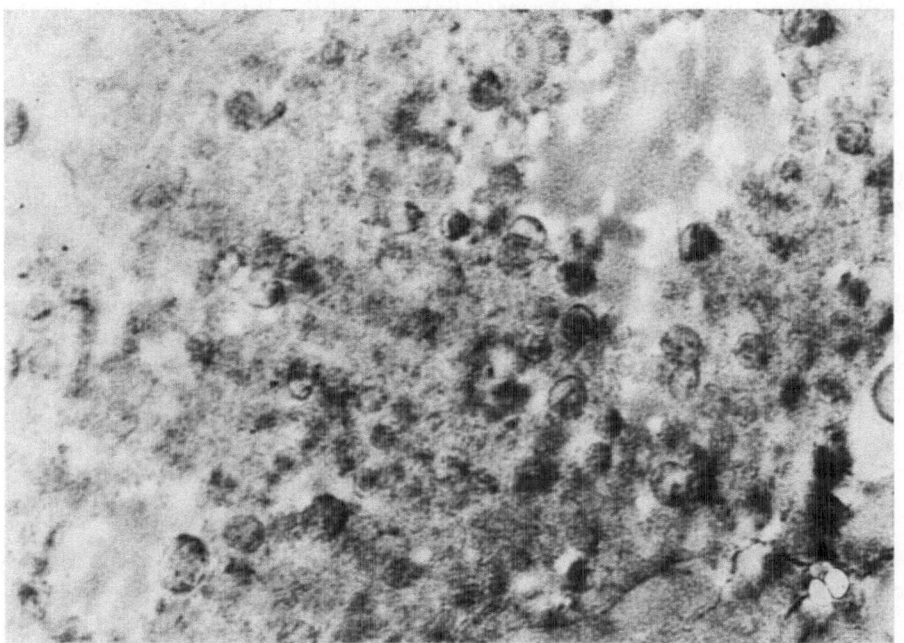

FIG. 4. A collection of sparsely granulated particles, presumably depleted granular vesicles, in the neuropile of a reserpinized cerebral ganglion

likewise diminished the fluorescence intensity in both areas. These observations suggest the presence of large amounts of primary catecholamine, presumably mainly DA, in the neuropile of *Spisula* ganglia, and that 5–HT is localized mainly within the cell bodies. Similar findings have been reported for the ganglia of *Anadonta piscinalis* (Dahl *et al.* 1966) and *A. cygnea* (Zs.-Nagy 1967). Since the granules were almost exclusively localized in the neuropile of *Spisula* ganglia, it seems that DA is more likely to be bound within them than 5–HT.

Further studies have been conducted to test the possibility that substances in addition to DA and 5–HT may give a positive reaction with the histochemical method of Wood as outlined above. Nineteen amino acids have been detected in *Spisula* ganglia (Cottrell and Seraphini–Fracassini, unpublished observation). Other related substances for which there is evidence for occurrence in the ganglia include histamine, ACh, and noradrenaline (NA) (Cottrell 1967), in addition to DA and 5–HT. *In vitro* experiments with glutaraldehyde and potassium dichromate reagents, prepared as above, and $5 \cdot 10^{-3}$ M solutions of the individual amines, ACh, and each of the amino acids, showed that only DA, NA, and 5–HT give a positive reaction with the histochemical method, i.e., produced a precipitate of chromium salt. However, when each of the different constituents, which formed a yellow color with the glutaraldehyde and/or were present in extremely large amounts, were tested in concentrations proportional to those found for the different substances in the tissue, the basic amino acid, arginine, in addition to DA and 5–HT, formed a precipitate. The color of the precipitate formed with arginine was light yellow, compared with the dark brown precipitate formed with DA and 5–HT. However, the amount of precipitate formed with the concentration of arginine used, about eight times greater than DA (same ratio as present in the ganglia), was about the same as that formed with the DA solution. This observation was made gravimetrically using filter sticks. None of the other constituents gave a positive reaction; noradrenaline at the concentration tested, about one-tenth that of DA, did not yield a perceptible deposit. Therefore, on the basis of concentration within the ganglia and reactivity with glutaraldehyde and dichromate of the natural constituents, the granular deposits most probably represent accumulations of DA, 5–HT, or arginine.

DISCUSSION

To summarize, the results of the reserpine experiments, fluorescence-microscopical studies, and *in vitro* experiments with different naturally occurring amines and amino acids suggest that DA is the most likely candidate for localization within the small granular vesicles described in this study. However, the possibility that the granules may also contain 5–HT cannot entirely be ruled out at present for two reasons: (1) It is possible that the intense green fluorescence in the neuropile of ganglia exposed to paraformaldehyde may mask smaller amounts of yellow fluorescence characteristic of 5–HT, and that 5–HT, therefore, may also be present in the neuropile region of the ganglia. (2) No particles other than the 400–1000 Å granules have been consistently observed so far in ganglia treated by the Wood histochemical technique which might be considered to bind 5–HT. Chromatographic evidence for the presence of comparatively small amounts of NA in *Spisula solida* ganglia (Cottrell 1967) further complicates the issue.

None of the other small molecular weight compounds containing an amino group which have been detected in the tissue, with the exception of arginine, appears to contribute to the formation of dense deposits in tissue treated with glutaraldehyde and dichromate. Furthermore, arginine only produced a precipitate in the *in vitro* experiments when tested at a concentration some eight times greater than that used for DA, i.e., in the same ratio as that found for the different substances in the ganglia, the averaged value for DA being 120 μg/g and that for arginine 969 μg/g (Cottrell and Seraphini-Fracassini, unpublished observation). Before it can be proposed that the dense granules represent the localization of arginine, it is first necessary to suggest that at least some of the arginine is not freely soluble in the cytoplasm of the neurones but is associated with subcellular particles, as has been shown with DA and 5–HT, as well as with ACh. If this is the case, the results of the *in vitro* experiments indicate that the granules would have to contain about eight times as much arginine as DA to give the same degree of electron density. If these assumptions are correct, then arginine must be considered as likely to be associated with the granules as DA, provided that arginine is concentrated in the neuropile regions of the ganglia. On reflection, the probability is not very great that the granules bind arginine alone.

Thus, at the present time, it is tentatively concluded that the small granular vesicles contain DA, but that further studies are required to determine whether NA, 5–HT, or arginine are not also, or alternatively, associated with the particles. A satisfactory procedure for further studies to test these possibilities may be to use a combination of the electron-microscopic technique of Wood (1965, 1966) or that of Coupland and Hopwood (1966) with homogenate fractions rich in the different catecholamines and with fractions rich in 5–HT.

The size of the granular vesicles observed in this study is somewhat similar to that of the sympathetic nerve granules that are thought to contain NA (Wolfe *et al.* 1962). However, whether all the granules observed in *Spisula* ganglia fixed in glutaraldehyde and incubated in dichromate solution have the same appearance in tissue fixed with osmium tetroxide in the conventional fashion is not known. Tranzer and Thoenen (1967), working with cat iris, have shown that the small granular vesicles of sympathetic nerves, seen when tissue is double fixed in glutaraldehyde and osmium tetroxide, appear hollow when fixed in osmic tetroxide alone. Thus, many of the granular vesicles seen in *Spisula* ganglia treated with glutaraldehyde and dichromate may appear hollow after osmium tetroxide fixation.

SUMMARY

Small granular vesicles ranging in size from about 300 to 1000 Å in diameter have been observed in *Spisula solida* ganglia prepared according to the method of Wood (1965, 1966) for the histochemical localization of amines with the electron microscope. The granulated vesicles were almost exclusively found in the neuropile region of the ganglia. *S. solida* ganglia are rich in DA and 5–HT. Results of depletion experiments with reserpine, histochemical studies at the cellular level, and *in vitro* experiments with 23 amino acids and amines as well as ACh (all of which have been detected in the ganglia), suggest that DA is the most likely candidate for localization within the granulated vesicles. However, the possibility that these granules may alternatively, or also, contain 5–HT, NA, or possibly arginine cannot be completely excluded at the present time.

Acknowledgments: This investigation was supported by grants from the British Medical Research Council and The Wellcome Trust. Thanks are due to Mr. B, Powell and Mr. J. Brown for technical assistance and to Ciba Laboratories Ltd., Horsham, Sussex, for supplying the soluble reserpine.

REFERENCES

BELL, C. E. and SOMERVILLE, A. R. (1966): A new fluorescence method for detection and possible quantitative assay of some catecholamines and tryptamine derivatives on paper. *Biochem. J.* **98** 1C—3C.

CARLSSON, A. and WALDECK, B. (1958): A fluorimetric method for the determination of dopamine (3-hydroxytyramine). *Acta physiol. scand.* **44** 293—298.

CORRODI, H. and JONSSON, G. (1967): The formaldehyde fluorescence method for histochemical demonstration of biogenic monoamines. A review on the methodology. *J. Histochem. Cytochem.* **15** 65—78.

COTTRELL, G. A. (1966): Separation and properties of subcellular particles associated with 5-hydroxytryptamine, with acetylcholine and with an unidentified cardio-excitatory substance from *Mercenaria* nervous tissue. *Comp. Biochem. Physiol.* **17** 891—907.

COTTRELL, G. A. (1967): Occurrence of dopamine and noradrenaline in the nervous tissue of some invertebrate species. *Brit. J. Pharmacol.* **29** 63—69.

COTTRELL, G. A. and MASER, M. (1967): Subcellular localization of 5-hydroxytryptamine and Substance X in molluscan ganglia. *Comp. Biochem. Physiol.* **20** 901—906.

COUPLAND, R. E. and HOPWOOD, D. (1966): The mechanism of the differential staining reaction for adrenaline- and noradrenaline-storing granules in tissue fixed with glutaraldehyde. *J. Anat.* **100** 227—243.

DAHL, E., FALCK, B., VON MECKLENBURG, C., MYHRBERG, H. and ROSENGREN, E. (1966): Neuronal localization of dopamine and 5-hydroxytryptamine in some Mollusca. *Z. Zellforsch.* **71** 489—498.

FALCK, B. (1962): Observations on the possibilities of cellular localization of monoamines by a fluorescence method. *Acta physiol. scand.* **56** Suppl. 197.

LOVELAND, R. E. (1963): Some aspects of cardio-regulation in *Mercenaria mercenaria*. Ph. D. Thesis, Harvard University.

MIROLLI, M. and WELSH, J. H. (1964): The effects of reserpine and LSD on molluscs. In: *Comparative Neurochemistry*, Pergamon Press, Oxford, 433.

QUAY, W. B. (1963): Differential extractions for the spectrophotofluorometric measurement of diverse 5-hydroxy- and 5-methoxyindoles. *Analyt. Biochem.* **5** 51—59.

SHORE, P. A. and OLIN, J. S. (1958): Identification and chemical assay of norepinephrine in brain and other tissues. *J. Pharmacol.* **122** 295—300.

SWEENEY, D. (1963): Dopamine: Its occurrence in molluscan ganglia. *Science* **139** 1051.

TRANZER, J. P. and THOENEN, H. (1967): Significance of "empty vesicles" in post-ganglionic sympathetic nerve terminals. *Experientia* **23** 123—124.

WELSH, J. H. (1958): Evidence for 5-HT granules in molluscan ganglia. *Acad. Rec.* **132** 516.

WELSH, J. H., MOORHEAD, M. (1960): The quantitative distribution of 5-hydroxytryptamine in the invertebrates, especially in their nervous systems. *J. Neurochem.* **6** 146—169.

WOLFE, D. E., POTTER, L. T., RICHARDSON, K. C. and AXELROD, J. (1962): Localizing tritiated norepinephrine in sympathetic axons by electron microscopic autoradiography. *Science* **138** 440—442.

WOOD, J. G. (1965): Electron microscopic localization of 5-hydroxytryptamine (5-HT). *Tex. Rep. Biol. Med.* **23** 828—837.

WOOD, J. G. (1966): Electron microscopic localization of amines in central nervous tissue. *Nature* **209** 1131—1133.

ZS.-NAGY, I. (1967): Histochemical demonstration of biogenic monoamines in the central nervous system of the lamellibranch mollusc *Anadonta cygnea* L. *Acta biol. Acad. Sci. hung.* **18** 1—8.

DISCUSSION

I. Zs.-Nagy: Your results are in agreement with ours as concern the localization of DA, but as regards the subcellular localization of 5–HT, there are some differences between us. We found, working with the differential centrifugation technique, that the majority of 5–HT was in the fractions, which contained many endoplasmic reticulum vesicles [Zs.-Nagy *et. al. J. Neurochem.* (1965) **12** 245]. At the same time, you demonstrated 5–HT in the perikarya histochemically, but by density-gradient centrifugation you have not obtained any evidence that 5–HT is localized in endoplasmic reticulum. Your pictures presented here show a considerable shrinking of elements, which is caused most probably by the hypertonic sucrose solutions you applied in density-gradient centrifugation. I think that this phenomenon may be one of the reasons for differences between your results and ours. What is your opinion about it?

G. A. Cottrell: I have not obtained any evidence that 5–HT is localized in 'endoplasmic reticulum vacuoles'. No 5–HT was found in the microsomal fraction of *Mercenaria* ganglia homogenate, as one might expect if the amine is simply trapped in 'vacuoles' of endoplasmic reticulum during homogenization. Further, the purest fraction of 5–HT particles that we obtained did not contain very many vacuoles which might be considered to have derived from the endoplasmic reticulum.

The 5–HT and ACh particles release their active substances in a quantitatively similar manner to changes in pH, reduction in osmolarity of the medium, and elevated temperatures. Thus, it seems that there are similarities in the binding mechanism for both substances. Although one can make hypotheses about the type of inclusions to which 5–HT is associated, I think the subcellular localization of 5–HT in molluscan ganglia is still really an open question at present.

D. A. Sakharov: I have some comment on the Falck–Hillarp method, which seems to have become very popular in invertebrate neurobiology. Those who worked with this method and demonstrated their results at this Symposium know how complicated the procedure is. Recently, Dr. A. V. Sakharova and myself succeeded in obtaining a considerable simplification of the fluorescent method for invertebrate nervous tissue, thus supporting the idea of Prof. N. Eränkö that aqueous solution of formaldehyde is capable of converting intraneuronal monoamines into strongly fluorescent compound.

We worked with ganglia of *Lymnaea stagnalis* which had been previously studied by means of the Falck–Hillarp method. We thus had the opportunity to compare results obtained by means of both procedures. The best results were obtained with ganglia immersed in 8% formalin for 7–15 min or in 4% solution for 45–60 min (formalin solutions were prepared in snail ringer and cooled with ice). Air-dried cryostat sections of such formalin-fixed ganglia gave excellent fluorescence of the same sensitivity and specificity as preparations obtained by means of the Falck–Hillarp procedure.

But what is really strange is the obvious discrepancy between histochemical and biochemical manifestations of monoamine depletion by reserpine. The histochemical method is so sensitive that 50 % of a monoamine remaining in the tissue after reserpine treatment should give a rather strong fluorescence. Nevertheless, we usually do not see any specific fluorescence in ganglia of the reserpinized snails. So, I wish to ask Dr. Cottrell what is his opinion about this discrepancy. Might it be that the fluorescent method reveals only a peculiar part of tissue monoamines, namely, a reserpine-sensitive part, at least in molluscs?

G. A. Cottrell: The response to reserpine in different molluscs may well vary. In our experiments with reserpine, the maximum depletion of dopamine achieved was about 50% of the total in the ganglia. In these experiments, the drug was added to the seawater bathing the clams in order to ensure uniform administration of the drug to sufficient animals for electron-microscopic and biochemical studies. Injection of reserpine into specimens of *Spisula* may possibly bring about a greater depletion. This is something we must try. Nevertheless, our reserpine experiments, although not giving 100% depletion of DA, have enabled us to observed differences in the number and appearance of granulated vesicles in relation to the content of DA and 5–HT in the different ganglia, thus providing additional evidence that DA is stored in these granulated vesicles.

J. Salánki: What do you think about the physiological role of the DA? Do you find any characteristic change in the behavior of the animal after reserpine treatment?

G. A. Cottrell: We did not closely follow the behavioral effects of reserpine on the animals because it would be difficult to interpret such observations in view of the fact that both 5–HT and DA are released, and also possibly NA, with reserpine administration.

INTEGRATION

Symposium on Neurobiology of Invertebrates 1967 (367—380)

THE REGULATING FUNCTION OF THE INVERTEBRATE NERVOUS SYSTEM

A. K. VOSKRESENSKAYA

Sechenov Institute of Evolutionary Physiology and Biochemistry
Academy of Sciences of the USSR
Leningrad, USSR

For a long time now (Pavlov 1922; Orbeli 1923), Russian physiological schools have been developing the idea that there are two kinds of central nervous system influence on effectors, of two essentially different integrative mechanisms: a starting mechanism (or functional mechanism, by the original terminology) and a mechanism for regulating the level of activity which was originally regarded as a trophic one. Later, Orbeli accentuated the adaptable significance of this regulating function of the nervous system and introduced the new term, adaptation-al-trophic function, into physiology.

Due to his evolutionary point of view in studying the functions of the animal nervous system, Orbeli was able to conceptually formulate the historical connection between these two kinds of nervous activity.

He considered the regulating action of the nervous system to be more ancient phylogenetically and to be particularly prevalent in some stages of evolutionary development. Afterwards, along with the development of specific starting mechanisms of nervous control, the regulating function also became connected with special and phylogenetically younger parts of the central nervous system. Orbeli believed the sympathetic adrenergic nervous system to be one of the most important pathways of the adaptational trophic influence on the peripheral effectors in vertebrates.

In the course of studying the structure and the functions of the animal nervous system the great significance of the so-called 'unspecific', 'tonic', prolonged forms of nervous control has been revealed; these do not initiate the action of the system, but secure a certain level of its activity and condition the strength and character of the reflex reaction.

The significance of this regulating role of nervous control may be so great that the normal function of the starting nervous mechanisms is possible only against the background of this 'unspecific' influence.

These 'unspecific' integrative mechanisms are very essential in understanding such complicated nervous functions as 'memory', 'learning,' 'training' and so on. These mechanisms can be analyzed from new angles due to the possibility that exists today of investigating physiological functions on the cellular level.

The invertebrates are very good preparations for studying the regulating function of the nervous system, its interconnections with starting nervous mechanisms, and its phylogenetic development.

For a long time in my research and that of my colleagues, attention has been paid to the regulating character of the nervous influence on the effectors in invertebrates. Similar findings in the papers of other investigators also drew our attention.

Let me pass to the discussion of different displays of the nervous system regulating function in some representatives of invertebrates in the line of Protostomia.

In their recently published works, del Castillo *et al.* (1963, 1967) established the fact that the rhythmic action potentials controlling the somatic muscle of *Ascaris* are of myogenic origin when generated at the muscle syncytium. The stimuli applied to the nerve cord caused the facilitation or inhibition of the muscle action potentials and altered the frequency of myogenic spikes by changing the membrane potential of syncytium. Therefore, the nerve cord produced the regulating influence on an autonomic peripheral mechanism of activity in somatic muscles of Nematoda.

From recent work in our laboratory on some representatives of Annelida (earthworm), it was shown that longitudinal muscles of body wall possesses spontaneous electrical activity appearing as periodic bursts of action potentials (Fig. 1). These spontaneous bursts of muscle action potentials were preserved after the elimination of the whole nerve cord or a part of it within 10–15 segments. After that, only some characteristics of the muscle electrical activity were changed: the duration of bursts, the intervals between them, and so on.

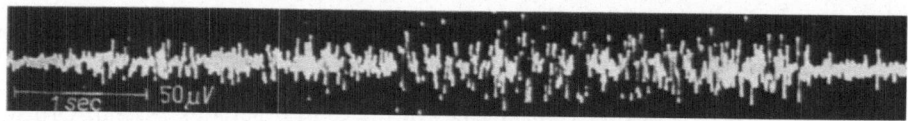

FIG. 1. The spontaneous electrical activity of the longitudinal muscle of earthworm with intact innervation

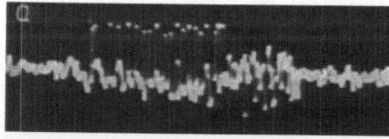

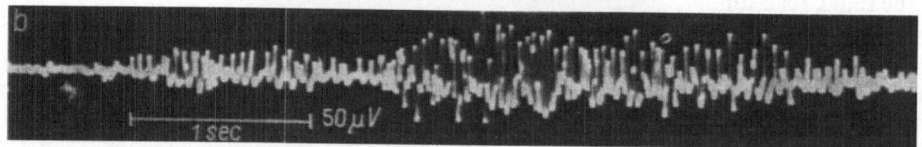

FIG. 2. Enhancement of electrical activity of the earthworm longitudinal muscle after the stimulation of the surface of the body by 1% solution of HCl (the nerve cord is eliminated).
a — control, b — after stimulation by 1% HCl

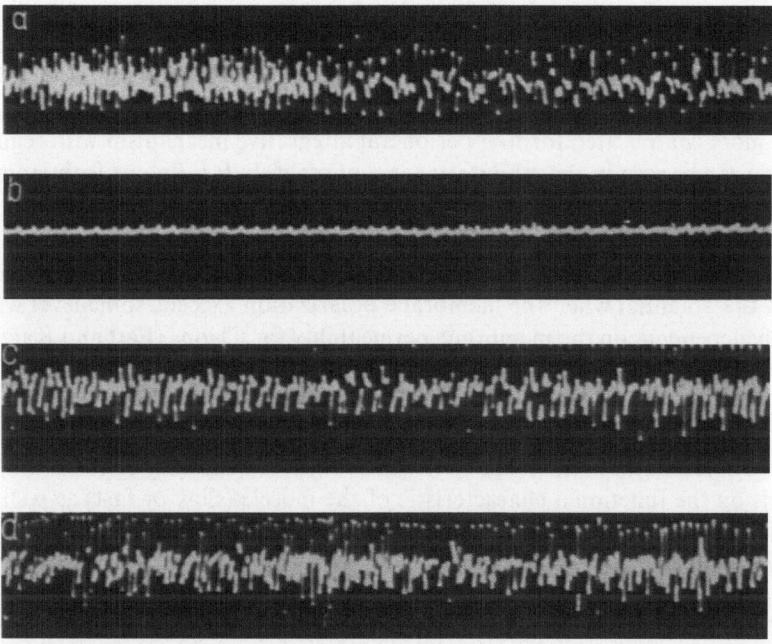

FIG. 3. The elimination of action potentials isolateind longitudinal muscle of earthworm after anaesthesia of the skin receptors. a — control; b — 1 min after putting 0.25 % solution of cocaine on the body surface; c — recovery of action potentials after washing cocaine off; d — action potentials of untreated segments (control)

Therefore, we can regard the mechanism of the initiation of the rhythmic action potentials as being of peripheral origin, with the nerve cord only exerting a regulating influence on it.

Further experiments showed that the electrical activity of the worm's longitudinal muscle depends on the peripheral afferent system. The bursts of action potentials became stronger when a chemical (HCl) or mechanical irritation was applied on the surface body of the worm, and the action potentials disappeared, but reversibly, in segments anaesthetized by a solution of cocaine (Figs. 2 and 3).

These data led us to the conclusion that the peripheral mechanism responsible for the initiation of the action potentials in the worm somatic muscle is not of myogenic, but is of peripheral neurogenic origin. It is based on the possibility of closing the reflex within receptor-sensitive cells – afferent axon - motor axon or muscle cell, without the participation of the central nerve cord.

Earlier, Prosser (1935) and Beritov (1950) showed it was possible to excite the somatic muscle of some Annelida by means of skin receptors. That possibility was also discussed by Bullock and Horridge (1965).

Further elucidation of the mechanism of action-potential initiation in the worm somatic muscle, as well as of the controlling nervous mechanism, peripheral or central, requires further and more precise experiments of the cellular level. However, the somatic muscles of different representatives of Vermes can surely serve as a valuable model for studying the peripheral autonomic reactions and the regulating function of the central nervous system.

The more complicated form of peripheral integrative mechanism with regulating function can be seen in the inhibitory nerve of crayfish. It is known from numerous investigations on this nerve that, as a rule, it has a hyperpolarizing effect on the membrane of muscle fibers or a repolarizing one if the membrane is depolarized by excitatory stimuli. However, it also can give some depolarizing effect or reverse the sign of the potential when the membrane polarization exceeds some level which is possibly dependent on the membrane permeability for Cl ions (Fatt and Katz 1952; Wiersma 1957; Hoyle and Wiersma 1958).

Our experiments showed that the so-called inhibitory nerve of crayfish *(Astacus)* can decrease as well as enhance the mechanical reaction of the muscle upon stimulation of an excitatory axon. The prevalence of the facilitating or inhibiting effect depends on the functional characteristic of the muscle (slow or fast) as well as on the current functional properties of the neuromuscular system (Voskresenskaya, Kuntsova and Svidersky 1959).

These facts were confirmed later by the works of Koshtoyants *et al.* (1962) and recently by Ovetchkin's experiments (1966).

Then in our laboratory Kuntsova (1962) showed that the inhibitory nerve can exert a dual effect on the mechanical reaction in different muscles of diverse representatives of Crustacea as well as on the same preparation in the same experiment (Figs. 4 and 5).

All these facts allowed us to regard the function of the crayfish inhibitory nerve as a regulating one, changing the functional state of the neuromuscular system in alternative ways depending upon the external and internal environmental conditions.

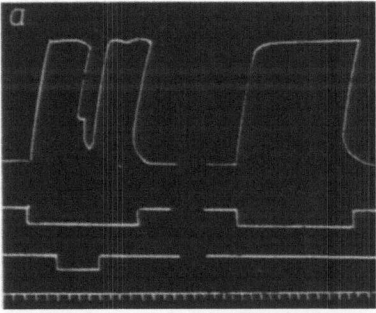 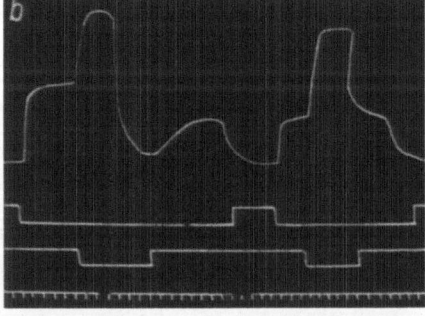

FIG. 4. Inhibiting and facilitating action of inhibitory nerve on the adductor muscle of the crayfish claw. a — inhibiting influence; b — stimulating influence. Reading downward: muscle contractions, the marks of motor-nerve stimulation, the marks of inhibiting nerve stimulation, time marks 1 sec

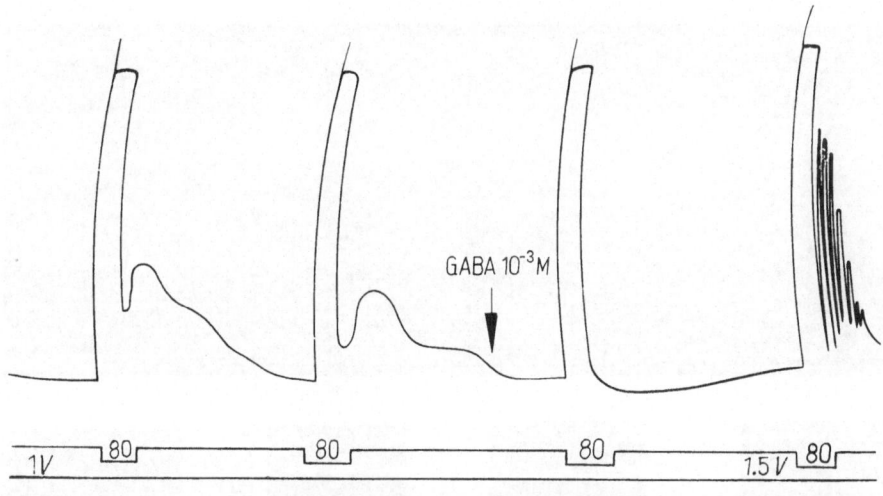

GABA 10^{-3}M

FIG. 5. The action of GABA on the mechanical reaction of the adductor muscle of crayfish
claw 2 days after cutting the inhibitory nerve. Same notation as Fig. 4

Earlier, Wiersma and Ellis (1942) and Hoyle and Wiersma (1958) mentioned
that they could hardly get an inhibition of mechanical reaction from the inhibitory
nerve in the fast muscle fibers.

Atwood, Hoyle and Smith (1965) proved that not each muscle fiber responds
by electrical reaction upon the stimulation of the inhibitory nerve.

Oniany (1964) published data suggesting that the inhibitory nerve inhibits the
contraction and decreases the local electrical response in the 'slow' muscle fibers
of the crayfish but does not decrease the spikes and tetanic contraction of the 'fast'
muscle fibers.

In the very interesting work by Usherwood and Grundfest (1965) the existence
of an inhibitory axon in some insect muscle fibers was established. They found
properties very similar to the crayfish inhibitory nerve in the axon S_2 discovered
earlier by Hoyle (1955) in the muscle extensor tibia of some Orthoptera jumping
legs *(Schistocerca gregaria* and *Romalea microptera)*. However, as Usherwood and
Grundfest showed, the function of an inhibitory axon was not found in the muscle
fibers innervated only by the fast motor axon. In these fast muscle fibers the inhibitory action of gamma-aminobutyric acid (GABA) was not revealed either.

It is known that the synaptic action of GABA is very similar to that of the crayfish inhibitory nerve (Takeuchi and Takeuchi 1966), though the transmitting role
of GABA is not proved (Florey and Chapman 1961). In this regard the data obtained by Kuntzova (1961) in our laboratory is of interest. When the inhibitory
nerve in the claw of *Astacus* was cut off in a chronic experiment, 2–5 days after
cutting, GABA failed to inhibit the contraction of the muscle upon stimulation of
the excitatory nerve (Fig. 5).

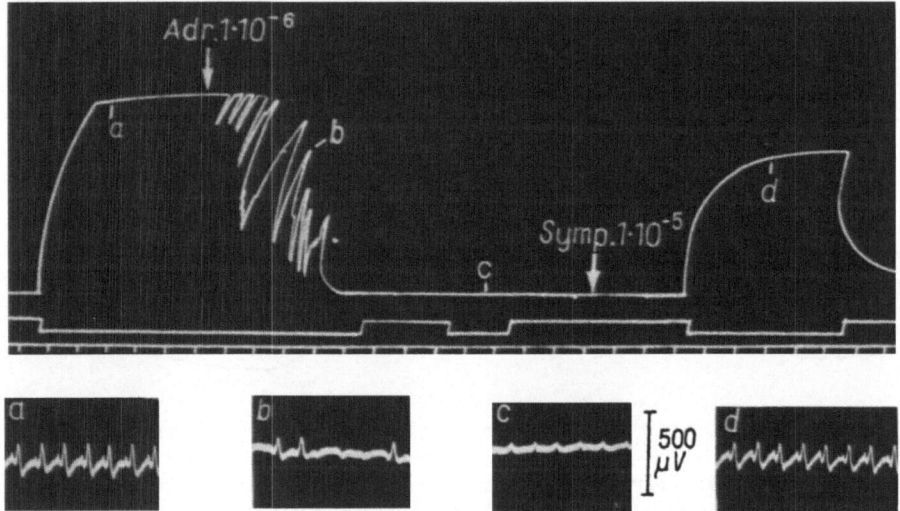

FIG. 6. The inhibitory action of adrenaline (10⁻⁶ g/ml) on the mechanical and electrical reaction of the claw muscle adductor of crab *Pachygrapsus marmoratus* is removed by sympatholytine (10^{-5} g/ml) a — normal; b, c — after injection of adrenaline; d — recovery of normal activity after injection of sympatholytine. Same notation as Fig. 4

Consequently the effect of GABA remains unrevealed in both cases: in the absence of inhibitory innervation (in insects) and after the degeneration of the inhibitory nerve terminals in the crayfish muscle fibers.

The fact that GABA fails to substitute for the inhibitory nerve but only assists it in its action, together with the dual regulating character of the inhibitory nerve action on the neuromuscular system of crayfish, led us to the assumption that nerve fibers of adrenergic nature can be involved in the transmission of inhibitory nerve influences. It was natural to suggest this possibility by analogy with numerous facts uncovered by us on insects and published previously. I will review some of these later.

In many experiments we managed to show that both the inhibiting and facilitating actions of the inhibitory nerve in the muscles of crayfish and crab claws can be imitated by some catecholamines (adrenaline, noradrenaline, euspiran, pedrolon) and the effects of these drugs, as well as the action of the inhibitory nerve, can be removed by different sympatholytics (Br-dibenamine, d. h. ergotamine, d. h. ergotoxin, aminazinum (Voskresenskaya, Kuntsova and Svidersky 1959; Kuntsova 1963). The effective concentrations of these drugs were 10^{-10}–10^{-4} g/ml (Fig. 6).

Figure 6 shows the inhibitory action of adrenaline 10^{-6} g/ml on the mechanical and electrical reaction of adductor muscle of the crab *Pachygrapsus marmoratus;* the effect was eliminated by sympatholytine (Br-dibenamine) 10^{-5} g/ml.

Sympathomimetics tested in these experiments exerted inhibition or facilitation (primarily the former) in varying degrees. For adrenaline the effect depended on the dose of the drug; when the dose was increased up to 10^{-6}–10^{-5} g/ml the inhibitory action became prevalent.

In most the experiments we managed to remove both the inhibiting and facilitating actions of the inhibitory nerve by sympatholytine (Fig. 7).

These facts were repeatedly confirmed on diverse representatives of Crustacea and they led us to the conclusion that the action of the inhibitory nerve is not uniform with regard to the transmitter substance. Together with GABA or some other specific inhibitory transmitter the compounds of the catecholamine group participate in the transmission of inhibitory nerve influences. These sympathomimetic substances can exert dual effects like the adaptational-trophic action of the sympathetic, adrenergic nerve on the somatic muscle of vertebrates (phenomenon of Orbeli–Ginetzinsky) as well as in insects, as our investigations showed.

We were greatly encouraged by the article of McLennan *et al.* published in 1961 in which they reported a strong inhibitory action of some monoamines in the slowly adapted stretch-receptor neurons of some species of crayfish and also suggested the occurrence of the two kinds of receptors sensitive to GABA and to the substances of the catecholamine group.

In this respect, the suggestion of Usherwood and Grundfest is of great interest that in insects, different inhibitory axons or even different branches of one axon exerting various effects can be correlated with the action of different amounts of transmitter or with the involvement in the process of some intermediar mechanism controlling the secretion of transmitters.

The multiform kingdom of invertebrates gives us a great amount of various autonomic peripheral systems, peripheral regulating, inhibiting, and stimulating forms of nervous influences, and shows the progressive centralization of the nervous control function and the differentiation of controlling nervous mechanisms.

The centralization of nervous mechanisms and the progressive subordination of the effectors to the central nervous system are closely connected with the development of the starting function of the nervous system and is typical for the fast locomotor apparatus. However, in the slow neuromuscular systems working with facilitation, there is found the initial regulating function of the nervous system which can be displayed somewhere in the peripheral inhibitory influence. In course of the development of the starting function, the regulating and generally improved functional properties can change into special nervous structures that are younger phylogenetically, as is the sympathetic nervous system in vertebrates.

When investigating the mechanisms of nervous control in some invertebrates we can see different stages of this process, and very often the complicated and multiform character of nervous influences can depend on incomplete differentiation of the nervous structures and functional mechanisms.

From that point of view the inhibiting action of the inhibitory nerve of Crustacea and perhaps that of insects in slow muscle fibers can belong to one kind of neuron, but the dual or mostly facilitating influence in the fast systems of the same animals can belong to another kind of neuron with another chemical.

The sympathetic nervous system of Decapoda Crustacea is not differentiated

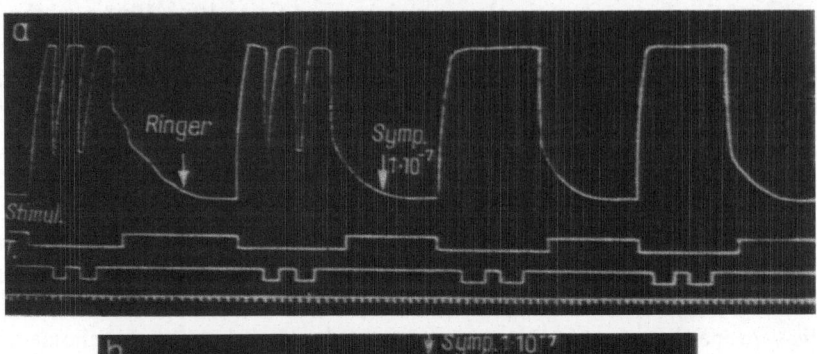

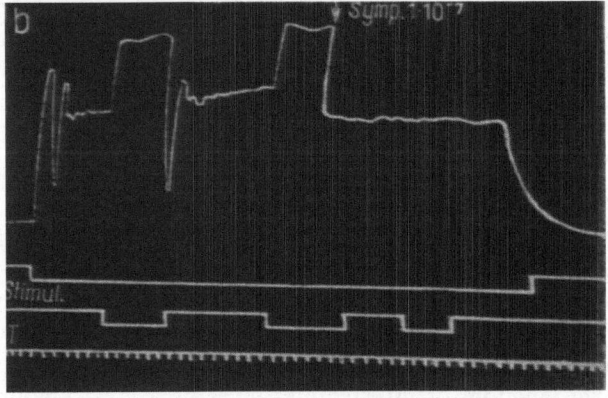

FIG. 7. Sympatholytine 10 7 g/ml removes the inhibitory (a) and facilitating (b) effect of the inhibitory nerve on the mechanical reaction of the crayfish claw adductor muscle

histologically. The histochemical fluorescent reaction in the crayfish nervous system showed the diffuse distribution of monoamines in the neuropile of ganglia (Plotnikova and Govyrin 1966). However, the marked influence of the sympatho-mimetic and sympatholytic substances on the neuromuscular system of these animals leads us to suggest the participation of adrenergic elements in their nervous functions.

The data recently published by Hoyle (1966a, 1966b) on the function of the inhibitory-conditioning axon innervating the insect leg muscles are of exceptional interest for our point of view. Earlier, it was noticed by Hoyle that the stimulation of the third axon S_2 in jumping legs of locust hyperpolarized the membrane of some part of muscle fibers but it did not inhibit the mechanical reaction, and even enhanced it.

In his more recent work, Hoyle managed to show the existence of the axon that did not have a starting but a regulating function, which sometimes inhibits and sometimes enhances both electrical and mechanical responses of the anterior coxal adductor muscle in two species of Orthoptera. This muscle can be regarded as neither 'fast' nor 'slow', because its muscle fibers show the intermediary properties and it has a single motor innervation.

Thus, from the works of Hoyle, Usherwood, and Grundfest we can judge that besides ordinary motor axons, the muscle fibers of insects are supplied with additional innervation which does not give any moving effect, but which changes the reaction of the muscle fiber and the whole muscle upon the excitation of the motor axon, i.e., it exerts the regulating influence. According to Usherwood and Grundfest, only the slow muscle fibers receive this additional innervation which always has an inhibitory influence. From Hoyle's experiments it follows that the influence of the additional axon can be both inhibiting and facilitating and either effect can prevail in different muscle fibers. Hoyle emphasized the dual, regulating action of this axon in its name (inhibitory-conditioning) and he regarded this additional axon as a somatic efferent axon.

The innervating influences in the insect neuromuscular system were also investigated for a long time in our laboratory. It was established that somatic muscles in the locomotor apparatus of legs and wings in locust and in some other insects were under double nervous control. Subordinated to the starting action of the central nervous system, these locomotor apparatus also receive additional regulating adaptable influences in the transmission of which the sympathetic system of the unpaired ventral nerve takes part.

According to investigations of Zawarzin's histological school the main part of the vegetative nervous system in insects is sufficiently differentiated and it forms the isolated nuclei in the nerve cord ganglia. These nuclei form the system of the unpaired ventral or median nerve which can be regarded by analogy with the vertebrate sympathetic nervous system.

At one time Svidersky showed that by cutting or destroying the median nerve conducting paths different changes could be demonstrated in the function of the locust jumping leg muscles. In 60% of his experiments, the muscles were released from some inhibition; after cutting, their functional abilities were improved. The other part of his experiments displayed some facilitating influence of the unpaired (median) nerve (Voskresenskaya 1950, 1959).

Consequently, the dual regulating influence of the sympathetic nerve was found in the muscles consisting of both slow and fast muscle fibers and innervated by both slow and fast motor axons.

In the numerous investigations performed in our laboratory we have studied the function of the sympathetic unpaired nerve in the fast muscles involved in frequent rhythmic movements and in the wing muscles of locust and of some other insects.

The rhythmic after-reaction, typical for the flight neuromuscular system, was a very suitable model to judge the functional ability of our preparations. After the cessation of the afferent nerve stimulation, both mechanical and electrical responses of wing muscles followed and lasted for tens of seconds (Fig. 8). This peculiar feature of the wing neuromuscular system could not help but attract our attention. It was found that this after-reaction is of central origin (Voskresenskaya and Svidersky 1960).

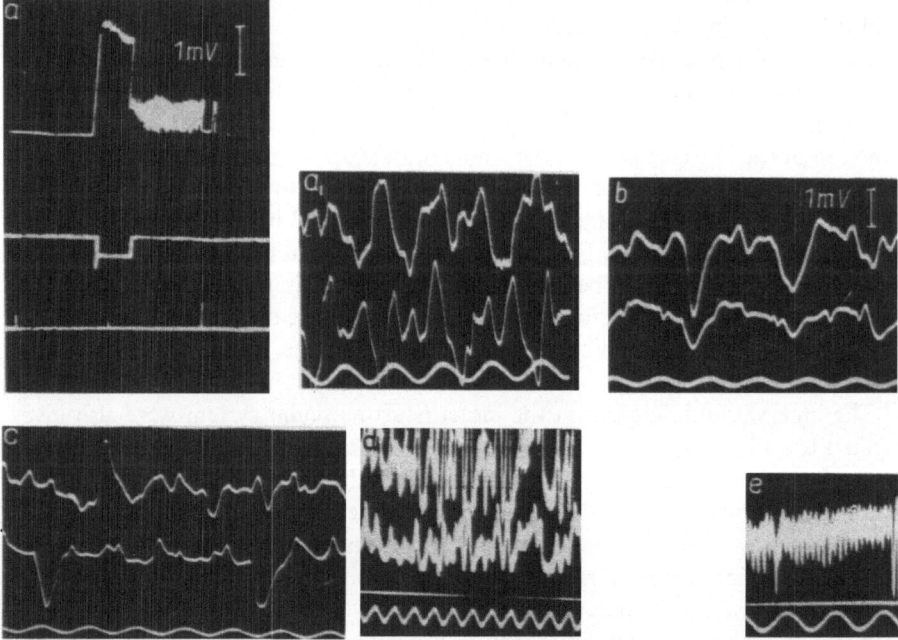

FIG. 8. After-discharges of insect wing muscles on 3–5 sec after the cessation of stimulation. (a — myogramm; a_1—e — electrograms.) a—a_1 — locust *(Locusta migrataria L.)*; b — dragonfly *(Aeschna carulea)*; c — bumblebee *(Bombus lucorum)*; d — bee *(Apis mellifera)* e — fly *(Calliphora erythrocephala)*

While examining the central mechanisms controlling the insect wing apparatus, Svidersky (1965, 1967) found that besides spontaneous discharging neurons in the ganglion there are some other neurons which start to discharge in response to the afferent stimulation, but that the rhythm of these discharges is regular, irrespective of the frequency of stimulation and corresponding to the rhythm of the wing muscle activity. These neurons seemed to be the starting mechanism of the wing system. They possess some inertness, as they continue discharging for several seconds after the cessation of the afferent stimulation. The after-discharges of these neurons are probably the mechanism of the rhythmic after-activity of the wing muscles.

However, the wing muscles reproduced these after-discharges only if the sympathetic innervation of this segment remained intact. When the unpaired nervous system was destroyed or some sympatholytic substance was injected into the muscle, the after-activity of the wing muscles ceased despite the fact that after-discharges continued to arrive from the neurons of the ganglion (Fig. 9). We considered this phenomenon as evidence of the adaptive influence of the sympathetic unpaired nerve which improves the wing muscle's functional ability.

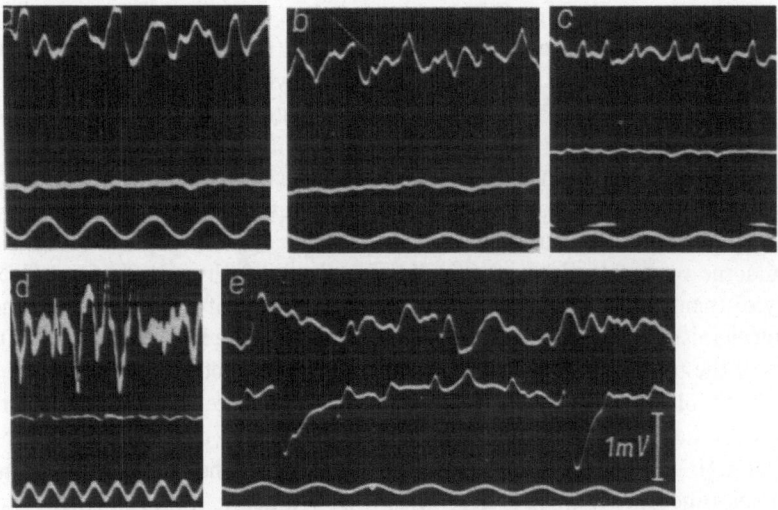

FIG. 9. Switching-off rhythmic after-reaction of wing muscles after cutting off locust unpaired nerve (a), after the injection of sympatholytine $(10^{-10}\,g/ml)$ into the muscle in (b) dragonfly, (c) bumblebee, and (d) bee. Restoration of after-potentials in the bumblebee muscle (e) after washing sympatholytine by some physiologic solution. All the observations were made during the 3rd second after the cessation of the irritation of the ganglion

In the fast neuromuscular systems of the wing we always observed the facilitating action of the sympathetic nerve, but not the inhibitory one.

For the present we failed to show histologically the sympathetic innervation of the insect muscle fibers, but Plotnikova (1968) recently found that some neurons, belonging to the nucleus of the unpaired nerve in Locust metathoracic ganglion, send their axons along with the somatic nerves. A similar fact had been established in the dragonfly larva by Tzvileneva (1950a, 1950b).

Further investigations must show the true histological and chemical nature of the regulating, conditioning influence of the nervous system in different somatic neuromuscular apparatus. They must also show in what cases these regulating actions are exerted by somatic motoneurons and in what cases they belong to a special nervous system with adaptable function.

SUMMARY

1. Over a long period of time in Russian physiological school the idea has developed that there are two kinds of central nervous system influences on the effectors – the functional, or starting, influence, and the regulatory, or trophic influence, which regulates the level of activity (Pavlov 1922).

Due to his evolutionary point of view in understanding nervous system functions, Orbeli ventured to express his conception on the historical connections between these two kinds of nervous control. He suggested that the regulating function of the nervous system is more ancient phylogenetically, but preserves its universal significance for all the organs and systems of living organisms. He considered the sympathetic nervous system to be one of the main pathways transmitting the regulating, adaptational-trophic influence in vertebrates.

The invertebrates are a good material for the investigation of the nervous system regulating (conditioning) function and its phylogenetic development.

2. In some representatives of Annelids (earthworm) the spontaneous electrical activity of their body muscles depends on the existence of some peripheral neurogenic mechanism. The central nervous system controls the level of this spontaneous activity of the muscles by changing its quantitative characteristics.

3. The inhibiting nerve of crayfish is a more complicated form of peripheral integrative mechanism. It is known that this nerve can show the effect of hyperpolarization as well as that of depolarization, depending on the initial level of muscle-fiber membrane potential. It is also known now that the inhibitory action of the inhibiting nerve is not revealed in 'fast' muscle fibers of crayfish. The data of our experiments on the effect of the crayfish inhibiting nerve on the mechanical and electrical reaction of a muscle allowed us to consider this nerve as the nerve with regulating (conditioning) function, and with adrenergic nervous elements participating in the transmission of this action.

4. In insects, the most specialized and 'fast' wing muscles and the muscles of some other rhythmic movements possess an original controlling nervous mechanism (starting the rhythmic movements). At the same time, our experiments proved the essential role of the sympathetic (adrenergic) nervous system, which in this case showed the stimulating action which improved the functional properties of muscles.

Some histological and histochemical investigations of the insect sympathetic nervous system were performed.

The existence of nerves with regulating (conditioning) function has recently been found in some other insect neuromuscular systems (Hoyle 1966).

5. The information now available allows us to draw some conclusions about the nature and evolutionary formation of the regulating function of invertebrate nervous systems.

REFERENCES

ATWOOD, H. L., HOYLE, G. and SMITH, T. J. (1965): Mechanical and electrical responses of single innervated crab muscle fibres. *J. Physiol.* **180** 449.

BERITOV J. S. (1950): On the origin of spontaneous activity and transmission in the nerve cord ganglion of the leech *(Hirudo medicinalis)*. *J. gen. Biol.* **11** 31 (in Russian).

BULLOCK, T. H. and HORRIDGE, G. A. (1965): *Structure and Function in the Nervous Systems of Invertebrates.* San Francisco–London, Vol 1, 755.

DAVID, O. F. (1968): The spontaneous electrical activity of longitudinal muscles of earthworm. In: Neurophysiology of Invertebrates. Suppl. to *J. evol. Biochem. Physiol.* (in press).

DE BELL, J., DEL CASTILLO, J. and SANCHEZ, V. (1963): Electrophysiology of the somatic muscle cells of *Ascaris lumbricoides. J. cell. Comp. Physiol.* **62** 159.

DEL CASTILLO, J., DE MELLO, W. C. and MORALES, T. (1967): The initiation of action potentials in the somatic musculature of *Ascaris lubricoides*. *J. exp. Biol.* **46** 2 263.

FATT, P. and KATZ, B. (1952): The action of inhibitory nerve impulses on the surface membrane of crustacean muscle fibers. *J. Physiol.* **118** 47.

FLOREY, E. and CHAPMAN, D. D. (1961): The non-identity of the transmitter substance of Crustacean inhibitory nerves and gamma-aminobutiric acid. *Comp. Biochem. Physiol.* **3** 92.

HOYLE, G. (1955): Neuromuscular mechanisms of Locust skeletal muscle. *Proc. roy. Soc B.* **143** 343.

HOYLE, G. (1966a): An isolated insect ganglion-nerve-muscle preparation. *J. exp. Biol.* **44** 413.

HOYLE, G. (1966b): Functioning of the inhibitory-conditioning axon innervating insect muscles. *J. exp. Biol.* **44** 429.

HOYLE, G. and WIERSMA, C. A. G. (1958): Inhibition at neuromuscular junctions in Crustacea. *J. Physiol.* **143** 426.

KOSHTOYANTS CH. S., SMIRNOVA, I. A. and ORLOVSKAYA, J. V. (1962): Theses of the report on the 3rd *Scientific Conference on the Evolutionary Physiology*. Leningrad 107.

KUNTSOVA, M. J. (1961): The action of gamma-aminobutiric acid on mechanical reaction of normal and denervated adductor muscle of the crayfish claw. *Bull. exp. Biol. Med.* **12** 8. (in Russian).

KUNTSOVA, M. J. (1962): On the regulating function of the crayfish inhibitory nerve. *Physiol. J.* **48** 7, 833 *USSR* (in Russian).

KUNTSOVA, M. J. (1963): The action of sympathomimetic amines on the crayfish and crabs neuromuscular systems. *Physiol. J. USSR* **49** 3, 370 (in Russian).

MCGEER, MCGEER, P. L. and MCLENNAN, H. (1961): The inhibitory action of 3-hydroxytyramine, GABA and some other compounds towards the crayfish stretch-receptor neurons. *J. Neurochem.* **8** 36.

ORBELI, L. A. (1923): *The sympathetic innervation of skeletal muscles*. In: *Selected Works*, II, 53, Moscow–Leningrad, 1962 (in Russian).

ONIANY, T. H. (1964): *The Problems of Comparative Physiology of the Neuromuscular System*. Tbilisi (in Russian).

OVETCHKIN, V. G. (1966): On the character of inhibitory nerve influence on the claw adductor-muscle of *Astacus*. *Bull. exp. Biol. Med.* **61** 10 (in Russian).

PAVLOV, I. P. (1922): *On the trophic innervation*. In: *Complete Works* **1** 402, Moscow–Leningrad 1940 (in Russian).

PLOTNIKOVA, S. I. (1968): The structure of insect sympathetic nervous system. In: *Modern Problems of the Insect Nervous System Structure and Function*. The works of Entomological Soc. 52 (in press) (in Russian).

PLOTNIKOVA, S. I. and GOVYRIN, V. A. (1966): Distribution of catecholamine-containing nerve elements in some representatives of Protostomia and Coelenterata. *Arch. anat. histol. embryol.* **50** 79 (in Russian).

PROSSER, C. L. (1935): Impulses in the segmental nerves of the earthworm. *J. exp. Biol.* **12** 95

SVIDERSKY, V. L. (1961): On the peripheral inhibition in insects. *Dokl. Akad. Nauk SSSR* **141** 1260 (in Russian).

SVIDERSKY, V. L. (1965): The activity of single neurons of thoracic ganglion of locust (*Locusta migrataria* L.). *Dokl. Akad. Nauk USSR* **164** 5, 1204 (in Russian).

SVIDERSKY, V. L. (1967): Central mechanisms controlling the activity of locust flight muscles. *J. Insect. Physiol.* **13** 899.

TAKEUCHI, A. and TAKEUCHI, N. (1966): A study of the inhibitory action of gamma-aminobutiric acid on neuro-muscular transmission in the crayfish. *J. Physiol.* **183** 418.

TZVILENEVA, V. A. (1950a): Nerve cord of *Aeschna* (using the materials of akad. A. A. Zawarzin). I. The structure of thoracic ganglions. *Proc. of Acad. Sci. of USSR*, B. 2, 91 (in Russian).

TZVILENEVA, V. A. (1950b): Again on the structure of sensitive neuropile in the nerve cord of *Aeschna*. *Dokl. Akad. Nauk USSR* **72** 161 (in Russian).

USHERWOOD, P. N. R. and GRUNDFEST, H. (1965): Peripheral inhibition in skeletal muscles of insects. *J. Neurophysiol.* **28** 497.

VOSKRESENSKAYA, A. K. (1950): On the sympathetic innervation of skeletal muscles of insects. *Physiol. J. USSR* **36** 176 (in Russian).

VOSKRESENSKAYA, A. K. (1959a): *The Functional Properties of the Insect Neuromuscular System*. Moscow–Leningrad (in Russian).

VOSKRESENSKAYA, A. K., KUNTZOVA, M. J. and SVIDERVKY, V. L. (1959b): Relationships between innervating systems in the nerve-muscle system of Crustacea. *Phyisol J. USSR* **45** 7, 831 (in Russian).

VOSKRESENSKAYA, A. K. and SVIDERSKY, V. L. (1960): Analysis of the nature of rhytmical after-reactions in the neuromuscular apparatus of the insect wing. *Physiol J. USSR* **46** 1050 (in Russian).

WIERSMA, C. A. G. (1957): Neuromuscular mechanisms. In: *Recent Advences in Invertebrate Physiology*. 143.

WIERSMA, C. A. G. and ELLIS, S. H. (1942): A comparative study of peripheral inhibition in decapod Crustaceans. *J. exp. Biol.* **18** 223.

ZAWARZIN, A. A. (1941): The essay on the evolutionary histology of the nervous system. *Selected Works III*, Moscow–Leningrad, 1950 (in Russian).

Symposium on Neurobiology of Invertebrates 1964 (381—390)

EXCITATION OF THE RECEPTOR CELLS OF THE CRUSTACEAN PD ORGAN

E. G. Boettiger and H. B. Hartman

University of Connecticut Storrs
Connecticut, U.S.A.

Movement of the appendages of Crustacea involves the cooperative activity of muscles in several joints. Usually, at each joint, movement is restricted to one plane and involves only two muscles. These muscles are innervated by a small number of motor excitatory and inhibitory cells. Such a simplified movement system is ideal material for the comparative physiologist.

The central control of muscles in the various segments during the complex behavioral movements of Crustacea requires information about the position, movement, and velocity of all segments. This information is supplied by a series of innervated strands and plates whose lenghts are determined by the relative positions of the segments united at a joint. The basic features of all these receptor organs are very similar, although differences in both organ structure and in the fine-structure of the nerve endings are present.

The best known of these receptor organs, the PD organ, bridges the propodite-dactylopodite joint. It can easily be located in the propodite (Burke 1954). The anterior and posterior surfaces of this segment are flattened and the dactyl is constrained to move in the dorso-ventral plane from the dorsal open position to the ventral closed position. The closer muscle is located ventrally in the propus. The PD organ originates on the closer tendon and runs obliquely across the joint to insert on the inner dorsal surface of the dactyl. The organ is revealed by removing the anterior surface of the propus. It is also flattened anterior-posteriorly especially near its origin on the closer tendon. The strand is elastic and contains connective tissue cells, collagen fibers, and a peripherally located amorphous material (Whitear 1962). The organ elongates about 1 mm when the joint closes (flexion) from the fully extended open position.

Axons leave the main leg nerve and proceed ventrally to the tendon end of the strand. The large cell bodies of bipolar cells are loosely attached to the strand and their dendrites insert either into the dorsal edge or the anterior surface (Hartman and Boettiger 1965, 1967). A group of small axons form a branch that innervates the distal region with a line of tightly adhering small bipolar cells. The axons of the receptor cells run centrally in distinct bundles that may easily be separated from the main nerve in the meropodite (Wiersma 1959).

The fine structure of the dendrite endings has been described by Whitear (1959). Distally the dendrites come together in pairs, so closely in one region, termed the ephapse by Whitear, that the separation is only about 200 Å. This possible low-resistance pathway between paired dendrites is of unknown significance. Beyond the ephapse, the paired dendrites pass into a space presumably filled with extracellular fluid and surrounded by a unique enveloping cell, the scolopale cell. Within this cell is a solid fibrous component of characteristic structure, the scolopale. This substance may guard the dendrite endings from external forces or contribute to the directional sensitivity of the cell by allowing these forces to act in certain planes only. The dendrite endings, where enveloped by the scolopale cell, have a ciliary-like structure. In most receptor strands one of the pair, the ciliary ending, is less modified than other, the paraciliary ending.

Only in the coxal-basopodite (CB) organ are the paired endings of the same ciliary structure. The degree of degeneration of the paraciliary ending is characteristic of the particular receptor strand. In the PD organ the paraciliary ending is well developed. The paired dendrites terminate together in a tube structure surrounded by part of the scolopale cell. Internal mechano-receptor elements of this type are called chordotonal organs and are common in arthropods.

The relationship of the paired endings and the enveloping scolopale cell to the connective tissue of the strand is not known. Whitear (1962) describes some structures that can be considered mechanical connections between collagen fibers and the scolopale cell surface. These may be important in determining how the cell is stimulated (Bush 1965b).

Interest in these receptor organs centers around three major problems: (1) what information they relay to the CNS; (2) how the information is used to control leg movements; and (3) how changes in strand length generate this information. In order to solve these problems it is necessary to analyze in detail the behavior and location of each receptor cell in the strand.

INFORMATION OBTAINED BY THE PD ORGAN

Wiersma and Boettiger (1959), recording the activity of single fibers isolated from bundles in the meropodite, found both tonic position cells and phasic movement cells. All cells, position and movement, were unidirectional, responding either to elongation of the strand, elongation-sensitive cells (ESC), or to relaxation of the strand, relaxation-sensitive cells (RSC). Thus basic types of receptor information channels are found in the PD organ; pure static or tonic position cells that are either ESC (called ESPC) or RSC (called RSPC) and movement cells that are either ESMC or RSMC. Pure movement cells and movement cells influenced by velocity and/or length of the strand (dynamic position) may also be differentiated. Similar cell types have been found in the other innervated strands of the leg (Bush 1965b; Wiersma 1959), although their relative numbers differ in the different organs.

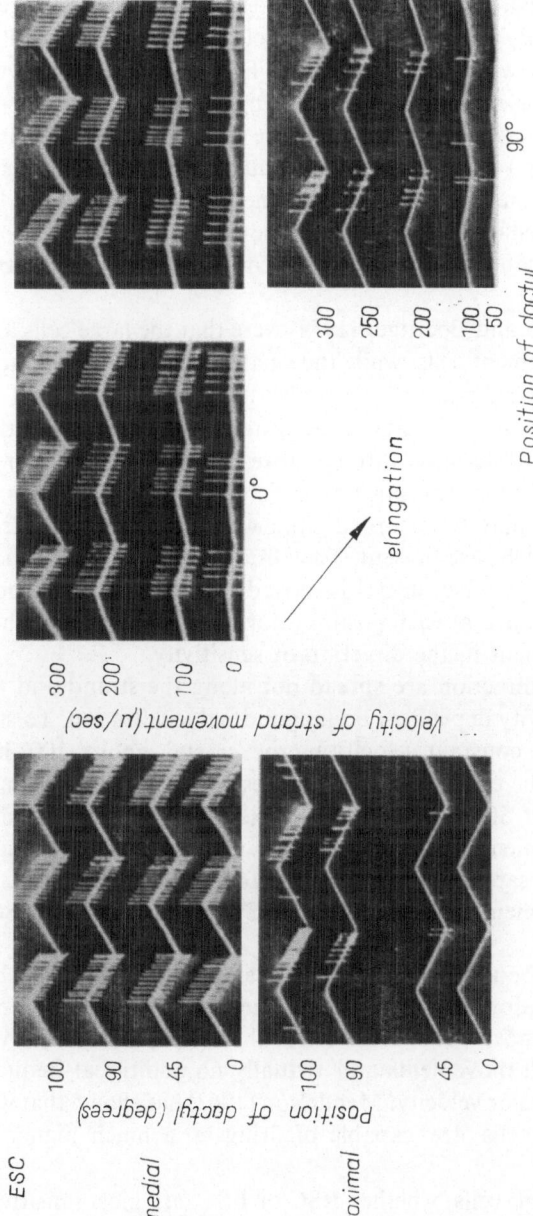

Fig. 1. Oscilloscope records of single movement elongation sensitive cells, ESC, during elongation (downward movement) and relaxation (upward movement of the beam). When the dactyl is fully extended, the joint position angle is 0°. Cells located medially in the group of receptors respond in the same way to small elongations regardless of dactyl position, although they are somewhat velocity-sensitive in the extreme positions. Proximal cells are quite velocity-sensitive and respond only to movement while the dactyl is flexed. Distal ESC are rare

Hartman and Boettiger (1965, 1967 and unpublished) prepared the PD organ of the rock crab, *Cancer irroratus*, *in situ* free of all muscle tissue and dissected so it could be viewed at high power with transmitted light. The cut end of an individual axon or axon plus cell body was sucked into a glass Ringer-filled microelectrode of such a size so as to make a close fit without injury to the cell. Axons could be drawn into the electrode and expelled repeatedly without change in response characteristics. The strand was cut free of the dactyl and held by fine forceps attached to a large crystal of Rochelle salts. Crystal movements in response to dc voltages from a function generator produced controlled constant velocity movements of the strand of up to 200 μ.

The conclusions of Wiersma and Boettiger (1959) were that the large cells at the tendon end are dynamic movement cells, while the small distal cells respond tonically to position. Movement RSC and ESC were also found to have distinct relations to the strand. Large bipolar cells whose dendrites enter the dorsal surface respond to relaxation (RSC), while those with dendrites entering the anterior surface respond to elongation (ESC). The position cells are small and their axons make a compact bundle that may be recorded from with the suction electrode. Both RSC and ESC position cells are present. They fire with increasing frequency over the whole movement range when dactyl is moved in a direction to produce excitation. The frequency attained at each position during the return is much less than that of the initial movement in the direction of sensitivity.

Movement cells for either direction are spread out along the strand and have characteristic response sensitivity dependent on longitudinal localization. To study this, the strand was moved at constant velocity by the crystal about 100–300 μ centered on three basic lengths equivalent to open, rest, and closed positions of the dactyl. Normal movement of strand is 1000 μ. The test of dynamic position sensitivity was made on movement cells in all locations. Choosing speeds of length change of 25 μ–300 μ/sec, the same cells were studied for velocity sensitivity. The typical results of these experiments are shown for ESC in Fig. 1 and for RSC in Fig. 2.

The type of information about mechanical events at the joint collected by a single cell depends on the location at which the dendrite enters the strand. Medial cells, especially ESC, are almost pure unidirectional movement cells showing frequency saturation for small movements and virtually no additional frequency change with changes in position or velocity. Mendelson (1963) has shown that when treated with nicotine, these cells are capable of firing at a much higher frequency.

Proximally located movement cells, whether RSC or ESC, are most sensitive to movement when the dactyl is closed (strand is stretched). Proximal RSC are somewhat position-sensitive, with a higher frequency to small movements in the closed position than in the open position. They are also velocity-sensitive. Proximal ESC are of two types – one type similar to the RSC, being position – and velocity-sensitive, responding over the whole movement range, and a second type responding primarily at the end of the flexion with a sharp increase in frequency.

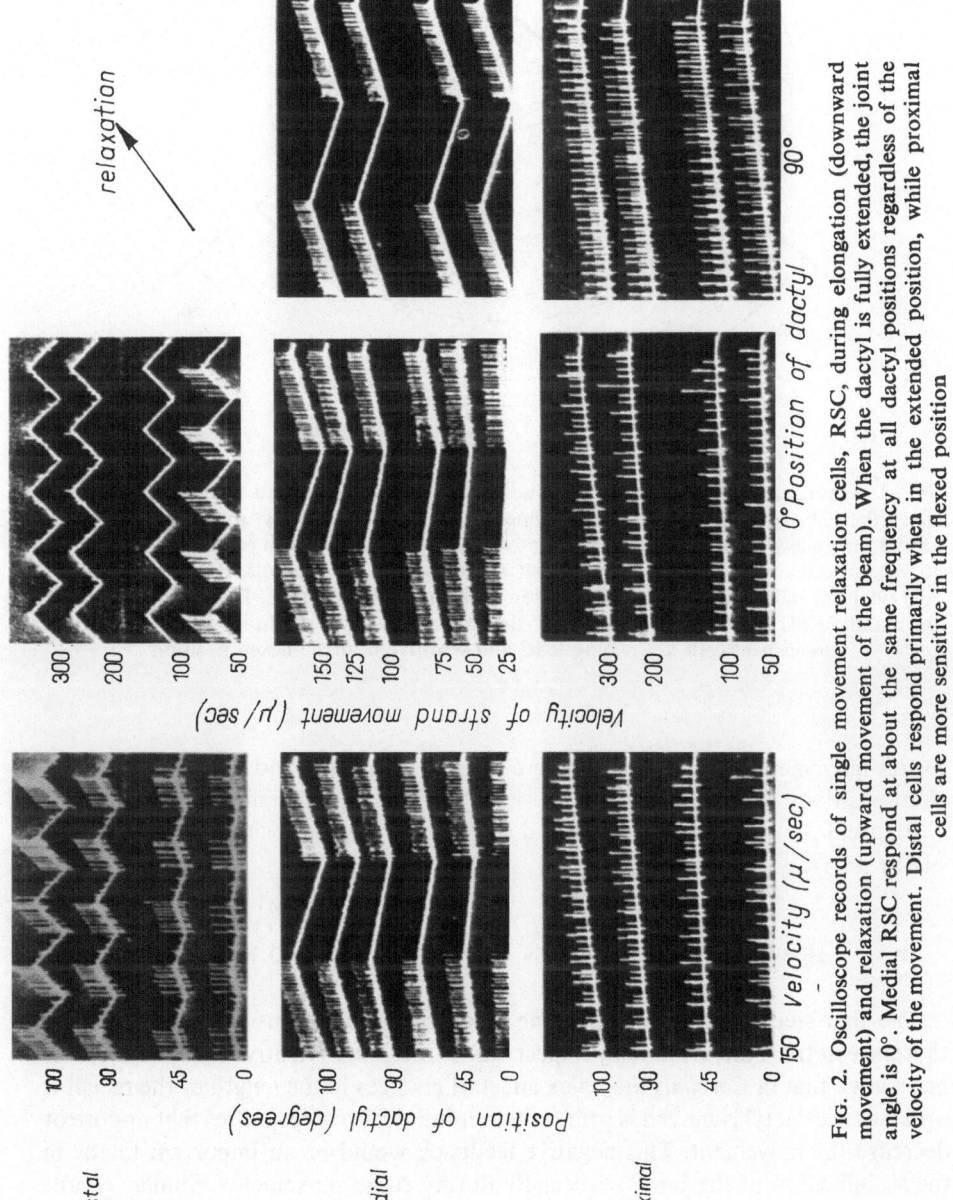

FIG. 2. Oscilloscope records of single movement relaxation cells, RSC, during elongation (downward movement) and relaxation (upward movement of the beam). When the dactyl is fully extended, the joint angle is 0°. Medial RSC respond at about the same frequency at all dactyl positions regardless of the velocity of the movement. Distal cells respond primarily when in the extended position, while proximal cells are more sensitive in the flexed position

E. G. BOETTIGER and H. B. HARTMAN

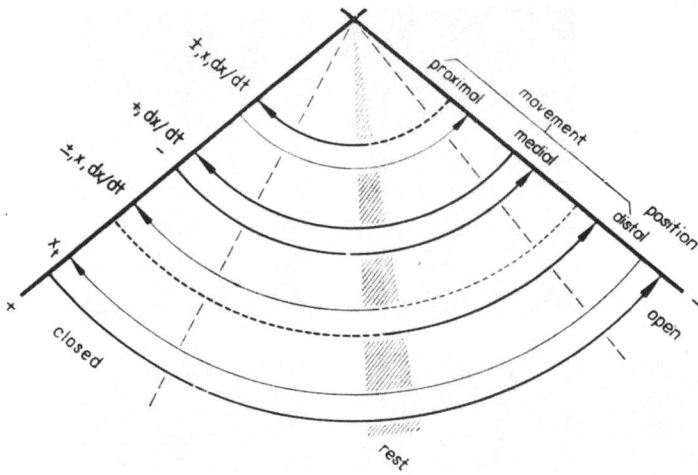

FIG. 3. Summary of response characteristics of receptor cells located in different positions along the PD organ. The widest lines define the boundaries of dactyl movement. The solid curved lines show the direction, range of movement, and, by the width of the line, the relative number of cells, with response characteristics indicated by the symbols. Position cells signal only position and are tonic X_t. RSC position cells are more numerous. Distal movement cells are mainly RSC and most sensitive near the extended (open) position while the proximal movement cells are mainly RSC and sensitive near the flexed position

Distal movement relaxation sensitive cells also have wide and narrow band position sensitivity and some velocity sensitivity. Distal ESC are rare.

Figure 3 summarizes the results and relates them to the position and movement of the dactyl.

HOW THE INFORMATION FROM THE STRAND IS USED

The first step in the analysis of the mechanical conditions in the strand, and therefore at the joint, is made by a spectrum of mechano-sensitive cells. Bush (1962) has shown that in *Carcinus* the reflex effect of changes in the length of the receptor organ as the dactyl is moved is primarily to initiate muscle responses that oppose or decrease the movement. This negative feedback would be an important factor in the stabilization of the joints, especially during rapid movements. Similar results were obtained in *Procambarus* by Muramoto (1965). Pure movement receptor feedback would be unbiased as to position or velocity and so maintain a steady damping on movement by keeping the antagonist muscle under the same tension at all positions and velocities. As more extreme positions are reached, some cells not previously active develop a sharp increase in frequency. For the opened positions, the narrow band distal RSC, and for the closed positions, the narrow band proximal ESC, would be most effective. The movement will be opposed by more receptor activity as the velocity sensitive cells are recruited.

Why is there a correlation between location on the strand and response characteristics? To be effective in coordinating movement, the various features of the mechanical conditions must be sorted out into different channels. Much of this analysis occurs in the strand, cells with similar characteristics being grouped together. This orderly arrangement of cells may be a means of obtaining orderly central connections.

SORTING OF INFORMATION INTO SEPARATE CHANNELS

The third problem concerns the means by which dendritic endings are made to respond to one or more of the mechanical components, direction, velocity, and position. Two major theories have been proposed to explain this: (1) That the response characteristics of the cells are basically the same and the analysis is due to the special transformer action of the strand causing stimulus for one cell to occur on strand relaxation and for another on strand elongation so as to give directional sensitivity and to confer on cells various degrees of movement, position, and velocity sensitivity; (2) that RSC cells are structurally, or at least physiologically, different from ESC, responding oppositely to the same stimulus with respect to direction.

The discovery that two different paired endings are enclosed by the same scolopale cell gave support to the idea that there are at least two distinct cell types. Whitear (1962) suggested that the ciliary type might respond to relaxation and the paraciliary to elongation. However, in the CB organ, where identical endings (ciliary) are paired, unidirectional responses are found. Hartman and Boettiger (1967) showed that in the PD organ, paired endings always respond to the same direction of movement, one cell being more sensitive than the other. These results do not disprove the idea of two or more cell types, for the significant physiological differences may not have morphological counterparts.

The separation of the position cells responding to relaxation from those responding to elongation is not obvious, in contrast to the case of movement cells. The RSC and ESC position cells are in a compact linear arrangement. The axons are too small for isolation, and so it has not been possible to determine functional localization. If the RSC dendrites and ESC dendrites have different mechanical arrangement within the strand they cannot be resolved with the light microscope. The position cells insert into the same material as the ESMC on the anterior surface of the strand. This material may compose the amorphous connective tissue described by Whitear (1962). It can easily be stripped away from the strand, bringing with it the ESMC and the position cells (Hartman and Boettiger 1967). Likewise, axons of the ESMC and position cells are found in the same bundle (Wiersma and Boettiger 1959).

The results of Hartman and Boettiger (1967; and unpublished) favor the idea that the receptor cells are the same but are subject to different transformer action by their location on the strand: for example, the specific planes of insertion of the RSMC and ESMC and the fact that paired cells respond to the same direction. Experiments have been carried out in which the strand was altered to see if the resulting changes in strand mechanics would alter the response characteristics of the cells. Mendelson (1963) cut the fanning connective tissue fibers at the origin of the strand that Wiersma and Boettiger (1959) thought might control the direction of forces on the strand. No effect was observed on the response, nor did they obtain dramatic changes when the strand was twisted $2^1/_2$ times and then stretched and relaxed. Hartman and Boettiger (unpublished results) made longitudinal slits to separate the RSMC and ESMC cells without affecting the responses. While recording from RSC, the origin of the strand at the tendon was severed in stages until only one bundle of strand fibers remained. Responses were normal. When ESC were studied in the same way, it was found that drastic reduction of the strand attachment increased somewhat the mechanical threshold for the ESC. Finally, elongation of the excised strand from the proximal or distal ends produced the same excitatory action on both RSC and ESC.

The insensitivity of the PD organ to all these manipulations suggests that the differential action is local in origin and that the fibers of the strand are closely bound together. There would then be little shearing action between the fibers.

Any theory describing how the cells on the strand respond differently to the same strand movement must take into account the complicated structure of the sensory and associated scolopale cells. The initial idea that the cells respond to stretch (Whitear 1962; Wiersma and Boettiger 1959) has been replaced by various explanations requiring bending of the excitatory structures (Bush 1965b; Mendelson 1963). In the leg receptors discussed above, the transformer action of the elastic strand sorts out the mechanical components by differential action on the cells. The flagella chordotonal organ, FCO, of the hermit crab studied by Taylor (1967a, 1967b) is composed of paired bipolar cells with an enveloping scopale, as in the case of the PD organ. However, the cells themselves, without any accompanying series of elastic elements, bridge the joint at the base of the flagella. Consequently, very little transformer action is possible. The adequate stimulus for the cells, demonstrated by direct observation, is a bending of the terminal structure. Symmetrical elongation of the endings (stretch) inhibits all discharge. Bending of the paired dendrites would produce a differential effect on opposite sides and some compression. When the complex scolopale structures are considered, the effect of bending on the enclosed dendritic endings might be complicated.

The response characteristics of the FCO cells include unidirectional cells for both movement and position. However, all movement cells are dynamic position- and velocity-sensitive. Only in one position range do they resemble pure movement cells. Movement cells also respond to static position, a condition not normally found in the PD organ. Thus, in the absence of the elastic strand, the peripheral analysis of the mechanical features is not as complete.

Taylor (1967a) has suggested that the unidirectional behavior of the strand receptors, such as the PD organ, can also be explained by the bending theory.

RSMC terminals are located in the elastic portion of the strand, allowing bending on relaxation to excite the cells. ESMC would traverse a nonelastic material to insert into the elastic portion of the strand. Differential length changes on elongation bend the dendrite. A similar conclusion has been reached by Thurm (1965) working with an insect mechanoreceptor.

The observations of Hartman and Boettiger (1967 and unpublished) seem to support this idea. While the RSMC are inserted at the dorsal surface into the collagen-rich area, the ESMC insert at the anterior surface into a very different material. It can easily be stripped away from the rest of the strand, carrying with it the ESC and the distal position cells without injury to the RSMC left on the strand.

Only the first of the problems outlined has been adequately studied. The response characteristics of all the movement cells have been determined. With this information as a basis, it should be possible to attack the other problems with some hope of success.

SUMMARY

The responsiveness of the receptor cells to direction, velocity, and range of movement as well as to position of the dactyl is related to the location of the cells on the PD organ. Most distal cells are small, and are attached to the strand in the anterior ventral region. They are unidirectional, tonic position cells, some responding to elongation and a greater number to relaxation of the strand. Large cells near the tendon end of the strand with dendrites inserted into the dorsal edge of the strand are relaxation sensitive cells. Similar cells entering the strand at the anterior surface are elongation sensitive cells. These large cells respond only during movement in one direction and to varying degrees to the range of the movement (dynamic position) and movement velocity. The medial large cells are pure movement cells with firing frequency independent of range or velocity. More of the proximal cells are elongation sensitive, affected both by range and velocity, while more of the distal cells are relaxation sensitive, also range- and velocity-dependent.

The PD organ analyzes the mechanical events at the joint and transmits various aspects of it over separate channels to the central nervous system.

REFERENCES

BURKE, W. (1954): An organ for proprioception in the legs of *Carcinus maenas* (L.). *J. exp. Biol.* **39** 89—105.

BUSH, B. M. H. (1962): Proprioceptive reflexes in the legs of *Carcinus maenas*. *J. exp. Biol.* **39** 89—105.

BUSH, B. M. H. (1965a): Proprioception by the coxo-basal chordotonal organ, CB, in legs of the crab, *Carcinus maenas*. *J. exp. Biol.* **42** 285—297.

BUSH, B. M. H. (1965b): Proprioception by chordontonal organs in the merocarpopodite and carpo-propodite joints of *Carcinus maenas* legs. *Comp. Biochem. Physiol.* **14** 185—199.

HARTMAN, H. B. and BOETTIGER, E. G. (1965): Localization of receptor types in the propodite-dactylopodite organ of the crab *Cancer irroratus*. *Say. Am. Zool.* **5** 651.

HARTMAN, H. B. and BOETTIGER, E. G. (1967): The functional organization of the propus-dactylus organ in *Cancer irroratus*. *Say. Comp. Physiol.* **22** 651– 663.

MENDELSON, M. (1963): Some factors in the activation of crab movement receptors. *J. exp. Biol.* **40** 157–169.

MURAMOTO, A. (1965): Proceptive reflex of the PD organ of *Procambarus clarkii* by passive movement and vibration stimulus. *Jour. Fac. Sci. Hokkaido U.* VI **15** 533–534.

TAYLOR, R. C. (1967a): The anatomy and adequate stimulation of a chordotonal organ in the antennae of a hermit crab. *Comp. Biochem. Physiol.* **20** 709–717.

TAYLOR, R. C. (1967b): Functional properties of the chordotonal in the antennal flagellum of the hermit crab. *Comp. Biochem. Physiol.* **20** 719–728.

THURM, U. (1965): An insect mechanoreceptor. Part 1. Fine structure and adequate stimulus. *Cold Spring Harbor Symp.* **30** 75–82.

WHITEAR, M. (1962): The fine structure of crustacean proprioceptors I. The chordotonal organs in the legs of the shore crab, *Carcinus maenas*. *Phil. Trans. Roy Soc. B.* **245** 291–295.

WIERSMA, C. A. G. (1959): Movement receptors in decapod Crustacea. *J. mar. Biol. Ass. U. K.* **38** 143–152.

WIERSMA, C. A. G. and BOETTIGER, E. G. (1959): Unidirectional movement fibers from a proprioceptive organ of the crab, *Carcinus maenas*. *J. exp. Biol.* **36** 102–112.

DISCUSSION

A. O. D. Willows: Have you tried to determine the functional role of these organs by removing or inactivating them in the whole animal?

E. G. Boettiger: We have not carried out such experiments, but now that we know the response characteristics we will try to destroy individual cells and note changes in reflex responses and behavior.

G. M. Hughes: To what extent are the differences in directional sensitivity of the different receptor cells dependent upon the integrity of the whole organ? Do they always maintain the same sensitivity after removing adjacent parts of the organ?

E. G. Boettiger: We have cut all but a few of the connective tissue strands connecting the PD organ to the tendon. Longitudinal cuts separating the ESC cells from the RSC cells were also made. These procedures did not alter directional responses except in some cases to decrease sensitivity to the same length change.

Symposium on Neurobiology of Invertebrates 1967 (391—411)

SENSITIZATION AND THE EVOLUTION OF ASSOCIATIVE LEARNING

M. J. WELLS

Department of Zoology, University of Cambridge
Cambridge, England

INTRODUCTION

It is probable that all animals are capable of changing their behavior as a result of individual experience. In many cases, it can be shown that these changes are not attributable to sensory adaptation or to muscular fatigue, and it is assumed that central nervous processes are responsible. Animals differ widely in the extent and rapidity with which they can alter their actions as a result of experience. At one end of the scale one finds creatures like ourselves and sophisticated invertebrate predators, like the cephalopods to be discussed below, animals that can learn to make fine distinctions between things that they see or touch within a few trials, and which probably base a good deal of their normal individual behavior on the results of trial and error experimentation with their environment. At the other extreme are the coelenterate polyps, organisms that undoubtedly show adaptive behavior, but for which there is as yet no unequivocal evidence to eliminate explanations based on muscular or sensory fatigue (Robson 1965; Ross 1965). Between the two extremes are a whole range of animals, from flatworms to annelids and arthropods, all of which appear at times to learn under the sort of conditions that would lead to central nervous changes in higher forms. More specifically, they can be shown to change their actions as a result of events consequent upon these actions. Where an event that is good or bad for the animal – a reward or a punishment – follows soon after some action taken by the animal, the animal will tend to adjust its behavior so that the probability of its repeating the action is adaptively enhanced or reduced. This associative learning dominates the individually acquired behavior of higher animals to such an extent that one tends to forget other forms of adaptive behavior that seem to be relatively important in the lower invertebrates. 'Sensitization', where experience alters the chances of a response regardless of the precise sequence of action and reward, can also lead to adaptive changes, and in many lower animals where equipment for sensory analysis and information storage is relatively simple, mechanisms that lead to changes in behavior irrespective of the sequence of events may provide the main means by which accumulated experience is used to increase the chances of individual survival.

In the discussion that follows, it will be argued that sensitization is an adaptive phenomenon developed before the evolution of sequence-dependent mechanisms capable of relating actions taken with the results that follow, and that the machinery of associative learning can be regarded essentially as a development from this preexisting adaptive mechanism.

EXPERIMENTS WITH CEPHALOPODS

1. *Trial and error learning*

Octopuses and cuttlefish can readily be taught to make discriminations under laboratory conditions, and a considerable literature now exists describing the conditions under which the animals learn, what they can learn, and which parts of the brain appear to be responsible (Wells 1962, 1966; Young 1961, 1964).

The results of a typical series of training experiments are summarized in Fig. 1, a record of the performance of an individual *Octopus vulgaris* successively trained to discriminate between two squares of different size, between horizontal and vertical rectangles, and between black and white discs, all tasks that the animal learned quite rapidly. In each of the three experiments, the octopus was rewarded with a piece of fish for attacking one object of the pair and punished by means of a small electric shock (9 V ac) for attacking the other. The figures were shown successively. The effect on the probability of attacking each member of the pair was progressive and opposite in sign, so that attacks on the 'positive' (the figure that

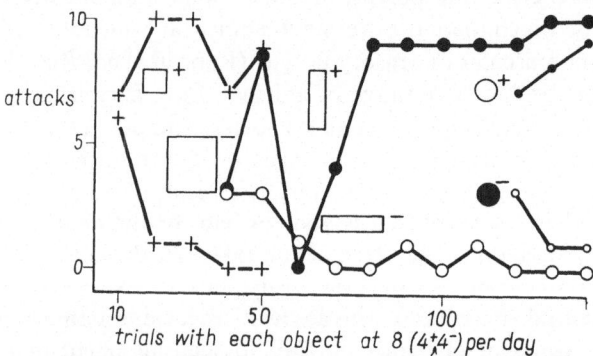

FIG. 1. Discrimination learning by *Octopus vulgaris*; results of a typical visual training experiment. The animal was trained by means of rewards and punishments to distinguish between (a) two squares, one with 4-cm sides, the other with 8-cm sides. (b) The same 10·2 cm rectangle, shown vertically and horizontally, and (c) two 3-cm diameter discs, one black and one white. All three discriminations were learned. In the original experiments, training to distinguish between the squares was continued thoroughout; the scores made in this part of the experiment have been omitted to make the figure less complex. (Replotted from Boycott and Young 1957, Fig. 5)

the animal was rewarded for attacking) became more and more frequent at the same time as attacks on the negative (which the animal was shocked for attacking) were made less and less often. This progressive alteration in performance does not take place if an octopus is fed or shocked before instead of after each trial (Young 1960).

Similar experiments have been made on the tactile responses of octopuses blinded by cutting the optic nerves; the animals can be trained to distinguish both the texture and the taste of objects that they grasp with the arms.

Once an octopus has been trained to make a visual or a tactile discrimination, it remembers. Octopuses will make distinct responses to the members of pairs of objects that they have been taught to distinguish three months previously (Saunders, Wells and Young 1968), and some preliminary observations on visual retention imply a similar duration for visual memories (Sutherland 1957). So far no attempt has been made to test discrimination following longer periods of retention. The internal representations of external events that ensure discrimination are not

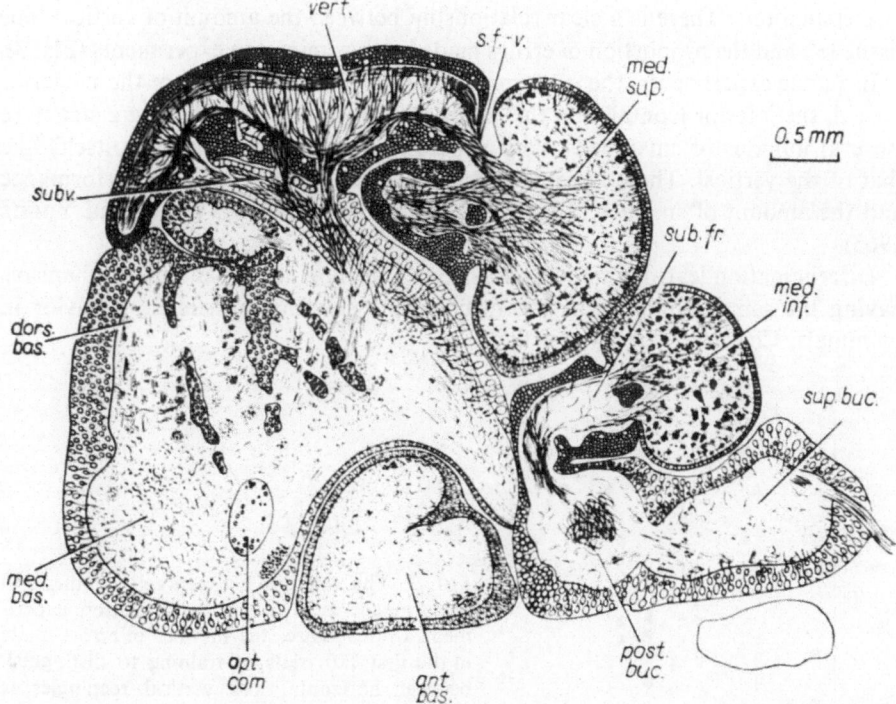

Fig. 2. A diagram of the brain of *Octopus*, to show the location of parts mentioned in the text. This figure shows a longitudinal vertical section through the supraoesophageal part of the brain only. Visual learning depends upon the optic lobes, which are attached to the brain at the level of the optic commissures, and on the integrity of the superior frontal and vertical lobes. Touch learning is slowed by removal of the vertical lobe and abolished by removal of a further small-cell area, the subfrontal, at the anterior end of the brain. The basal lobes are higher motor centers, not apparently concerned in sensory analysis or learning

eliminated by anaesthesia or by Faradic stimulation of the brain (Boycott and Young 1955).

The effect of these training procedures is specific. There is, indeed, some transfer, so that training on one set of visual figures or tactile objects affects the chances of positive and negative responses to further test objects. But the degree of transfer depends on the similarity of the test figures to those that the octopus has already learned to recognize and is, in any case, only detectable as a statistical effect on the probability of attacks on shapes that fairly closely resemble the originals (Sutherland 1960).

The parts of the brain responsible for learning have been identified by means of brain lesion experiments in which regions of the central nervous system were exised before or in the course of training. In visual experiments, damage to the vertical, superior frontal, or optic lobes is critical (Fig. 2). If the vertical and/or the superior frontal lobes are removed, learning is considerably slowed; even with simplest tasks, animals damaged in this way require more training trials to achieve the same standards of performance as before, and they may fail altogether in difficult discriminations. There is a clear relationship between the amount of vertical lobe tissue left and the proportion of errors made in discrimination experiments (Fig. 3).

In tactile experiments, the corresponding regions of the brain are the posterior buccal, the inferior frontal, and the subfrontal lobes, the latter having a structure (several millions of small cells, with processes ending within the lobe itself) like that of the vertical. There again seems to be a relationship between performance and the amount of small cell tissue left after brain operations (Wells and Young 1965).

Discrimination learning in octopuses seems, in short, to depend on mechanisms having the same properties as the mechanisms determining similar behavior in mammals. Characteristically, we find:

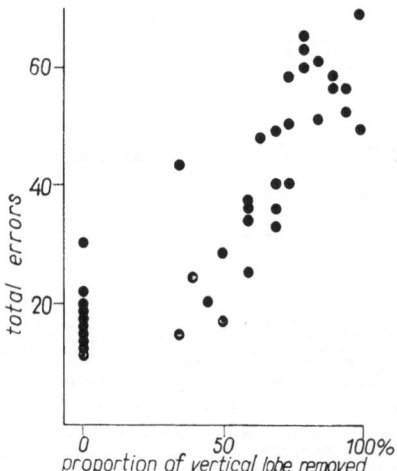

FIG. 3. The relation between cell number and performance is a visual discrimination experiment. In this figure, the number of errors made in the first 120 trials of training to distinguish between horizontal and vertical rectangles is plotted against the amount of the vertical lobe removed. Intact, this region is composed of about 25 million small cells with processes that do not leave the lobe, plus about 80,000 larger cells which carry its output downwards into the rest of the supraoesophageal brain (see Fig. 2) (after Young 1958, Fig. 5)

1. The animals learn when reward or punishment *follows* some action taken by the animal but not if food or shocks are given *before* the action.

2. The representations established in the memory store of the animals are *specific;* they alter subsequent responses to particular objects or figures only.

3. The representations *last for a long time*, months anyway, and possibly much longer. They are not eliminated by disrupting the pattern of electrical activity of the brain.

4. They seem to depend upon the integrity of regions of the brain having very large numbers of small cells.

2. *Sensitization phenomena in octopuses*

If an octopus is given an electric shock or fed, and then tested soon afterwards, the probability of its responding positively toward anything that it sees or touches is found to be altered. This is an unspecific effect, affecting responses to *all* objects regardless of whether they have previously been associated with shocks or food.

Figures 4 and 5 show visual experiments in which feeding and shocks were used to elevate and depress the probability of attacks on a vertical rectangle. In the sea, octopuses tend to live in cavities in the rocks; in the laboratory, they will settle in a 'home' of bricks placed at the end of a tank. In the experiments, a white 10 · 2 cm rectangle was shown and moved up and down at the end of the tank distant from the octopus watching from its home in the bricks. If the animal approached and touched the rectangle within 15 sec, it was recorded as having attacked.

Figure 4a (replotted from Young 1960) illustrates a straight-forward sensitizing effect. For this experiment, each octopus was fed 5–10 g of fish in the home at zero time and then shown the rectangle at the far end of the tank at intervals over the next 24 h. The animals were not rewarded for attacking on these occasions. The zero time level for the responses was established as the average from a series of unrewarded presentations made before the start and after the end of the experiment, several hours after any feeding.

Figure 4a shows quite clearly that feeding increased the chances of an attack on the rectangle despite the fact that the animals were never fed for attacking or just after seeing the figure, and never fed at the far end of the tank where the figure was shown. They had no previous training to attack the rectangle, and one must conclude that feeding itself increases the probability of an attack. The second half of Fig. 4 shows subsequent experiments with the same animals on two successive days that began with a showing of the rectangle paired with feeding. On this occasion, the animals were fed immediately *after* seeing the rectangle whether they attacked it or not. This was the only' training' that the animals got; subsequent showings of the rectangle were unrewarded. Figure 4b again shows that feeding is followed by a rise and then a fall in the probability of attack, and shows that a single training trial has very little effect on the probability of attack — no more than would be expected from sensitization effects alone. The differences between

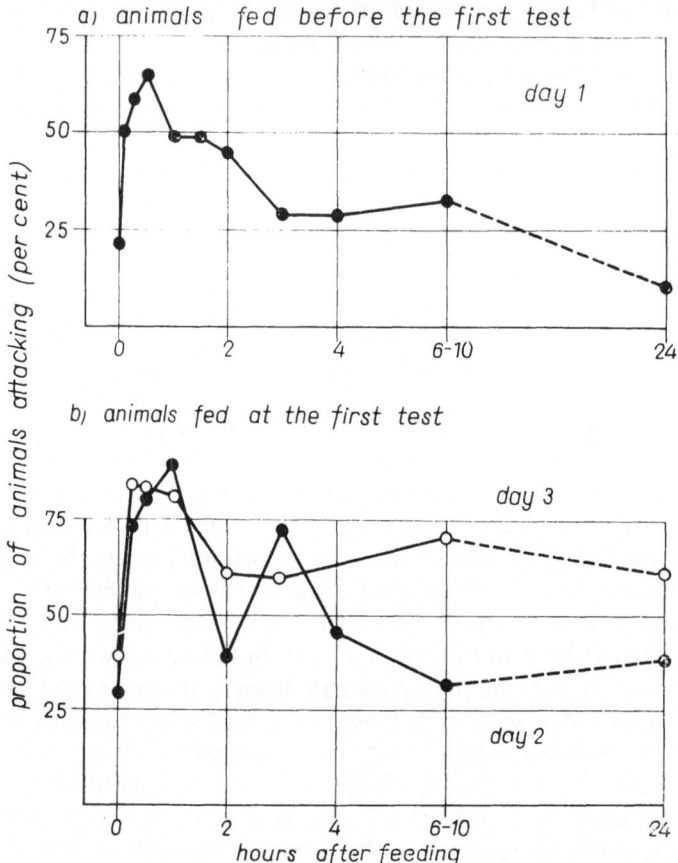

FIG. 4. Sensitization and conditioning in *Octopus*. In the first
part of this experiment, a) the animals were fed in their homes
and then shown a previously unseen vertical rectangle at the
opposite end of their tanks at intervals over the next 24 h; attacks
increased and then declined. Fifty-four animals were used, half
of them to establish the attack level at each time after feeding.
Twenty-four hours after feeding, the level of response was
lower than at the start of the experiment. In b) the same animals were fed immediately *after* showing the rectangle at the
first trial and thereafter tested as in a). The results are shown for
2 days of this treatment, which led to a progressive and relatively
permanent rise in the probability of attacks on the rectangle
(replotted from Young 1960, Fig. 3)

the 'rewarded' and the unfed animals begins to become apparent, however,
after the second 'rewarded' trial, in that the level of attack rises and stays up.
There is a cumulative effect of training that lasts for more than 24 h.

A similar, longer experiment (also from Young 1960) is shown in Fig. 5. During
the first four days, the 10 animals concerned in this experiment were never rewarded

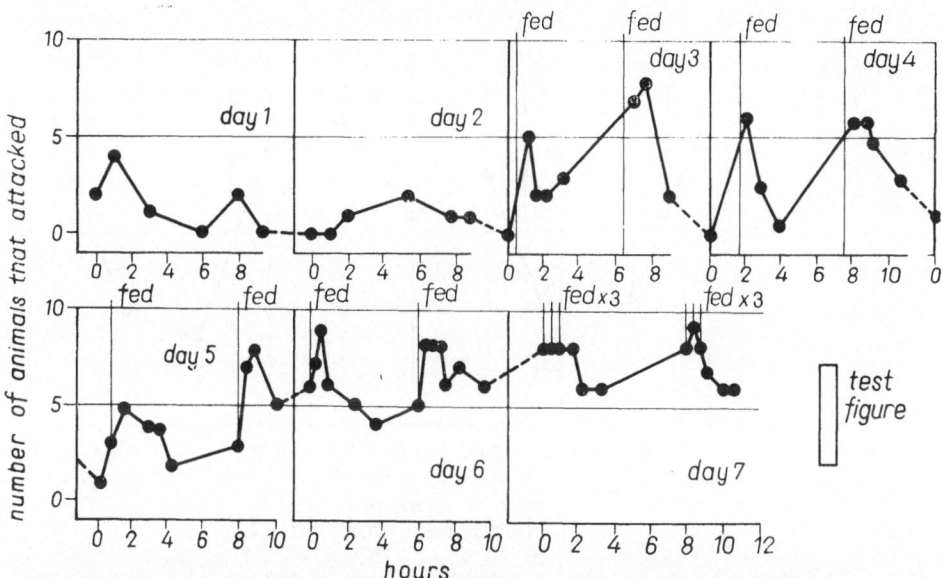

FIG. 5. A longer-term experiment in the manner of that shown in Fig. 4. Ten animals were used. On days 1 and 2, they were not fed for attacking the test rectangle. On days 3 and 4, they were fed immediately *before* the second showing of the rectangle. On each occasion, the response increased but declined to a similar base level, with only one or two of the animals attacking. On days 5, 6 and 7, the animals were fed immediately *after* showing them the rectangle on the occasions indicated. The behavior of the animals changed rapidly until most of them were attacking the rectangle at every presentation. (Replotted from Young 1960, Fig. 2)

for attacking the rectangle; feeding produced the same transient effects as in Fig. 4a. On days 5 and 7, the animals were fed immediately *after* showing them the rectangle on the occasions indicated; there was a steady increase in the probability of attacks that was maintained from one day to the next. This is in contrast to the 'sensitized' elevation in attack level on days 3 and 4, an effect that disappeared altogether within a few hours.

Figure 6 shows a slightly more complex example, in which shocks were used to depress attacks as well as feeding to increase them. The same figure (a white rectangle) was used throughout, with animals already trained to attack, so that their initial level of response was higher than in the experiments already discussed. These eight animals were tested 11 times per day at intervals of 40 min or more. After the first five unrewarded tests on each day, the rectangle was shown once and the animal shocked (11 V ac) or fed, whether or not it attacked, the shock probe and the food being introduced into the home if necessary. The animals were then given five further unrewarded tests. Figure 6 shows the number of animals attacking on each day in the 5 morning tests and the number attacking in the 5 evening tests. Shocks and food clearly alter the probability of response in this training experiment, and on each occasion, some part of the effect lasted

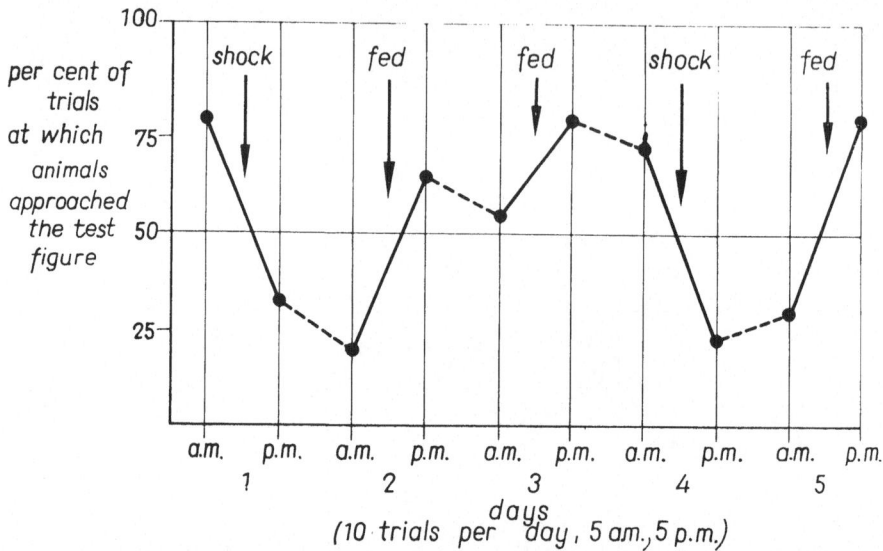

FIG. 6. The effect of shocks and food on the level of attack by *Octopus*. Eight animals were trained to attack a vertical rectangle and then tested as follows: In the morning of each day, the rectangle was shown five times to each animal at intervals of 40 min or more. In the middle of the day, each animal was given food or a shock in the home following a single showing of the rectangle, whether it attacked or not. A further five unrewarded tests with each octopus was made in the afternoon (From Young 1960, Fig. 6.)

until next day. The probability of attack, however, nearly always returned to a mean value, so that some of the elevation or depression was temporary. Figure 6, in short, illustrates a point that must be relevant to all training experiments. What we observe is a change in the probability of an attack; this probability is in part determined by long-term changes brought about by stimulus-reward pairings and in part due to relatively 'direct' short-term effects of pain or food.

It should be noted that these relatively short-term effects are reversible, and that under certain conditions they may well mask a capacity to discriminate that is less readily reversed but which remains concealed if the overall level of attack is particularly high or low. Manipulation of these short-term effects can be used as a tool to reveal concealed capacities for discrimination. An example of this is given in Fig. 7, taken from a series of retention test given by Wells and Wells (1958). In tactile experiments, octopuses are rewarded or punished for grasping test objects and passing these under the interbranchial web to the mouth. This is how the animals normally feed; untrained animals take most small objects in this way and try to eat them. Animals that have been trained to reject by giving them shocks for taking an object will revert to positive responses if left untrained. The animal whose performance is summarized in Fig. 7 took all of the objects presented to it when tested 10 days after the end of a period of tactile training; one might reasonably have concluded that all the effects of training had been lost.

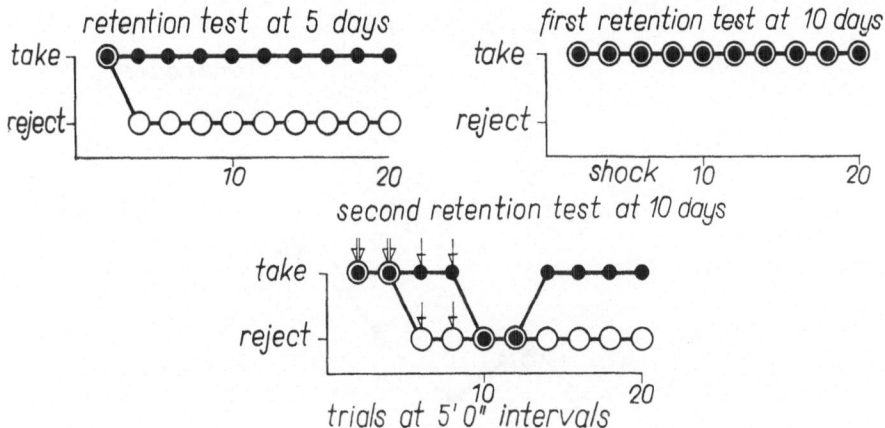

FIG. 7. A series of retention tests made 5 and 10 days after the end of training to discriminate by touch between rough and smooth plastic cylinders. At 10 days, the blinded octopus took both cylinders indiscriminately, any capacity to discriminate being masked by an overwhelming tendency to take. When the threshold for this positive response was raised by giving the animals a succession of small electric shocks (irrespective of which object was presented, or whether it was taken or not), the positive response to the object that the animal had once learned to reject declined first, revealing a persistent capacity to discriminate (from Wells and Wells 1958)

When, however, its overall level of response was depressed by shocks (given, note, irrespective of the objects presented or the responses made), the animal proved quite capable of distinguishing between the test cylinders.

In summary, it would seem that both long-term conditioned responses and short-term sensitizing effects are detectable when *Octopus* is trained by means of rewards and punishments. The latter are particularly noticeable in the early phases of experiments, but since they are transient and the long-term effects of conditioning are cumulative, 'sensitization' becomes relatively unimportant under most conditions. The early phases of training experiments with octopuses tend to be characterized by rather violent fluctuations in the level of response that are progressively ironed out as the system becomes more permanently altered by experience.

EXPERIMENTS WITH ANNELIDS

A somewhat different state of affairs is found in annelids, where the difficulty is not so much to detect sensitizing effects in a learning system dominated by long-term changes as to show that classical associative learning occurs at all. There is no doubt that it does occur, at least in oligochaetes (Jacobsen 1963, Rattner 1965), and it is the probable explanation of the performance of polychaetes in T-mazes (Evans 1966). More important in the present context is the observation that nonassociative processes undoubtedly play a very large part in the adaptive behavior of worms.

M. J. WELLS

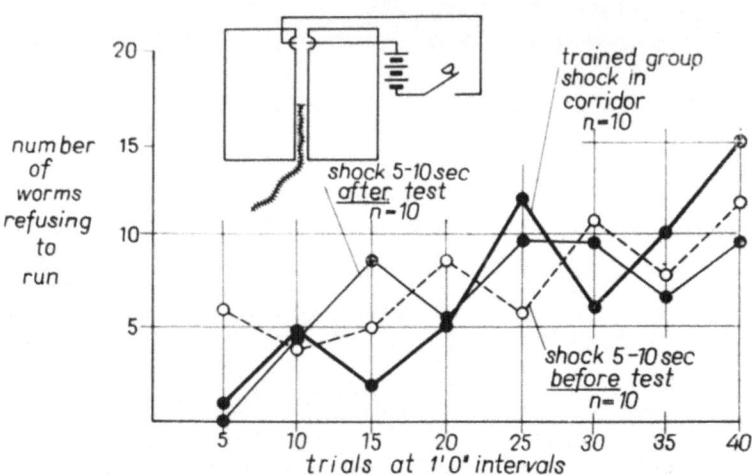

Fig. 8. Three experiments with *Nereis diversicolor*. Untrained *Nereis* will explore a narrow channel such as that in the apparatus shown, between two Perspex blocks. Initially, they take 9 or 10 sec to reach the far end and crawl out; a control group (not shown here) continued of run at this rate for 40 trials. If the worms are given a 6 V dc shock as they reach the far end of the channel, they soon begin to refuse to run up the channel. But the same effect is produced if they are given shocks after leaving the apparatus, or before they enter it, so their change in behavior is no indication of associative learning. (Data from Evans 1966a)

The following examples are taken from the work of Evans (1965, 1966a, 1966b) who, experimenting with nereids, has repeatedly pointed out how easy it is to confound sensitization with associative learning.

Figure 8a shows an 'avoidance learning' experiment. *Nereis diversicolor* lives in tubes that it makes in mud and muddy sand between the tidemarks. If it is placed in a dish it will explore any tube or narrow channel that it encounters. The apparatus used in the experiment shown in Fig. 8 was a simple channel between two Perspex blocks, a little wider than the worms used, equipped with a pair of electrodes recessed into the walls of the channel close to the 'exit' end. The whole was placed in seawater. Worms were directed toward the entrance to the channel with a soft brush. Once the head entered the channel, control worms always ran to the exit, an exercise that typically took them 9 or 10 sec. This response was not detectably altered by 40 repetitions at one-minute intervals.

An experimental group, given 6-V shocks as they arrived at the exit, quickly slowed down; running time increased progressively to 15–20 sec by the end of the first 20 trials and from there to 25 sec or more by the end of the experiment. An increasing proportion of the worms refused to run at all, reversing in the channel, turning round, and reemerging at the entrance end, or simply refusing to move after inserting the head and the first few segments into the entrance. Figure 8

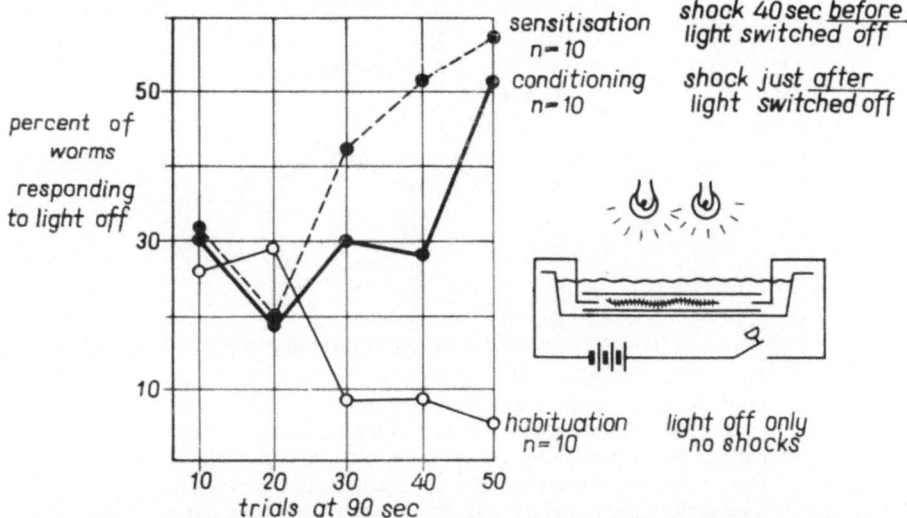

FIG. 9. Further experiments on sensitization and conditioning with *Nereis*. The animals, living in glass tubes, were given 6 V dc shocks following a sudden decrease in illumination. The proportion of worms responding to the stimulus increased in this conditioning experiment. A second 'sensitization' group, shocked between trials, showed the same progressive changes in behavior. A third group of worms, tested without shocks, became progressively less responsive. (From Evans 1965, Fig. 1)

plots the number of worms that failed to complete a run to the exit within 40 sec of the start of each trial.

This looked very much as if the worms had learned to associate the act of running up the channel with the electric shock following their arrival at the far end. That this is not the explanation was, however, quickly shown by two further experiments, in which worms were given shocks outside the apparatus. In the first of these, the worms were given shocks *after* they had crawled out of the far end of the passage. In the second, they were given shocks just *before* each trial. The results of these are summarized together with the results of the 'training' experiment in Fig. 8. The worms from all three groups became increasingly reluctant to run as trials were continued; there was no significant difference between the performance of the animals given shocks outside the apparatus and the performance of the 'trained' group.

A similar result was obtained in an experiment designed to condition the giant fiber response of *Nereis*, summarized in Fig. 9. Again the response was enhanced by giving the worms shocks, regardless of whether the shocks followed or preceded the sudden decrease in light intensity to which the worms were responding – again a case of sensitization rather than associative learning.

Responses to positive stimuli could also be enhanced by feeding. In a further ex-

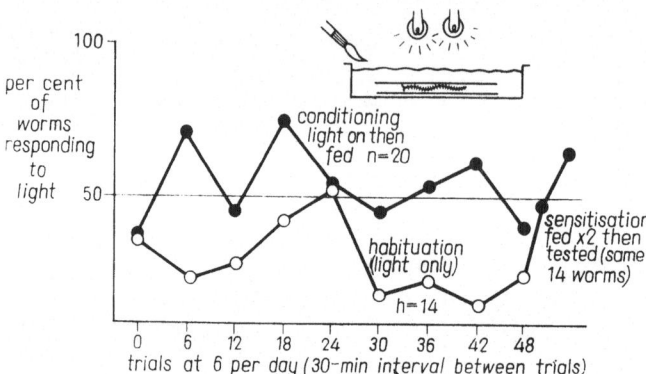

FIG. 10. Sensitization and conditioning in *Nereis*. The plots
show the proportion of worms responding by coming to the
ends of their tubes within 2 min of a sudden increase in illumination. The conditioning group (n = 20) were fed almost
immediately after the start of each trial; the plot here shows
their performance at every sixth trial, when they were tested
but not fed. The habituation group (n = 14) were not fed
during the first 48 trials. They were then fed twice (no change
in illumination) and tested again, the food apparently having
made them much more responsive to light. (Data from Evans
1966b)

periment, described by Evans (1966), 19 *Nereis diversicolor* were trained to run
to the ends of their tubes when the light intensity was suddenly increased. Initially,
only four out of the 19 worms responded. After a single feeding on wheat germ
extract, 12 of the 19 responded in this way, crawling to the ends of their tubes
in a mean time of 81 sec, a significantly faster time than the 106 sec averaged by
the four worms responding to the prefeeding trial. Similar results were obtained
with worms trained by pairing light with food and by a trace-conditioning procedure, where the worms were fed 5 min after the end of the light stimulus. There
were no cumulative effects when training was continued for eight days at a rate
of six trials per day, although the level of response remained consistently higher
than at the first prefeeding trial (Fig. 10).

DISCUSSION

Under many conditions, sensitization mimics associative learning. Yet, it clearly
depends upon some simpler mechanism, since it is not dependent upon a particular
sequence of events. The effects of sensitization are comparatively short-lived,
enduring for a few hours at the most, whereas the results of classical stimulus-
reward pairing are typically more or less permanent.

Since associative learning would appear to be a more certain and reliable means of achieving adaptive responses, and since nearly all animals can be shown to learn by association under some conditions, it is not obvious why the apparently more primitive system should have been preserved at all. If the animal alters its responses regardless of the stimulus-reward sequence, it will never be able to determine which of its actions produce desirable and which produce undesirable effects. It cannot hope to make progressive long-term improvements in its performance. It would appear that every advantage must lie with adaptive mechanisms capable of discriminating between events in relevant and irrelevant sequences.

But this view presupposes that the animals concerned have sensory and analytical machinery capable of defining the stimulus in the first place. It is no good learning to respond or not to respond to a stimulus unless the stimulus is rather precisely defined. Indeed, if similar (but to the animal, indistinguishable) stimuli are likely to signal both favorable and dangerous situations, it may be disadvantageous for the animal to commit itself in this way. Associative learning, as we have seen, tends to be cumulative and enduring. A run of occasions on which a given stimulus turned out to be favorable might well commit an animal to a series of mistakes, where a sensitized animal would make only one or two errors before reversal when conditions changed, regardless of the length of its previous favorable experience. An animal like *Nereis*, living in a burrow in the mud, probably has no means of determining whether the shadow passing the end of its tube is a 'good' or a 'bad' signal. It may mean food (*Nereis* feeds on fragments of algae and the like, which it collects from the end of its burrow), or it could signal the approach of a predator; *Nereis* has no means of knowing. The animal can, however, improve upon its chances of making a correct decision by taking advantage of the fact that favorable and unfavorable events rarely occur at random. Food comes in batches; predators appear, wait for a while, and go away. The worm that remains withdrawn into its burrow for a while after it has been hurt is less likely to be missing a meal than one that has just fed. Sensitization ensures that the animal responds appropriately in the light of the prevailing conditions; it is a means of optimizing responses when the animal lacks machinery for distinguishing between the stimuli signaling forthcoming good and bad events.

Formally, the behavior of *Nereis* can be described in the terms shown in Fig. 11a. The probability of a positive response rises as the probability of a negative response falls. The two are altered by the recent past history of the animal, irrespective of the order of events. Figure 11b shows how such a system might be built up from units having the properties of neurons. A piece of neurological machinery with these properties would achieve better than randomly correct responses in any two-choice situation where the sequence of 'good' and 'bad' events occurred in runs of two or more. It is a basic model of sensitization. It is, however, clearly too simple. For one thing, it supposes only two possible responses to a given sensory input; more output channels could be postulated without altering its essential characteristics. More important, it presupposes that the external sensory input is unclassified, which it quite clearly is not in any real sense. At very least, animals distinguish between visual and tactile stimuli, and

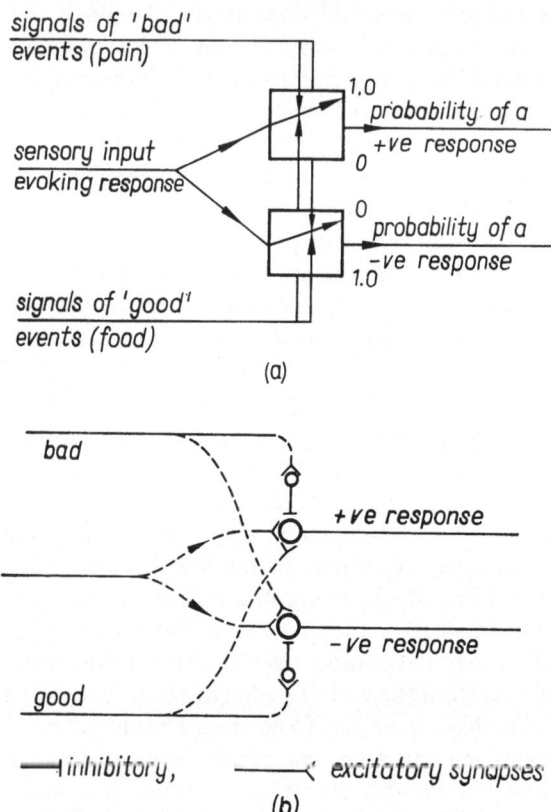

signals of 'bad'
events (pain)

1,0
probability of a
+ve response
0

sensory input
evoking response

0
probability of a
-ve response
1.0

signals of 'good'
events (food)

(a)

bad

+ve response

-ve response

good

⊣ inhibitory,   ⟨ excitatory synapses

(b)

FIG. 11. A model of the behavior of animals that sensitize. (a) Represents the state of affairs in an animal such as *Nereis* or *Octopus* which may become more or less responsive as a result of food or shocks irrespective of the sequence of these events and the sensory input evoking a response. A nervous machine with these properties would ensure better than randomly correct responses, even though the animal was unable to distinguish between stimuli, provided only that the events that these indicated were not strictly alternating or at random; (b) shows how such a machine might be constructed from units having the properties of neurons

generally between different intensities of each modality. So one must suppose that there is some classifying device channeling the stimuli to the units which determine the probability of response to the stimulus concerned. The second model (Fig. 12) incorporates this modification. It also includes recurrent circuits by which the mere performance of an action reduces the chances of its repetition in the absence of other 'good' or 'bad' signals. All animals habituate; an alternative model might postulate habituation by path failure in the manner that Horne

(1967) had suggested − it is not an important feature of the present series of models and will not be discussed here.

One must suppose that in life the stimulus-analyzing device is in the first instance innate, determined by selection in the past. In more sophisticated developments of the model, one might postulate means by which the probability units feed back to alter the flow of information through the stimulus-analyzing device. At all events, the stimulus-analyzing stage in the processing of the input would be complex, channeling the input to tens or thousands of alternative outlets.

In the model shown in Fig. 12, only two pairs of outlets are shown. Even such an elementary system could produce results that would mimic many of the features of discriminant learning, provided only that the two pairs of units determining the probability of responses were not equally sensitive to the 'good' and 'bad' input signals. If one were more sensitive than the other (A than B in Fig. 12), repeated good or bad signals could produce progressive differences in the degree of response to two input signals. An experiment with an animal organized in this manner could produce results in the pattern shown in Figs. 13a and 13b, results that look very like learning to discriminate. But it should be noted that the responses to the two signals are alike in kind. They differ only in degree. Such a

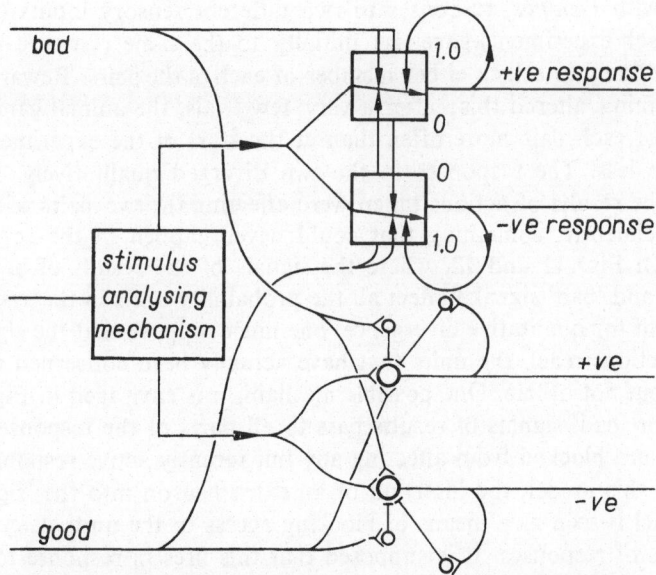

FIG. 12. A further development of the model proposed in Fig. 11. All animals classify incoming stimuli to some extent, and all animals so far investigated will habituate. A machine constructed in the manner shown above, with the pairs of units unequally sensitive to the 'good' and 'bad' inputs, could produce results like those shown in Fig. 13, which mimic associative learning to discriminate. The lower pair of units is shown in neuron form, as in Fig. 11b

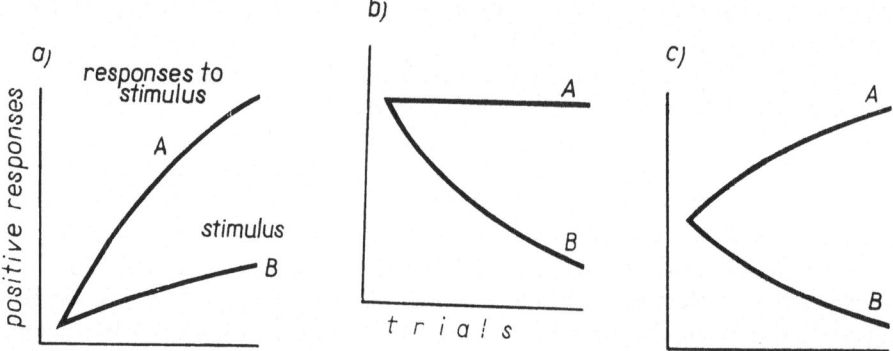

FIG. 13. Possible results of 'training' a machine with the properties of the model shown in Fig. 12. Progressively divergent responses a, b are possible with a sensitizing device. Note, however, that the direction of the divergent changes will always be the same so that such a system can never produce results of the type specified in c), where the responses to two stimuli initially evoking the same response diverge toward responses of different sign

mechanism could never give rise to the sort of divergent responses recorded in Fig. 13c, or (to pick a real case) the results shown in Fig. 1. In this series of experiments with *Octopus*, responses to two different sensory inputs (the pair of figures in each experiment) gave rise initially to the same response—the animal was equally likely to attack either member of each of the pairs. Reward and punishment training altered this; after a very few trials, the animal came to attack one figure of each pair more often than at the start of the experiment, and the other figure less. The responses to the two diverged qualitatively, that is, the signals of the results of actions taken were affecting the two pairs of probability units independently, something that could never happen in the sort of model postulated in Figs. 11 and 12, where the signals of the results of actions taken (the 'good' and 'bad' signals) affect all the probability units in the system.

To account for qualitative divergence, one must suppose that the signals of the results of actions reach the units that have actually been concerned in ordering the action but not others. One possible mechanism is envisaged in Fig. 14; here, the 'good' or 'bad' signals of results pass to all parts of the response command system, but are blocked from affecting any but recently active response-ordering neurons. In this model, the insertion of an extra neuron into the 'signals of results' channel is seen as a means of blocking access to the units determining the probabilities of responses. It is supposed that this fires in response to 'good' or 'bad' signals only if facilitated as a result of preceding activity on the part of the response unit that it supplies. No activity, no pass (Fig. 14).

Such a model would have many of the properties observed as necessary conditions for association learning. Rewards and punishment would have specific effects. The signals indicating the results of actions taken would only penetrate if they arrived quite soon after an action had been taken. The model would also, since it implies the existence of delay circuits and feedback loops not required

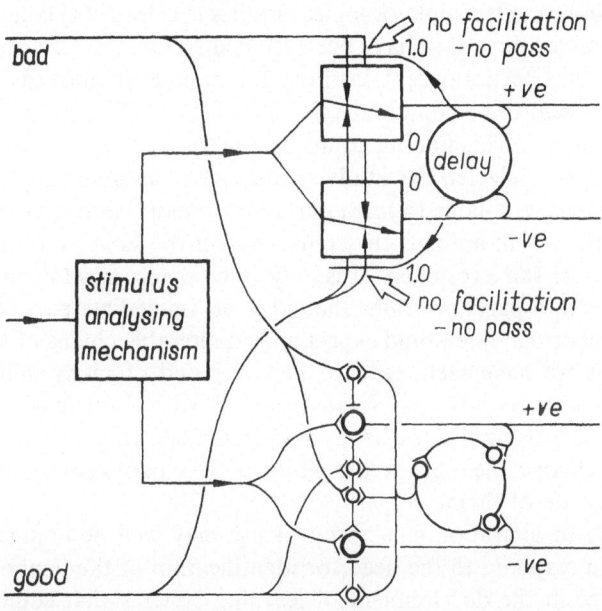

FIG. 14. Further development of the model shown in Figs. 11 and 12. In order that the machine shall be able to associate action with its consequences, some system must be encorporated to ensure exclusion of signals of events occurring *before* the action has taken place. In this model, the units determining the probability of responses to a specific stimulus are screened from the signals of 'good' and 'bad' events until activated, whereupon they become exposed for a while to these inputs. One way of achieving this is shown; active units stimulate delay circuits that for a while facilitate pathways leading to themselves. In this model, recurrent 'habituating' circuits have been omitted. The screening of irrelevant inputs allows the evolution of lability along the input–output pathways

in the simpler sensitization machine, require relatively large numbers of cells, again a constant feature of brains that learn by association.

With the isolation of the neurons actually responsible for ordering responses, one finds that enduring 'memory traces' appear, 'permanently' altering the probability of specific responses to specific stimuli. These alterations presumably depend upon changes at the synapses along the sensory-input–response-output line. They could also be achieved by altering the rate of spontaneous firing by neurons on this line. At all events, it implies that the state of the units ordering the response can be permanently changed as a result of the excitatory and inhibitory inputs penetrating to them; the units or their connections must be labile.

Long-term changes do not usually occur when chains of neurons are made active. In general, nerve cells and their connections must not be permanently altered by their inputs. Electrophysiological studies (Eccles 1964) confirm this, and in most of the preparations that have been studied so far, the effect of even thousands of impulses has been transient, lasting a few minutes in most cases. Relatively long-term effects seem to be comparatively rare (Tauc 1966; see also Chalazonitis in the present volume). This finding could be due to the selection of preparations used. Large neurons, selected for study because it is feasible to penetrate them with microelectrodes, are liable to form parts of common motor paths, reflex arcs and the like, pathways in animals that must remain the same; the same stimulus must give rise to the same response. It is only when one comes to consider central nervous effects within parts of brains that play no immediate part in the organization of movement that one would expect to find alterable chains of neurons, and these regions, as we have seen, tend to be composed of many million cells, at present or until very recently too small to attack with electrophysiological techniques. It is not surprising that we have so few records from neurons and synapses that appear to change their properties more or less permanently as a result of the use that is made of them.

Neurons liable to alteration as a result of use may well be limited to parts of brains evolved in response to the need for identification of the temporal sequence of events, a stage in the development of learning systems that could occur with advantage only after the evolution of analyzing machinery capable of separating the sensory input into precise categories.

To summarize we can say that associative learning machinery is seen as having evolved in a number of stages, as follows:

1. Animals first developed the capacity to sensitize, altering the probability of innately determined responses in the light of prevailing conditions. The animal was able to make better than chance responses, even if quite unable to analyze its sensory input (stage 1 in the model, Fig. 11).

2. Means for classifying the sensory input developed.

3. This allowed the possibility of becoming more readily sensitizable to some stimuli than to others and permitted forms of discrimination learning limited to quantitative differences in qualitatively similar changes in response to the stimuli distinguished (stage 2 in the model, Fig. 12).

4. With the development of increasingly precise means of defining the inputs from exteroceptors, the possible survival value of more permanent changes in response increased and synaptic lability was evolved.

5. With this arose a need to ensure that the signals of results of actions taken affected only the units actually concerned in the organization of the relevant response. One way in which this may have been achieved is suggested in the final stage (Fig. 14) of the model developed.

SUMMARY

Cephalopods can be taught to make visual and tactile discriminations under conditions that leave little doubt that they are learning by association. The central nervous changes due to the training procedures cannot be eliminated by anaesthetics, they may last for months, and they are specific to the test situations. Their probable location can be discovered from the results of brain lesion experiments.

Together with these more or less permanent changes, one finds temporary alterations in the probability of responses to objects seen or touched that are also dependent on feeding or punishment. In the normal course of events, these are superimposed upon the effects of conditioning procedures, but they can be distinguished from these because they take place even when the rewards and punishments precede rather than follow presentation of the test objects.

Sensitization phenomena are more obvious in annelids and among the nereid polychaetes, at least, they account for many of the changes of behavior that at first sight look like associative learning.

The role of sensitization in the organization of adaptive behavior is discussed. Sensitization is a means by which animals with only primitive apparatus for sensory analysis and information storage can optimize their responses by taking advantage of the nonrandomness of favorable and unfavorable naturally occurring events.

A series of models is presented to describe the role of sensitization in relation to learning and to show how associative machinery may have evolved from more primitive adaptive mechanisms.

REFERENCES

BOYCOTT, B. B. and YOUNG, J. Z. (1955): A memory system in *Octopus vulgaris* Lamarck. *Proc. roy Soc. B* **143** 449—480.

BOYCOTT, B. B. and YOUNG, J. Z. (1957): Effects of interference with the vertical lobe on visual discriminations in *Octopus vulgaris* Lamarck. *Proc. roy. Soc. B.* **146** 439—459.

CHALAZONITIS, N. (1968): Synaptic properties of the oscillatory neurons. In *"Neurobiology of Invertebrates"*. 201—225.

CLARK, R. B. (1960): The learning abilities of Nereid Polychaetes and the role of the supra-oesophageal ganglion. *Anim. Behav. Suppl.* 1, 89—100.

ECCLES, J. C. (1964): *The Physiology of Synapses.* Springer-Verlag, Berlin.

EVANS, S. M. (1965): Learning in the Polychaete *Nereis. Nature (Lond.)* **207** 1420.

EVANS, S. M. (1966a): Non-associative behavioural modifications in Nereid Polychaetes. *Nature (Lond.)* **211** 945—948.

EVANS, S. M. (1966b): Non-associative behavioural modifications in the polychaete *Nereis diversicolor. Anim. Behav.* **14** 107—119.

HORNE, G. (1967): Neuronal mechanisms of habituation. *Nature.*

ROBSON, E. A. (1965): Adaptive behaviour in cnidaria. *Anim. Behav. Suppl.* 1, 54—59.

ROSS, D. M. (1965): The behaviour of sessile coelenterates in relation to some conditioning experiments. *Anim. Behav. Suppl.* 1, 43—52.

SAUNDERS, G., WELLS, M. J. and YOUNG, J. Z. (1968): *in preparation.*

SUTHERLAND, N. S. (1957): Visual discrimination of orientation by *Octopus. Brit. J. Physiol.* **48** 55—71.

SUTHERLAND, N. S. (1958): Visual discrimination of shape by *Octopus:* Squares and triangles. *Quart. J. exp. Psychol.* **10** 40—47.

SUTHERLAND, N. S. (1960): Visual discrimination of shape by *Octopus:* Open and closed forms. *J. comp. Physiol. Psychol.* **53** 104—112.

TAUC, L. (1966): Physiology of the nervous system. In *Physiology of Mollusca.* Vol. **II.** Ed.: K. M. Wilbur and C. M. Yonge, 387—454.

WELLS, M. J. (1962): *Brain and Behaviour of Cephalopods.* Heinemann.

WELLS, M. J. (1966): Learning in the Octopus. *Symp. Soc. exp. Biol.* **20** 477—507.

WELLS, M. J. and WELLS, J. (1958): The effect of vertical lobe removal on the performance of octopuses in retention tests. *J. exp. Biol.* **35** 337—348.

WELLS, M. J. and YOUNG, J. Z. (1965): Split brain preparations and touch learning in the octopus. *J. exp. Biol.* **43** 565—579.

YOUNG, J. Z. (1958): Effect of removal of various amounts of the vertical lobes on visual discrimination by *Octopus. Proc. roy. Soc. B.* **149** 441—462.

YOUNG, J. Z. (1960): Unit processes in the formation of representations in the memory of *Octopus. Proc. roy. Soc. B.* **153** 1—17.

YOUNG, J. Z. (1961): Learning and discrimination in the octopus. *Biol. Rev.* **36** 32—96.

YOUNG, J. Z. (1963): Some essentials of neural memory systems. Paired centres that regulate and address the signals of the results of action. *Nature (Lond.)* **198** 626—630.

YOUNG, J. Z. (1964): *A Model of the Brain.* Oxford University Press.

DISCUSSION

G. Ádám: I would like to ask two questions in relation with Dr. Wells' excellent paper: (1) People working in the field of behavior and the physiology of conditioning are familiar with the possibility of elaborating a conditional response by associating the two stimuli in the inverse manner; i.e. by using UCS–CS sequence. This is the well-known 'backward conditioning' procedure. I wonder why the octopus does not learn in such a backward reinforcement situation? (2) The term 'sensitization', in my opinion, is not necessary to be introduced into neurophysiology, since the term 'summation reflex' in the classical Pavlovian terminology covers exactly the same phenomanon.

M. J. Wells: Perhaps the answer to the first question is that we have not really tried. As I understand it, to get backward conditioning, the unconditioned stimulus must precede the conditioned stimulus by quite a short time. If the UCS is more than a few minutes before the CS, backward conditioning does not happen. In the octopus experiments I quoted, the animals were fed half an hour or so before testing. We simply have not tried a shorter period. As regards terminology, I am not introducing a new term; 'sensitization' is already widely used in relation to experiments with flatworms and annelids and other invertebrates. It seems to me to have the advantage that it does not presuppose any particular sort of neurological explanation.

I. Madarász: Have you any experience concerning the behavior of the animals when punishment was 'overdosaged', i.e., if the animal receives too much or too strong electric shocks? Does the sensitization process remain unaltered or not?

M. J. Wells: If you give too many or too great electric shocks, octopuses refuse to collaborate at all. They simply stop taking objects presented to them. So the sensitization process is altered; the whole behavior swings negative. The voltage we use in most training experiments (6–9 V ac) has been arrived at by trial and error; it is strong enough to act as a deterrent (3 V ac was not, at least in some experiments) but not so unpleasant for the animal that he gives up attacking altogether.

W. C. Corning: The work of Lashley and a number of others indicates that 'the engram', whatever it is, is delocalized in brain tissue. Do you think there is a specific locus for information storage in the octopus, or are you perhaps interfering with some output path rather than with the storage process itself?

M. J. Wells: In the octopus we seem to have the situation that Lashley thought he had found for the cortex in rats; you can chop at the vertical lobe or the subfrontal lobe of an octopus and the animal's performance in discrimination experiments declines in proportion to the damage done. There seems to be no question of specific sites for the engrams of specific events. This means that we cannot localize memories within those areas. But I think we can make a case for suggesting that the structural changes really are in these regions, somewhere. The small cell areas of the subfrontal, for example, are essential to touch learning. As you remove the last traces of this tissue from the subfrontal-posterior buccal region touch learning fails altogether. The effects of past experience are lost, and the animal is not reteachable. Of course, one might be cutting a pathway going to or from a storage area elsewhere. But it seems unlikely on structural grounds. The subfrontal is at the end of the line; it is the point at which the connections with the arms turn around and go back, the highest point of the brain for this set of circuits. It is composed of $5^1/_2$ million small cells, which seems an awful lot for a pathway. Its removal upsets nothing else but learning, so far as we can discover. Strictly speaking, one can never prove that a particular part of the brain is a site of memory storage, short of pin-pointing a structural change that can be identified as the cause of a change in behavior. Nobody has done this yet. We would like to hink that in the octopus we are beginning to know where to look.

Symposium on Neurobiology of Invertebrates 1967 (413—421)

LEARNING OF ISOLATED GANGLIA OF THE MOLLUSC
LYMNAEA STAGNALIS

B. N. VEPRINTSEV and S. I. ROSANOV

Institute of Biological Physics, Academy of Sciences of the USSR
Putchino on Oka, Mocow region
Moscow, USSR

In the present work we attempt to demonstrate the possibility of the formation of long-term memory traces in the isolated ganglia of *Lymnaea stagnalis*. In setting up the experiments we started from the following facts and empirical generalizations:

1. Molluscs easily develop conditioned reflexes (Fischel 1931; Sokolov 1959).

2. All information from receptors is transmitted to the brain as electrical impulses. The main portion of information sent by the brain to effectors is also transmitted in the form of nervous impulses.

3. Nerve cells in the isolated ganglia retain their electrical activity for about two days. Twelve to sixteen hours after preparation, cells are capable of responding to stimulation by increasing RNA synthesis as well as protein synthesis (Veprintsev and Dyakonova 1967).

4. The ability to form long-term memory must be one of the intrinsic properties of the nervous system.

Accordingly we started with the idea that since intensive and sufficiently lengthy stimulation of two or three nervous trunks of the isolated brain of *Lymnaea stagnalis* evokes ortho- and antidromic excitation of neurons and consequently activates the synthesis of RNA as well as of proteins, this should cause the rebuilding of interneuronal connections and a steady change in the electrical activity of the isolated brain.

To conduct the experiments, we isolated the oesophageal ring with the main nervous trunks but without buccal ganglia (Fig. 1). The isolated ganglia were placed in perspex camera perfused by a physiological solution (Fig. 2). For the registration of activity and stimulation, the nerves were sucked in capillary glass electrodes filled with physiological solution (Fig. 3).

The electrical activity was registered in the following nerves: n. pallialis inferior, n. pallialis dexter, n. columellus, n. pedalis medius, and n. pedalis posterior, and the following nerves were stimulated: n. tentacularis, n. opticus, and n. pedalis posterior (Fig. 1). Registration was made by means of multi-channel electro-encephalograph ('Sanei').

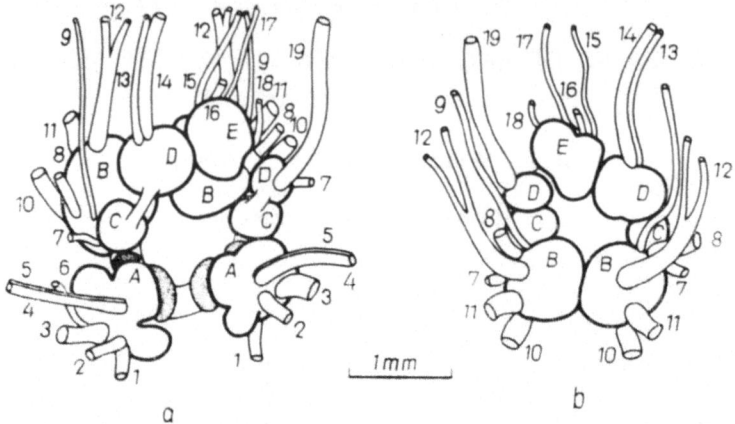

FIG. 1. Scheme of oesophageal ring of *Lymnaea stagnalis* (by courtesy L.S. Botcharova). a — dorsal view; b — ventral view. *A* — cerebral ganglion; *B* — pedal ganglion; *C* — pleural ganglion; *D* — parietal ganglion; *E* — visceral ganglion. 1 — cerebrobuccal connective; 2 — n. frontolabialis; 3 — n. labialis medius; 4 — n. opticus; 5 — n. tentacularis; 6 — n. penis; 7 — n. cervicalis anterior; 8 — n. cervicalis posterior; 9 — n. collumelus; 10 — n. pedalis anterior; 11 — n. pedalis medius; 12 — n. pedalis posterior; 13 — n. pallialis inferior (radix parietalis); 14 — n. pallialis dexter; 15 — n. pallialis inferior (radix visceralis); 16 — n. aortalis; 17 — n. genitalis; 18 — n. cutaneus pallialis; 19 — n. pallialis sinister

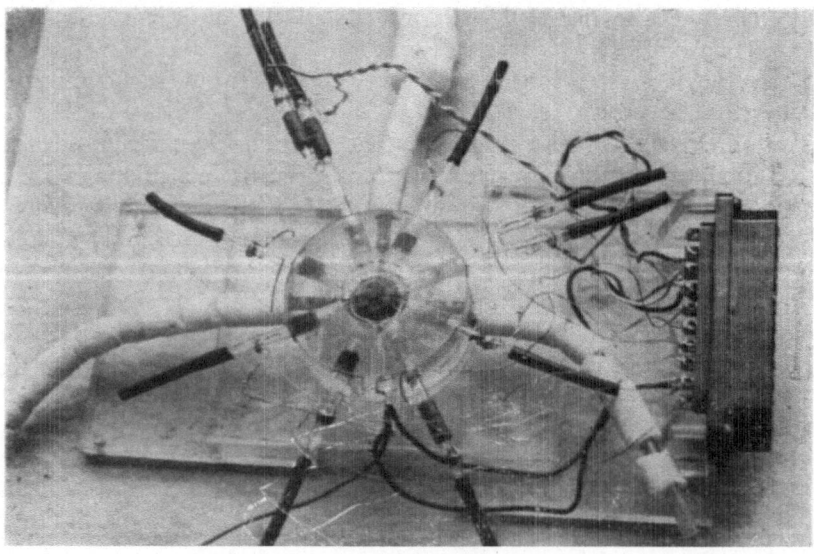

FIG. 2. The perspex chamber with glass suction electrodes for stimulation of nerve trunks and registration of their activity

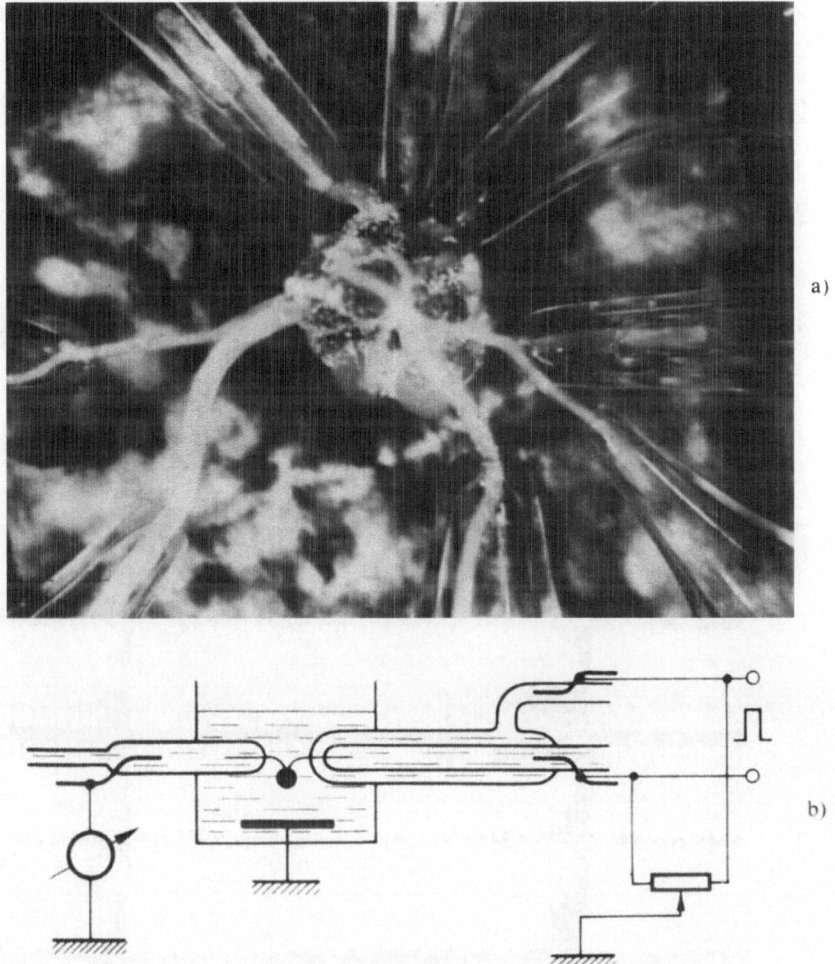

FIG. 3. a) Photo of ganglia with suction electrodes; b) scheme of
registration of activity and stimulation of nerves

The spontaneous activity from the single nervous trunk is shown in the irregular
sequence of nervous impulses with amplitudes from 20 μV to 1 mV (Fig. 4a).
Action potentials registered from the nervous trunk reflect the activity of single
axons. One may frequently observe identical patterns of impulses in different nerve
trunks (Fig. 5). Such impulses apparently reflect the activity of the different
branches of the same axon in different trunks, which is common in the nervous
system of gastropod molluscs (Hughes and Tauc 1962; Tauc and Hughes 1963).

The nerve trunks were stimulated by supermaximal impulses of 20-msec duration with amplitudes three to four times above threshold. An increase in stimulation did not affect the response of nerve trunks. The interval between stimulating

B. N. VEPRINTSEV and S. I. ROSANOV

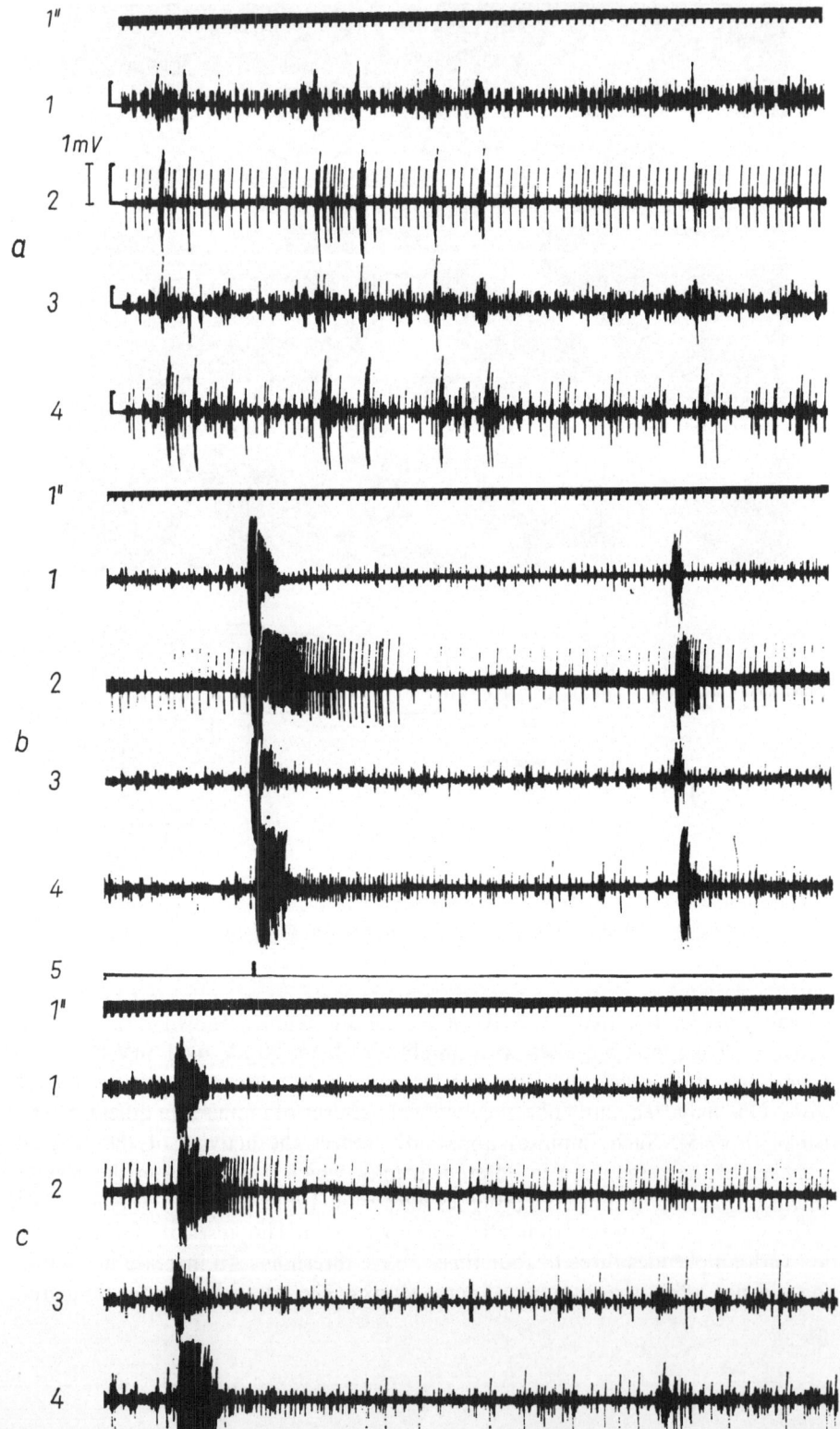

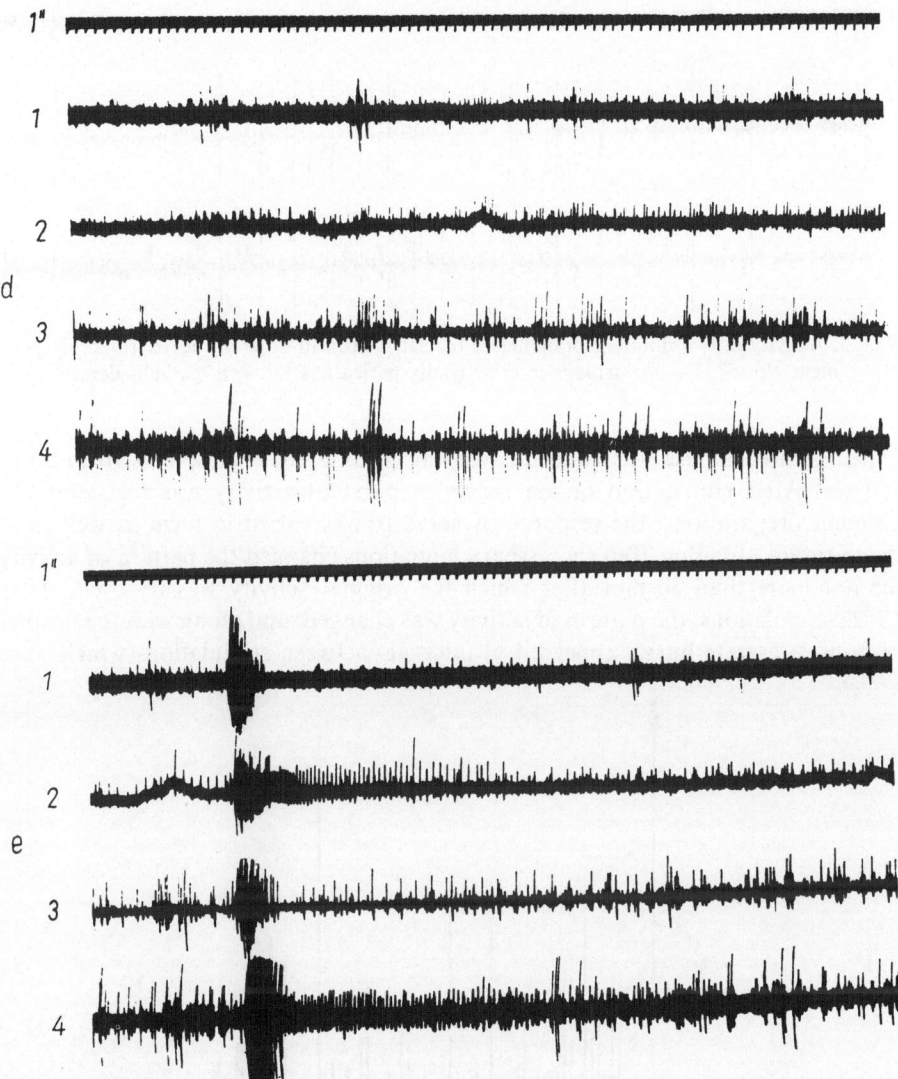

FIG. 4. Electrical activity in nerve trunks in the process of 'learning'. a— spontaneous activity 5 h after preparation; b — the response to stimulation and one of the first synchronous bursts of activity appearing at intervals between stimulating impulses; c — one of the synchronous bursts of activity 90 min after cessation of stimulation; d — activity of spontaneous type 12 h after cessation of stimulation; e — restitution of burst activity after 10 stimulations; 13 h after cessation of first stimulations; 1 — n. pallialis inferior (rad. visceralis); 2 —n. pallialis inferior (rad. parietalis); 3 — n. pallialis dexter; 4 — n. pallialis sinister; 5 — marking of stimulation.
Experiment No. 67

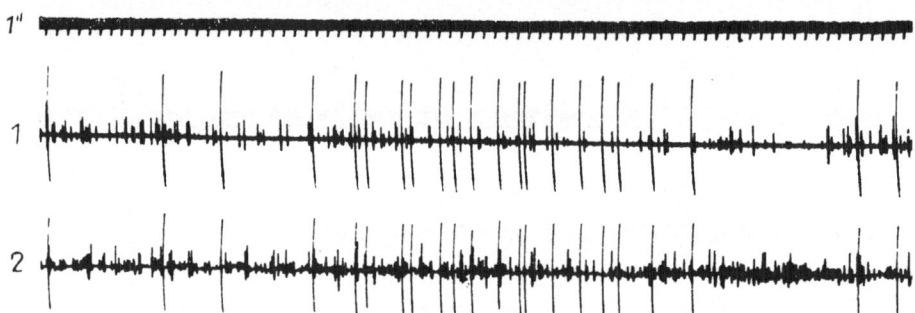

FIG. 5. Activity in two different branches of the same axon in different nerve trunks. Experiment No. 57. 1 — n. pallialis inferior (radix parieatalis); 2 — n. pallialis dexter

impulses in each case was constant, varying in different experiments from 90 to 250 sec. After stimulation of the nerves, a burst of activity was registered. In different preparations, the response of nerve trunks varied in form as well as in the duration of fading. Ten successive stimulations changed the pattern of activity for not more than 20 min after which the original activity was restored. After 15–20 stimulations, the pattern of activity was changed, and along with the normal responses discrete bursts appeared at intervals between stimulations which were

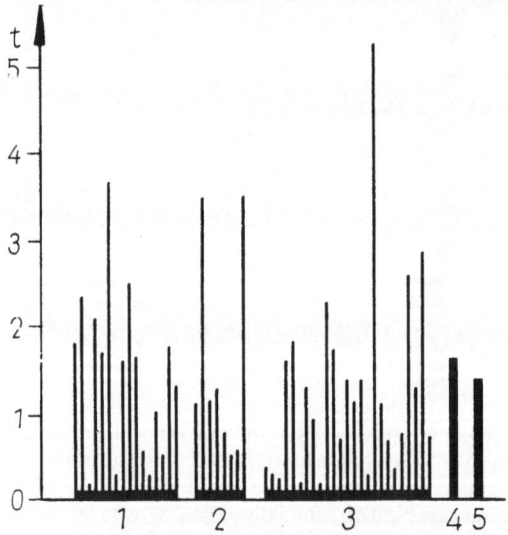

FIG. 6. The succesive intervals between bursts of activity. 1 — Immediately after cessation of first stimulations; 2 — 80 min later; 3 — after 10 additional stimulations; 13 h after cessation of first stimulations; 4 — interval between stimulating impulses; 5 — mean interval between bursts. Ordinate — time in min

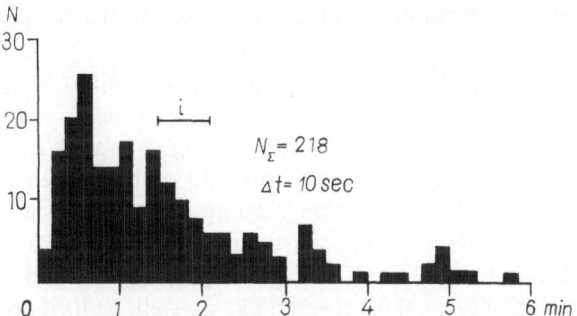

FIG. 7. Histogram of intervolley intervals after cessation
of stimulation, summing up the results of seven experiments with the same intervals between stimulating impulses (i)

weaker than the responses to stimulation (Fig. 4b). From 40 to 50 stimulations
changed the pattern of activity steadily. Some time after the stimulation was
stopped, powerful bursts of activity similar to the responses to stimulation were
observed at irregular intervals (Fig. 4c). In a few cases we registered such synchronous bursts of activity in all trunks 20–24 h after stimulation was stopped. We are
unable to trace any regularity in the intervolley intervals (Fig. 6). The histogram
of the intervolley intervals, which summarizes the series of experiments, does
not show any assimilation of the rhythm of excitation (Fig. 7). In the majority
of experiments, 10 or 12 h after stimulation was stopped, the bursts of activity
were followed by periods of 10–30 min of restoration of original activity. These
periods had a tendency to become longer and were followed by periods of burst
activity. However, after the restoration of the original activity (Fig. 4d) it was
sufficient to give 5 or 10 stimulations to restore the burst pattern of activity
(Fig. 4e).

Apparently, the change in the pattern of impulse activity in the isolated brain
of *Lymnaea stagnalis* after stimulation reflects steady changes in the nature of
interneuronal connections as a result of the systematic stimulation. This phenomenon may be regarded as analogous to learning. The following facts bear out
this conclusion:

1. The dynamics of formation of steady changes of activity as a result of stimulation are similar to the dynamics of the formation of a conditioned reflex.
The changed pattern of activity that disappears some time after stimulation is
easily restored by several additional stimulations, representing an analogy of the
confirmation of a conditioned reflex.

2. The number of stimulation required for a steady change in activity is comparable with an average number of inputs required to produce a reflex in invertebrates (Sokolov 1959).

3. The absence of a clearly defined periodicity in the bursts of activity, i.e. the
histogram of intervolley intervals (Fig. 7), fits a Poisson distribution, and speaks

for a steady change in the interneuronal connections rather than for the formation of impulse circulation caused by facilitation.

4. Changes in activity evoked by training are preserved or easily activated after 12 or more hours.

Unfortunately, not all ganglia possess 'learning capability'. In 40 experiments where isolated ganglia were stimulated, only 20% showed such capability. The difference between ganglia capable of learning and those incapable of learning remains obscure. Apparently, there must exist a better 'criterion of learning' independent of the individuality of the given preparation and manifesting itself in the change of activity of the nerve trunks as well. Still, it seems that the registration of these changes when they take place after stimulation is feasible only with the help of a computer.

SUMMARY

An attempt was made to demonstrate the possibility of the formation of long-term memory traces in the isolated ganglia of the mollusc *Lymnaea stagnalis*. Two nerve trunks of isolated oesophageal ganglia ring were stimulated. The nervous activity was registered in four other nerve trunks. After 40–50 stimulations of the nerves during 1.5–2 h, the original pattern of activity was changed, and some time later, powerful bursts of activity similar to the responses to stimulation but appearing at irregular intervals were observed. In a few cases, such bursts of activity were registered 12–24 h after stimulation was stopped. After this kind of activity faded, it was sufficient to give 5–10 stimulations to restore the burst pattern of activity. It seems the change in the pattern of the impulse activity in the isolated brain of the *Lymnaea stagnalis* after stimulation reflects steady changes in the nature of interneuronal connections as a result of systematic stimulation. This phenomenon may be regarded as analogous to learning.

REFERENCES

FISCHEL, W. (1931): Dressurversuch mit Schnecken. *Z. vergl. Physiol.* **15** 50.

HUGHES, G. M. and TAUC, L. (1962): Aspects of the organization of central nervous pathways in *Aplysia depilans*. *J. exp. Biol.* **39** 45—69.

SOKOLOV, V. A. (1959): Conditioned reflex in Gastropod mollusc *Physa acuta*. *Vestnik Leningrad Univ.* 9, 2. ser. Biol. (in Russian).

TAUC, L. and HUGHES, G. M. (1963): Modes of initiation and propagation of spikes in the branching axons of molluscan central neurons. *J. gen. Physiol.* **46** 533—549.

VEPRINTSEV, B. N. and DYAKOVA, T. L. (1967): Cytological changes in neurons of nudibranchia mollusc *Tritonia* in the process of action potential generation. *Biophysics* **12** 104—113 (in Russian).

DISCUSSION

A. O. D. Willows: You said that you failed to obtain classical conditioning in these animals. What conditioning situation did you try?

B. N. Veprintsev: We tried to obtain a conditioned reflex model with preparation of isolated ganglia of *Lymnaea stagnalis*. The stimulation of the pedal nerve was used for reinforcement, while stimulation of the optical nerve was used for conditioning. The effect was estimated by the response of n. collumelus, which governs the retraction of the leg. It was expected that combined stimulation of the optical and pedal nerves would in time change the pattern of activity in n. collumelus during the stimulation of the optical nerve alone. Unfortunately, the results obtained were rather ambiguous.

A. O. D. Willows: Which works do you know of which refer to conditioning in Gastropoda?

B. N. Veprintsev: I know of two works, that of W. Fischel, (1931), and that of V. A. Sokolov (1959) from Leningrad, who obtained classical conditioning in *Physa acuta*.

M. J. Wells: Have you tried to cool the preparation? Did it not help to reduce learning?

B. N. Veprintsev: We tried cooling to 10–12 °C, but this did not interfere with learning. Of course, it is necessary to use further cooling to reduce all activity in ganglia. I hope to do this in the future.

Symposium on Neurobiology of Invertebrates 1967 (423—441)

THE LEFT AND RIGHT GIANT NEURONS (LGC AND RGC) OF
APLYSIA

G. M. HUGHES

Department of Zoology, University of Bristol
Bristol, England

INTRODUCTION

Mainly because of the large amount of physiological and pharmacological work being done with the nervous system of *Aplysia*, the large size of its neurons is now very well known. However, apart from the work of Tchou Si-Ho (1942) and Arvanitaki and Tchou Si-Ho (1942), few quantitative observations seem to have been recorded on the relative size of its nervous system or on the numbers and sizes of neurons in individuals of different sizes. There has, in fact, been relatively little work of this kind carried out on molluscs with the exception of *Octopus* (Young 1963), *Anodonta* (Gubicza 1965) and *Tritonia* (Dorsett 1967). As part of a study on recognizable cells in the ganglia of *Aplysia*, it was decided to make some measurements of the two most easily distinguishable neurons in the CNS of a series of specimens.

The left and right giant cells (LGC and RGC) contained in the left pleural and right side of the abdominal ganglia respectively, were also selected because of the interest in these cells from the point of view of their anatomy (Hughes and Tauc 1963; Hughes 1967), their biophysical properties (Fessard and Tauc 1956; Kandel and Tauc 1965), their pharmacological properties (Tauc and Gerschenfeld 1962) and their learning capabilities (Bruner and Tauc 1966).

The results obtained gave further support to the view that these were a pair of homologous and symmetrical cells which have become asymmetrically placed during the course of development (Hughes and Tauc 1961; Hughes and Chapple 1967).

(A) MEASUREMENTS ON A POPULATION OF *APLYSIA*

The animals used in this study were taken from several hundred *Aplysia fasciata* collected at Naples during a few days in April, 1967. Figure 1 shows the weight distribution in this collection, which proved to be continuous, in spite of an impression, before all the animals had been weighed, that this might not be so. A cumulative percentage curve for this population falls almost entirely along a single straight line, and therefore supports the view that the population investigated was part of a single generation (compare Carefoot 1967). The animals were

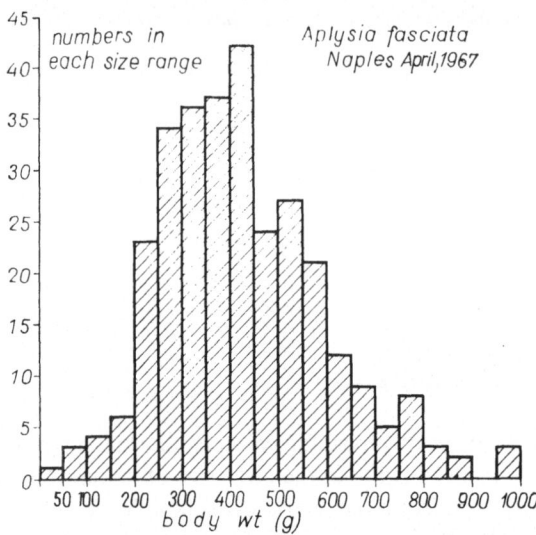

FIG. 1. *Aplysia fasciata*. Histogram to show distribution of body weights among 300 individuals of a population studied at Naples

taken from the aquarium, and the external water removed by draining and absorption on towels. About 15% of the specimens were dissected and the major part of their central nervous ganglia and connectives removed. All nerves were cut close to the ganglia, but the connectives remained between the cerebral, pleural, pedal, buccal, and abdominal ganglia. This formed the central nervous system, which was dried on blotting paper and weighed. No attempt was made to remove the connective tissue which forms a sheath to the ganglia and connectives, but all other connective tissue was dissected away. The result of weighing these specimens (Fig. 2) showed that there was an increase in weight of the CNS with increasing body weight, but there was a considerable scatter in the data, especially with the larger weights. Two main reasons can be suggested for this scatter, in addition to normal variations in size of the nervous tissue: (a) variations in the degree of hydration of the ganglia and connectives themselves; and (b) variations in the hydration of the whole of the body of the animal, leading to variations in the body weight taken. The existence of the first of these factors was only judged by the appearance of the ganglia and sheaths, but the second was tested for a sample of the animals. This was done by weighing the remaining part of the animal following the removal of the central nervous ganglia and connectives. The major drop in weight was due to the loss of body fluid and, as can be seen from Fig. 2, most of the values obtained are close to 50%* of the original body weight (mean for 18 specimens = 49.4%; range 26–70%). There were some notable exceptions to this, however; for example, the animals weighing 800 and 810 g, respectively. In these two specimens, the body weight after dissection was 210 g. It is notable

---

* Similar measurements on a sample of *A. fasciata* collected at Arcachon in September 1967 gave values of about 40% (average of 15 animals = 40.3%; range 29–50%).

FIG. 2. (a) Weight of the central nervous ganglia and connectives of animals from a range of body weight. The individual measurements are plotted as filled circles and a full line shows the general increase in CNS weight with body size. (b) The weights of *Aplysia* following dissection are plotted as crosses, and a dashed line is drawn to show where these points would fall if the animal lost 50% of its body weight as body fluid during dissection

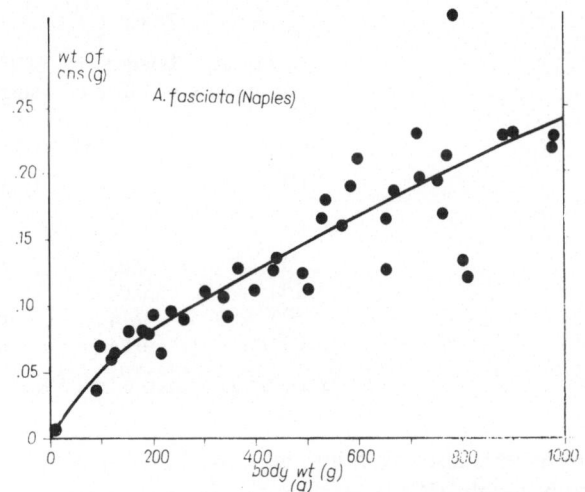

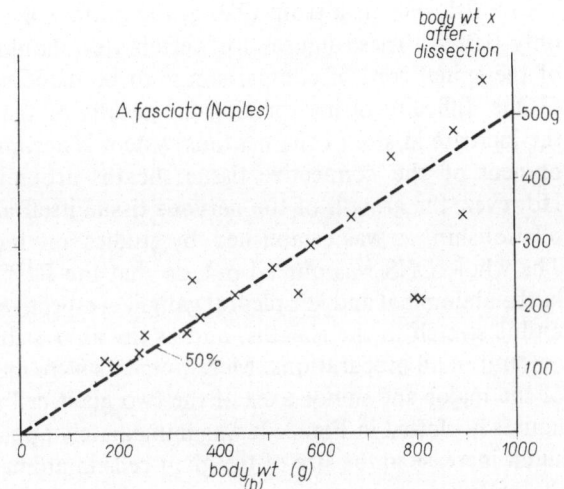

that the weight of the CNS was exceptionally low (0.133 and 0.121 g) in both cases. This suggests that their body weight under conditions of hydration comparable to the other specimens would be about 400 to 500 g, and if this were the case their CNS weight would be very close to that of others in this weight range.

From the curve showing the relationship between body weight and CNS weight, it is apparent that not only is there a general increase in weight of the nervous system with increasing size, but also that its percentage of the body weight decreases, as summarized in Table 1.

But perhaps the most remarkable feature is the relatively low proportion of the body weight which the CNS comprises in all specimens. There are relatively few figures of this kind but for man's brain it is 2.5% and in *Octopus* it is probably about 1% for the brain alone, but if the weight of the nerve cords in

G. M. HUGHES

TABLE 1

A. fasciata. Relative size of CNS
in animals of different weights

Body weight g	CNS weight	
	g	% of body wt
50	0.03	0.06
100	0.05	0.05
500	0.14	0.028
1000	0.24	0.024

Mean: 492 g   Mean: 0.139 g   Mean: 0.0282%

the arms is also taken into account then it may approach as much as 5% of the body weight (Wells, personal communication).

In a fish such as a trout (220 g) the brain is 0.19% and in a cod of 10 kg, it is only 0.05%. These figures for vertebrates should be at least doubled, because of the spinal cord, if comparison is to be made with the invertebrate CNS.

One difficulty in interpreting the results of the *Aplysia* measurements is that the increase in size of the nervous system is accompanied by an increase in development of the connective tissue sheaths around the ganglia and connectives. However, the growth of the nervous tissue itself also appears to follow a similar relationship, as was established by studies on individual cells of the ganglia. The whole CNS was pinned out, so that the RGC and LGC were clearly visible in the abdominal and left pleural ganglia, respectively. Care was taken not to apply undue stretch to the sheaths, and as far as possible this was done to a constant amount in all preparations. Measurements were made with a micrometer eyepiece of the major and minor axes of the two giant cell somata, and the mean of these figures is plotted in Fig. 3. It was immediately apparent that there was a relatively slight increase in the size of the giant cells in animals above 100 g. This confirmed the impression obtained during dissections of the nervous system when making physiological preparations that the two giant cells are more conspicuous in smaller specimens. This factor is also aided by the more transparent nature of the sheaths surrounding the ganglia, since these two cells are usually darkly pigmented. In animals above 100 g, the two cells fell within the size range 300–600 μ mean diameter. It is clear from this plot that the two giant cells are very similar in their diameters, but, if anything, the right cell is slightly smaller in most cases. This is hardly significant, however, when all the data is taken into account, as seen in Fig. 3 where lines have been drawn to show two standard deviations from the best straight line for the data obtained for each cell. Close similarity in the slope of these lines and the overlap in the areas they enclose once more emphasizes the similarity between the two cells. It is of interest that measurements obtained for giant cells of *A. fasciata* collected in the Bassin d'Arcachon, France, in Sep-

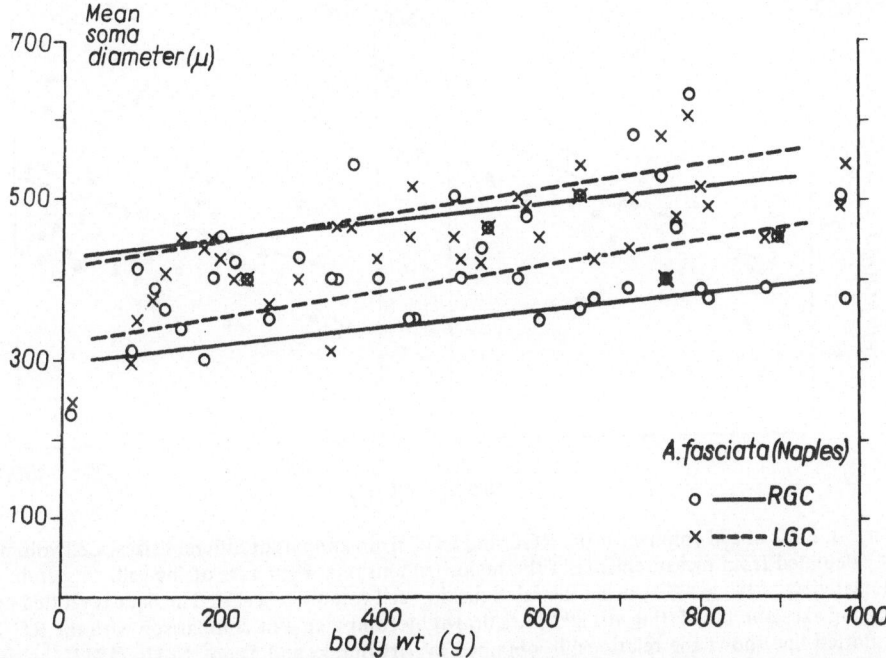

FIG. 3. Mean diameter of the somata of the RGC and LGC in the central nervous ganglia of animals from a range of body sizes. For each cell, a pair of lines has been drawn by a computer to show the spread of the data for one standard deviation on each side of the best-fitting straight line

tember, 1967 showed that 60% of the points fell above the upper standard deviation lines shown in this figure. This confirmed the impression that the cells of *A. fasciata* at Naples were relatively smaller than those observed previously during physiological experiments at Arcachon.

From these measurements it was later decided to obtain some measure of the cell volume by the same method used by Arvanitaki and Tchou Si-Ho (1942). The area of the apparent surface (S_a) of the cell approximated to a rectangle was obtained from the two linear dimensions measured and the assumption that the cell volume was proportional to the 3/2 power of the apparent surface area, i.e., $S_a^{3/2}$. The figures obtained produced a more scattered appearance, and it was not readily apparent on a log/log plot which was the most preferable straight line. A computer was used to determine the best straight line to the data, as shown in Fig. 4. The slope of these two lines is fairly close and provides evidence for the similar growth characteristics of the somata of these two cells.

These findings are also interesting in relation to the work of Arvanitaki and Tchou Si-Ho (1942) on *A. depilans*. They found that the slope of the line varied from cell to cell, but of great interest was the way in which the slope showed a marked change at a given weight for a particular cell (Fig. 5).

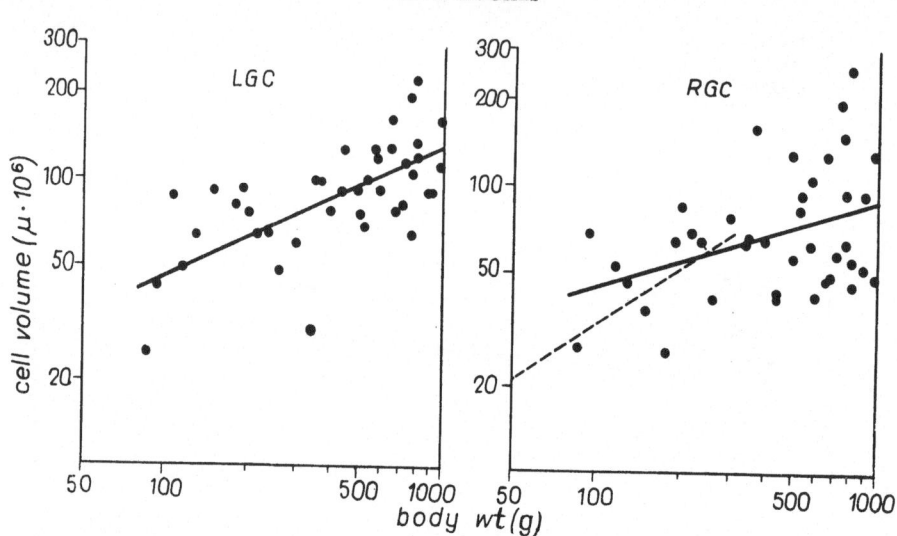

Fig. 4. Plots of cell volume for the RGC and LGC from animals of different sizes. Cell volume is calculated from measurements of the major (x) and minor (y) axes of the cell, as seen in a dorsal view of the ganglia, using the relationship cell volume $=(xy)^{3/2}$. The data is plotted cn log/log axes and best-fitting straight lines drawn by computer. For comparison with the RGC, a dotted line shows the relationship obtained by Arvanitaki and Tchou Si-Ho (1942) for cell A in *Aplysia depilans*

In their paper, Arvanitaki and Tchou Si-Ho (1942) drew attention to the special growth characteristics of their cell A from the right upper quadrant of the abdominal ganglia. This cell did not show a sudden change in slope of the log/log plot, and, furthermore, its gradient was least steep. The latter feature seemed to be associated with the large size of this neuron, and there is little doubt that neuron A is the RGC of *A. depilans* (Arvanitaki, personal communication). Although there are differences in the slope of the line obtained by Arvanitaki and Tchou Si-Ho for cell A (0.65 compared to 0.47 and 0.30 for the LGC and RGC), it is perhaps most significant that in all three cases this value is positive, and there is no evidence for a change in slope, i.e. the growth constant α in the equation $y = bx^{\alpha}$ remains the same at all sizes. Thus, in spite of the wider scatter of the data on the log/log plot obtained in the present work, it seems to conform to the same general growth characteristics as the RGC of *A. depilans*.

In conclusion, then, these observations on the LGC and RGC, together with the fact that they are usually the most darkly pigmented cells on each side, support the electrophysiological evidence discussed below suggesting that these two giant cells are homologous.

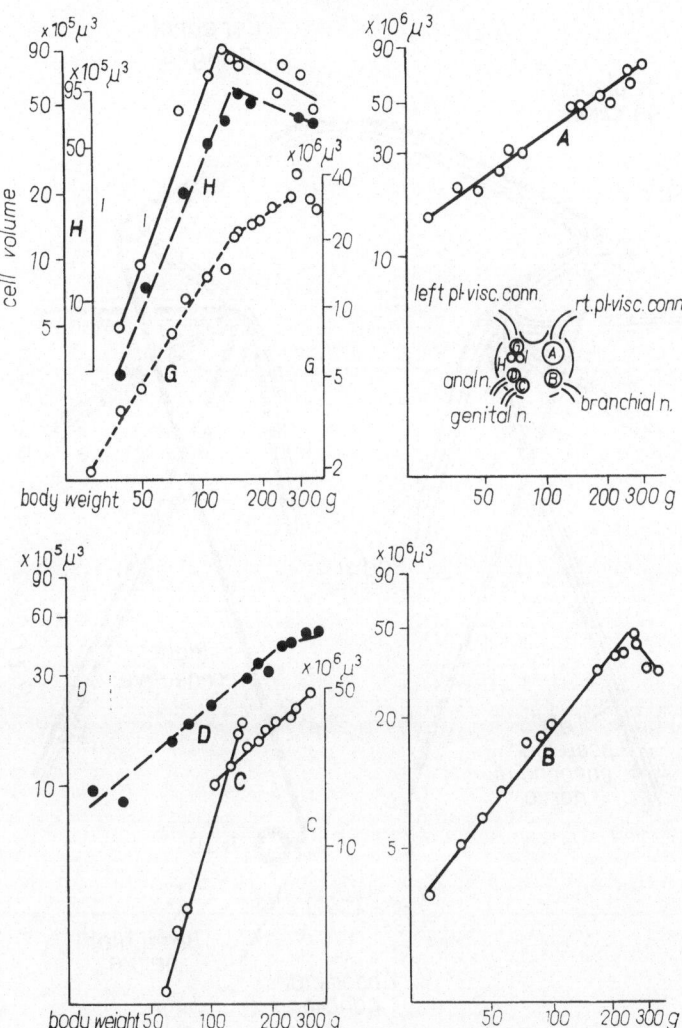

FIG. 5. Increase in size of the soma of 7 identified neurons from the abdominal ganglia of *Aplysia depilans*. The data is plotted on log/log coordinates in relation to body weight. The position of the cell is shown in the inset diagram of a dorsal view of the ganglia (after Arvanitaki and Tchou Si-Ho 1942)

(B) ELECTROPHYSIOLOGICAL ANATOMY OF THE LGC AND RGC

During work on the physiology of the right giant cell, many workers have shown that a large axon of this cell enters the right connective. Thus, each time the cell fires, a large potential is recorded by extracellular electrodes hooked under the connective. Stimulation of the connective is followed by the recording of an

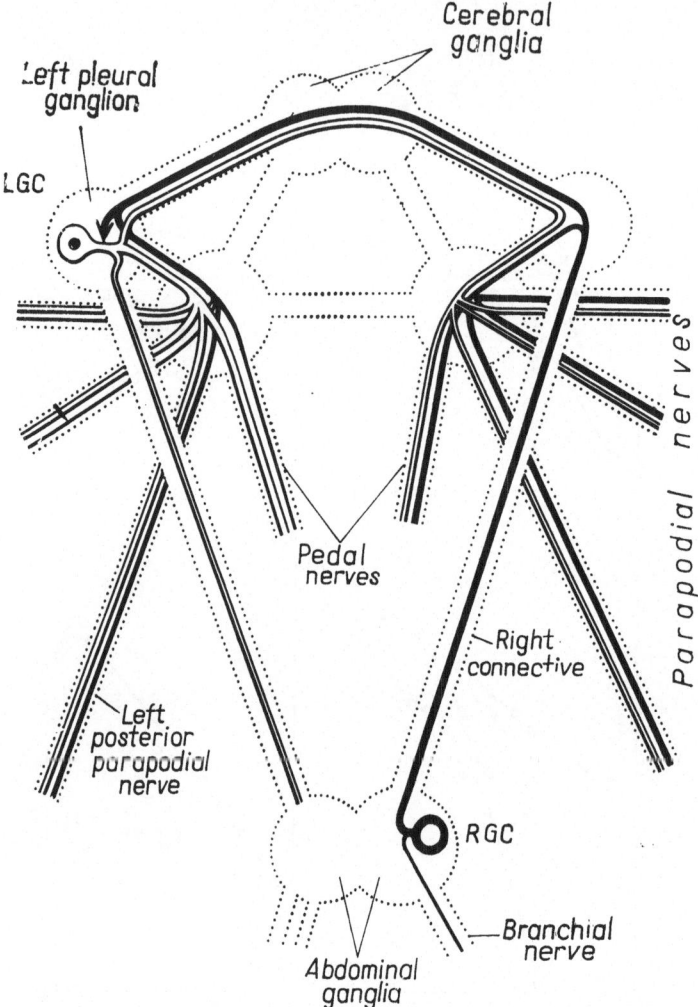

FIG. 6. *Aplysia fasciata*. Diagram, based upon electrophysiological evidence, of the main central nervous ganglia and
connectives showing the typical branching pattern of the LGC
and RGC. Some indication of the probable relative sizes of the
branches can also be deduced from the experimental data

antidromic potential in the RGC soma, and detailed investigations have been
made on this potential and also on the effect of synaptic input to the giant cell
(Tauc 1957, 1962). We began to investigate the function of the RGC by experiments in which activity was recorded intrasomatically and stimulation carried out
through a microelectrode in the cell soma of isolated and whole animal preparations (Hughes and Tauc 1961, 1963). It was soon discovered that branches of
the main axon in the right connective enter many of the nerves which innervate

rt. mid. parapodial n.

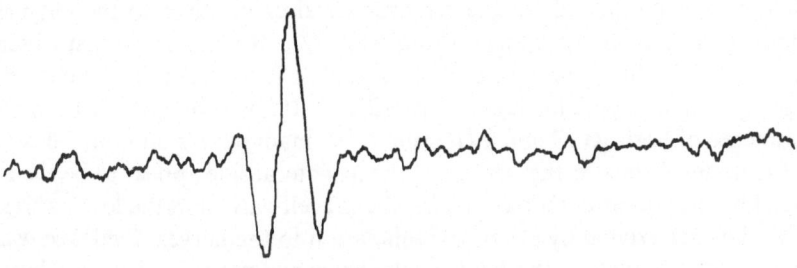

left connective

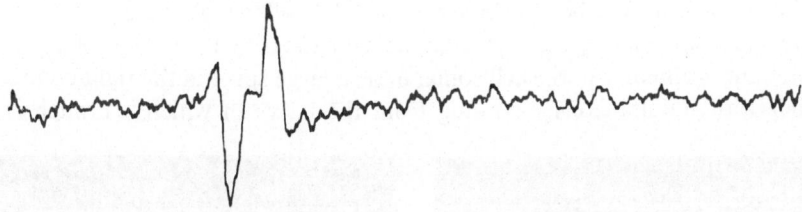

rt. post. parapodial n.

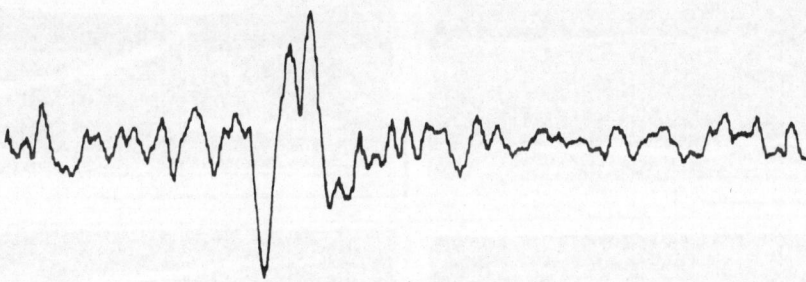

branch of rt. ant. parapodial n.

50 msec

FIG. 7. Recordings obtained by means of a BIOMAC 1000 computer showing the effect of averaging 256 sweeps. Each sweep is triggered by the spike in the LGC soma and recordings are made from different nerves. The clear presence of an action potential at a definite interval (conduction time) after the cell has fired is shown for the right middle and posterior parapodial nerves and for the left connective. In this preparation, no evidence was obtained for an LGC axon in this particular branch of the right anterior parapodial nerve. Time calibration, 50 msec

the foot and parapodia on the right side of the animal. It was also shown that the main branch continued via the cerebropleural connectives to the left pleural ganglion. It was during the course of this work that the LGC was first observed, and a study of its branching pattern showed that it has axonal branches in the main parapodial and pedal nerves of the left side. It has subsequently been shown that branches of both RGC and LGC run in the main nerves on both sides of the animal (Fig. 6). Axons in the nerves on the ipsilateral side appear to be the larger as judged by the size of potentials recorded extracellularly and the lower threshold at which they are excited by electrical stimulation to the nerves. Evidence was obtained for a fine branch of the RGC in the branchial nerve, as has also been observed in *Aplysia californica* by Strumwasser (1967). Up to the present time, all the evidence for the branching of these neurons is based on electrophysiological experiments which may be considered under the following headings.

1. *Direct stimulation of the cell soma*, accompanied by recordings of the action potential intracellularly in the cell soma, or as a large spike in the right connective, together with simultaneous recording from the nerve in which a branch is sus-

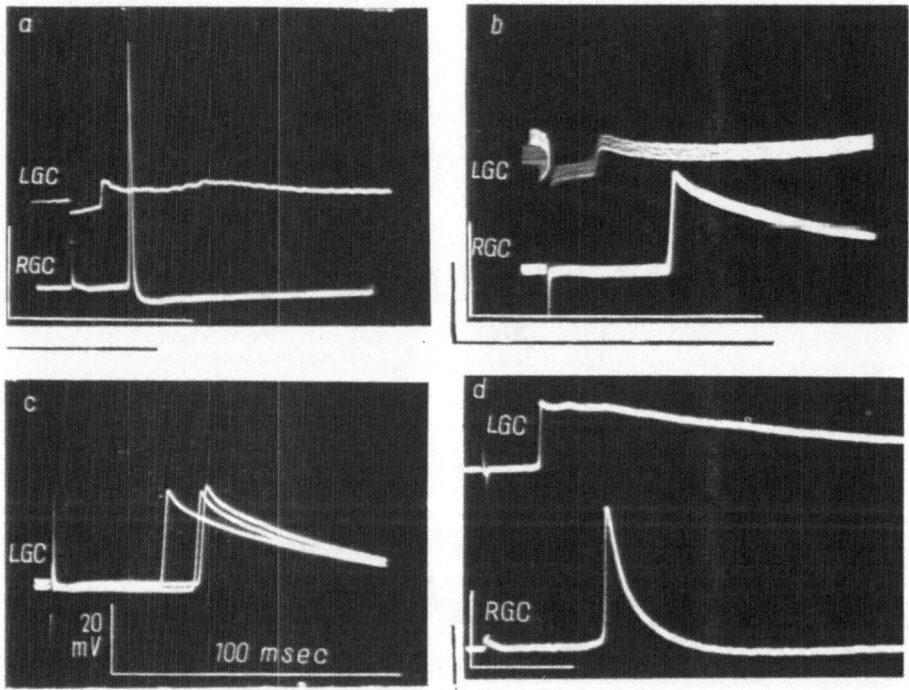

FIG. 8. Simultaneous recordings from the right and left giant cells following stimulation of the pedal (a), middle (b) and posterior (d) parapodial nerves on the right side. In these recordings an *A* spike is recorded in both cells and indicates the presence of a branch of the axon of both giant cells in these nerves. In (b) many sweeps are superimposed and show the constant delay. (c) Identical *A* spikes in the LGC following stimulation of the posterior and middle parapodial nerves and the pedal nerve of the left side. Note different delay times

pected. When the cell is fired alone by electrical stimulation, it is more often possible to distinguish the action potential in the nerve concerned, especially by the use of a time base triggered by the somatic spike and superimposing many sweeps. However, because of the spontaneous activity present in the parapodial nerves and the relatively small size of the action potential in the giant cell axon, it is often very difficult to observe this spike. This work has very recently been facilitated by the use of an averaging computer, and confirmation has been obtained for branches of the LGC in the right parapodial nerves, which previously had proved very difficult using photographic integration. The averaging computer also enabled us to show the presence of a branch of the left giant cell in the left connective (Fig. 7), although such a branch had not been suspected in experiments using the techniques mentioned below (Hughes, Weevers and Hartley, unpublished).

2. *Antidromic stimulation.* Electrical stimulation of a nerve which contains a branch of the giant cell leads to the establishment of an action potential which, when it arrives at the cell soma, produces a characteristic A or AS spike which can usually be distinguished from a synaptically evoked spike by studying the effects of hyperpolarization and high-frequency stimulation (Tauc 1957; Hughes and Tauc 1962). Following such stimulation, action potentials were also recorded

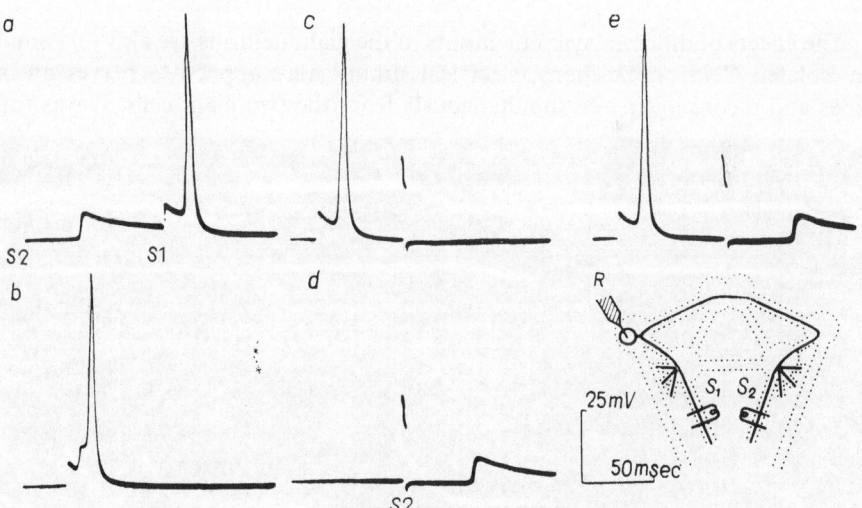

FIG. 9. *Aplysia fasciata.* Demonstration of the presence of a branch of the LGC in both the right and left pedal nerves by collision experiments. The A spike from the left side is followed by invasion of the soma, whereas stimulation of the right nerve (S2) is not. Successive recordings are shown in which the interval between these two shocks is changed. In (a) and (b), the S2 spike precedes S1 and there is no blocking action, but in (c), when the S1 spike precedes the S2 spike, the A spike resulting from S2 is absent, (d) shows the effect of stimulating S2 alone at the same delay as in (c). In (e), S2 is sufficiently delayed after S1 that the impulse producing the A spike is not blocked by collision with that arising following stimulation of S1

from other branches of the giant cell axon by extracellular electrodes on the other parapodial nerves. It should be noted that the delay between the recording of a spike in the cell soma and in one of the parapodial nerves, corresponds to the delay between a stimulus and the recording of an *A* spike in the cell soma and remains constant during repetitive stimulation (Fig. 8).

3. *Impulse collision experiments.* Where two nerves are believed to contain branches of the same axon, it is possible to produce blockage of one because of the collision of impulses if stimuli are applied to the two nerves within certain intervals of one another. Where such collisions occur, only one *A* spike will be recorded in the cell soma. In many ways, this is perhaps the most unequivocal technique, and the experiments illustrated in Fig. 9 clearly demonstrated the existence of branches of the LGC in the right pedal nerve, as had been suspected by recording of an *A* spike in the cell soma following stimulation of both nerves.

From investigations of this type, the electrophysiological anatomy of the right and left giant cells has been elucidated and is shown diagrammatically in Fig. 6. Confirmation of this type of branching of giant neurons into many nerves on both sides of the body has recently been obtained in *Tritonia* by Dorsett (1967) using similar techniques.

(C) SYNAPTIC CONNECTIONS OF THE LGC AND RGC

The effects of different synaptic inputs to the giant neurons are also very similar. In isolated CNS preparations, electrical stimuli were applied to nerves on both sides and recordings made simultaneously from the two giant cells. It was found

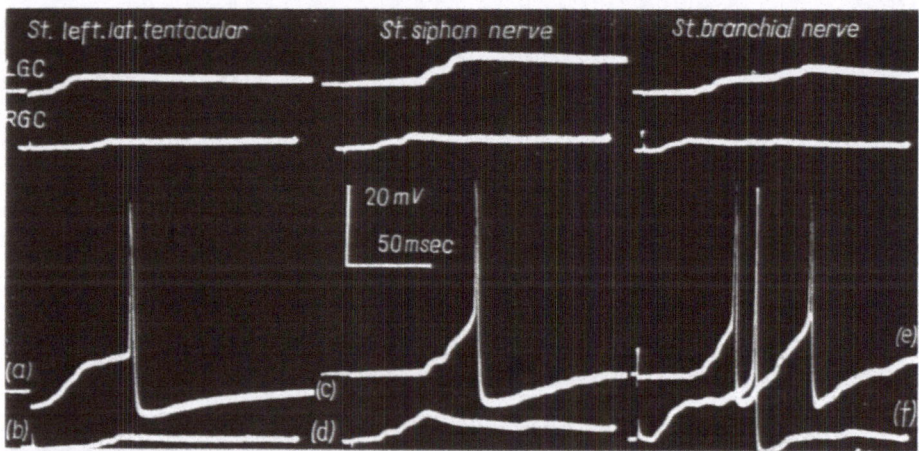

FIG. 10. Recordings showing the similar synaptic effects on the LGC and RGC which result from stimulation of different nerves (a) and (b), left lateral tentacular nerves; (c), (d), stimulation of the siphon nerve; (e), (f), stimulation of branchial nerve. In each case, stimulation at two different intensities is shown — the upper row of recordings are subthreshold for the initiation of a propagated spike, whereas the compound EPSP's are sufficient to do this in the lower row of recordings, at least for the LGC. (After Hughes 1967)

that the amplitude and time-course of EPSP's following a shock to a given nerve were often very alike (Fig. 10). This was a little unexpected, as it might have been supposed that the LGC would be more responsive to stimulation of nerves from the left and anterior part of the animal and *vice versa* for the RGC. This was not the case, but the LGC tended to have the lower threshold and responded more readily to all types of synaptic input. In whole animal preparations, mechanical stimulation of any part of the body has been known to affect the RGC, stimulation in the siphon and gill regions being most effective. Recent recordings from the LGC in whole-animal preparations (Hughes, Weevers and Hartley, unpublished) have shown that this cell is also responsive to mechanical stimuli all over the body. In most preparations, the threshold was generally lower to stimulation of the left parapodium than the right parapodium. These preparations also confirmed the observation made with electrical stimulation that the LGC tends to be more excitable than the RGC, but this varied from preparation to preparation. Both giant cells show a decline in their responsiveness with a repetition of the stimulus, which appears to be a form of habituation (Hughes and Tauc 1963; Bruner and Tauc 1966).

Simultaneous recordings from the LGC and RGC have given good evidence that they both receive input from many common interneurones (Hughes 1967), and has been confirmed in whole-animal preparations. It seems likely that the general similarity in form of the compound EPSP's recorded from the two cells is related to this feature of the organization of their synaptic connections. It has also been shown (Hughes and Tauc 1965, 1968) that the two cells have a direct synaptic connection between them, because each time the RGC fires, a biphasic postsynaptic potential (BPSP) is recorded in the LGC after a delay of about 0.15 sec. This type of synaptic connection seems to be typical of the two giant cells, because in a unique preparation both cells were in the pleural ganglia, and it was found that a BPSP could also be recorded in the RGC following stimulation of the LGC. In this unique preparation, the branching pattern of the two cells was symmetrical, which suggests that normally they may be symmetrical early in development.

(D) BIOPHYSICAL PROPERTIES

Because of the relative ease with which two or three microelectrodes may be inserted into the soma, these cells have frequently been used in experiments to determine membrane resistance, capacitance, and other properties. A summary is given in Table 2 and shows their close similarity for many of these parameters. Both cells are hyporpolarized by acetylcholine and show anomalous rectification in their current–voltage curves. The very long time-constant of both cells is of particular interest, and no doubt is related to their very large size, which may therefore have some significance in relation to the integrative function of these cells, as it means that inputs arriving during relatively long periods can be summed. The

G. M. HUGHES

TABLE 2

Membrane properties of somata of the RGC and LGC in *Aplysia fasciata* (after Hughes 1967)

Position in CNS	RGC Abdominal ganglia	LGC Left pleural ganglion
Diameter (μ)	500–700	600
Resistance	Slope $10 \cdot 10^5 \Omega$* and $2.2 \cdot 10^5 \Omega$	$3.55 \cdot 10^5 \Omega$ and $2.33 \cdot 10^5 \Omega$
Specific membrane resistance ($K\Omega$/cm^2)	11.3–2.5*	4.01–2.63
Time constant (msec)	100–200	160
Specific membrane capacity (F/cm^2)	12.4–56*	40–60
Pacemaker sensitivity $\dfrac{\text{Impulses/sec}}{A \cdot 10^{-9}}$	0.3–0.4	0.7–0.9
Rheobase ($A \cdot 10^{-9}$/cm^2)		177
Acetylcholine	Hyperpolarize	Hyperpolarize

* Kandel and Tauc (1965).

specific membrane resistance and capacitance of these cells are very high, and it has been suggested (Hughes 1967) that this may be related to the membrane structure of the cell soma, which is known to be folded a great deal. The infoldings have generally been supposed to help satisfy their nutritional and respiratory requirements (Chalazonitis and Arvanitaki 1966), but may also be related to their function in integration if they increase the membrane capacitance and resistance.

(E) DISCUSSION

The data on the growth of the right and left giant cells reported in this paper supports the electrophysiological evidence for the hypothesis that the two giant cells are a pair of symmetrical cells which during development have become displaced as a result of morphogenetic changes in the visceral loop. Figure 11 shows a hypothetical ancestral condition with a visceral loop containing pairs of parietal and intestinal ganglia and a single terminal visceral ganglion. It suggests that the RGC and LGC somata were originally in the paired parietal ganglia and that during developement there is a coalescence of the ganglia on the visceral loop such that the right parietal and intestinal ganglia have become fused with the visceral ganglion and also the main part of the two ganglia on the left side, thus producing the abdominal ganglion complex. However, on the left side, that part of the parietal ganglion containing the LGC appears to become fused with the left pleural ganglion. Thus, the abdominal ganglion in *Aplysia* is made up of the visceral ganglion together with two or more other ganglia. It is for this reason that perhaps the more neutral term 'abdominal ganglia' is to be preferred, rather than equating this ganglion with one of the more morphologically defined ganglia of the typical visceral loop of the gastropods.

It is clear that before a complete understanding is obtained of the structure and function of neurons such as the LGC and RGC (after Hughes 1967), it is essential to take into account as many features of their organization as possible, with the growth characteristics being just as important as the biophysical properties.

This study has shown that the two giant cells are already very large in animals nearly 100 g in weight. It therefore suggests that the cells are of importance to the animal at all stages of its life cycle. Unfortunately, no work has yet been possible on the post-veliger stages, which would give best evidence as to the disposition of these two cells in the earlier stages of development. The large size of the somata of these and other molluscan neurons, has aroused the interest of many neurobiologists, but as yet no completely satisfactory explanation is available as to its significance. The cell soma is generally considered to be the nutritive part of the neuron, and perhaps their size is related to the very extensive axonal branching. However, neurons with very long and branching axons are known in other animals, but they do not always have large somata. It has generally been presumed that the characteristic infolding of the soma in molluscan neurons is related

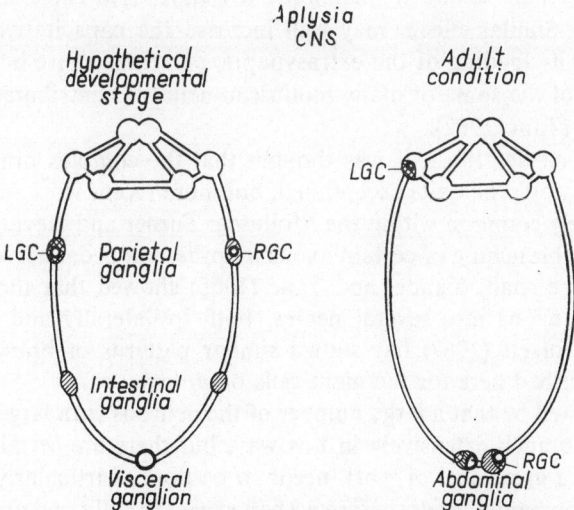

FIG. 11. Diagrams to illustrate the hypothesis that the asymmetrical position of the RGC and LGC somata in the central nervous ganglia of *Aplysia* has resulted from asymmetries in the distribution of the principal ganglia of the visceral loop during development. It can be seen that the abdominal ganglia of *Aplysia* is composed of several of the primary ganglia, including the visceral ganglia

to their large size and the need to increase the surface area through which the soma may receive nutrients (Amoroso *et al.* 1964; Bullock 1961; Borovyagin, Sakharov and Veprintsev 1964). The axons also have such folded surfaces (Schlote 1957; Batham 1961) and the glial elements both here and in the intuckings of the soma contain granules of glycogen (Rosenblueth 1963). It has been noted, however, that the characteristic infolding is not present in axons below a diameter of about 2μ, and for somata of *Archachatina* they are less evident in smaller neurons ($<60\mu$) (Amoroso *et al.* 1964). It would be of interest to compare the growth characteristics of neuron somata of different sizes in relation to the degree of development of the infoldings of the somatic surface membrane. One might expect a decrease in the apparent rate of growth, as judged by the method used here and by Arvanitaki and Tchou Si-Ho. The observation of such changes by Arvanitaki and Tchou Si-Ho, especially for the smaller cell somata, may well be related to sudden increases in the infoldings.

Another effect of the infolding is to increase the extent of the cell membrane and hence to give false values for the specific membrane resistance and capacitance as normally calculated (Fessard and Tauc 1956). The effect of similar intracellular processes which provide alternative channels for current flow between the inside and outside of frog and crustacean muscle fibers, has recently been discussed by Falk and Fatt (1964). They suggest that these changes may contribute to the apparently high values of membrane resistance and capacitance that have been measured. Similar effects may well increase the capacitative effect of the cell soma and its loading of the extrasynaptic currents. There is little doubt that the large size of the soma of many molluscan neurons contributes to their integrative functions (Tauc 1962).

When first described, it was thought that the complex branching of the RGC and LGC would be quite exceptional, but more recent work has shown that it may be relatively common within the Mollusca. Turner and Nevius (1951) had already shown the branching of certain axons in ipsilateral pedal nerves of the slug *Ariolimax*. In the snail, Kandel and Tauc (1966) showed that the metacerebral giant cells send axons into several nerves, both ipsilaterally and contralaterally, and recently Dorsett (1967) has shown similar patterns of branching in *Tritonia* to those described here for the giant cells of *Aplysia*.

It may well be that a large number of the neurons with large giant somata in the Mollusca branch extensively in this way, but there are certainly exceptions. It is clear that a great deal of work needs to be done, particularly at the biochemical and developmental levels, before we can expect to fully understand the significance to the size of molluscan neurons and the function of the LGC and RGC.

SUMMARY

1. A survey is given of some of the growth characteristics, anatomy, synaptic connections, and biophysical properties of the left and right giant cells of *Aplysia*, which are the largest cells on each side of the CNS. The LGC is contained in the

left pleural ganglion and the RGC is found in the right upper quadrant of the abdominal ganglia.

2. The relative size of the main central nervous ganglia and connectives has been determined for a population of *Aplysia fasciata* over the weight range 50–1000 g. It is shown that the main ganglia and connectives usually constitute about 0.05 % of the body weight in smaller specimens, but 0.025 % in larger animals. Some variations in the data obtained were due to differences in the amount of body fluid of the animal, which was usually about 50 % but in certain specimens was as great as 75 %.

3. Measurements of the size of the LGC and RGC showed that they are normally from 300 to 600 μ in diameter for this range of body sizes, and are relatively larger in the smaller specimens.

4. The cell volume of the LGC and RGC increases with a growth constant (α in the relationship $y = bx^\alpha$) of 0.47 and 0.30, respectively.

5. An account is given of the present views on the extensive branching of these two neurons in the main nerves and connectives of the CNS based upon electrophysiological evidence.

6. Both giant cells are relatively inactive in whole-animal preparations but respond to stimuli at any place on the body surface. Electrical stimuli to nerves have similar synaptic effects on the two giant cells, and these observations may be related to the presence of numerous common interneurons.

7. It is concluded that the two cells originated as a symmetrical pair in ganglia of the visceral loop which during development have coalesced in such a way that they are now asymmetrically placed although their branching pattern, etc., is almost symmetrical.

Acknowledgments: I wish to thank the Royal Society for the financial assistance to help me carry out at Naples some of the work described here and also to attend this Symposium. It is also a pleasure to record my thanks to Dr. P. Dohrn and Professor R. Weill of the Stazione Zoologica di Napoli and the Institut de Biologie Marine at Arcachon, respectively, for the accommodation they provided, and to their staffs for collecting the animals. The use of the IBM 1300 computer was generously made available by Hydronautics Incorporated, Laurel, Maryland, U.S.A. Some of the work at Arcachon was supported by a grant from the Science Research Council.

REFERENCES

AMOROSO, E. C., BAXTER, M. I., CHIQUOINE, A. D. and NISBET, R. H. (1964): The fine structure of neurons and other elements in the nervous system of the giant African land snail *Archachatina marginata*. *Proc. roy. Soc. B.* **160** 167–180.

ARVANITAKI, A. and TCHOU SI-HO (1942): Les lois de la croissance relative individuelle des cellules nerveuses chez l'Aplysie. *Bull. Histol. Appliguée* XIX 244–256.

BATHAM, E. J. (1961): Infoldings of nerve fibre membranes in the Opisthobranch mollusc *Aplysia californica*. *J. biophys. biochem. Cytol.* **9** 490–492.

BOROVYAGIN, V. L., SAKHAROV, D. A. and VEPRINTSEV, B. N. (1964): Satellites of nerve cells in gastropod ganglion. *Proc. 3rd Europ. Reg. Conf. on Electron Microscopy*. Prague, 273.

BRUNER, J. and TAUC, L. (1966): Long-lasting phenomena in the molluscan nervous system. *Symp. Soc. exp. Biol.* **20** 456—475.

BULLOCK, T. H. (1961): On the anatomy of the giant neurons of the visceral ganglion of *Aplysia*. From "*Nervous Inhibition*". Pergamon Press, London.

CAREFOOT, T. H. (1967): Studies in a sublittoral population of *Aplysia punctata*. *J. mar. biol. Ass. U. K.* **47** 335—350.

CHALAZONITIS, N. and ARVANITAKI, A. (1966): L'"espace gliosomatique" dans les ganglions d'*Aplysia* et diffusibilité de l'oxigène. *C. R. Soc. Biol.* (Paris) **160** 1897.

DORSETT, D. A. (1967): Giant neurons and axon pathways in the brain of *Tritonia*. *J. exp. Biol.* **46** 137—151.

FALK, G. and FATT, P. (1964): Linear electrical properties of striated muscle fibres observed with intracellular electrodes. *Proc. roy. Soc. B.* **160** 69—123.

FESSARD, A. and TAUC, L. (1956): Capacité, resistance et variations actives d'impédance d'un soma neuronique. *J. Physiol. (Paris)* **48** 541—544.

GUBICZA, A. (1965): Relation of body size, ganglions and neuron dimensions in a freshwater mussel *Anodonta cygnea* L. *Annal. Biol. Tihany* **32** 3—9.

HUGHES, G. M. (1967): Further studies on the electrophysiological anatomy of the left and right giant cells in *Aplysia*. *J. exp. Biol.* **46** 169—193.

HUGHES, G. M. and CHAPPLE, W. D. (1967): The organisation of nervous systems. From "*Invertebrate Nervous Systems*" Ed.: C. A. G. Wiersma, University of Chicago Press, Chapter 14.

HUGHES, G. M. and TAUC, L. (1961): The path of the giant cell axons in *Aplysia depilans*. *Nature (Lond.)* **191** 404—405.

HUGHES, G. M. and TAUC, L. (1962): Aspects of organisation of central nervous pathways in *Aplysia depilans*. *J. exp. Biol.* **39** 45—69.

HUGHES, G. M. and TAUC, L. (1963): An electrophysiological study of the anatomical relations of two giant nerve cells in *Aplysia depilans*. *J. exp. Biol.* **45** 469—486.

HUGHES, G. M. and TAUC, L. (1965): A unitary biphasic postsynaptic potential (BPSP) in *Aplysia* 'brain'. *J. Physiol. (Lond.)* **179** 27—28P.

HUGHES, G. M. and TAUC, L. (1968): A direct synaptic connexion between the left and right giant cells in *Aplysia*. *J. Physiol.* **197** 511—527.

KANDEL, E. R. and TAUC, L. (1965): Mechanisms of heterosynaptic facilitation in the giant cell of the abdominal ganglion of *Aplysia depilans*. *J. Physiol.* **181** 28—47.

KANDEL, E. R. and TAUC, L. (1966): Anomalous rectification in the metacerebral giant cells and its consequences for synaptic transmission. *J. Physiol.* **183** 287—304.

ROSENBLUETH, J. (1963): The visceral ganglion of *Aplysia californica*. *Z. Zellforsch.* **60** 213—236.

SCHLOTE, F. V. (1957): Submikroskopische Morphologie von Gastropodennerven. *Z. Zellforsch.* **45** 543—568.

STRUMWASSER, F. (1967): Types of information stored in single neurons. From "*Invertebrate Nervous Systems*" Ed. C. A. G. Wiersma, University of Chicago Press, Chapter 23.

TAUC, L. (1957): Stimulation du soma neuronique de l'Aplysie par voie antidromique. *J. Physiol. (Paris)* **49** 973—986.

TAUC, L. (1962): Site of origin and propagation of spike in the giant neurone of *Aplysia*. *J. gen. Physiol.* **45** 1077—1097.

TAUC, L. and GERSCHENFELD, H. M. (1962): A cholinergic mechanism of inhibitory synaptic transmission in a molluscan nervous system. *J. Neurophysiol.* **25** 236—262.

TCHOU, SI-HO (1942): *Contribution a l'étude de la physiologie des cellules nerveuses chez l'Aplysie*. Bose et Rion, Lyon.

TURNER, R. S. and NEVIUS, D. B. (1951): The organization of the nervous system of *Ariolimax columbianus*. *J. comp. Neurol.* **94** 239—256.

YOUNG, J. Z. (1963): The number and sizes of nerve cells in *Octopus*. *Proc. Zool. Soc. Lond.* **140** 229—254.

DISCUSSION

A. O. D. Willows: In regard to your statement that the left and right giant cells have axons together in certain trunks, I want to ask how you made certain that the spikes recorded in the cell bodies were indeed antidromic and not due to orthodromic input?

G. M. Hughes: In general, potentials recorded intrasomatically following orthodromic and antidromic inputs can be distinguished by the effect of repetitive stimulation which results in variations in the delay of the orthodromically elicited spike, whereas the antidromically elicited one is not so variable. By hyperpolarizing the soma, it is also possible to distinguish the antidromic potential from the synaptic potential, as the former decreases in size whereas the EPSP tends to increase in size at greater polarizations. However, as I mentioned, the best way of recognizing that a spike recorded in the soma is antidromically produced is by showing that the action potential producing it can be collided with a spike elicited in another branch of the cell, as described in the collision experiments. The method of recording the antidromic potential simultaneously in both cells is not always satisfactory because of the differences in threshold, presumably because of the disparity in size of the RGC and LGC axons in a given nerve trunk. However, it should be emphasized that it is by the combined results of the various electrophysiological methods in many preparations that the anatomy of these cells has been made out.

G. A. Kerkut: Did you record from the cell *in situ* in a swimming animal?

G. M. Hughes: No, we have not yet recorded from either of the giant cells of a swimming animal. One of the difficulties of course is that in an animal of this size and flexibility, it is difficult to implant an electrode in the cell soma. We have hopes, however, of recording from the right connective where it should be possible to pick up the large spike of the RGC in a free-moving animal.

V. D. Gerasimov: What do you think about the functional role of these giant neurons?

G. M. Hughes: The short answer is that we still do not yet know their function, although this was the objective when Dr. Tauc and I started this work. Whatever their function, it would appear to result from the firing of these cells at relatively low frequencies and involve the whole animal. A neurosecretory function of some of the giant neurons cannot be excluded.

N. Chalazonitis: The disagreement between Arvanitaki and Tchou Si-Ho with your own results concerning plottings of neuronal diameters versus weight of *Aplysia*, may be assigned to the unavoidable uncertainties when measuring the neuronal diameter through the ganglionic capsule of connective tissue.

G. M. Hughes: I am sorry that the impression was given that there is any disagreement between the results obtained by Arvanitaki and Tchou Si-Ho and myself. The only significant difference would appear to be in the amount of scatter in the results obtained by me. Certainly, it is difficult to measure accurately the neuronal diameter through the connective tissue sheaths.

Symposium on Neurobiology of Invertebrates 1967 (443—461)

CORRELATION OF BEHAVIOR WITH THE ACTIVITY OF SINGLE IDENTIFIABLE NEURONS IN THE BRAIN OF *TRITONIA**

A. O. D. WILLOWS and G. HOYLE

University of Oregon
Oregon, U.S.A.

With few exceptions, the nerve cell bodies in all members of the animal kingdom are less than 25 μ in diameter. This, coupled with the fact that most nerve cells are practically indistinguishable from their immediate neighbors in live preparations, makes analysis of the unit level ganglionic events controlling behavior a difficult task. The problem is aggravated by the fact that many of the large nerve cell bodies in insect and crustacean ganglia appear to be electrically inexcitable. This is particularly unfortunate, since the notable progress achieved in studies of arthropod sensory and neuromuscular mechanisms leads one naturally to attempt to explore the neuropile where most integrative processes occur.

Gastropod molluscs, on the other hand, frequently have large identifiable nerve cells whose somata are electrically excitable. However, the absence of any articulated skeletal support and the reduced dependence upon discrete anatomical muscles have restricted attempts to make quantitative correlations between behavioral activity and electrical events in the central nervous system in these animals. It has been possible to overcome these restrictions to a degree by taking advantage of favorable anatomical features in the nudibranch *Tritonia gilberti*, and in this report, we wish to describe experiments which establish details of the input and output to over fifty identifiable nerve cells and to correlate activity in certain of these cells with the behavioral activities of the intact animal.

NEURAL TOPOGRAPHY IN THE ISOLATED GANGLION

The central nervous system of gastropod molluscs is composed of a number of more or less dispersed ganglia interconnected by commissures and connectives. In

*A premilinary report of this work appeared in *Science* **157** 570-574.

The work was supported in part by N.S.F. Research Grant GB 3160 to G. Hoyle, and by Marine Sciences Training Grant, GB 3386 to the Friday Harbor Laboratories. Supported also by Training Grant 2 TI GM 336 from National Institutes of Health U.S.P.H.S.

The work was conducted mainly at the Friday Harbor Laboratories of the University of Washington, and the authors wish to thank the Director, Dr. Robert Fernald, for making available the excellent facilities of that laboratory.

nudibranchs, a high degree of centralization has taken place and one finds a number of ganglia aggregated in a ring around the oral tube (Fig. 1). The cerebral, pedal, and pleural ganglia of *Tritonia gilberti* are centralized in this way, and, furthermore, since the parietal, visceral, and abdominal ganglia are fused within the pleurals, this cerebral-pleural-pedal (c.p.p.) ganglion complex comprises the vast majority of the animal's central nerve cell bodies. In common with certain other nudibranchs, *Tritonia* has a number of large, brightly pigmented nerve cell bodies on the surfaces of its ganglia (Veprintsev, Krasts and Sakharov 1964; Dorsett 1965, 1967; Sakharov, Borovyagin and Zs.-Nagy 1965; Willows 1965). There are about 500 of these somata, ranging in diameter from 40 to over 800 μ and in color from deep orange to crystalline white.

The apparatus used in studies of the neural geography consisted of two stimulating-recording intracellular microelectrodes and thirty stimulating-recording polyethylene suction electrodes (Fig. 2). The intracellular micropipettes were constructed in the usual way, filled with 3 M KCl and direct-coupled to a dual beam Tektronix oscilloscope through cathode-follower amplifiers. Stimulating currents were provided by Grass S–4 stimulators with stimulation isolation units. To reduce signal losses through the stimulators to ground, 40 MΩ resistors were placed in the stimulating circuits. Appropriate amplitude corrections were made in the calibration of the amplifiers to compensate for this resistance in parallel with the approximately 10 MΩ electrode resistance.

To permit stimulation and recording from the various nerve trunks, an array of 30 suction electrodes was constructed. Lengths of polyethylene tubing (PE 90) on

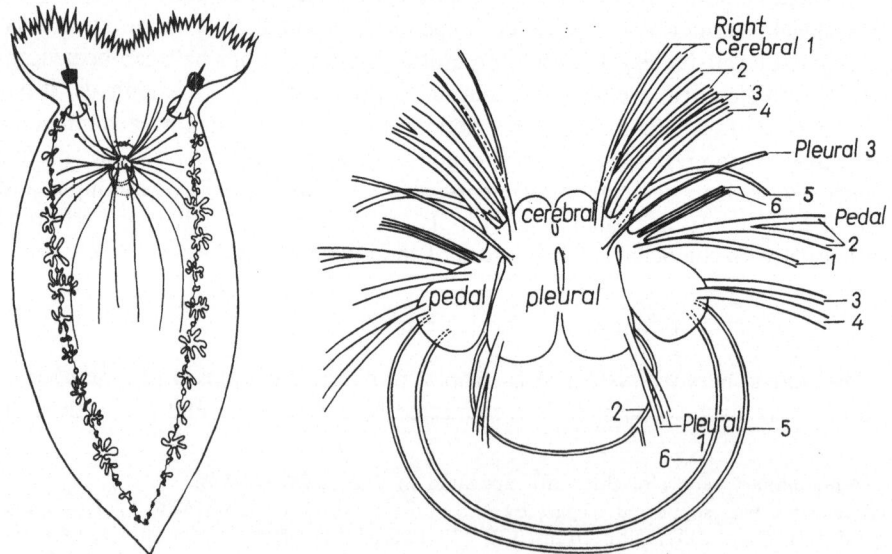

FIG. 1. (Left) The location of the cerebral-pleural-pedal ganglion complex in the intact animal. (Right) The pleural ganglia are fused anteriorly to the cerebrals and the pedals are attached across short connectives

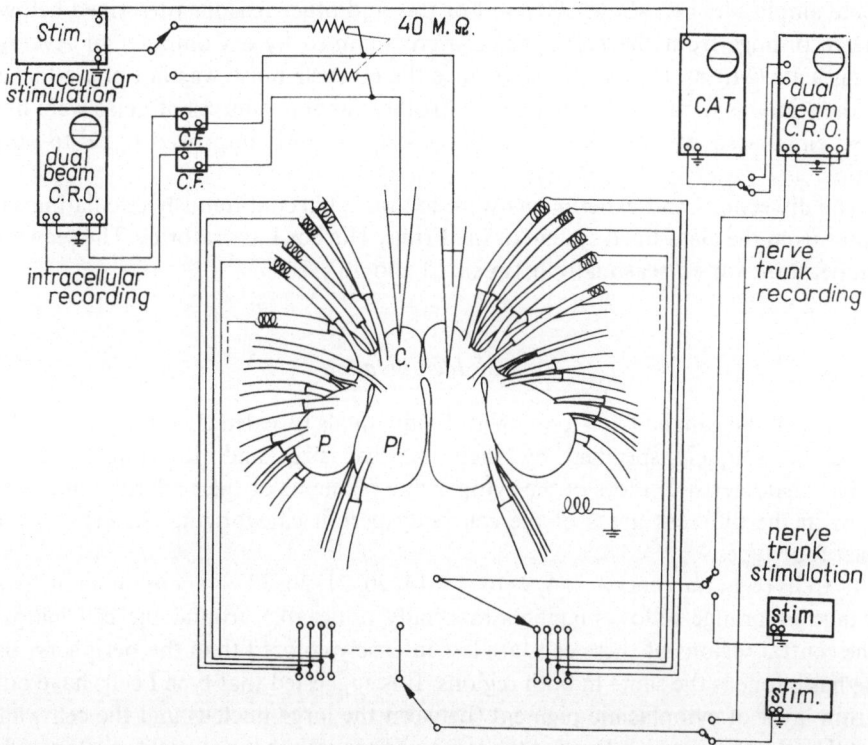

FIG. 2. Stimulating and recording apparatus. Intracellular microelectrodes were used for stimulating or recording from the nerve cell bodies. A shielded high resistance in the stimulating circuit prevents loss of the recorded signal to ground. Each nerve trunk may be connected either to a stimulator or to recording apparatus through seawater-filled suction electrodes. A computer of average transients (Mnemotron CAT) is used to average successive stimulus-response cycles to extract information phase-locked to the stimulus and reject random or stray 60-cycle noise and spontaneous spikes (Copyright 1967 by the American Association for the Advancement of Science)

20 gauge hypodermic needles were attached to 1 ml plastic syringes. The open ends of the tubes were heated near a soldering iron and then pulled out to reduce the inside diameter to approximately that of each of the various nerve trunks. A close fit is desirable to improve the signal-to-noise ratio for recording purposes and to reduce the amount of stimulating current required to initiate impules in the cut nerve ends. Electrical connections were made by inserting short lengths of chlorided silver wire into the bases of the syringes. The leads from these silver wires were soldered to the contacts of two 30-position rotary switches. The armatures of the switches could then be connected to stimulators or recording equipment, permitting simultaneous stimulation or recording from any two nerve trunks chosen by the position of the armatures.

Since the lengths of tubing were quite long, and both tubing and switches were unshielded, the induced 60-cycle noise level was about the same as the typical spike amplitude, i.e., about 500 μV. For this and other reasons (described below), the recordings from the nerve trunks were summed by a Computer of Average Transients (Mnemotron CAT), and since the 60-cycle noise was not phase-locked to the stimulus it was averaged to zero. All other random sources of noise were likewise eliminated by this means, yielding a significantly improved signal-to-noise ratio.

The dissections and experiments were conducted in continuously circulating seawater from the glass-lined system of the Friday Harbor Laboratories. The seawater temperature varied seasonally between 10 and 13 °C.

Visual Identification

Study of the ganglia from a number of individuals revealed that many cell bodies could be uniquely specified by their relative sizes and locations (Fig. 3).

The characteristic colors of the cytoplasmic pigments or their relative concentrations in the different areas of the cell body permit categorizing the cells into at least four types.

I. Cells (e.g., numbers 2, 3, 4, 9, 10, 13, 14, 20, 21, 30, 33, 34) containing brilliant orange or orange-yellow pigments especially noticeable around the cell margin. The central regions of these cells are less intensely colored than the periphery, but the basic tone is the same in both regions. It is suggested that type I cells have only a thin layer of cytoplasmic pigment (between the large nucleus and the cell wall), but that it appears more dense at the margins because the tangential aspect includes a greater depth of pigment.

II. Cells (e.g., numbers 1, 15–18, 24–29) similar to type I in the color of pigment but which have a thicker layer of pigment in the central regions. There is, as a result, less contrast between the central and peripheral pigment intensities, the whole cell appearing more uniformly orange.

III. Cells (e.g., numbers 5, 6, 11, 12, 31, 32) containing the orange pigments of types I and II cells, but having in addition, a whitish hue that is especially apparent in the central regions of the cell. The relative amounts of the orange and white pigments vary, so that in some cases these cells are almost pure white, while and in others, they appear similar to type I cells.

IV. Cells (e.g., number 22) containing large amounts of crystalline white pigment with little or no coloration of any other hue.

Input and Output Connections

The nerve trunks carrying synaptic input to any particular cell were ascertained by stimulating specific trunks through the suction electrodes while recording intracellularly from selected cell bodies. The presence of an axon in a trunk could be detected by stimulating the appropriate cell body intracellularly while recording in the nerve trunk.

A shock delivered to a whole nerve trunk produced one of three responses in any particular cell body. Examples of these responses are shown in Fig. 4.

(i) Often, a single excitatory or inhibitory postsynaptic response or a burst of such synaptic responses was seen.

(ii) Less often, this last mentioned burst was either preceded or accompanied by a spike or an abortive response, which indicated that a spike had occurred some distance away in the axon.

(iii) Occasionally, no response whatever was seen.

If either of the first two alternatives occurred, the stimulating and recording connections were reversed, and depolarizing shocks were applied to the cell body while a record of the electrical responses in the nerve trunk was made. If there was no response in the nerve trunk to the intracellular shocks, it was assumed that the cell sent no axonal branches in that trunk. This observation coupled with (i) or (ii) above were taken as evidence that the trunk contained one or more axons with synapses on the dendrites of the cell in question. If, on the other hand, a response phase-locked to the stimulus could be elicited in the trunk by intracellular stimulation, it was assumed that the trunk carried one or more of the axons from the cell involved. In this regard, to improve the signal-to-noise ratio in the extracellular record, and to emphasize any response in the nerve trunk that was phase-locked to the stimulus, a series of stimulus-response cycles were averaged in the CAT. Such averaging served to retrieve extremely small action potentials (produced by small or damaged axons) from beneath the noise level and to separate phase-locked responses from random background spike activity (Fig. 5). Secondary responses produced in the trunk as a result of other cells being driven by the stimulated cell were also rejected, since the gradually increasing synaptic delay constantly changed the phase of the action potential seen in the trunk.

Just under 20,000 measurements were made of the input-output connections associated with identifiable cells in the ways described above, and the results for the 30 nerve trunks and 54 cells are summarized in Table 1.

Cell-cell Interactions

Using two intracellular microelectrodes placed in different cell bodies, the effect of a stimulus applied to one cell could be observed in another, and the presence or absence of synaptic connections between them could be inferred. Such interactions could also, in a few cases, be deduced from an analysis of the on-going spike activity in pairs of cells. Detectable synaptic interactions were rare. However, one group of 25–30 cells, located adjacent and anterior to cells 11 and 12 on the left pleural ganglion (collectively referred to as cell 10), as well as another similar group on the right pleural ganglion near cell 22 (collectively referred to as cell 21) were found always to make reciprocal excitatory synaptic connections between all members of the group in question (Fig. 6). These same cells never made synaptic contact with other identifiable cells nearby, such as 11–18 or 22–28.

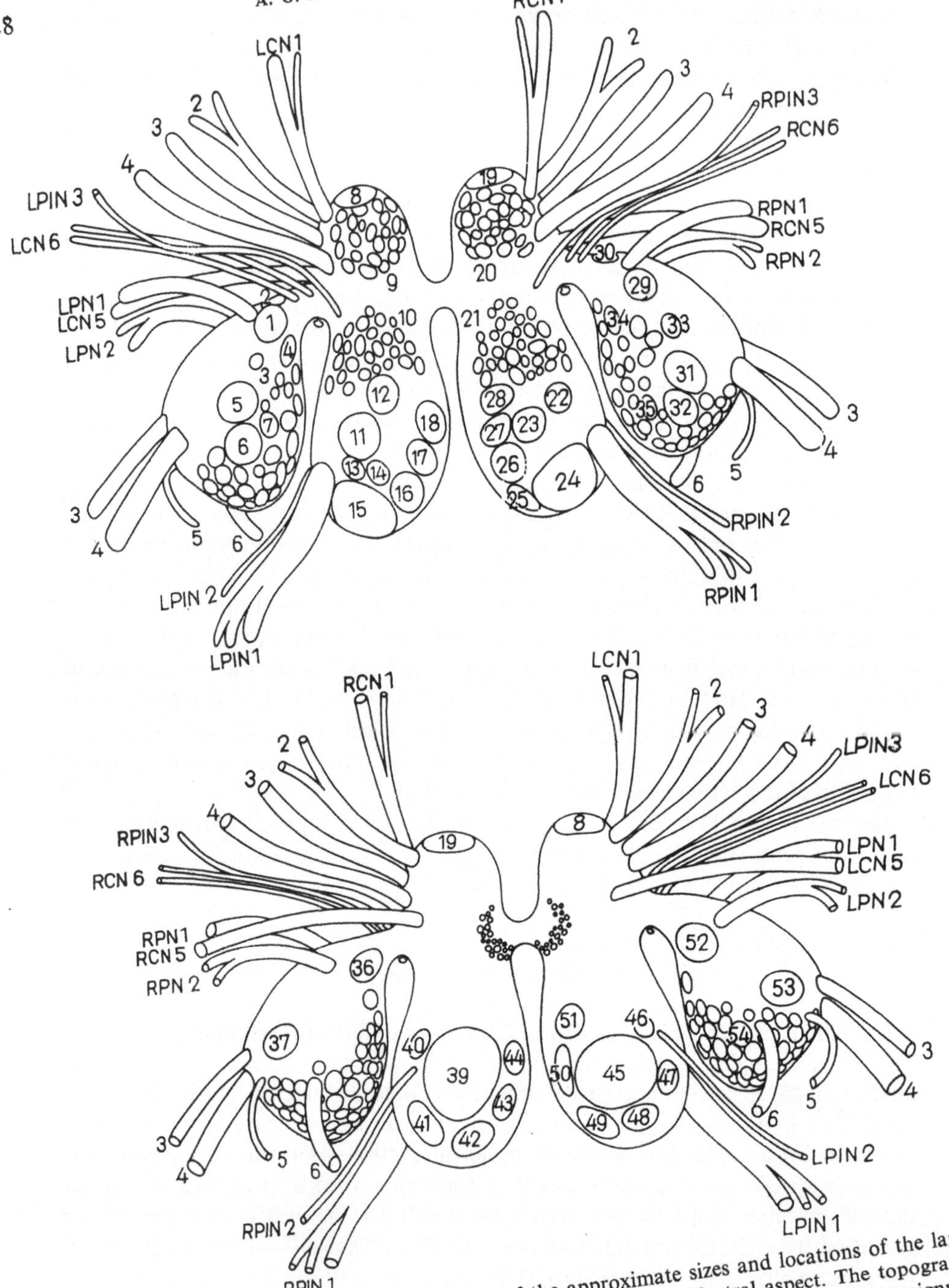

FIG. 3. Semidiagramatic representation of the approximate sizes and locations of the larger cells of the c.p.p. ganglion complex. (a) Dorsal aspect. (b) Ventral aspect. The topographic locations of many smaller cells vary sufficiently from animal to animal to prohibit assignment of unique identifying numbers. There are several hundred smaller cells (50–100 μ) not represented here. The scale may be judged by the fact that the entire complex is 6–7 mm in width. Abbreviations: C — cerebral, L — left, N — nerve, P — pedal, Pl — pleural, and R — right

Fig. 4. Intracellular responses to stimulation of selected nerve trunks through suction electrodes. a, b— Excitatory synaptic inputs. c — A depolarizing wave probably resulting from the electrotonic invasion of the soma by the spike blocked in the hillock region. After about 70 msec, the wave has increased (due to delayed synaptic inputs) sufficiently to trigger an abortive spike in the soma. d — The delayed synaptic input was sufficient to trigger a spike in the soma after several hundred msec. e — A burst of spikes on a depolarizing wave. The cell (1) receives a large amount of input from the trunk (RP1N1) but may be shown not to have an axon in the trunk. f —The antidromic spike from an axon in the trunk abolishes any synaptic input that may also have impinged on the cell from that trunk

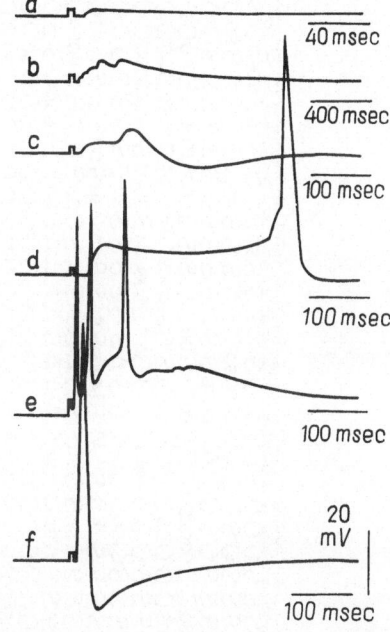

Common Input Sources

In many instances, two or more cells received excitatory synaptic input in common. The intracellular records of on-going activity in such pairs had sequences of synaptic input that were synchronized in time and of the same relative amplitudes. Cells that appeared, on the bases of location and pigmentation, to be members of a bilaterally symmetric pair, such as numbers 1 and 29, often received such inputs (Fig. 7). Individual cells within group 10 and others within group 21 also had common sources of input.

Background or 'Spontaneous' Activity

Many cell bodies in the isolated ganglion preparation had either continuous or bursting, on-going spike activity. Examples are shown in Fig. 8. This long-term rhythmic firing is to be differentiated from the transient spike trains often observed for a short period immediately following microelectrode penetration of the cell. The site of initiation of the long-term spiking is, most probably, within the cell body and not in one or another of the axons. Axonal spikes sufficient in amplitude to trigger a spike in the soma invariably reset the pacemaker, whilst spikes that were blocked in the hillock region (seen as pseudospikes in the soma), did not (Fig. 9). The implication is that the pacemaker for such 'spontaneous' firing is in the soma, since it is not affected by spikes originating elsewhere.

TABLE 1

Input—output connections. A — Indicates that the cell has an axonal branch in that trunk. S — Indicates that the cell receives synaptic input from that trunk. O — Indicates that neither of the above apply. A pair of letters at an intercept indicates that the situations implied by both letter are seen in different animals sufficiently frequently that neither may be unquely assigned

FIG. 5. Electrical responses in the nerve trunks recorded and averaged in the Computer of Average Transients. The trace is triggered by the intracellular stimulus. a — Five stimulus-response cycles averaged from RPN4. The large axon from cell 31 produces an easily recognized spike in the trunk. b — Fifty cycles were necessary to retrieve this very small response from the background noise level. c, d — The small wave appearing after 25 msec is produced by a gradually fatiguing cell that is being driven by the stimulated cell. A check on the repeatability of this response in d increases confidence that such a weak response is the result of a stimulus-related event rather than random. e — One hundred cycles were insufficient to recover any phase-locked response in this trunk. The transients in the first 5 msec and the declining baseline are artifacts of stimulation. Calibration mark = 18 msec

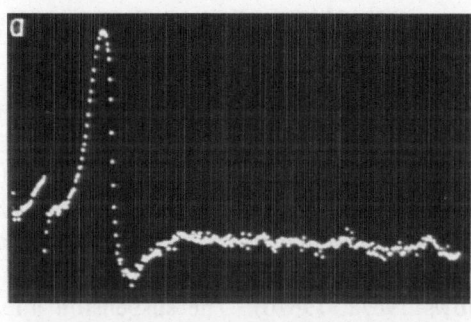

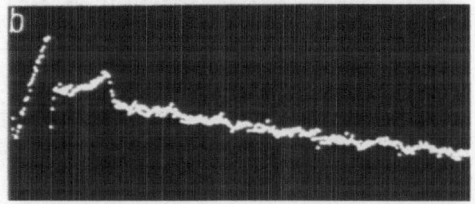

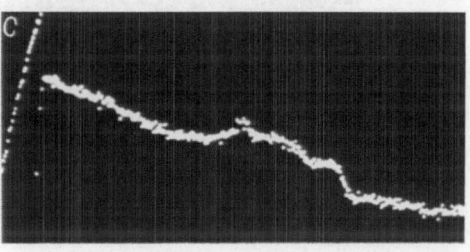

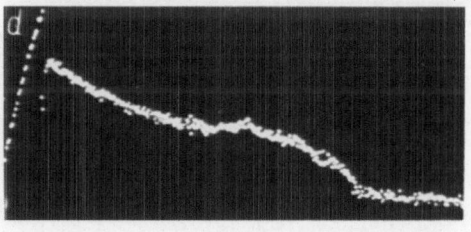

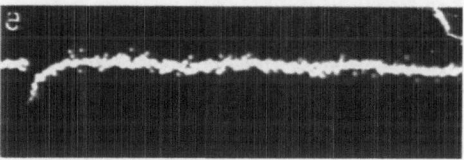

FUNCTION OF SINGLE CELLS IN THE CONTROL OF BEHAVIOR

Since individual cells could be identified on the basis of topographical evidence, and since, furthermore, cells identified by these criteria appeared to have reasonably predictable input and output connections, the question quite naturally arose as to whether the behavioral roles played by individual cells remained constant from animal to animal. In order to answer this question, a means was devised by which the animal could be suspended in a tank of circulating seawater with its brain immobilized and accessible to intracellular microelectrodes (Fig. 10) (Willows 1967a, 1967b). The suspension apparatus interfered minimally with nor-

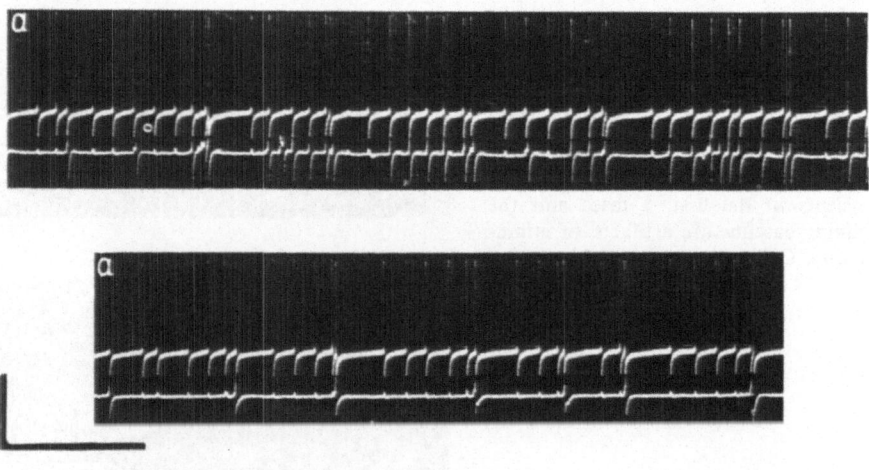

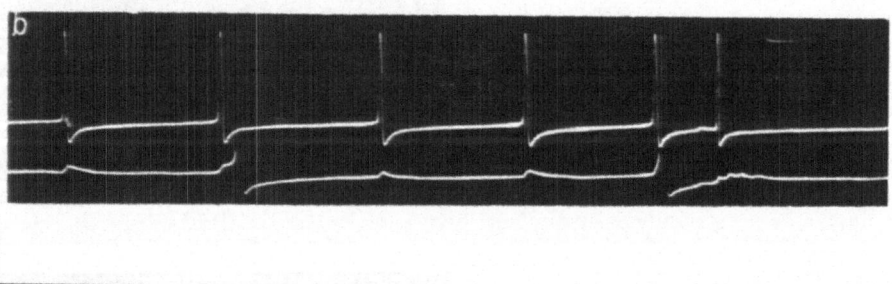

FIG. 6. Excitatory synaptic connections between a pair of cells in group 21. a — Positive synaptic feedback within the network generates recurring bursts with a 5–10 sec period. In (b), a single burst is shown on a faster time base. Note the e.p.s.p.'s produced in the cell on the lower trace by spikes in the other cell. Several spikes elsewhere in the network near the end of the trace evoke e.p.s.p.'s in both cells. A 1–2 sec period of depression follows the burst. Calibration: (a) upper trace, 20 mV, 10 sec; lower trace, 10 mV, 10 sec, (b) upper trace, 20 mV, 1 sec; lower trace 10 mV, 1 sec

mal motor activities and the surgical procedures seemed to disturb the animal only temporarily.

Results of these experiments are summarized below:

(1) Type III and IV cells evoke no obvious motor responses when stimulated, nor do they respond electrically to moderate levels of peripheral tactile or chemoreceptor stimulation.

(2) Shocks applied directly to certain specific type I and II cells elicit predictable and stereotyped motor responses ranging from local contractions of isolated body wall muscles and selective withdrawal of the branchial tufts on either side to turning of the whole body to the left or to the right.

(3) A number of type II cells evoke a generalized increase in body tonus. On occasion, a single spike in one of these cells may trigger a prolonged sequence of swimming activity normally released by contact with certain echinoderms.

(4) Cells in the left pedal ganglion receive synaptic input primarily from the left-hand side of the animal, and elicit contractions primarily in muscles of the left side of the body. Likewise, cells of the right pedal ganglion are prominent in functions of the right-hand side. Cells of the pleural ganglia tend to exert influences on both sides equally.

(5) Reflexive withdrawals of local skin regions in response to weak noxious stimuli appear to occur without the involvement of central nervous cells. Only when the strength of the stimulation increases does sufficient sensory input impinge upon the relevant central units to evoke an electrical response.

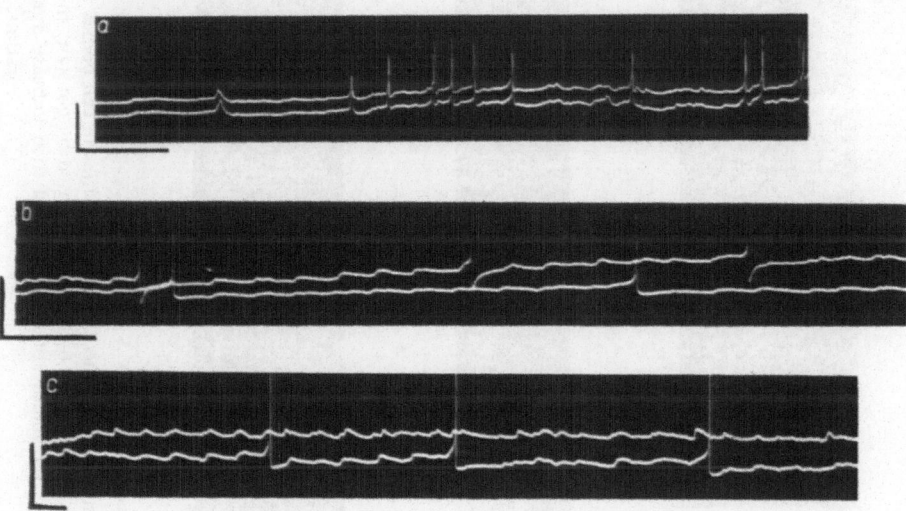

FIG. 7. Pairs of cells with common input sources. a — Two cells in group 10, firing nearly in phase, also appear to be receiving synchronous inputs; b, c — common synaptic volleys to cells 1 (upper trace) and 29. Note that the cell bodies of 1 and 29 are 2–4 mm apart on the l. and r. pedal g., respectively. Calibration: a — 40 mV and 1 sec; b — upper, 25 mV and 1 sec: lower, 50 mV and 1 sec. c — 25 mV and 1 sec

A. O. D. WILLOWS and G. HOYLE

FIG. 8. Ongoing or 'spontaneous' spike activities in cell bodies of the isolated ganglia. a — Regular pacemaker pattern in cells 15 (upper trace) and 24. b — Trains of 6–8 spikes separated by constant intervals seen in cell 15. c — Bursts of 6–8 spikes separated by intervals that decrease to the middle of the burst and then increase again. Cell 22. Similar to parabolic bursting of cell 3 in *Aplysia* (2). Calibration: 40 mV and 2 sec

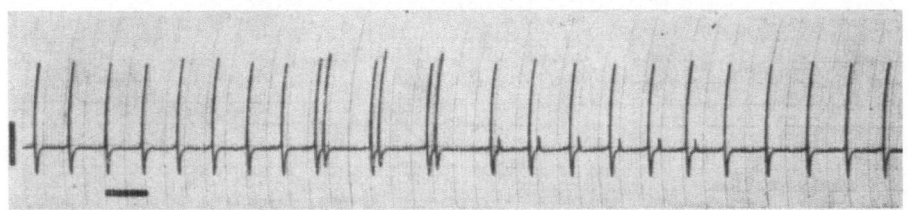

FIG. 9. Pacemaker activity in cell 45 is reset (after 'spontaneous' spikes 9–11) by antidromic stimulation of sufficient amplitude to discharge soma. Antidromic spikes that are blocked in the hillock region (after 'spontaneous' spikes 12–17) do not affect pacemaker rhythm. Site of origin of this pacemaker activity is therefore likely to be the soma, rather than the axon

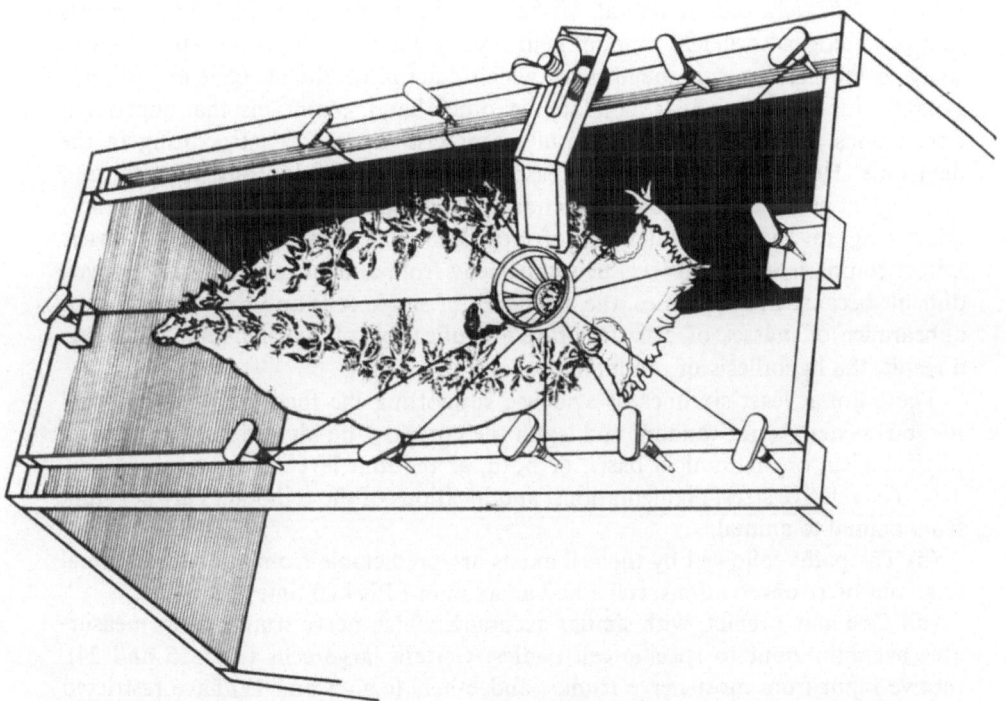

FIG. 10. *Tritonia* suspended in a tank filled with sea water by threads hooked into the edges of an incision over its brain. The hook ring, rigidly attached to the side of the tank by a bolt and heavy spring, immobilizes the animal's brain (Copyright 1967 by the American Association for the Advancement of Science)

(6) The background or 'spontaneous' activity commonly observed in cells of isolated preparations is not apparent in cells of the intact animal. Spiking usually occurs in response to overt stimulation or during the execution of motor activities. At other times, most cells are silent except for low-level, apparently random synaptic input, and the occasional spike or burst of spikes.

DISCUSSION

Topographical and Functional Constancy

The profuse branching of nerve processes seen in histological sections of invertebrate neuropile and in vertebrate cortical areas encourage the view that synaptic connections between cells and the paths followed by individual axons may be random. Studies of the theoretical properties of neuronal networks often assume random connectivity (Allanson 1956; Smith and Davidson 1962). In one such study, Beurle (1956) has demonstrated the theoretical capability of masses of randomly-connected interneurons to act as decision units and to perform learning functions. However, others attribute the complex features of sensory-motor relations to relatively precise neural wiring. Horridge (1961), for instance, suggests that physiological order in random neuropile is due to a selective chemical sensitivity in the postsynaptic membranes which determines the effectiveness of each synaptic input. Kennedy (1966), on the other hand, maintains that functional connections are more likely the result of precise anatomical structuring in the neuropile. He argues that in some cases, there is supporting physiological and histological evidence for such connections, while Horridge's alternative, chemical addressing, involves large numbers of ineffectual and therefore wasted synapses. Direct comparison of unit-level neuroanatomy from animal to animal has proven difficult because in most cases, the small size of single cells and the homogeneous appearance of masses of cells frustrate identification of specific nerve units. As a result, the hypothesis of randomness is difficult to test.

There are at least six lines of evidence supporting the theory that this central nervous system is not random and is, on the contrary, highly ordered, on either a physiological or anatomical basis, or both, at the unit level.

(i) The relative sizes, pigmentations, and locations of the cell bodies are constant from animal to animal.

(ii) The paths followed by the cell axons are predictable from animal to animal (e.g., out of 21 observations, cell 5 had an axon in LPN4 20 times).

(iii) One may predict, with similar accuracy, which nerve trunks carry measurable synaptic input to specific cell bodies. Certain large cells (e.g., 15 and 24), receive input from most nerve trunks, and others (e.g., 8 and 19) have restricted synaptic input fields.

(iv) Certain specifiable cells regularly make excitatory synaptic contacts with other specific cells. Such cells do not synapse on all nearby cells, i.e., connections are limited to an anatomically defined group.

(v) Certain cells receive synaptic input in common with others even when the cell bodies are separated by two connectives and a commissure, a distance of 2–4 mm (Cells 1 and 29).

(vi) The functional roles of individual cells in the intact animal are constant. Stimulation of cells such as 1, 29, 3, and 33 evokes repeatable motor responses (e.g., tuft retraction, turning, or withdrawal) while a single stimulus in others may trigger a prolonged behavioral sequence. The repeatability of these phenomena implies a high degree of structural constancy in different animals.

While one can only guess at the ontogenetic forces that produce such constancy, some of its implications are reasonably clear:

(a) It has been suggested that concepts such as sensitization, habituation, conditioning, and learning could be accounted for if the synaptic connections between central cells were highly mobile and in continuous flux. This kind of plasticity is not, however, compatible with the observations of relative stability of synaptic and axonal connections described above. It might be argued that the detailed synaptic interactions of the neuropile are not always detectable by observations made several hundred microns away in the soma. This argument does not take two experimental observations into account, however. First, while an unknown number of synaptic potentials decay to the noise level between the neuropile and the cell bodies, a considerable amount of measurable activity does reach the somata. And this activity is not observed to change, either in time or over different animals, as would be expected if synapses were being made and broken continuously. Secondly, the functional constancy described in (vi) above is difficult to understand if the input–output relations of these cells are not relatively precise and constant both over time and from animal to animal within the species.

(b) These data support the hypothesis that the bases of patterned behavior, as well as reflex activity, lie in the static geographical details of connectivity between cells and cell groups, as opposed to a description in terms of dynamic properties of interacting nerve cells.

(c) None of the above experiments provides evidence which precludes either precise anatomical connectivity or a system of chemical addressing as the means providing central nervous constancy at the unit level. However, the former alternative seems more likely. If nerve cells are connected randomly, and the effectiveness of synaptic transmission depends upon the chemical sensitivity of the postsynaptic membrane, then one is forced to propose that each and every motor function that can be specified by a particular cell must be characterized by sensitivity to a different chemical transmitter. This seems unlikely since it would require an unrealistic number of different transmitters – at least several hundred, and perhaps many thousands.

Cell Interactions

Studies in primitive nerve nets (Pantin 1952; Horridge 1954) and even of the integrative properties of the mammalian nervous system (Sherrington 1906) have demonstrated that a high degree of synaptic interconnectivity underlies nervous integration. These classical studies have been extrapolated to central nervous functioning in general, leading to the commonly held assumption that most, if not all, central nervous systems contain large numbers of interneurons whose function it is to interconnect cells more or less at many levels.

In this study, on the other hand, there is compelling evidence indicating that the sensory-motor integrative channels for activities differing only slightly are relatively well insulated from one another.

In the isolated ganglion preparation, where a systematic search was made, and in the intact animal, where a smaller number of cells could be explored, cell-cell interactions were rare. In fact, cells in groups 10 and 21 were the only ones which regularly interconnected synaptically. One would expect that if any significant number of interneurons exist, at least a few intraganglionic synaptic interactions would have been detected. In all, an examination of over 1000 pairs of cells, including approximately 100 different cells, uncovered six clearly-defined synaptic connections (not including the observations in groups 10 and 21). It may be argued that the individual synaptic contacts in the neuropile produce only negligible currents in the distant somata, and, therefore, may not be measurable by the techniques used here. But if interconnectivity is the rule, and if large numbers of cells are in fact internuncial in function, one could reasonably expect a higher percentage of detectable synapses. This is especially true since synaptic input from the nerve trunks, including individual e.p.s.p.'s, was routinely observed in cell bodies.

It might also be expected that interneurons would be confined to the ganglia, i.e., not have axons in any of the trunks leaving the c.p.p. ganglion complex. A search of Table 1 does not, however, reveal any cell, of the 54 listed, confined in this way. In fact, all cells have at least one axonal process leaving the ganglia.

Evidence on this same point (scant interconnectivity) was obtained in the intact-animal preparation. It was noted, for instance, that different cells sometimes perform overlapping motor functions and yet did not appear to interact synaptically. Spikes in cell 1 cause left branchial tuft retraction, spikes in cells 15 or 24 (both of which have multi-branched axons) cause retraction of all branchial tufts, including those on the left-hand side. Yet, neither high-frequency intracellular stimulation nor spikes induced by peripheral stimulation in cell 1 produced any measurable electrical effect in cells 15 or 24. Neither did spiking in 15 or 24 cause measurable electrical change in 1. Even cells 15 and 24, whose functions appear identical, do not interact synaptically.

Single Cells and Behavior

The responses elicited by intracellular stimulation of individual nerve cells fell into three categories:

(i) The type III and IV cells produced no obvious motor responses whatever. These same cells were particularly unresponsive to sensory stimulation. A group of similarly pigmented cell bodies in *Aplysia* have recently been studied by Coggeshall *et al.* (1966), and on the bases of electrophysiological and ultrastructural evidence these authors have suggested that such cells are probably neurosecretory. We propose that this same function is served by the type III and IV cells in *Tritonia*.

(ii) Many type I and II cells elicited motor activities in either isolated or larger coordinated groups of muscles. However, these responses appeared to be driven by the stimulation, since they continued only so long as the stimulation.

(iii) A few cells (almost invariably type II, such as cells 15–18 or 24–28) occasionally triggered a prolonged escape response. The behavior, however, did not depend for its maintenance upon stimulation, since a single shock of just suprathreshold strength could release the sequence.

SUMMARY

The nudibranch *Tritonia* has several hundred large, pigmented nerve cell bodies on the surface of its cerebral-pleural-pedal ganglion complex. Many of these cells are identifiable from animal to animal by visual inspection. Studies of the input and output relations of these cells in the isolated ganglia indicate that the paths of axons and the cell-cell synaptic interactions are constant from animal to animal. A technique is described by which the ganglia of the intact animal may be stabilized for intracellular stimulation and recording on a rigid suspension of hooks, leaving the animal free to engage in normal behavioral activities. Stimulation of single identifiable brain cells elicits behavioral acts that range from reflex withdrawal of gill tufts to a sequence of coordinated swimming movements that is normally evoked by the proximity of certain echinoderms.

REFERENCES

ALLANSON, J. T. (1956): *Information Theory*. Third London Symposium. Ed.: C. Cherry London, Butterworth's Scientific Publications, 303.

BEURLE, R. L. (1956): Properties of a mass of cells capable of regenerating pulses. *Phil. Trans. B*. **240** 55—94.

COGGESHALL, R. E., KANDEL, E. R., KUPFERMAN, I. and WAZIRI, R. (1966): A morphological and functional study of cluster of identifiable neurosecretory cells in the abdominal ganglion of *Aplysia californica. J. cell. Biol.* **31** 363—368.

DORSETT, D. A. (1965): The brain of the sea slug. *Rep. Challenger Soc.* **3** XVII, 24.

DORSETT, D. A. (1967): Giant neurons and the axon pathways in the brain of *Tritonia. J. exp. Biol.* **46** 137—153.

HORRIDGE, G. A. (1954): The nerves and muscle of Medusae. I. Conduction in the nervous system of *Aurellia Aurita Lamark. Nature* **171** 400.

HORRIDGE, G. A. (1961): *Nervous Inhibition*. Ed.: E. Florey, Oxford, Pergamon Press, 395.

KENNEDY, D. (1966): *Advances in Comparative Physiology and Biochemistry*. Ed.: O. Lowenstein, New York, Academic Press, 117.

PANTIN, C. F. A. (1952): The elementary nervous system, *Proc. roy. Soc. B*, **140** 147—168.

SAKHAROV, D. A., BOROVYAGIN, V. L. and ZS.-NAGY, I. (1965): Light fluorescence and electron microscope studies on neuro-secretion in *Tritonia diomedia* Bergh (Mollusca, Nudibranchiata). *Z. Zellforsch.* **68** 660—673.

SHERRINGTON, C. S. (1906): *The Integrative Action of the Nervous System*. 2nd ed.: New Haven, Yale University Press, 1947.

SMITH, D. R. and DAVIDSON, C. H. (1962): *Biological Prototypes and Synthetic Systems*. Ed.: E. E. Bernard and M. R. Kare, New York, Plenum Publishing Corp. V. i, 148.

Strumwasser, F. (1965): The demonstration and manipulation of a circadian rhythm in a single neuron in *Circadian Clocks.* Ed.: Jurgen Aschoff, Amsterdam, North Holland Publishing Co., 442—462.

Veprintsev, B. N., Krasts, I. V. and Sakharov, D. A. (1964): Nerve cells of the nudibranchiate mollusc *Tritonia diomedia* Bergh. *Biofizika.* **9** 327—336 (in Russian).

Willows, A. O. D. (1965): Giant nerve cells in the ganglia of nudibranch Molluscs. *Comp. Biochem. Physiol.* **14** 707—710.

Willows, A. O. D. (1967a): *PhD. Thesis,* University of Oregon.

Willows, A. O. D. (1967b): Behavioral acts elicited by stimulation of single, identifiable nerve cells. *Symposium on Physiological and Biochemical Aspects of Nervous Integration.* Prentice Hall (in press).

DISCUSSION

G. A. Cottrell: I believe you said that a vigorous behavioral response lasting 30 sec results from a single electrical stimulus through a micropipette to one particular cell. Is it possible that the initial electrical stimulus just initiates the response and that the continued movement lasting $^1/_2$ min results from a mechanical effect of the electrode on the moving neurones?

A. O. D. Willows: I cannot entirely exclude the possibility that slight movements of the animal's brain with respect to the electrode during the swimming movements produce activity in the cell body which in turn serves to further excite the cell and thus drive the activity. Since the stimulating microelectrode is also used for recording, however, one can monitor the intracellular electrical activities during the swimming and, as a rule, detect such damage responses by noting gross changes in the rest potential.

B. G. Boettiger: Does the cell continue to fire after this swimming activity has ended?

A. O. D. Willows: The cell appears normal after such activity and spikes may be elicited in it by the usual means.

A. Arvanitaki-Chalazonitis: From what depth are these animals obtained?

A. O. D. Willows: Between 25 and 35 m.

A. Arvanitaki-Chalazonitis: Could not the shock of being rapidly raised from such a depth cause abnormal behavioral responses?

A. O. D. Willows: Yes, this is a valid criticism. For 24 h or even 36 h after collection, the animals behave very abnormally, and some do not survive the shock. But then they appear to return to normal and are commonly seen eating, exploring the tank, copulating, or performing other activities. Such animals may be maintained for several months in the laboratory.

N. Chalazonitis: Any comparison of neuronal electrical properties in isolated and *in situ* ganglia implies the establishment of identical values on the environmental factors: such as temperature (T), pO_2, pCO_2, etc. For instance, in many *Aplysia* neurons, a $\Delta T = 1 °C$, or an intracellular $\Delta pO_2 = 3$ mmHg, are determinant of a $\Delta(MP) = 1$ mV which is highly significant on firing neurons.

On the other hand, the synchronous bursts of many *Tritonia* neurons you demonstrated by simultaneous recording are extremely important as an example of pacemaker synchronization. In *Aplysia* such pacemaker synchronization has also been recorded, although the duration of each burst is not always identical.

A. O. D. Willows: The experiments on intact and isolated nervous systems were done under identical conditions insofar as the pO_2 and temperature were concerned. That is, in both cases, continously changing seawater from the glass-lined water system of the laboratory perfused the brain.

B. N. Veprintsev: In our studies on *Tritonia diomedia* we have frequently observed that stimulation of one cell in a bilaterally symmetric pair causes a spiking response in the opposite member of the pair, although the excitatory postsynaptic potentials from the first cell were not evident. It could be that you have not seen many cell-cell interactions because the electrical activities in the neuropile are not always conducted to the cell bodies.

A. O. D. Willows: Yes, I agree that this is possible.

G. A. Kerkut: Dr. Willows, are you suggesting that the studies of Strumwasser on circadian rhythm in giant *Aplysia* cells may be artifacts, since the cells tend to be less active in the intact animal?

A. O. D. Willows: Since Dr. Strumwasser's elegant analyses indicate electrical activity rhythms which are functions of the daily cycle and which are the same in different animals, his results cannot be attributed to such artifacts. My results merely suggest that long-term rhythmic spiking in cell bodies of isolated ganglia should not be assumed to be meaningful in terms of the whole animal until such spiking is demonstrated for the cell involved in the intact animal. Quite clearly, nervous systems must contain elements that serve as a time base to which the activities of the animal may be referred, but I am suggesting that the internal clock may manifest itself as a subtle change of cell excitability or rest potential and not as soma membrane spiking.

Symposium on Neurobiology of Invertebrates 1967 (463—477)

BEHAVIORAL AND NEUROPHYSIOLOGICAL INVESTIGATIONS OF *LIMULUS POLYPHEMUS*[*],[+]

W. C. CORNING[*] and R. VON BURG[†]

*Department of Psychology Fordham University, U.S.A.
†Department of Biology College of New Rochelle, U.S.A.

INTRODUCTION

Anyone who frequents the beaches along the East Coast of the United States is most likely well acquainted with one of the ocean's more bizarre inhabitants, the horseshoe crab or *Limulus polyphemus*. The large size of this arthropod (specimens up to 75 cm long are quite common), its unusual morphology, and its purported phylogenetic antiquity would all seem to make it a fascinating and popular zoological subject. Yet, aside from extensive and well-known research on its large lateral eyes and a moderate amount of work on other receptors, on the heart, and on the neuromuscular apparatus, there is a paucity of information available on the neurological integrative mechanisms and little more than speculation as to how central nervous system properties relate to the transactions the animal must carry on with the world in order to survive.

Our research on the *Limulus* began with what were initially considered to be two independent investigations, one concerned with the neural regulation of cardiac rhythms and the other with the central processing of visual input. In the course of these investigations we learned that in the central nervous system of this animal these two systems were not independent and that, as in higher animals, we must consider the total integrative activity of the nervous system when attempting to comprehend how specific functions are regulated.

LIMULUS: GENERAL DESCRIPTION

The *Limulus* is found along the eastern coast of North and Central America, and three other species are located in the Indo-Pacific region (Schuster 1960). The optimum time to observe and procure the horseshoe crab is from April through June, during its annual mating season. The females, with males attached, climb onto the beaches at high tide to deposit their eggs. This behavior usually occurs during the high tides of the new-moon or full-moon phases, but it can be affected by a number of other variables such as water temperature, clear skies, and calm seas. For example, this year in Long Island Sound the weather was unusually cold

* Portions of the research reported in this paper were supported by Grant NSG 475, National Aeronautics and Space Administration.
+ Some of the data presented in this paper are part of a PhD. dissertation of R.V.B.

during April and May and egg deposition did not occur until June, during the first week of warm, clear weather which happened to fall between the full moon and new moon. The beach migration is a hazardous undertaking, exposing the seemingly impervious *Limulus* to predators such as gulls and grackles, as well as to fishermen who wantonly kill them, believing they are the cause of the depleted oyster and clam beds. However, as Rood (1967) has so aptly pointed out, the *Limulus* and its prey have been able to survive together for millions of years prior to the arrival of man and his waste products.

The *Limulus* is not really a crab but is probably more related to the scorpion, its ancestors being the Eurypterids or sea scorpions of 425 to 230 million years ago. During its larval stage it closely resembles the trilobites. It is the only aquatic survivor of the once widespread Merostomata (Tiegs and Manton 1958). Some peculiar features of this animal are the presence of a spine, the development of a new set of lenses after molting, the presence of book-like gills in the second part of the animal, and the fact that it chews food by means of barbs on its 'shoulders' (Milne 1965). These characteristics are attributed to fossils of 230 million years of age, and because of this the *Limulus* is commonly referred to as a 'living fossil'.

There was a time when it appeared that the horseshoe crab's only value to modern man was as a source of fertilizer. With the discovery of Hartline and Graham (1932) that the animal's large lateral eye and associated optic nerve were suitable for single-cell electrophysiological recording, the *Limulus* became a valuable preparation in sensory research. Other general attributes of the animal are its large size, hardiness, and availability. As an unusually large arthropod, it offers the behaviorist the possibility of studying an animal whose behavioral repertoire is easily observed and recorded, and for neurophysiological investigation its correspondingly large nervous system permits ready location of components for stimulation, recording, and lesioning purposes. From a phylogenetic point of view, research on this preparation may indicate the kinds of capacities and neurobiological properties that were present in ancient animals. Perhaps some insight may be gained as to the precursors of properties found in more complex systems.

VISUAL REFLEXES: CHARACTERISTICS AND PLASTICITY

Detailed information on the significance of visual information for various functional modes the animal engages in is unavailable. Cole (1922) reports that the *Limulus* is normally positively phototropic, although the tropism could be changed by starvation and handling as well as other factors. By covering or removing the median and lateral eyes unilaterally, Cole obtained evidence that the main pathway of leg activation by visual input is contralateral with some ipsilateral effect. This asymmetry in activation causes the movement toward the light source. When the *Limulus* is given a choice of two light fields flickering at different rates, it will select the faster of the two fields, showing that in these arthropods as well there is a strong reaction to flicker (Wolf and Wolf 1937).

Attempts to demonstrate plasticity in response systems of the animal have so far been limited to classical conditioning and habituation. Smith and Baker (1960) attempted to establish a more reliable light response by conditioning a telson movement. Their basic procedure was to pair a 10-sec light stimulus with a shock delivered across the posterior carapace. The shock regularly elicited a telson reflex, and after an average of 220 trials the movement was elicited by the light at a high rate (criterion was 17/20 responses for three consecutive days). A replication of this study (Corning, unpublished data) showed that not all animals are capable of being trained, and the response modification obtained was not as consistent over time as that reported by Smith and Baker (1960). Some subjects actually showed a drop in telson movements toward the end of conditioning.

Because of the rather general nature of the information available on visual reflexes in the *Limulus*, it was necessary to obtain a clearer definition of innate responses to light prior to initiating any attempt at demonstrating plasticity in system organization. When the *Limulus* was clamped in a position affording good visualization of the ventral side, consistent reflexes were observed when the lateral eye was unilaterally excited (see Fig. 1). The telson swings toward the light source, the flap-like endings at the end of the fifth contralateral walking leg close at the onset of light and open at its cessation, and walking movements occur mainly in contralateral appendages. These findings concur with the observations of Cole concerning the asymmetry of activation (Table 1); in all appendages, the contralateral activation was greater than the ipsilateral. The flap-closing reflex was *always* contralateral to the source of light, whereas with the other leg movements there was always some degree of ipsilateral activation.

In addition to a delineation of visual reflexes, a response hierarchy was also noted. Certain reflexes displayed a higher reaction to light than others, with the flap-closing being the most consistently elicited, followed by activation of the 5th, 4th, 3rd, 2nd, and 1st legs. Some animals deviated somewhat from this pattern during the habituation study which will be described below — at times, the 3rd and 4th leg movements appeared to be temporarily dominant, indicating that the response hierarchy is probably not immutable.

To ascertain the degree of plasticity in the response topography observed during light stimulation, a group of 13 animals was given habituation trials. During the trials, one eye of the subject was covered with black tape or a lampblack and wax

TABLE 1

Response Asymmetry In *Limulus**

Appendage	1	2	3	4	5
Contralateral	30.6	35.8	38.5	48.1	56.5
Ipsilateral	20.5	21.4	19.9	18.5	21.8

* Scores represent the average percent response. The flap-closing reflex of the 5th walking leg was always contralateral to the light source. For a summary of statistical analyses on these data, see Table 2.

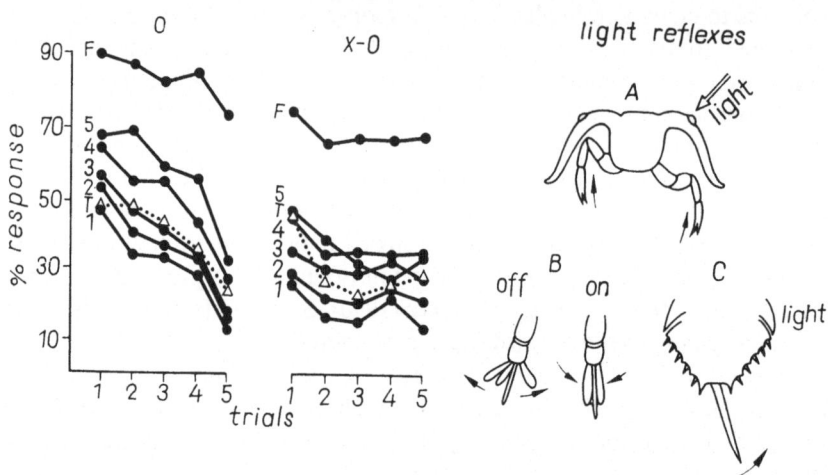

FIG. 1. Habituation of visual reflexes in *Limulus*. O indicates average per cent response of original eye; X–O indicates average per cent response during cross-optic tests

mixture and the light stimulus delivered to the opposite, exposed eye. The intensity of this light source was varied from animal to animal; an intensity was chosen which elicited consistent responses for 10 test trials. The subjects were presented 20–40 habituation trials per day. Each trial consisted of a 10-sec light period, and the intertrial interval varied from 90 to 120 sec. All animals were given these treatments until the total number of leg movements was five or less during the last 10 trials of two consecutive habituation sessions. This level is, on the average, close to the lower limit for the crab under the particular experimental conditions.

As can be seen from the data represented in Fig. 1, all of the reflexes displayed a diminution in activation with repeated trials. A summary of statistical analyses may be found in Table 2. The response hierarchy persists throughout habituation, with the flap-closing reflex of the 5th leg being the most resistant. Fatigue and sensory adaptation may both be ruled out as factors being responsible for the drop in response frequency. When a shock is delivered across the posterior carapace, all the appendages respond. A total of 400 shock trials does not produce any response decrease, and since it took an average of 200 trials to habituate, fatigue was not involved. Furthermore, the carry-over of habituation from one day to the next would also rule out sensory adaptation.

Following achievement of criterion, the opposite, 'naive' eye was exposed and the animals subjected to the same treatments to assess the degree of cross-optic transfer of the habituation. These data, presented in Fig. 1, demonstrate that there is some central storage of visual information. For all but two appendages, the degree of savings was beyond the 0.05 probability level (see Table 2).

Concomitant with these behavioral analyses, investigations of the neural pathways subserving the visual reflexes were undertaken. Although far from complete,

TABLE 2

Summary of Statistical Analyses

A. Asymmetry: Contralateral vs   Ipsilateral reflexes ($n = 13$)

Appendage	p (sign test)
1	0.003
2	0.002
3	0.001
4	0.001
5	0.001

B. Hierarchical organization: Appendage comparisons ($n = 13$)

Leg 1 vs leg 2	0.001 (sign test)
Leg 2 vs leg 3	0.172
Leg 3 vs leg 4	0.001
Leg 4 vs leg 5	0.001

Leg 1 > leg 2 > leg 3 > leg 4 > leg 5: $p = 0.001$

C. Habituation (L test for differences; O's pooled)

Original contralateral: $p = 0.001$
Cross-optic test: $p = 0.01$

D. Cross-optic transfer ($n = 9$)

Appendage	p (sign test)
1	0.002
2	0.002
3	0.090
4	0.020
5	0.090
Tail	0.011
'Flap-closing'	0.001

they indicate that ipsilateral activation of the lateral eye does not produce any activity in units of the contralateral forebrain. To relate this to the habituation data of above, it indicates that the savings observed in the cross-optic tests must be due to central storage elsewhere in the system, not in the forebrain structures.

CARDIAC RHYTHMS: PERIODICITIES AND SENSORY MODIFICATION

In the course of these investigations of the *Limulus*, methods were developed for the chronic monitoring and stimulation of physiological events (Corning *et al.* 1965). Techniques have so far been developed for the chronic implantation of electrodes in the lateral eye optic nerve, in the ventral cord, and dorsally for EKG recordings. Quite by accident it was discovered that visual input could readily

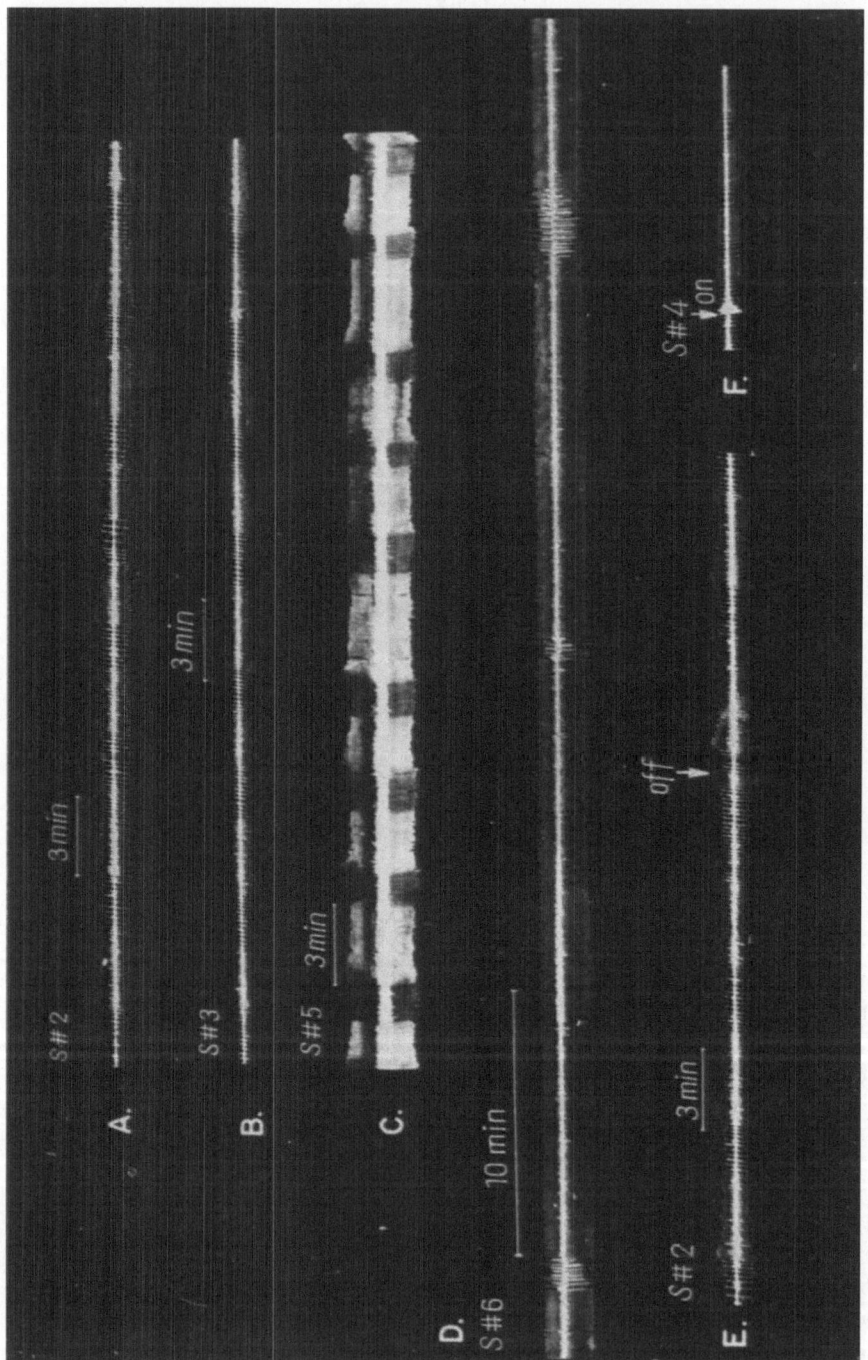

FIG. 2. Periodicities in heart rate of *Limulus*

modify ongoing cardiac rhythms, and interest expanded to the central regulation of heart function.

The EKG of the *Limulus* shows definite periodicities. These are not clearly emergent until a chronically implanted animal has been 24–36 h in a tank relatively free from stimulation. Some examples of the kinds of periodicities observed may be seen in Fig. 2. The most typical is an acceleration–inhibition cycle of approximately 3 min in duration. Other rhythms of longer duration (15 to 20 min) have also been observed. The neurophysiological bases of these periodicities are as yet undetermined, but it is interesting to note that in one animal, lesioning the optic lobes caused the immediate appearance of the 3-min periodicity, while in another animal, complete removal of the forebrain produced a periodicity of a longer duration. Our suspicion is that sensory input may desynchronize elements in the system which, under reduced input conditions, tend to operate in unison.

Larimer and Tyndel (1966) have reported on sensory modification of cardiac rhythms in the crayfish. We find that in the *Limulus* as well, these rhythms are remarkably sensitive to changes in incoming information. Variations in illumination, temperature changes, and other forms of stimulation have all proved effective. In Fig. 2 it can be seen that in one case (F), the presentation of light led to the onset of periodic fluctuation in rate which eventually diminished, while in the other example (E), the cessation of light during an acceleratory phase caused the cycle to 'reset' itself.

At this point it was decided that the central mechanisms influencing cardiac rhythms should be included in our studies. Since the initial phase of this portion of the research is completed, these data will be discussed in the following section.

CENTRAL REGULATION OF CARDIAC RHYTHMS

(A) Previous research

The cardiac rhythm in *Limulus* is neurogenic in origin, i.e., it is dependent upon the cardiac ganglion which lies dorsal and medial along the heart. Other portions of the nervous system have also been implicated in cardiac regulation. Nerves 7 and 8 emerging from the accessory brain have been found to be inhibitory in nature, although Pax and Sanborn (1964) report that acceleration occurs at high frequencies of stimulation. The role of the ventral cord ganglia is less clear. Carlson (1905) has obtained acceleration of the rhythm when stimulating the abdominal commissures. The dorsal nerves of the ventral cord ganglia are supposed to carry acceleratory influences. We have found some aspects of this picture to be oversimplified, while other aspects are not at all correlative with our findings.

(B) Forebrain involvement

The previously mentioned finding of sensory modification of cardiac rhythms hinted strongly at forebrain involvement, particularly since the effect was quite rapid. The effects on the rhythm of a series of lesions in the forebrain are represent-

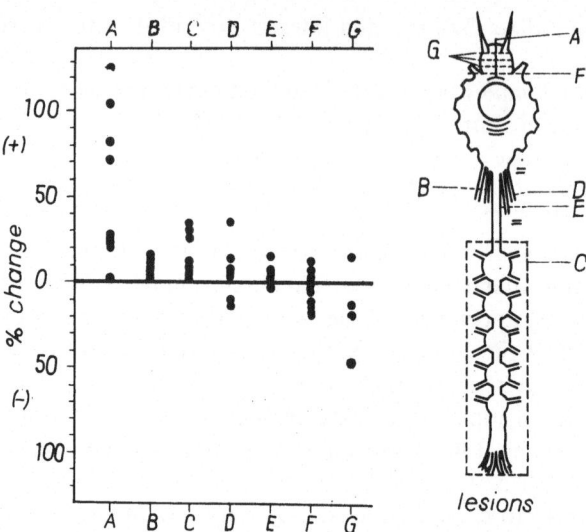

FIG. 3. Forebrain lesions in *Limulus*. The data in
columns A–G represent the percent change from
prelesion rates. The types of lesions performed are
represented in the diagram at the right

ed in Fig. 3. The critical lesion with respect to changes in beat frequency is a longitudinal split which must involve the commissures of the 'association' area of Hanstrom (1926). Lesions anterior or posterior to this region do not lead to any significant changes. The direction of change is always acceleration in the intact animal.

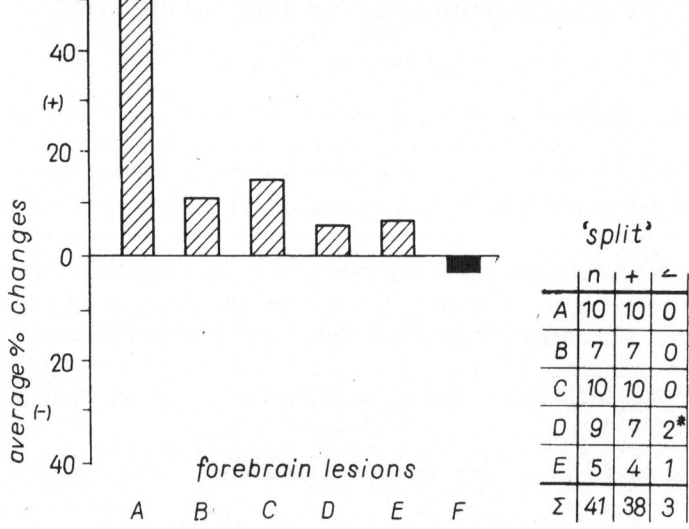

FIG. 4. Summary of forebrain lesion effects (see Fig. 3 for description
of lesions A–F)

When the ventral cord is removed, (lesion C), splitting the forebrain still results in acceleration, although not as much. If the cord is left intact and the output of the accessory brain removed by severing nerves 7 and 8, a split again produces acceleration. For this lesion, however, there were two exceptions — two subjects displayed a slight inhibition in rhythm frequency (see lesions B, O and E, Fig. 3).

Bilateral removal of the forebrain (lesion F) produces mixed effects, indicating that in the intact animal, the centers whose effects are released by splitting the forebrain are probably held in check either by mutual inhibition or by having their effect opposed by centers of opposite function. In other words, if there were only

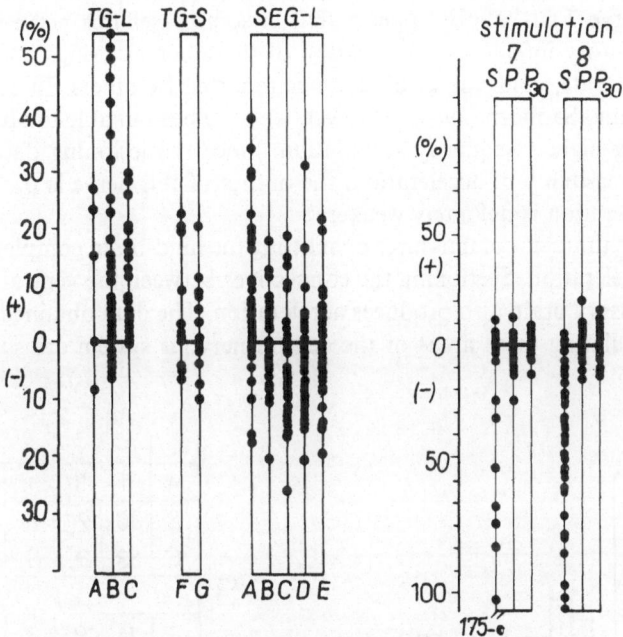

FIG. 5. Lesioning and stimulation results in various central nervous system regions. The data in the columns represent the percent change from base rates taken before the lesion or before stimulation

TG–L: A — Cutting nerves 7 and 8 together. B — Cutting nerve 7. C — Cutting nerve 8.

TG–S: F — Cutting commissures between two halves of thoracic region. G — Splitting hindbrain region completely.

SEG–L: A — Cutting connectives between accessory brain and ventral cord ganglion 1.

B — Cutting connectives between ventral cord ganglion 1 and 2. C — Cutting connectives between 2 and 3. D —Cutting connectives between 3 and 4. E — Cutting connectives between 4 and 5.

Stimulation: S — During stimulation. P — Immediately after stimulation. P_{30} — 30 sec after stimulation

one type of center in each lobe of the forebrain, each with a contralateral effect elsewhere in the nervous system, the bilateral removal ought to have the same effect as splitting. Since the data of Figs. 3 and 4 suggest otherwise, we must postulate dual mechanisms of opposite function at the forebrain level.

(C) Thoracic ganglia and accessory brain

Lesions in neuromeres 1 through 6 have failed to bring about any consistent changes in cardiac rhythms. In the accessory brain, lesioning of nerves 7 and 8 produces acceleration in most animals (Fig. 5). Exceptions were noted in animals with only nerve 7 cut, and it appears that in some cases this nerve does mediate weak acceleratory influences. Stimulation of these nerves brings about a marked diminution in rate, confirming the findings reported by others. In contrast to the work of Pax and Sanborn (1964), however, we have been unable to obtain acceleration by using higher frequency stimulation. Since the lesioning data on nerve 7 did show a possibility of acceleration, the output of this nerve is probably mixed, but the acceleration is definitely weaker.

Cutting the thoracic commissures or splitting the hind-brain completely generally produces acceleration. Sectioning the connectives between the ventral cord ganglia and the accessory brain also produces acceleration. The data obtained from lesioning and stimulating these areas of the central nervous system are summarized in Figs. 5 and 6.

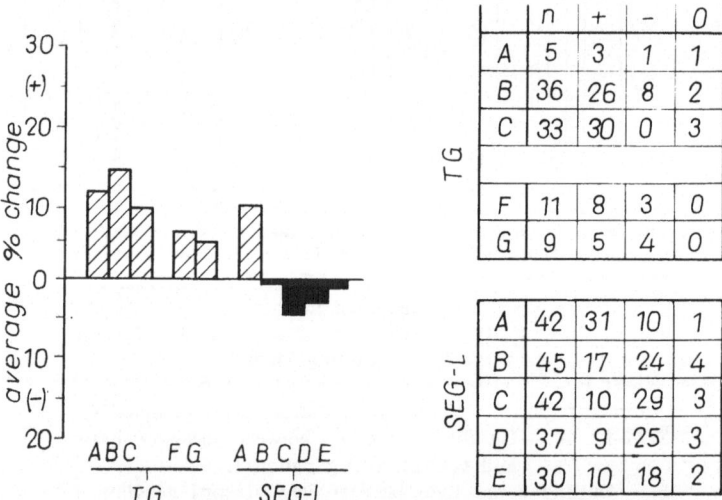

TG

	n	+	−	0
A	5	3	1	1
B	36	26	8	2
C	33	30	0	3
F	11	8	3	0
G	9	5	4	0

SEG-L

	n	+	−	0
A	42	31	10	1
B	45	17	24	4
C	42	10	29	3
D	37	9	25	3
E	30	10	18	2

FIG. 6. Data summary for various lesions and for stimulation of nerves 7 and 8 (see legend of Fig. 5 for description)

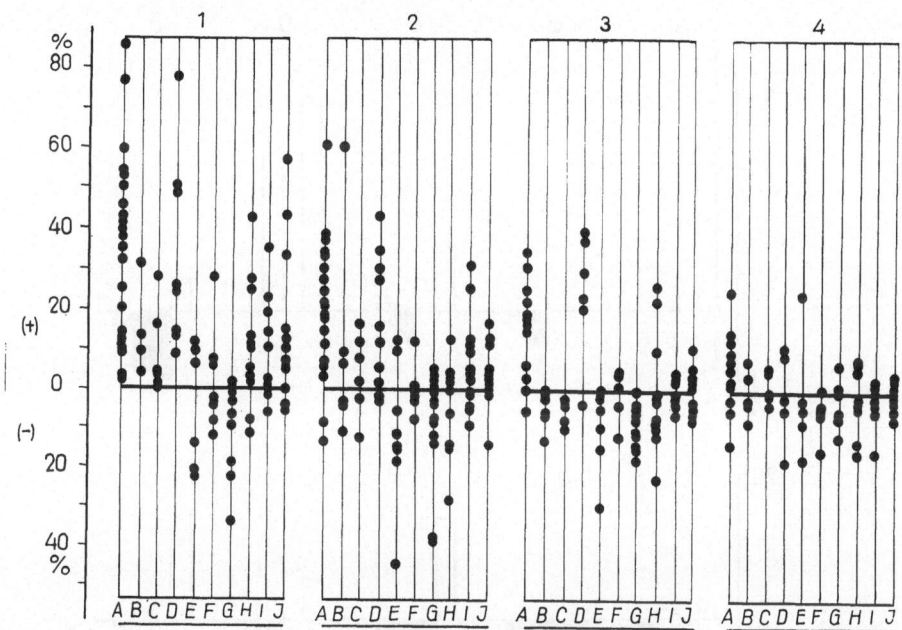

FIG. 7. Lesion effects in ventral cord ganglia

A — Splitting intact ganglion. B — Splitting ganglion with dorsal and ventral nerves severed
C — Splitting ganglion with dorsal nerves severed. D— Splitting ganglion with ventral nerves
severed. E — Cutting dorsal and ventral nerves post-split. F—Cutting ventral nerves post-split.
G — Cutting dorsal nerves post-split. H — Cutting dorsal and ventral nerves prior to split.
I — Cutting ventral nerves prior to split. J — Cutting dorsal nerves prior to split

(D) Ventral cord

A wide variety of lesions were performed in the ventral cord region. Only through
the complex array described in Fig. 7 were we able to learn anything about the
functional arrangement of these ganglia.

Lesioning the connectives between ganglia resulted in an inhibition of rhythm
for a majority of animals, although there were numerous exceptions. The types of
relationships existing between ganglia are most likely complex and possessing
tracts of mixed function (see Fig. 6).

The most important lesion with respect to alteration of ongoing rhythms was a
longitudinal split of the ganglion. The effect of this lesion was acceleration in all
ganglia, but the degree of acceleration decreased from ganglion 1 to ganglion 4
(see Figs. 7 and 8; lesion A). As in the forebrain, a critical function within the
ganglion is mediated by a contralateral pathway. Contralateral effects are also
excerted on other ganglia. If the outputs of ganglion 1 are cut and then a split is
performed (lesion B), acceleration is observed, indicating a drop in inhibitory influence on the remaining ganglia. The same effect is observed in ganglion 2, demonstrating an effect on the output of the others. The effect is not observed in 3 or 4.

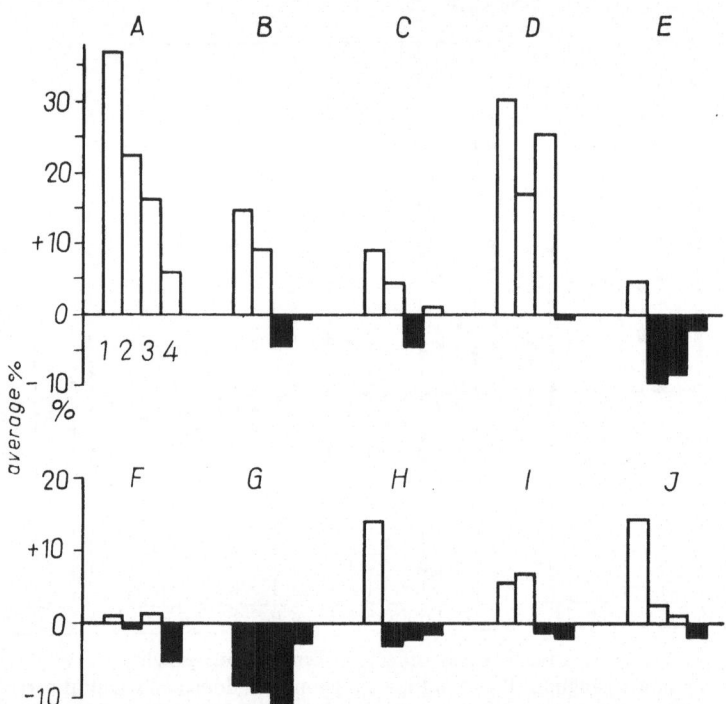

FIG. 8. Summary of lesion effects in ventral cord ganglia. (See legend of Fig. 7 for description of lesions A–J)

Splitting the ganglia with the dorsal roots cut has little effect (lesion *C*), but if a split is performed with the ventral roots cut, marked acceleration is observed in ganglia 1 through 3 (lesion *D*).

Of interest is the effect of cutting the dorsal and ventral roots before and after a split (lesions *E* through *J*). In particular, a comparison of lesions *G* and *J* is illuminating. Prior to splitting, cutting the anterior roots led to acceleration in ganglion 1 and little change in 2 through 4. After a split, cutting the anterior roots resulted in inhibition. The influence of the ganglion appears to change when the contralateral connectives are severed. Ganglion 1 lesions produce the most dramatic changes in rate. Furthermore, the main effects, both acceleratory and inhibitory, are carried by the dorsal roots.

These results indicate that the ventral cord can both drive and inhibit the cardiac rhythm. The direction of change is dependent upon the integrity of the contralateral pathways. They also demonstrate that ganglia 1 and 2 are most critical, with 3 and 4 somewhat involved. Lesions involving ganglia 5 through 8 produced little change in rate.

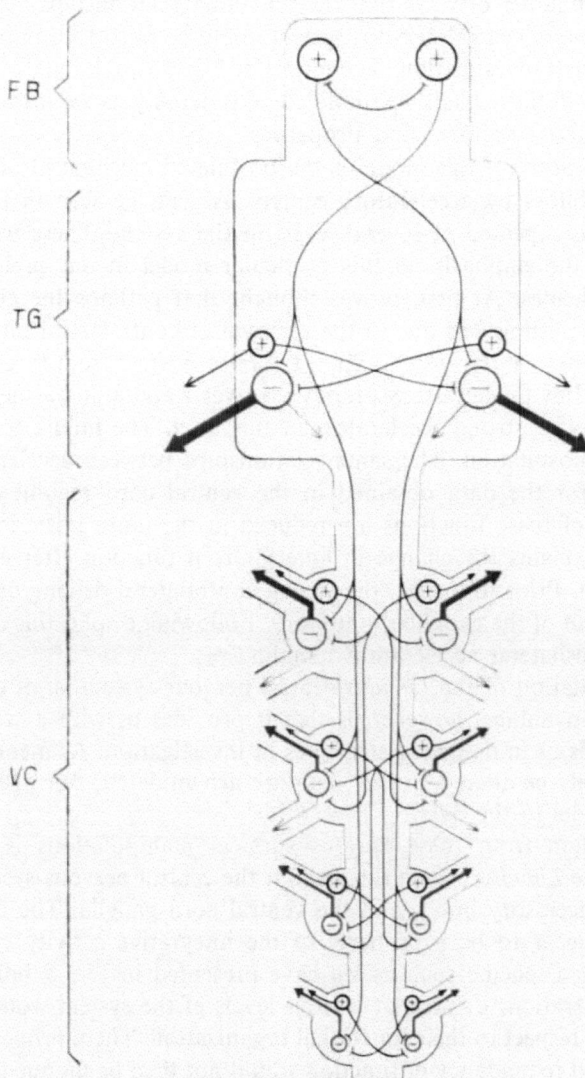

FIG. 9. A model of cardiac regulation in the central nervous system of *Limulus*. FB — forebrain. TG — thoracic ganglia and accessory brain region. VC — ventral cord

(E) A tentative model

There are a number of ways that the functional relationships between components of the *Limulus* central nervous system might be portrayed in order to fit the data of the present investigation. The model of Fig. 9 is only one of several possibilities, and it will most likely be modified considerably as we move toward finer analyses of neural structures and properties.

The critical feature of this model is the postulated mechanism of contralateral driving of inhibitors by acceleratory centers. As can be seen in the schematic, this operation is repeated at several levels in the system. There were a number of reasons for the emphasis on this particular model in our preliminary array of potential schemes. At first, it was thought that perhaps the effect observed with forebrain splitting was due to the removal of contralateral inhibition of accelerators elsewhere in the system. Since the release effect from forebrain splitting can be mediated by the accessory brain via nerves 7 and 8, it was necessary for us to find evidence for strong accelerators at this level. The failure to do so led us to the present postulation. The same relationships between accelerators and inhibitors work for the data obtained in the ventral cord region, providing the relative effects of these functions are reduced in the more posterior ganglia.

The model explains the change in anterior root function after a ventral cord ganglion is split. Prior to a split, the strong contralateral driving of the inhibitor makes the output of the ganglion inhibitory. Following a split, this driving is lost, and only the ipsilateral acceleration remains.

This representation of the *Limulus* central nervous system is, of course, highly simplified. Its advantage, however, is that it provides us with a scheme that we can test and aids us in designing strategies of investigation. As mentioned before, it will most likely be altered as new data are accumulated; our prime obligation is to fit the model to the data.

What is certain from these various surgical manipulations is that cardiac regulation in the *Limulus* occurs throughout the central nervous system involving the forebrain, accessory brain, and the ventral cord ganglia. The control of the heart would appear to be a property of the integrative activity of the system. Furthermore, if a scheme such as we have presented in Fig. 9 holds true, then the kinds of operations existing at various levels of the system would seem to be redundant with respect to their functional organization. These properties of multilevel control and redundancy of function would not then be unique to the *Limulus* — they are a basic principle of organization in higher brains.

SUMMARY

Considerable information has been accumulated concerning the properties and structure of the lateral eye and the heart of the horseshoe crab, but little is known about the central mechanisms mediating the functions of these systems. Our interests in the two systems were stimulated by the finding that visual input could alter cardiac rhythms. With respect to the functional significance of visual information and its processing by central structures, data are reviewed which demonstrates (1) that there are specific reflexes elicited by visual input; (2) that certain reflexes habituate and display a response hierarchy; (3) and that there is cross-optic transfer of habituation. Electrophysiological evidence indicates that the storage of visual information is not in the forebrain but is most likely in the circumoesophageal ring. Visual input can also accelerate or inhibit cardiac rhythms. Examples of endogenous rhythms and their modification by sensory stimulation are presented. An extensive series of stimulation and lesioning studies have attempted to clarify the role of the central nervous system in controlling the heart rate. Results suggest a multilevel control with considerable redundancy of operation. With regard to cardiac control, we cannot adhere to a 'segmental programming', but instead must consider the integrative action of the nervous system. A model of the possible interrelationship between various components of the nervous system is presented.

REFERENCES

CARLSON, A. J. (1905): *Amer. J. Physiol.* **13** 217—240.
COLE, W. H. (1922): *J. gen. Physiol.* **5** 417—428.
CORNING, W. C., FEINSTEIN, D. A. and HAIGHT, J. R. (1965): *Science* **148** 394—395.
HANSTROM, B. (1926): *Acta Univ. Lund. Aud.* **22** 1—79.
HARTLINE, H. K. and GRAHAM, C. H. (1932): *J. cell. comp. Physiol.* **1** 277—295.
LARIMER, J. L. and TYNDEL, J. R. (1966): *Anim. Behav.* **14** 239—245.
MILNE, B. (1965): *The Crab That Crawled Out of the Past.* New York, Atheneum.
PAX, R. A. and SANBORN, R. (1964): *Biol. Bull.* **126** 133—141.
ROOD, R. (1967): *Audubon,* **69** 38—42.
SCHUSTER, C. N. (1960): In: *Encyclopedia of Science and Technology.* New York, McGraw-Hill, vol. **14** 563—567.
SMITH, J. C. and BAKER, H. D. (1960): *J. comp. physiol. Psychol.* **53** 279—281.
TIEGS, O. W. and MANTON, S. M. (1958): *Biol. Rev.* **33** 255—333.
WOLF, E. and WOLF, G. Z. (1937): *J. gen. Physiol.* **20** 767—776.

Symposium on Neurobiology of Invertebrates 1967 (479—485)

AFFERENT MECHANISMS OF THE MAINTENANCE OF FLIGHT IN LOCUSTS

V. L. SVIDERSKY

Sechenov Institute of Evolutionary Physiology and Biochemistry,
Academy of Sciences of the USSR
Leningrad, USSR

The locust flight is a complex locomotory process involving a great number of nervous and muscular units. In recent years a great deal of attention has been paid to central mechanisms controlling the work of the locust wing muscles. It has been established that special rhythm-generating neurons or pacemakers play an important role in that control. These neurons possess a high level of automatism and are initiated by afferent impulses from definite sites of the insect body. In the present communication the receptor system of the head maintaining the work of wing muscle neurons and capable of starting them in a suspended insect is discussed.

According to Weis-Fogh (1956), on each side of the locust head there are five groups of receptors represented by sensillae. The stimulation of these receptors by an air current may initiate and maintain a flight. We consider the character of activity typical of sensillae of the second area containing up to 30–40 large hairs in the species *Locusta migratoria*.

Our primary task was to study the character of nerve impulses arising in a single receptor in response to stimulation and then entering the central nervous system of the locust. For this purpose we registered action potentials of single sensilla (Fig. 1) using tungsten microelectrodes with a tip diameter of about 1 μ. The microelectrode was introduced at an angle of 45° (Thurm's method, 1962) into the edge of an articular hair membrane.

The receptor was stimulated by an air current blown at a given rate (mainly 3.5 m/sec) through a thin hypodermic needle. The direction of the current is important, as receptors of the second area respond only to bending from side to side or backwards, which seems to be connected with the aerodynamic conditions of the locust flight.

The response of a single receptor in the second area regarded at various rates and bending angles did not on the whole differ from those of other areas, particularly of the area 1 (see Svidersky 1967). I shall try to give here a general picture of the response observed in a single receptor during prolonged air stimulation (Fig. 1). After a period of high-frequency activity (attaining 140 impulses/sec) for the first moment of stimulation, there is a decrease, and 20–30 sec of the continuos stimulation the discharge frequency of the receptor declines to 30–40 impulses/sec.

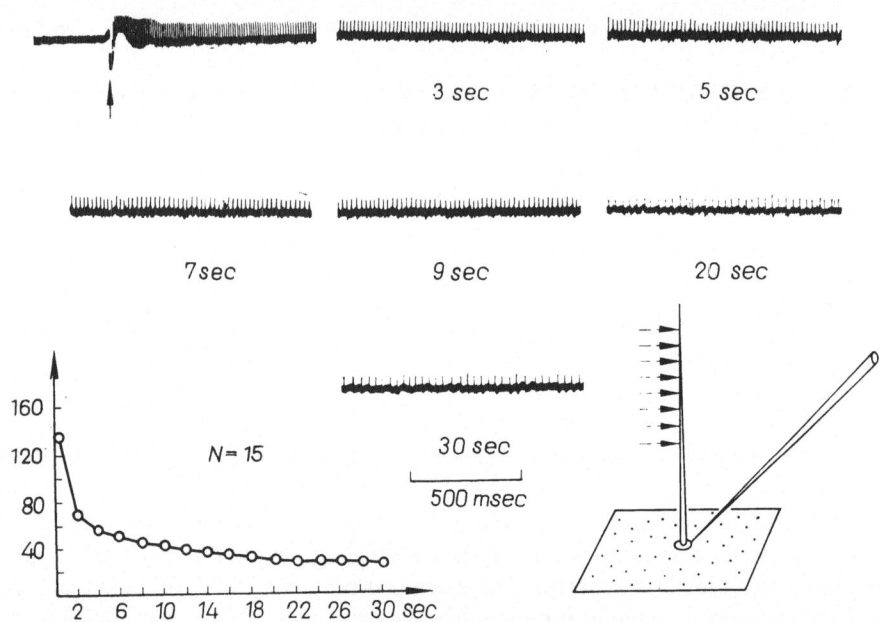

FIG. 1. Response of a single receptor of the second area during prolonged stimulation by an air current (numbers denote corresponding seconds of stimulation). The arrow marks the onset of stimulation. The time line represents 500 msec. Below at the left is a graph summing up observations of this type. The ordinate gives the frequency of the impulses, the abscissa the time in seconds. At the right is a scheme representing recordings of nerve impulses in a single sensilla

This 'final' frequency may be maintained stable for a very long time (dozens of minutes). The initial discharge frequency may vary within a considerable range, depending primarily on the bending rate of a hair (Svidersky 1967). In addition, the discharge frequency of receptor depends on the bending angle of a hair only at angles which do not exceed 25°. At winder angles the amplitude of bending is not significant for the discharge frequency. Therefore, both the rate and the bending angle (30°) were constant throughout all the experiments.

Hence, the receptors investigated behaved as typical mechanoreceptors with a very slow incomplete adaptation. Two phases can be distinguished in response of a single receptor: short-term phasic and tonic ones. Since these receptors can act to trigger and to maintain the work of generating neurons it would be only natural to suggest that it is the first, phasic part of the reaction that is used to initiate neurons whereas the 'final' frequency is important to maintain the activity of those neurons during the starting flight. This might have been the case had we been sure that the impulse activity arising in receptors undergoes no changes, i.e., it enters the locust thoracic ganglia and activates central neurons without any synaptic processing whatever.

Therefore, our second task was to decide, on the basis of electrophysiological data, whether the pulsed activity of receptors entering thoracic ganglia is processed, and, if so, in what way this occurs. For this purpose we registered a response arising at various levels of the nerve cord when hairs of the second area are stimulated by an air current. In the first series of experiments the recording was made with the aid of an external tungsten hook-like electrode from the connectives linking the ganglia. In the second series of experiments a glass microelectrode with the tip diameter 1 μ was introduced into mesothoracic or metathoracic ganglia near the anterior connective.

EXTERNAL RECORDING FROM CONNECTIVE

A response in the connectives when hairs of the second region are stimulated by air is identical at all the levels of the nerve cord beneath the suboesophageal ganglion, which shows the impulse activity undergoes no processing at this level. Following Guthrie (1964) we showed that in the head ganglia (in the suboesophageal ganglion, according to Guthrie, and possibly even higher) there is an intersection of nerve paths transmitting impulses of receptors to thoracic ganglia.

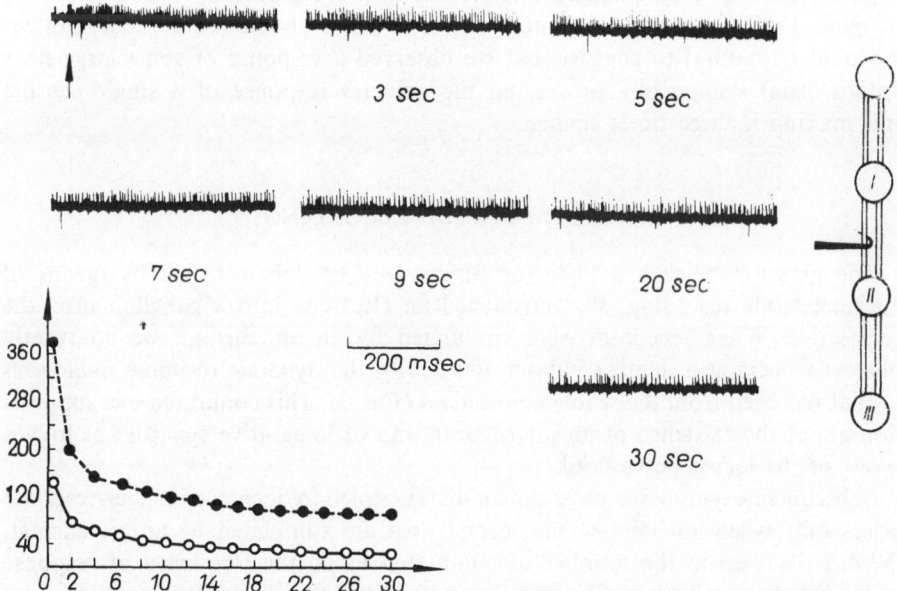

FIG. 2. Response registered in the I–II connective when receptors of the second region are stimulated by an air current (external recording). The time line represents 200 msec. The rest of the designations are the same as in Fig. 1. Below at the left is a graph of response recorded in connectives (upper curve) and in single receptors (lower curve). Below at the right is a graph of the impulse recordings from the connective. I, II, III — the locust thoracic ganglia

Thus, when hairs of the left region are stimulated, impulses are registered not only in the left connectives of the thoracic ganglia, but also in the right connectives, and the same holds for the stimulation of the right region. It is important that responses in the right and left connectives are actually identical.

Figure 2 shows a response arising in a connective between the first and second ganglia when hairs of the second area are stimulated by an air current.

The diagram sums up the similar data. The lower curve is a response recorded in a single receptor with the aid of a microelectrode.

It is seen that the two curves are similar in shape. However, in the case of the summed recording the frequency of response is about three times higher than in a single sensilla. It might have been supposed that the recording from connectives is a result of the summation of responses obtained from about three asynchronously discharged fibers. Each fiber is discharging as shown by the lower curve. However, the character of response (high-voltage and resembling a recording from single structures) and the fact that, according to Guthrie, sensory fibers in connectives have very small diameters (1–2 μ) reject such a possibility. In fact, it is least probable, for purely physical reasons, that a response of such fibers might be registered by an external electrode in a comparatively thick connective (Hughes 1965): the action potentials of such fibers should have been shunted. It would be natural to suggest that we observed a response of some large fiber (interneuron) synaptically processing the primary response of a single sensilla and making it three times higher.

MICROELECTRODE RECORDING

The presence of such a fiber (or fibers) may be demonstrated by means of microelectrode recording. We introduced an electrode into a ganglion near the connective. When receptors were stimulated by an air current, we constantly observed fibers at a depth of about 100 μ with the dynamic response analogous to that recorded from the whole connectives (Fig. 3). This confirmed our supposition about the existence of an interneuron and of integrative synapses at higher levels of the locust nerve cord.

Subsequent experiments have shown that a tripled frequency of a fiber-response arises only when all hairs of the second area are stimulated by an air current. With a decrease in the number of stimulated hairs, the frequency of response in the interneuron lowers. Figure 4 shows the character of interneuronal response observed when about one-third of the hairs are stimulated (stimulation of the upper part of the area).

With increasing number of stimulated hairs, a gap between the two curves (Figs. 3 and 4) might have been filled up (see also Fig. 5), although we have not observed a straight dependence between the number of stimulated hairs and the frequency of a fiber's response. It might be suggested that such a dependence should not have been linear, since in our experiments the discharge frequency

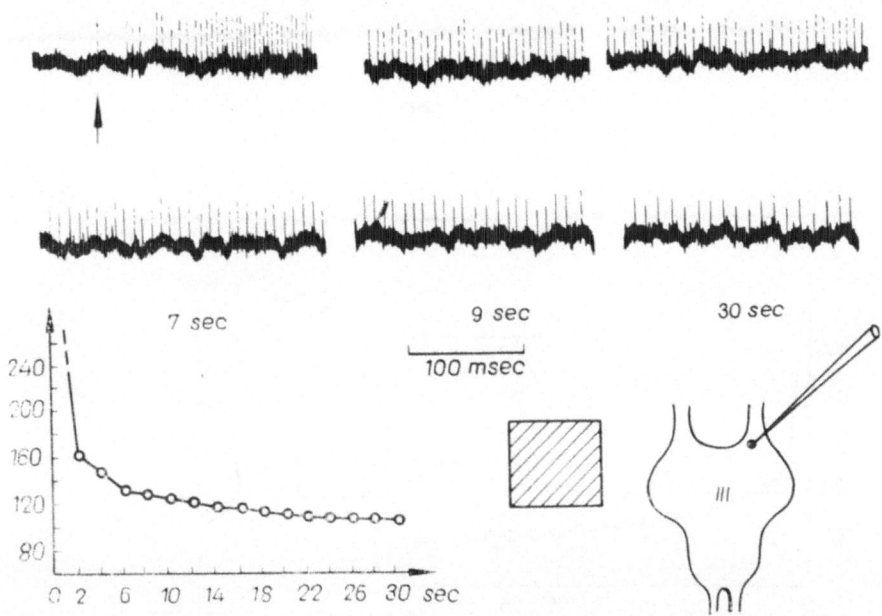

FIG. 3. Response recorded by means of a microelectrode (the recording scheme is presented below at the right) in the third thoracic ganglion of the locust when all the receptors of the second area are stimulated by an air current (completely hatched square). The time line represents 100 msec. Below at the left is a graph summing up observations of this type. The designations are the same as in the previous figures

of fibers does not significally differ for cases when one-third and one-half of all the hairs of the second region were exposed to air.

At present it is obvious that there is at least one synapse which is not a 1 : 1 type in the path of afferent impulses from the head receptors to thoracic ganglia regulating the locust flight. The impulses undergo a synaptic process which mostly affects the impulse frequency (it may vary considerably depending on the number of hairs stimulated). Such processing much less affects (or does not affect at all) the distribution of impulses in time (a picture of time-distributed discharges of central neurons resembles that given by single hairs, with phasic and tonic components of a response present). This integrating synapse is located in one of the head ganglia (perhaps in the suboesophageal ganglion) of the locust.

A general picture of the frequency and time relation of impulses that are propagated in neurons controlling the work of the locust flight muscles is suggested by the diagram in Fig. 5, although it concerns only one of the head regions maintaining the insect flight, the second region.

The above evidence does not preclude the possible existence of additional pathways between hair sensillae of the head and neurons of the flight muscles. For example, in the region of the connective, another type of responses

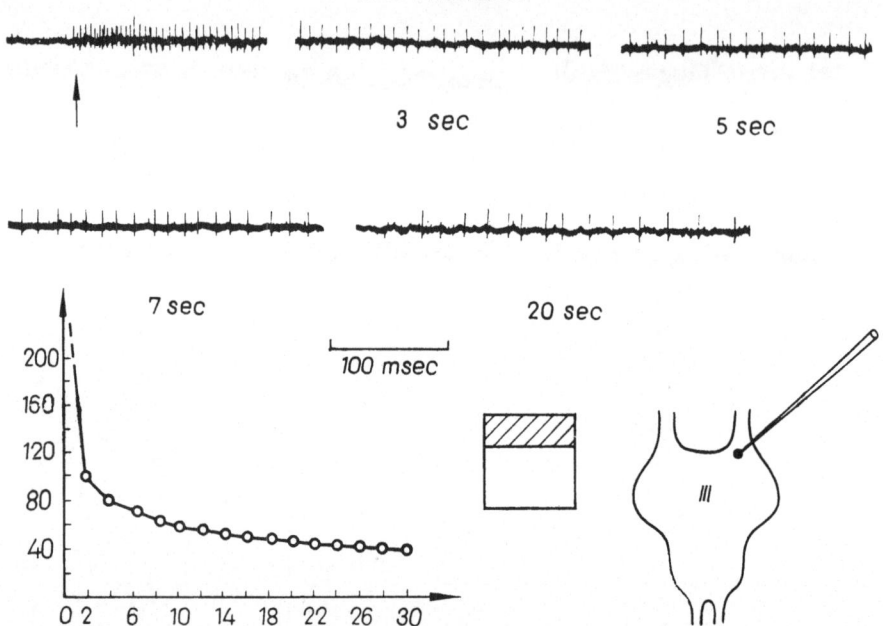

FIG. 4. Response recorded with the aid of a microelectrode (the recording scheme is shown below at the right) in the third thoracic ganglion of the locust when about one-third of the receptors of the second area are stimulated by air currents (one-third of the square is hatched). Below at the left is a graph of the response. The designations are the same as in the previous figures

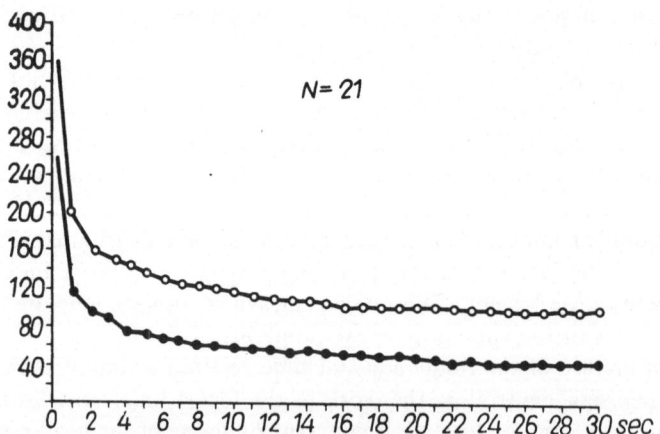

FIG. 5. Graph of the response in receptors of the second area when all hairs (upper curve) and about one-third of all the hairs (lower curve) are stimulated by air currents. The rest of the designations are the same as in the previous graphs

to continuous bending of hairs, i.e., phasic one may be recorded in the ganglion. Phasic response of interneurons sometimes was recorded even with external electrode located on the connective However, the activity of only phasic-tonic interneurons is presumably closely associated with prolonged activation of wing muscle neurons.

SUMMARY

The receptor system of the head maintaining the work of wing muscle neurons and capable of starting them in a suspended insect is discussed.

In CNS of the locust afferent impulses of head receptors undergo synaptic processing. There are two functional types of interneurons – phasis-tonic and phasic ones. The functional role of phasic-tonic neurons during flight of locust.

The impulses in the phasic-tonic interneurons undergo a synaptic process which mostly affect the impulse frequency (it may vary considerably depending on the number of hairs stimulated). Such processing affects much less the distribution of impulses in time (a picture of time-distributed discharges of central neurons resembles that given by single hairs with phasic and tonic components of a response present).

This integrative synapses are located in the head ganglia of the locust.

REFERENCES

HUGHES, G. M. (1965): Neuronal pathways in the insect central nervous system. in: *The Physiology of Insect* CNS. Ed.: J. C. Treherne and J. W. L. Beament, 79.

GUTHRIE, D. M. (1964): Observations on the nervous system of the flight apparatus in the locust *Schistocerca gregaria. Quart. J. micr. Sci.* **105** 183.

SVIDERSKY, V. L. (1967): Electrical activity of receptors maintaining the flight in locusts. *Dokl. Akad. Nauk USSR* **172** 1230 (in Russian).

THURM, U. (1962): Ableitung der Rezeptorpotentiale und Nervenimpulse einzelner Cuticula-Sensillen bei Insekten. *Z. Naturforsch.* **176** 285.

WEIS-FOGH, T. (1956): Biology and physics of locust flight: IV. Notes on sensory mechanisms in locust flight. *Phil. Trans. B.* **239** 553.

DISCUSSION

E. G. Boettiger: Were the responses arising in the left and right connectives identical?

V. L. Svidersky: Yes, they were practically identical on frequency.

A. Arvanitaki-Chalazonitis: Were the responses in the left and right connectives synchronous?

V. L. Svidersky: I did not record the responses in both of connectives simultaneously.

Symsposium on Neurobiology of Invertebrates 1967 (487—492)

CONTRIBUTION TO THE STUDY OF THE FUNCTION OF THE NERVOUS SYSTEM IN *NEMATODA*

B. A. SHISHOV

Helminthological Laboratory, Academy of Sciences of the USSR
Moscow, USSR

The study of the physiology of the nervous and muscular systems of nematodes is at present based mainly on the swine ascarid as a model. Other representatives of *Nematoda* have not been investigated as much in this respect.

Comparative studies on the structure and functions of the nervous system of *Ascaridia galli* and *Ascaris suum* are being carried out at the Helminthological Laboratory of the Academy of Sciences of the USSR. The study of these helminths is of definite interest, as both species are parasitic in the small intestine (though their hosts are rather remote systematically) and possess no special fixation organs. This allows us to believe that the location of these helminths is secured by their locomotor activity.

The observations on the locomotor activity of *Ascaris suum* have given the following results (Shishov 1961a, 1961b, 1962–1963, 1965):

After artificial introduction of the worms into different sections of the host's small intestine the nematodes actively orientate their head ends against the flow of the chyme and move to the upper parts of the intestine, i.e., they occupy the position corresponding to their usual location. Data on the orientation of intestinal nematodes against the chyme flow serve as a good reason for supposing that the leading role in development of adaptive movements of helminths is fulfilled by the anterior section of the nervous system.

A chemical or mechanical stimulation of the region of exteroceptor location causes a response contraction. In addition, changes in the worm's motion as a result of changes in the chemical composition of the environment have been observed in the course of other investigations (Krotov 1961, etc.)

Together with data on the participation of the ascarid exteroceptors in the regulation of movements, it has been established that when their functions are cut off the contractile activity does not stop. (In our experiments the disturbance of the nervous structure function was accomplished by applying ligatures on the corresponding sections of the nematode body. Subsequently the tied-up sections were cut off.)

Serpentine contractions in an intact worm usually become periodic. The periodicity is disturbed by the ablation of the cephalic nervous structures (Fig. 1). In

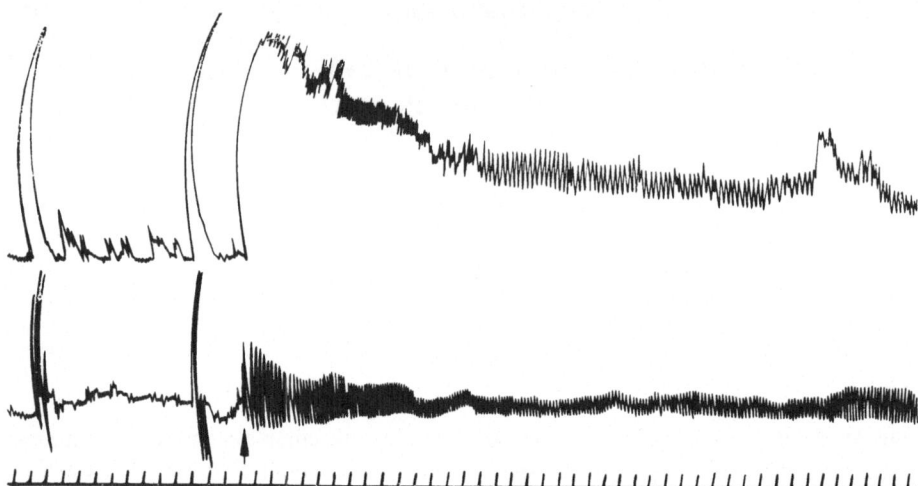

FIG. 1. Change in the locomotor activity of the ascarid after ablation of the cephalic end (↑) Upper line — registration of total contraction of the worm. Lower line — registration of serpentine movements along its body length. The time line indicates 3 min

this case serpentine movements continue constantly. This indicates that the function of the anterior end nervous structures of ascarids is not so much associated with the occurrence of the locomotion act as with its regulation. Thus, the processes of inhibition and regulation, which are carried out with the participation of the cephalic nervous structures, control the activity of the underlying locomotor sections and attach a more economic, periodic character to the helminth's movements. The inhibition of spontaneous and provoked contractions of the ascarid is also observed when irritating its cephalic area (Goodwin and Vaughan Williams 1963).

In contrast to the cephalic nervous structures, it is not evident that the anal ganglia exert a significant influence on the formation of the ascarid's contractions. The locomotor activity of nematodes does not noticeably change after their removal. Our experiments show that the ascarids retain their mobility after disturbance of the functions of anterior and posterior sections of the nervous system. These observations conform to the data of different authors on the contractile activity of the body fragments of these worms (Toscano-Rico 1926, Baldwin 1943, Baldwin and Moyle 1947).

The ascarid nerve cells form accumulations of ganglia in the anterior and posterior body sections; in addition, separate neurons or small groups of them are disposed along the nervous trunks. On the basis of this we can assume that the serpentine contractions of intact nematodes as well as of worms with ablated cephalic and anal sections of the nervous system are formed with the participation of these neurons.

The above-mentioned type of movements is interfered with by injection of curare-like and ganglion-blocking substances (0.1–0.2 ml, 0.2–2% diplacine, tubocurarine, nanophine solutions). The experiments were performed on preparations made by the method of Baldwin and Moyle (1947), in order to confirm the suggestion that inhibition of the influence of neurons takes place under the action of the above-mentioned substances.

Ventral fragments in whose nervous trunks we succeeded, by histological examination, to detect neurons, react on diplacine ($1 \cdot 10^{-5}$–$4 \cdot 10^{-5}$ g/ml), tubocurarine ($1 \cdot 10^{-4}$ g/ml), pentamine ($2 \cdot 10^{-4}$–$5 \cdot 10^{-4}$ g/ml), and nanophine ($4 \cdot 10^{-5}$ g/ml). Parallel with this, a modification of the rhythm of contractions takes place. It is characteristic for the ventral fragments that the muscles are in the state of contraction and relax periodically for short periods only. Under the influence of the above-mentioned substances the preparations relax considerably in most cases, and afterwards the frequency of contractions and relaxations increases (Fig. 2). In this case, the rhythm of contractions of the ventral preparations becomes similar to the rhythm of dorsal preparations, which is characterized by relatively fast alternation of contractions and relaxations. Further increases in the concentrations of the substances (up to 5–10 times) do not alter the effect. The substances tested do not influence the activity of dorsal preparations in spite of the increase (up to 6 times) of the concentration in comparison with concentrations acting on the ventral preparations. In some cases, an insignificant decrease of the amplitude of contractions has been observed. A histological control could not reveal neurones in the fragments of dorsal nervous trunks.

The study of the fragments allows us to draw some conclusions. The data suggest that the substances tested eliminate the influence of neurons controlling the locomotor activity, but at the same time do not significantly disturb the contractile properties of the muscles. In connection with this, we can consider the cessation of serpentine movements of ascarids under the action of ganglion-

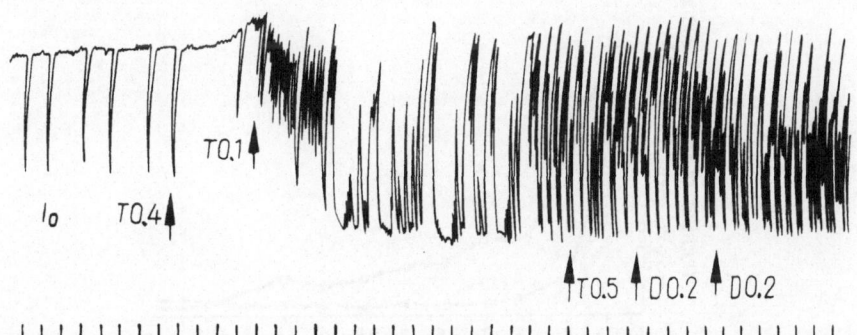

FIG. 2. The action of curare-like substances on a preparation of the ventral part of an ascarid. Contractions of the preparation were registered by means of the upward motion of the recorder. The arrows show the addition (in ml) of solutions to the experimental medium (volume 50 ml). T — tubocurarine, 1% sol. D — diplacine 2% sol. The time lines indicate 2 min

blocking and curare-like substances as contractions of neurogenic nature formed with the participation of neurons of the trunk part of the nervous system.

The fact that the ventral fragments are in a more contracted state then the dorsal ones indicates that the nerve cells of the trunks evidently exert a tonic influence on the somatic muscles.

Consideration of data on the different contractile activities of ventral and dorsal fragments, on morphological differences in their structures (the presence and absence of neurons in nervous trunks, respectively), and on their different reactions to the pharmacologic substances tested, indicates that ascarids may execute movements of neurogenic nature as well as contractile activity which does not depend on motor neurons and has a myogenic origin. A bioelectric rhythmic activity of myogenic nature has been recorded in the ascarid's muscles (de Bell *et al.* 1963; del Castillo *et al.* 1963, 1967).

The comparison of data on the locomotor activities of *Ascaris suum* and *Ascaridia galli* has revealed similarities as well as differences in the significance of nervous structures in the formation of the worm's movements. *Ascaridia galli*, like *Ascaris suum*, retain the contractile activity after the removal of the exteroceptors and the cephalic and anal ganglions. However, in contrast to *Ascaris suum*, we could not find in *Ascaridia galli* the participation of the nervous cephalic structures in the inhibitor processes regulating the motion. Kymograms of *Ascaridia galli* movements, which we obtained both under relatively normal conditions and under conditions of disturbed function of exteroceptors and ganglionic structures, are similar to the data obtained by Saenz-Beltran and Guevara Pozo (1965).

The significance of the nervous system in the regulation of functions of the helminth's organs and tissues (except for the regulation of locomotor activity) is obscure, though this problem is of great interest.

Investigations on the role of nervous structures in regulating the egg output of Nematoda are being carried out at present in our Helminthological Laboratory.

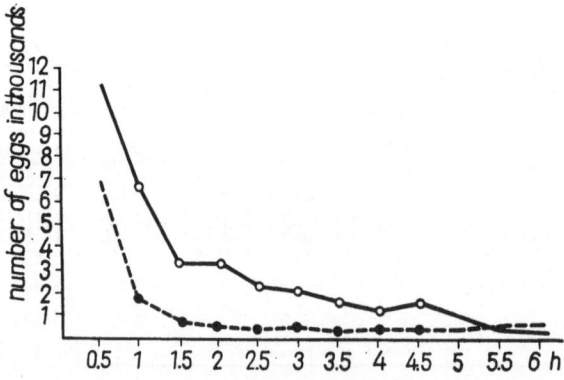

FIG. 3. Dynamics of egg-layings of *Ascaris suum*.
———— Under control conditions.
— — — After ablation of the cephalic nervous structures

Studies on the dynamics of egg production in *Ascaris suum* and *Ascaridia galli* have shown that these nematodes display *in vitro* great individual variations in intensity of egg-laying. Therefore, in order to gain average values, the total amount of eggs obtained from 10 specimens was determined. The experiments performed by Voronyuk have shown that the egg production continues after the disturbance of the function of the cephalic and anal ganglions. However, the removal of the anterior end results in a significant lowering of the egg output level in *Ascaris suum* (Fig. 3). The graph presents average values of the egg output of one worm, obtained from the data of 10 experiments, i.e., from one hundred specimens. We could not observe statistically significant differences in egg-laying after the same operation made on *Ascaridia galli*.

Thus, our experiments have shown that in the nematodes studied the cephalic nervous structures possess significant differences concerning the regulation of two leading functions of the worm's organism: the locomotor and the sexual functions. The functional significance of the ganglia in *Ascaridia galli* is not yet clear.

The fact that the cephalic sections of the nervous system of ascarid females control the activity of the underlying motor structures and take part in the regulation of the sexual system activity indicates a higher functional development of nervous structures in *Ascaris suum* than in *Ascaridia galli*.

SUMMARY

The role of nervous structures in regulating the activity of locomotor and sexual systems of females of *Ascaris suum* and *Ascaridia galli* was studied. Removing the peripharyngeal nerve ring and adjoining ganglions broke the periodicity of *A. suum* movements. In this case serpentine movements continued constantly. This indicates that cephalic nervous structures participate in the formation of inhibition of *A. suum* locomotor activity. The same operation did not cause essential changes in the contractile activity of *A. galli*. Disturbance of the function of exteroceptors and anal ganglions of both *A. suum*, and *A. galli* did not significantly influence on movements. Serpentine contractions of *A. suum* were interfered with by injections of curare-like and ganglion-blocking substances. They also changed the character of contractions of *A. suum* body ventral fragments, where neurons were found, and did not influence on the activity of dorsal preparations, where we did not find neurons. These data suggest that *A. suum* may have contractile activity of neurogenic and myogenic origin. The investigation of the dynamic egg-laying of nematodes *in vitro* showed that this process takes place both in intact helminths and in helminths with removed cephalic and anal ganglia. However, the level of the egg production of *A. suum* is considerably reduced when the head nerve structures are interfered with. This phenomenon is not observed in *A. galli*. These data show that *A. suum* has a higher functional development of the nerve structures than *A. galli*.

REFERENCES

BALDWIN, E. (1943): An *in vitro* method for chemotherapeutic investigation of anthelminthic potency. *Parasitology* **35** 89—111.

BALDWIN, E. and MOYLE, Y. (1947): An isolated nerve-muscle preparation from *Ascaris lumbricoides. J. exp. Biol.* **23** 277—291.

DE BELL, J. T., DEL CASTILLO, J. and SANCHEZ, V. (1963): Electrophysiology of the somatic muscle cells of *Ascaris lumbricoides. J. cell. comp. Physiol.* **62** 159—177.

DEL CASTILLO, J., DE MELLO, W. C. and MORALES, T. (1963): The physiological role of acetylcholine in the neuromuscular system of *Ascaris lumbricoides. Arch. Int. Physiol. et de Biochimie* **71** 741—757.

DEL CASTILLO, J., DE MELLO, W. C. and MORALES, T. (1967): The initiation of action potentials in the somatic musculature of *Ascaris suum. J. exp. Biol.* **46** 263—279.

GOODWIN, L. D. and VAUGHAN WILLIAMS, E. M. (1963): Inhibition and neuromuscular paralysis in *Ascaris lumbricoides. J. Physiol.* **168** 857—871.

KROTOV, A. J. (1961): *Experimental Therapy of Helminthiasis.* Medgiz, Moscow (in Russian).

SAENZ-BELTRON, Z. E. and GUEVARA POZO, D. (1965): El metodo quimgrafico utilizando *Ascardia galli* y sus resultados con alqunos antihelminticos y otras substancias. *Rev. ibér. Parasit.* **25** 131—163.

SHISHOV, B. A. (1961a): Movements of Ascarids as an indicator for their localization in the intestine. *Reports of the Helminthological Laboratory Acad. Sci. USSR* **2** 353—355 (in Russian).

SHISHOV, B. A. (1961b): On the regulation and mechanism of the contractive activity of *Ascaris suum. Helmintologia* **3** 299—310 (in Russian).

SHISHOV, B. A. (1962—1963): The nature of the contractive activity of the *Ascarids. Helmintologia* **4** 446—455 (In Russian).

SHISHOV, B. A. (1965): Motor activity of helminths and its regulation. *Reports of the Helminthological Laboratory Acad. Sci. USSR* **15** 232—237 (in Russian).

TOSCANO RICO, J. (1926): Sur l' emploi de l'*Ascaris lumbricoides* comme reactif pharmacologique. Action de la nicotine. *C. R. Soc. Biol. (Paris)* **94** 918—920.

Symposium on Neurobiology of Invertebrates 1967 (493—501)

ROLE OF CEREBRAL GANGLIA IN THE REGULATION
OF ACTIVITY IN FRESHWATER MUSSEL (*ANODONTA CYGNEA* L.)

J. Salánki

Biological Research Institute of the Hungarian Academy of Sciences
Tihany, Hungary

The accumulation of data about the nervous system and neural mechanisms in Pelecypods is proceeding at a considerably slower rate than in other groups of Molluscs. This may be explained partly by the fact that unlike Gastropods and Cephalopods, Pelecypods have neither giant cells nor giant axons. Furthermore, since their sense organs and motor activity are less well developed, it is not so convenient to work on them in the laboratory. During the last twenty years there have only been about a dozen publications dealing with functional problems of the nervous system in mussels and it is because of the small number of investigations that our knowledge in this field is inadequate. Thus, inevitably, even in newer monographs the data seem to be partly or completely out of date.

These were some of the reasons we began to investigate the functional and morphological properties of the nervous system in Pelecypods. The main problems we are interested in, are the neural regulation of rhythmic and periodic activity, including spontaneous activity and the reflex function of the nervous elements. We are also investigating the problems of long- and short-term interneuronal and neuromuscular transmission, which are of importance in the regulation of behavior.

In this paper I would like to summarize some of our results in connection with the significance and role of the cerebral ganglia in the regulation of muscle activity, to deal with some characteristic feature of metabolic processes in the ganglia, and to discuss the possible connection between humoral factors participating in regulation within the central nervous system and also on the neuromuscular level.

The functioning of the adductor muscles in Pelecypods is a basic life-phenomenon, since the animal communicates with the environment only during the time when the valves are open. The valves are opened during the relaxation of the adductor muscles, while contraction of the adductors can close the shells completely. In such a way the whole situation of the animal depends on the activity and position of the adductor muscles. For this very reason the activity of the adductors, i.e., the movement of the valves, seems to be a good indicator when investigating some properties of the nervous system and the neural regulation of complex behavioral processes.

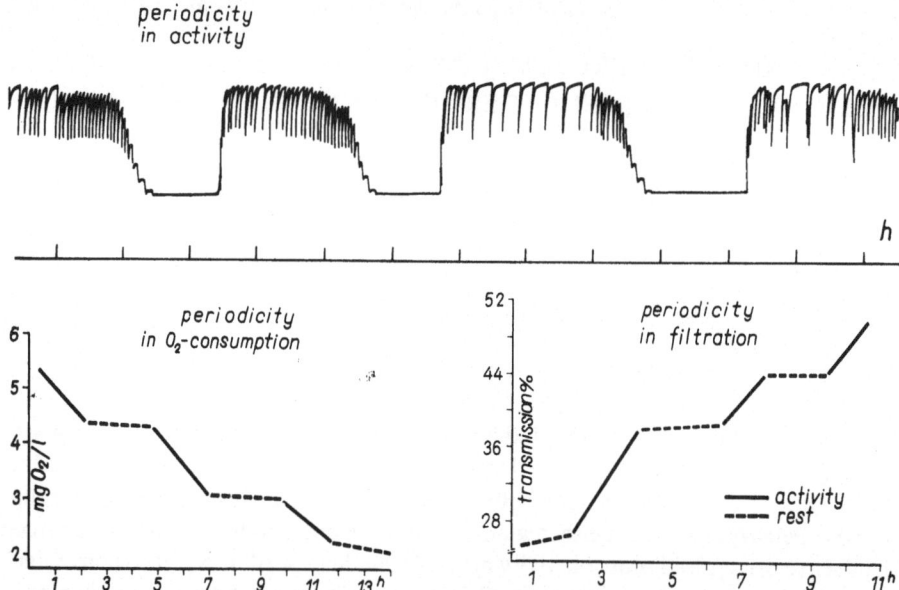

FIG. 1. Upper curve: periodicity in motor activity (actogram). Left curve: periodicity in O₂ consumption. Right curve: periodicity in filtration activity

According to earlier investigations (Marceau 1909; Barnes 1955; Koshtoyants and Salánki 1958) the adductors show two different phenomena: (1) there is an irregular rhythmic activity in the relaxed state, and (2) there is a periodic alteration of active and rest periods, the adductors being in tonic contraction during the latter (Fig. 1, upper curve).

I recently carried out investigations together with F. Lukacsovics to determine whether the periodicity is manifested only in the function of the adductors or whether other life processes also show periodicity in the freshwater mussel *Anodonta cygnea*. The oxygen consumption and filtration activity of mussels were investigated parallel with the recording of shell movement. Suspended neutral red particles served to measure the rate of filtration. The O_2 consumption of animals was determined with Winkler's method on samples taken from the paraffin-covered water in the aquaria.

The upper curve of Fig. 1 demonstrates the periodicity of the shell movement, showing the alteration of active and rest periods, respectively. The second curve indicates that there is also a similar periodicity in the O_2 consumption, and the third curve demonstrates that the same is true for the filtration rate. Thus, it is evident that the periodicity of shell movement is the indicator of several processes, and that O_2 uptake and water cleaning have identical periodicities.

The periodic activity contains four basic components: (1) rhythmicity of adductors during activity; (2) transition of adductor muscles into tonic contraction; (3) tonic contraction lasting for hours; and (4) relaxation of adductors after the rest period. It is quite obvious that these different processes are regulated by the nervous system. When one consider that in the adduction process two muscles situated far from each other are involved, it is evident that the nervous system has to insure not only a general regulation, but also the coordination between the adductors. Since in mussels the components of the central nervous system, namely the cerebral, visceral, and pedal ganglia, are situated far apart, the question arises, to what degree do they participate in the regulatory processes? Pavlov's investigations at the end of the last century emphasize the distinctive role of the cerebral ganglia. He assigned to them the role of controlling the relaxation of both adductors. This assumption was taken as our starting point in our earlier investigations, where different bioactive agents were tested for their effect on the neural regulation of the muscles (Salánki 1963).

At that time we found that substances known to be stimulating agents in molluscs (for example, serotonin) induced both relaxation and increased rhythmic activity of the posterior adductor when applied to the cerebral ganglia. These findings emphasized the presence of the relaxing center in the cerebral ganglia.

On the other hand, we found a definite connection between the basophilia of the nerve cells and the state of activity of the animal (Salánki, Zs.-Nagy and H.-Vas 1965). It seemed possible that changes occurring in the RNA content or in nucleic acid metabolism were responsible for the function of cerebral ganglia regulating the active and rest periods of the animal. Of course, the changes in the nucleic acid metabolism are realized through other enzymatic processes connected with the metabolism of activating or inhibitory substances.

I therefore carried out investigations together with I. Zs.-Nagy to see if there was a significant modification of the animal's activity after local treatment of the cerebral ganglia with inhibitors of nucleic acid metabolism, and, if there were a difference, to determine what type of cytological changes could be observed in the nerve cells. Figure 2 shows the results. In most cases, after actinomycin and chloramphenicol treatment of the cerebral ganglia, there was, within 4–6 h, a gradual decrease in the amplitude of the relaxations, until finally the tonic contraction of the posterior adductor took place. This corresponds to the physiological rest period. Now the cerebral and visceral ganglia were excised and after cresyl violet staining the content of the basophilic substance in the nerve cells was investigated. Only a few cells were found to have many Nissl granules in their cytoplasm; the number of cells which did not contain Nissl substance was markedly increased. As controls we used untreated animals which were in the active period. The difference was well expressed in the cerebral ganglia and only of a smaller degree in the visceral. The picture in the cerebral ganglia is in full agreement with our earlier findings that the beginning of the rest period and the low nucleic acid content in the ganglia are coincident (Zs.-Nagy, Brodsky and Salánki 1966).

The hypothesis seemed to be correct in its supposition that cerebral ganglia themselves play a decisive role in the regulation of adductor activity, since (1) their activation leads to the relaxation and to the increased activity of the muscle in tonic contractions; (2) by inhibiting their protein metabolism the inhibition of the animal's whole activity can be reached independent of the condition of the visceral ganglia.

Now, what is the role of the visceral ganglia in the regulation of closer muscles? According to the scheme of Pavlov, the visceral ganglion takes part only in the contraction of the posterior adductor and has nothing to do with the anterior adductor (Pavlov 1885). However, there are some observations which make this suggestion questionable. Pavlov himself reported that when stimulating the CVC (cerebro-visceral connective), the anterior adductor responds, even if only with tonic contraction, thus arguing against the presence of relaxing fibers.

To resolve this problem, experiments were performed on preparations containing a part of the CVC as well as the cerebral ganglia and the anterior adductor (Fig. 3). The CVC was stimulated by a series of square wave pulses. Investigations undertaken with T. Pécsi and E. Lábos on the effect of a wide range of parameters showed that in most cases the response of the anterior adductor was really of a tonic character. However, by stimulating the CVC under certain values of parameters, an expressed phasic response and a marked decrease in the level of the tonus could be obtained, indicating that there are different fibers present leading from the visceral to the cerebral ganglia: some of them increase tonus, but others

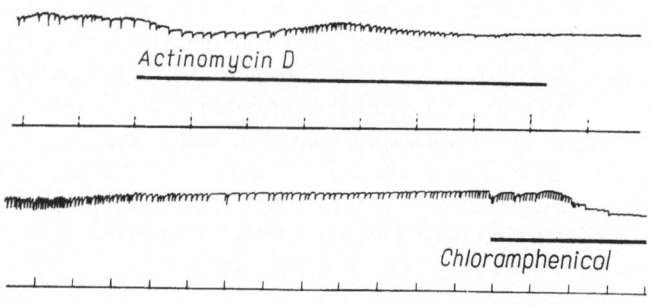

Neurons with Nissl-substance

	Control	Actinomycin D	Chloramphenicol
full	43%	4%	7%
transitory	23%	17%	26%
empty	34%	79%	67%

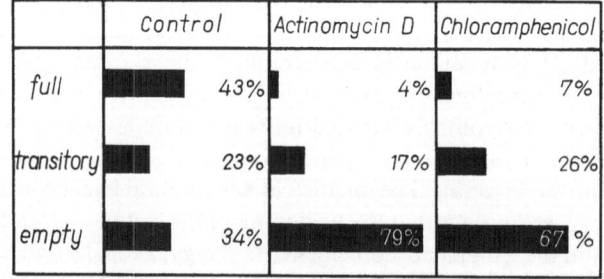

FIG. 2. Effect of actinomycin D and chloramphenicol applied to the cerebral ganglia on the activity of the posterior adductor muscle (actograms) and on the amount of the Nissl substance in nerve cells of the cerebral ganglia (table)

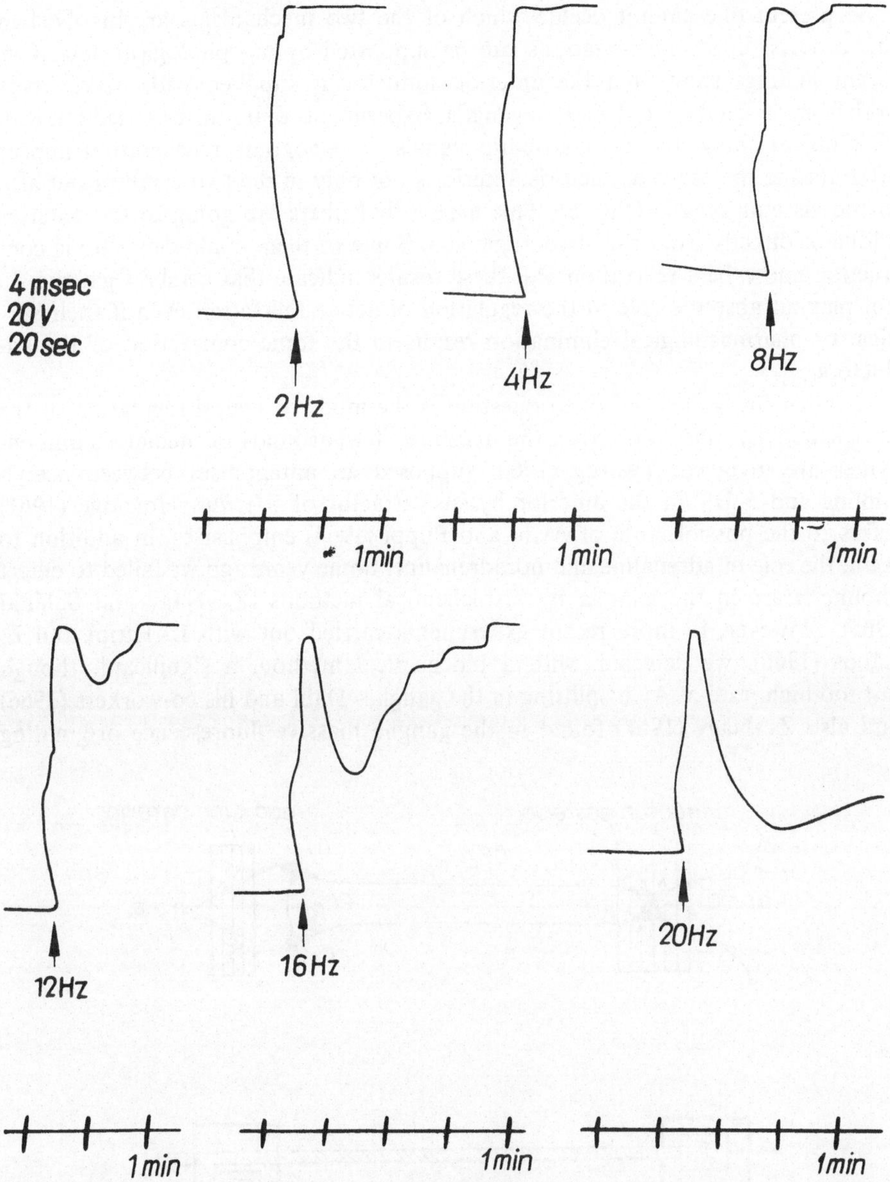

4 msec
20 V
20 sec

2 Hz

4Hz

8Hz

1min   1min   1min

12Hz

16Hz

20Hz

1min   1min   1min

FIG. 3. Effect of different parameters on the response of the anterior adductor muscle after stimulation of the cerebro-visceral connective

cause relaxation. Thus, Pavlov's theory must be supplemented, insofar as the visceral ganglion has the property to induce both tonic and relaxing effects on the anterior adductor. This influence is realized either directly on the muscle or through stimulation of the corresponding center in the cerebral ganglia (Fig. 4).

At present one cannot decide which of the two mechanisms are involved in this process. Both interpretations can be supported by morphological data. Our recent findings show that the anterior adductor is supplied with nerve fibers both from the cerebral and visceral ganglia. Experiments that I have carried out with A. Gubicza show that cytoplasmatic signals of retrograde regeneration appear after cutting the nerve adductoris anterioris not only in the two cerebral but also in the visceral ganglia (Fig. 5). This means that fibers are going to the anterior adductor directly from the visceral ganglia. Some of them could cause tonic contraction and others relaxation. All these results indicate that cerebral ganglia do not play an absolute role in the regulation of active relaxation even if their surgical or pharmacological elimination results in the tonic contraction of the adductors.

A rather intriguing and open question is the mode of neural regulation of the muscle activity, and, moreover, the question of what kinds of mediators and enzymes are involved. Twarog (1960) supposed an antagonism between acetylcholine and 5-HT in the anterior byssus retractor of *Mytilus*. Horridge (1961) refers to the possible role of ACh, and Puppi (1964) emphasizes, in addition to ACh, the role of adrenaline and noradrenaline. Some years ago we failed to detect cholinesterase in the ganglia by histochemical methods (Zs.-Nagy and Salánki 1965). However, in more recent experiments carried out with L. Hiripi and E. Lábos (1966), we detected, with a biochemical method, a significant, though not too high, rate of ACh splitting in the ganglia. Dahl and his co-workers (1966) and also Zs.-Nagy (1967) found in the ganglia massive fluorescence originating

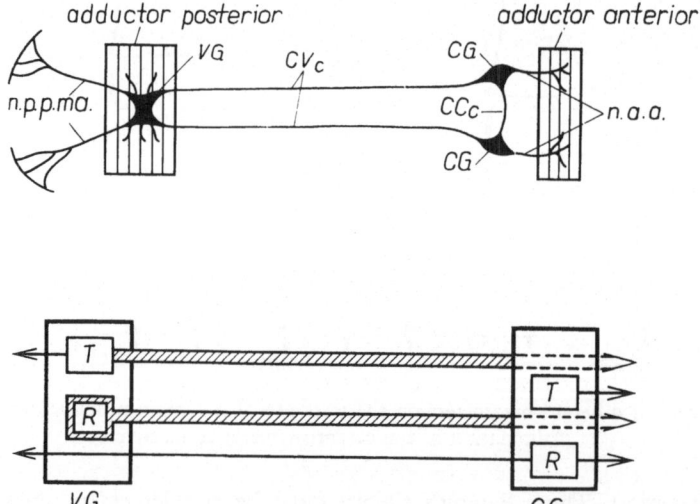

Fig. 4. Scheme of possible centers in the cerebral and visceral ganglia. ————— Pavlov's suggestion. ▨ Our additional suggestion

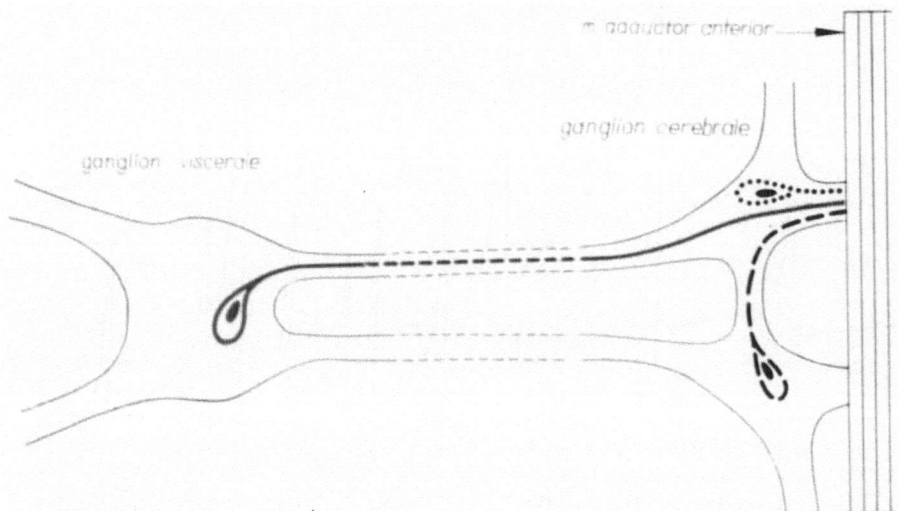

FIG. 5. Situation of nerve cells of the axons present in the nerve adductoris anterioris

from catecholamines and serotonin. They supposed the catecholamine to be dopamine. The question is, which of these substances takes part in the regulation of the adductor activity in the ganglia, and what is the type of effect they evoke at the neuromuscular level? The fact is that applying any one of these substances to the cerebral ganglia causes some increase in activity, and none of them results in tonic contraction of the adductors. It is known that the presence of the active agents in the ganglia does not only indicate that they take part in intraganglional processes, but also that they are transported along axons and perhaps released at junctions. For this very reason we investigated the effect of these agents applied directly to the adductor muscle.

The materials were injected into the posterior adductor during spontaneous rhythmic activity. It was found rather unexpectedly that ACh does not cause a lasting effect even in a 500-μg amount. Dopamine similarly showed no effect, even though the ganglion contains it in large quantities. Injection of 5–HT in 50-μg quantity resulted significantly in high relaxation, while adrenaline and noradrenaline resulted in gradual tonic contraction within 10–15 min (Fig. 6).

From these experiments it may be concluded that 5–HT and catecholamines play antagonistic roles in the regulation of adductor activity in *Anodonta*. Since 5–HT is present in the ganglia it may be supposed that there is a direct transport of it to the adductors, while the dopamine present in the ganglia could serve as a precursor for adrenaline or noradrenaline. It seems that ACh, notwithstanding the high cholinesterase activity of the adductors, plays an insignificant role in the regulatory processes of long-term phenomena.

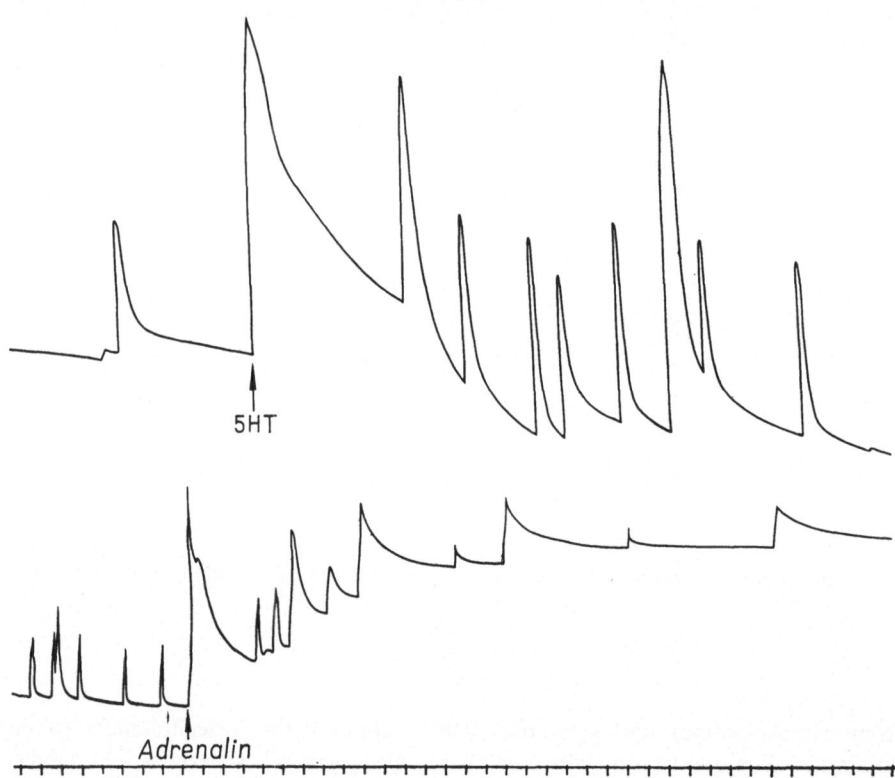

FIG. 6. Effect on the muscle tone of serotonin and adrenaline injected into the posterior adductor, time scale: 1 min

SUMMARY

It is believed that the cerebral ganglia play an important role in the opening of the shells needed for rhythmic and periodic activity in Pelecypods, but there are no exact data about the operation of this mechanism. Investigations were carried out showing that:

1. The periodicity of shell movement is an indicator of several processes (periodicity in O_2 uptake and in water cleaning).

2. By separately affecting the cerebral ganglia with drugs blocking RNA synthesis, the amount of the Nissl-substance in the nerve cells decreases and the relaxed state of the adductors turns to tonic contraction. This corresponds to our earlier findings; i.e., the activity runs parallel with the amount of the Nissl material of the nerve cells.

3. The character of the adductor response to electrical stimulation of nerves depends very much on whether cerebral ganglia are involved or not. The role of these ganglia differs in this respect from that of the visceral ganglia, but the relations and nerve pathways are more complex than suggested by earlier investigations.

4. ACh, 5–HT, and catecholamines, suggested or detected in the ganglia, cause different actions when applied directly to the cerebral ganglia or into the adductors. On the basis of these experiments indolalkylamines and catecholamines are suggested as antagonistic agents in the neural control of adductor activity in *Anodonta*.

REFERENCES

BARNES, G. E. (1955): The behaviour of *Anodonta cygnea* L., and its neurophysiological basis. *J. exp. Biol.* **32** 158—174.

DAHL, E., FALCK, B., VON MECKLENBURG, C., MYHRBERG, H. and ROSENGREN, E. (1966): Neuronal localization of dopamine and 5-hydroxytryptamine in some Mollusca. *Z. Zellforsch.* **71** 489—498.

HORRIDGE, G. A. (1961): The centrally determined sequence of impulses initiated from a ganglion of the clam *Mya*. *J. Physiol.* **155** 320—336.

KOSHTOYANTS, CH. S. and SALÁNKI, J. (1958): On the physiological principles underlying the periodical activity of *Anodonta*. *Acta biol. Acad. Sci. hung.* **8** 361—366.

MARCEAU, F. (1909): Recherches sur la morphologie, l'histologie et la physiologie comparées des muscles adducteurs des Mollusques acéphales. *Arch. Zool. Exp. Gen.* **2** 295—469.

PAVLOV, J. (1885): Wie die Muschel ihre Schaale öffnet. *Pflügers Arch. ges. Physiol.* **37** 6—31.

PUPPI, A. (1964): New contributions to the tone-inhibiting effect of catecholamines in the fresh-water mussel. *Experientia* **20** 620—621.

SALÁNKI, J. (1963): The effect of serotonin and catecholamines on the nervous control of periodic activity in fresh-water mussel *(Anodonta cygnea)*. *Comp. Biochem. Physiol.* **8** 163—171.

SALÁNKI, J., HIRIPI, L. and LÁBOS, E. (1966): Cholinesterase activity in the central nervous system of *Anodonta cygnea* L. *Annal. Biol. Tihany* **33** 143—150.

SALÁNKI, J., ZS.-NAGY, I. and H.-VAS, É. (1965): Change of the Nissl-substance in connection with the periodic activity in the central nervous system of fresh-water mussel *Anodonta cygnea* L. *Annal. Biol. Tihany* **32** 11—116.

TWAROG, B. M. (1960): Effects of acetylcholine and 5-hydroxy-tryptamine on the contraction of a molluscan smooth muscle. *J. Physiol.* **152** 236—242.

ZS.-NAGY, I. (1967): Histochemical demonstration of biogenic monoamines in the central nervous system of the lamellibranch mollusc *Anodonta cygnea* L. *Acta biol. Acad. Sci. hung.* **18** 1—8.

ZS.-NAGY, I. and SALÁNKI, J. (1965): Histochemical investigations of cholinesterase in different molluscs with reference to functional conditions. *Nature* **206** 842—843.

ZS.-NAGY, I., BRODSKY, V. J. and SALÁNKI, J. (1966): Fluctuation of the nucleic acid concentration in the central nervous system in connection with the periodic activity of freshwater mussel *(Anodonta cygnea* L.). *Acta biol. Acad. Sci. hung.* **17** 43—49.

DISCUSSION

B. Glaizner: Could you tell us something more about the coordination of the two adductors?

J. Salánki: Yes. It is really a very interesting problem and I only touched on it in my talk. As we found, in most of cases there exists some degree of coordination and it operates through the cerebro-visceral connective. But in other cases, namely, when the contraction appears as a reflex answer to the stimulation of the mantle, the adductors contract at different times, one after the other, depending on the path from the place of the stimulation. In this case, the contractions of the two adductors are near each other in time, but this 'coordination' operates without the participation of the cerebro-visceral connective.

Responsible for publication: Gy. Bernát, Director of
the Publishing House of the Hungarian Academy of
Sciences and of the Academy Press
Responsible editor: A. Kiss. Technical editor: Á. Garamvölgyi
Jacket and cover: É. Prácser
Set in: New Times Roman 10/11 point